Bioremediation Science
From Theory to Practice

Editors

Amitava Rakshit
Department of Soil Science & Agricultural Chemistry
Institute of Agricultural Sciences
Banaras Hindu University
Varanasi, UP
India

Manoj Parihar
Crop Production Division
ICAR-Vivakananda Parvatiya Krishi Anusandhan Sansthan
Almora, Uttarakhand
India

Binoy Sarkar
Lancaster Environment Centre
Lancaster University
Lancaster
UK

Harikesh B. Singh
Retired Professor of Mycology and Plant Pathology
Institute of Agricultural Sciences
Banaras Hindu University
Varanasi, UP
India

Leonardo Fernandes Fraceto
São Paulo State University
Sorocaba, São Paulo
Brazil

CRC Press is an imprint of the
Taylor & Francis Group, an **informa** business

A SCIENCE PUBLISHERS BOOK

Cover credit: Image taken by editors from Pixbay and modified by them.

First edition published 2021
by CRC Press
6000 Broken Sound Parkway NW, Suite 300, Boca Raton, FL 33487-2742

and by CRC Press
2 Park Square, Milton Park, Abingdon, Oxon, OX14 4RN

© 2021 Taylor & Francis Group, LLC

CRC Press is an imprint of Taylor & Francis Group, LLC

Reasonable efforts have been made to publish reliable data and information, but the author and publisher cannot assume responsibility for the validity of all materials or the consequences of their use. The authors and publishers have attempted to trace the copyright holders of all material reproduced in this publication and apologize to copyright holders if permission to publish in this form has not been obtained. If any copyright material has not been acknowledged please write and let us know so we may rectify in any future reprint.

Except as permitted under U.S. Copyright Law, no part of this book may be reprinted, reproduced, transmitted, or utilized in any form by any electronic, mechanical, or other means, now known or hereafter invented, including photocopying, microfilming, and recording, or in any information storage or retrieval system, without written permission from the publishers.

For permission to photocopy or use material electronically from this work, access www.copyright.com or contact the Copyright Clearance Center, Inc. (CCC), 222 Rosewood Drive, Danvers, MA 01923, 978-750-8400. For works that are not available on CCC please contact mpkbookspermissions@tandf.co.uk

Trademark notice: Product or corporate names may be trademarks or registered trademarks and are used only for identification and explanation without intent to infringe.

Library of Congress Cataloging-in-Publication Data

```
Names: Rakshit, Amitava, editor.
Title: Bioremediation science : from theory to practice / editors, Amitava
   Rakshit, Department of Soil Science & Agricultural Chemistry, Institute
   of Agricultural Science, Banaras Hindu University Varanasi, UP, India,
   Manoj Parihar, Crop Production Division, ICAR-Vivakananda Parvatiya
   Krishi Anusandhan Sansthan, Almora, Uttarakhand, India, Binoy Sarkar,
   Lancaster Environment Centre, Lancaster University, Lancaster, UK,
   Harikesh B. Singh, Retired Professor of Mycology and Plant Pathology,
   Institute of Agricultural Sciences, Banaras Hindu University, Varanasi,
   UP, India, Leonardo Fernandes Fraceto, São Paulo State University,
   Sorocaba, São Paulo, Brazil.
Description: First edition. | Boca Raton, FL : CRC Press, [2021] | Includes
   bibliographical references and index.
Identifiers: LCCN 2020039958 | ISBN 9780367343965 (hardcover)
Subjects: LCSH: Bioremediation.
Classification: LCC TD192.5 .B5577 2021 | DDC 628.5--dc23
LC record available at https://lccn.loc.gov/2020039958
```

ISBN: 978-0-367-34396-5 (hbk)

Typeset in Times New Roman
by Radiant Productions

Thankfully and lovingly dedicated to our wives and our children who motivated us to fly towards our dream

Preface

The introduction of human civilization and their progressive development has placed enormous pressure on environmental integrity. Unscientific disposal of waste released from industries, agriculture and domestic sources is posing a great threat to natural biodiversity and its existence. Therefore, there is an urgent need of diminution of hazardous pollutants and xenobiotics using economical, effective and viable alternative. In this regard, bioremediation offers a greater possibility to manage the wide range of potential toxic pollutants. Bioremediation includes plants, microbes and their enzymes in order to detoxify or degrade the heterogeneous and toxic substances to an innocuous state. Due to natural attenuation and cost effective feature, this approach provides an edge over other conventional methods. However, the long duration involved and inappropriate levels of residual contaminants render this technique ineffective, but technical expertise and considerable experience with appropriate design of bioremediation program can become the game changer that modern world needs. The continuous changes and evolution of new metabolic pathways have facilitated these microbes to degrade or detoxify the resistant and various heterogeneous pollutants present on this planet. In the last few years, research on bioremediation has received much attention and increased dramatically to develop a more effective and applied bioremediation process. However, this rapidly advancing field requires a multidisciplinary approach with up-to-date knowledge to broaden the perspective of researcher engaged in this field. In this book, we have tried to cover the current research development along with some new approaches, methods and management strategies for the abatement of contaminants present in soils, sediments, surface water and aquifers for the environmental conservation in a sustainable way. It offers new insights and perspectives for novices and experts as for students, academicians, teachers, researchers, environmentalists and other individuals interested in the field of bioremediation.

This book provides state of the art description of various approaches, techniques and some basic fundamentals of bioremediation to manage a variety of organic and inorganic wastes and pollutants present in our environment. The scope of this book extends to environmental/agricultural scientists, students, consultants, site owners, industrial stakeholders, regulators and policy makers with a holistic and systematic approach. It covers new development and recent advances in the field of bioremediation research within relevant theoretical framework to improve our understanding of the cleaning up of polluted water and contaminated land. The book is easy to read, and the language can be readily comprehended by aspiring newcomer, students, researchers and anyone else interested in this field.

Amitava Rakshit
Manoj Parihar
Binoy Sarkar
Harikesh B. Singh
Leonardo Fernandes Fraceto

Acknowledgement

We bow our appreciation to several enthusiastic hearts due to whose promise and devotion in the subject, we have been able to accomplish this gigantic job. Our profound sense of veneration and sincere thanks to the contemporaries and students who assisted us in our venture in bringing this book to light. Our sincere thanks also go to Prof. Mankombu Sambasivan Swaminathan for his immense impact, unwavering back-up and support. Thanks to everyone on my publishing team. Special thanks to Raju Primlani and Danielle Zarfati. Last but not the least we should thank our family, immediate and extended, who always encouraged us to continue the enormous task. Finally, we thank all the people who have supported us to complete the task directly or indirectly.

Amitava Rakshit
Manoj Parihar
Binoy Sarkar
Harikesh B. Singh
Leonardo Fernandes Fraceto

Contents

Preface   v

Acknowledgement   vii

1. **Bioremediation: Concepts, Management, Strategies and Applications**   1
   Alexandre Marco da Silva, Fátima Piña-Rodrigues, Debora Zumkeller Sabonaro, Ivonir Piotrowski, José Mauro Santana, Lausane Soraya Almeida, Lucas Hubacek Tsuchiya, Marco Vinicius Chaud and *Vanderlei Santos*

2. **Bioremediation: Current Status, Prospects and Challenges**   15
   Ruby Patel, Anandkumar Naorem, Kaushik Batabyal and *Sidhu Murmu*

3. **Integrative Approaches for Understanding and Designing Strategies of Bioremediation**   37
   Shiv Prasad, Sudha Kannojiya, Sandeep Kumar, Krishna Kumar Yadav, Monika Kundu and *Amitava Rakshit*

4. **Ecological Tools for Remediation of Soil Pollutants**   57
   Nayan Moni Gogoi, Bhaswatee Baroowa and *Nirmali Gogoi*

5. **Phytoremediation: A Green Approach for the Restoration of Heavy Metal Contaminated Soils**   79
   Sivakoti Ramana, Vassanda Coumar Mounissamy and *Jayant Kumar Saha*

6. **Soil Heavy Metal Pollution and its Bioremediation: An Overview**   92
   Swagata Mukhopadhyay, Swetha R.K. and *Somsubhra Chakraborty*

7. **Mechanism of Heavy Metal Hyperaccumulation in Plants**   103
   Supriya Tiwari

8. **Biological Indicators for Monitoring Soil Quality under Different Land Use Systems**   121
   Bisweswar Gorain and *Srijita Paul*

9. **Aromatic Plants as a Tool for Phytoremediation of Salt Affected Soils**   138
   B.B. Basak, Smitha G.R., Anil R. Chinchmalatpure, P.K. Patel and *Prem Kumar B.*

10. **Microbial Mediated Biodegradation of Plastic Waste: An Overview**   154
    Rajendra Prasad Meena, Sourav Ghosh, Surendra Singh Jatav, Manoj Kumar Chitara, Dinesh Jinger, Kamini Gautam, Hanuman Ram, Hanuman Singh Jatav, Kiran Rana, Surajyoti Pradhan and *Manoj Parihar*

11. **Agrochemical Contamination of Soil: Recent Technology Innovations for Bioremediation** — 170
    Suryasikha Samal and C.S.K. Mishra

12. **Bioremediation of Pesticides with Microbes: Methods, Techniques and Practices** — 180
    Rakesh Kumar Ghosh, Deb Prasad Ray, Ajoy Saha, Neethu Narayanan, Rashmita Behera and Debarati Bhaduri

13. **Compost-assisted Bioremediation of Polycyclic Aromatic Hydrocarbons** — 202
    N.S. Bolan, Y. Yan, Q. Li and M.B. Kirkham

14. **Petroleum Hydrocarbon-Contaminated Soils: Scaling Up Bioremediation Strategies from the Laboratory to the Field** — 212
    José A. Siles

15. **Heavy Metal Pollution in Agricultural Soils: Consequences and Bioremediation Approaches** — 227
    Abdul Majeed

16. **Arsenic Toxicity in Water-Soil-Plant System: An Alarming Scenario and Possibility of Bioremediation** — 240
    Ganesh Chandra Banik, Shovik Deb, Surajit Khalko, Ashok Chaudhury, Parimal Panda and Anarul Hoque

17. **Bioremediation of Fluoride and Nitrate Contamination in Soil and Groundwater** — 252
    Lal Chand Malav, Gopal Tiwari, Abhishek Jangir and Manoj Parihar

18. **Soil Degradation in Mediterranean Region and Olive Mill Wastes** — 267
    Victor Kavvadias, Evangelia Vavoulidou and Christos Paschalidis

19. **Membrane Bioreactor for Perchlorate Treatment** — 277
    Benny Marie B. Ensano, Sivasankar Annamalai and Yeonghee Ahn

20. **Nanobioremediation Technologies for Clean Environment** — 298
    B. Chakrabarti, P. Pramanik, S.P. Mazumdar and R. Dubey

21. **Biochar—An Imperative Amendment for Soil and Environment** — 310
    Sumita Chandel, Ritika Joshi and Ashish Khandelwal

22. **Endophytic Microorganisms from Synanthropic Plants—A New Promising Tool for Bioremediation** — 318
    Olga Marchut-Mikolajczyk and Piotr Drozdzynski

23. Bioremediation of Chlorinated Organic Pollutants in Anaerobic Sediments	**330**
Archana V., Salom Gnana Thanga Vincent and *Thava Palanisami*	
24. Bioremediation of Wastewater by Sulphate Reducing Bacteria	**337**
Panchami Shaji, Salom G.T. Vincent and *Thava Palanisami*	
Index	**345**
About the Editors	**347**

1

Bioremediation

Concepts, Management, Strategies and Applications

Alexandre Marco da Silva,[1,]* *Fátima Piña-Rodrigues,*[2]
Debora Zumkeller Sabonaro,[2,3] *Ivonir Piotrowski,*[2] *José Mauro Santana,*[2]
Lausane Soraya Almeida,[2] *Lucas Hubacek Tsuchiya,*[1] *Marco Vinicius Chaud*[3]
and *Vanderlei Santos*[4]

1. Introduction—Environmental degradation and approaches for environmental repairs

An ecologically healthy and equilibrated environment is the foundation of human life. It provides us with the goods and services that we need to survive and prosper. However, the planet is becoming more and more degraded (Diehl 2018, IPCC 2019). Environmental degradation is any process that reduces the aptitude of a given ecosystem to sustain life. This process is related to biological and/or physical vicissitudes that affect ecological stability.

Such alterations usually modify natural fauna and flora, sometimes causing biodiversity loss, in terrestrial or aquatic systems (Figures 1 and 2). Although they may occur due to natural factors, problems concerning environmental degradation are habitually associated with anthropogenic actions, modifying the trajectory of the evolution of the environment (Tripathi et al. 2017).

Aiming to maintain the ecological health and the quality of the ecosystems services provided by the forests, oceans, rivers, and others ecosystems, currently, there are two options: (i) conserving the remaining original, pristine ecosystems (natural capital), and (ii) restoring the degraded ones (Silva and Rodgers 2018, Arponen 2019).

Several techniques and approaches have been developed in order to fulfill the second option (restore or repair degraded ecosystem): stop the degradation process and/or repair the degradation by means of interventions that might restore the original ecological conditions of the degraded ecosystem, reclaim it, or rehabilitate it. The conceptual differences between these three approaches are depicted in Figure 3.

---

[1] Department of Environmental Engineering – Institute of Sciences and Technology of Sorocaba – São Paulo State University (UNESP).
[2] Department of Environmental Sciences – Federal University of São Carlos – Campus Sorocaba.
[3] ITEPEC Enterprise – Environmental Technology and Consulting.
[4] University of Sorocaba.
* Corresponding author: alexandre.m.silva@unesp.br

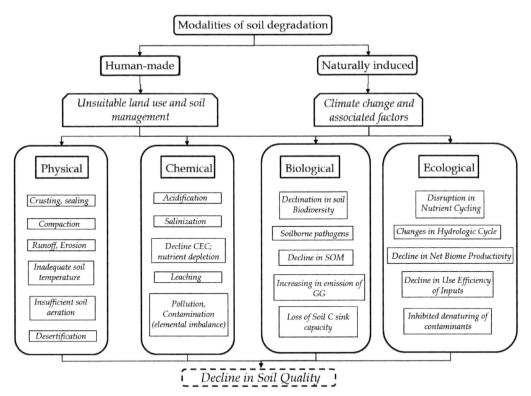

Figure 1. Types of degradation in soil-related systems. Source: (Lal 2015) – modified.

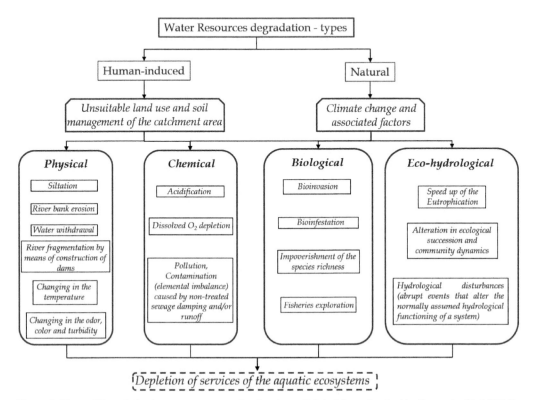

Figure 2. Types of degradation in water recourses-related systems. Original figure inspired by the work of Lal (2015).

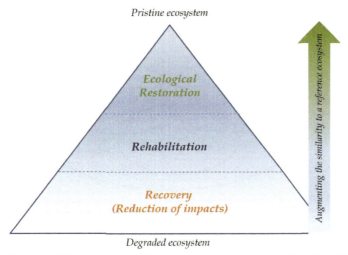

Figure 3. Conceptual differences between the three approaches. Source: modified from (SER 2017).

Activities that aim to repair damaged ecosystems may range from (a) local to (b) regional scale, and from (i) efforts of benevolent volunteers to (ii) logistical projects of the multi-agencies. We find interventions varying from (a) the "do nothing" attitude (i.e., just removing the degradation factor(s) and allowing the natural succession of the environment) to (b) a variety of abiotic and biotic interventions designed at speeding up or shifting the course of ecosystem recovery (Trujillo-Miranda et al. 2018, Rydgren et al. 2019). However, even in very resilient ecosystems, when degradation is severe, advanced or prolonged (or both), the ecosystem may be impotent to entirely recover on its own. This is when restoration practitioners can step in Aronson et al. (2016).

One of the most important options for repair degraded ecosystems is a set of techniques and approaches named bioremediation. Such set of techniques consist chiefly in using biological organism (several species of plants, as well as numerous species of microorganisms) as an agent of extraction, accumulation, and/or transformation (complexation or degradation) of chemical composites, in order to diminish or eradicate the toxicity of the compost. Recovery of contaminated soils, effluent and waste treatment, and cleaning of pipelines and equipment, constitute some examples of the wide application of the bioremediation.

2. Concept and categories of bioremediation

The central point of the bioremediation process is the mechanism of transformation of a contaminant performed by a microorganism or plant (Varjani et al. 2018). Bioremediation embraces a set of biotreatment processes that cover diverse types of biochemical mechanisms that may lead to a humification, target's mineralization, the partial transformation of a composite or altered redox state for metallic elements, for example (Bharagava and Saxena 2020). It is viewed as the safest method to combat some kinds of degraded environments with anthropogenic composites in ecosystems (Paliwal et al. 2012). Environmentally responsive and advantageous cost-saving feature are amongst the major advantages of bioremediation related to both chemical and physical approaches of remediation (Azubuike et al. 2016).

The primal role in bioremediation is that of the interplay of metabolic features of the plant or microbial communities living within that hampered ecosystem (Paliwal et al. 2012). Nonbiological remediation technologies (e.g., excavation, pump-and-treat systems) and bio/phytoremediation might complement each other and they're not mutually exclusive (Pilon-Smiths 2005).

The central difference between bio, phyto, and phycoremediation is the category of living organisms used in each method (Adams et al. 2015, Biswas et al. 2015, Azubuike et al. 2016). Normally the literature considers as bioremediation the microbiological-related processes of

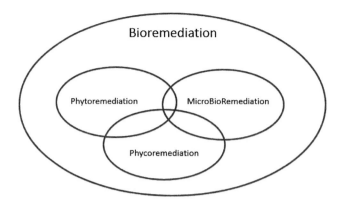

Figure 4. Graphical depiction of the concept of bioremediation and its sub-groups. Diagram elaborated with data provided by Velázquez-Fernández and Muñiz-Hernández (2014).

remediation, and due to this, the phytoremediation and phycoremediation are placed in a different category. However, the term *bioremediation* is here considered as the overall set of techniques that might be sub-divided into three categories: phytoremediation and phycoremediation and micro bioremediation (Figure 4).

We have four major biological agents in bioremediation: (i) vegetation, especially the root system of vascular plants, and the microbiological community, especially (ii) bacteria, (iii) algae, and (iv) fungi. Especially in opened sites and *in situ* techniques (concept explained ahead) the vegetation has been considered, under the variability of environmental conditions, as an agent of acceleration of the process of degradation of organic chemical residues in soils normally in association with a microorganism community (Burges et al. 2018).

2.1 In situ and ex situ techniques

Currently, we have techniques and approaches designed to remediate both terrestrial and aquatic environments considered degraded (Lal 2015, Shishir et al. 2019). In general terms, the techniques are categorized as *ex situ* and *in situ* (Gomes et al. 2013, Lal 2015, Azubuike et al. 2016) (Figure 5).

In situ bioremediation technologies encompass the treatment and manipulation of the contaminants in the local itself (Wadgaonkar et al. 2019). Amidst the most common techniques

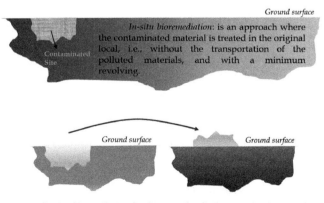

Figure 5. Illustrations of the two major groups of remediation technologies.

categorized as *in situ*, we mention the passive remediation or monitored natural attenuation, also named as "do nothing" approach as mentioned earlier. Furthermore, another major group of *in situ* techniques is constituted by a set of techniques that are embraced in a category named "enhanced techniques" (Table 1).

Ex situ techniques are the actions and treatments that eliminate contaminants at a distinct and separate treatment facility (Wadgaonkar et al. 2019). In works involving the *ex-situ* bioremediation approach, we might cite the bioreactors as a usual technique (Table 2). Also, nutrients may be added in order to accelerate the chemical or physical decomposition of environmental pollutants. For instance, the *ex situ* remediation of heavy metals in soil is further improved by the

Table 1. Explanations of the two main categories of techniques of *in situ* bioremediation.

Technique		Description
Monitored, natural attenuation		This technique establishes a manner of reducing the mass or mobility, as well as the level of toxicity of a contaminant without human influence (EPA 1999). It is based on the premise that under favorable circumstances, certain contaminants may be transformed, degraded, immobilized, as well as detoxified naturally, without any human interference (Sayler et al. 1995). The attenuation occurs due to the chemical, physical and biological processes of degradation of the contaminant (Scow and Hicks 2005). Among the main features and assumptions of this technique are: (i) no site manipulation is required, (ii) we assume that the process of contamination is controlled and finished (no more contaminant is launched into the local environment), (iii) some local environmental features (i.e., presence or absence of sunlight, level of soil moisture, soil pH, aeration and/or level of oxygen of the water (if waterbody), environmental temperature, among others) may favor transformations of the contaminant, (iv) nature and level of concentration of the contaminant. This technique or approach presents as main advantages the low cost, the no necessity of interference, and the no or minimal necessity of human contact with the contaminant and/or contaminated environment (Naeem and Qazi 2019). On the other hand, as main disadvantages, we cite the long-term to adequately transform the contaminant into a non-toxic product, and the dependence of specific site (or local) conditions to execute the work.
Enhanced techniques	Bioslurping and Bioventing	Constitutes a mix of two processes: bioventing and vacuum-enhanced free-product recovery. Bioslurping is an excellent alternative in *in situ* treatment as it also remediates floating waste elements on top of the groundwater. This method applies a vacuum to extract, water, soil vapor, and additional free products from the subsurface. Next, such products are separated and then treated for biodegradation (Varshney 2019). On the other hand, the process of bioventing complements bioslurping, once it involves the injection of air into the contaminated environment (pumps the air only into the unsaturated or vadose zone) at a rate established to maximize local biodegradation and minimize (or eliminate) the off-gassing of volatilized elements to the atmospheric environment. This process is also capable of degrading less volatile organic molecules. Due to the reduced volume of air required, it allows for the treatment of less permeable soils (Khan et al. 2004).
	Biosparging	It involves pumping air and, if necessary, nutrients into the saturated zone (Varshney 2019). This technique is very similar to bioventing in that air is injected into soil subsurface to activate microbial activities, in order to stimulate pollutant elimination from polluted sites. However, different from bioventing, a volume of air is injected in the saturated layer or zone. Such process may cause upward dislocation of volatile organic molecules to the unsaturated zone to facilitate the process of biodegradation (Maitra 2018).
	Phytoremediation	It means the use of plants and, habitually, their associated microbes for environmental cleanup, whatever the kind of substrate: solid, liquid or gaseous (Pilon-Smiths 2005). Rhizobacteria constitutes an important group of microorganisms that live in the plant root system and the interaction with the plants is usually mutual or symbiotic (i.e., they are not pathogenic microorganisms), since the microorganisms produce and liberate some plant growth products as hormones.

Table 2. Description of the major groups of techniques of *ex situ* bioremediation.

Technique	Description
Windrow (or Biopiles) composting	It is a technique for purifying contaminated soils or dried sediments and consists of using appropriate bacteria communities to eliminate pollutants. We can cite the main advantages of the biopile technology: (a) windrow systems are relatively easy to design and make; (b) remediation can be finished in a reasonably short time (3 to 6 months); (c) windrows may be cost-competitive with landfilling and are preferred over landfilling; (d) windrow technology is effective on organic contaminants that are difficult to desorb. On the other hand, the technology presents as the chief limitation the fact that it may not be effective for high contaminant concentrations of some kind of contaminants (Gomes et al. 2013). There are two basic categories of biopiles for bioremediation of contaminated soils, where the main difference is related to the aeration system: (i) static windrow, where the aeration process is forced by means of perforated pipes installed on the bottom of the pile. They are connected to a blower (vacuum pump or an exhaust-driven wind are normally considered too); (ii) dynamic windrow, where the aeration is executed by means intervallic soil tillage, analogous to the procedure applied to the composting windrows (Lopes et al. 2014).
Windrow composting	This technique has been performed using the following steps: (i) initially, contaminated material (soils) is excavated and sieved to remove large-sized debris; next (ii) the sieved material is transported to a composting pad with a provisional protected structure to provide protection from climate extremes. Additional materials, such as straw, manure, agricultural wastes, and wood chips, may be used as bulking facilitators and also as a supplemental carbon. Hence, the materials (soil and amendments) are arranged in long piles, named windrows. The windrow is systematically mixed by turning with an adequate tractor. Indicators such as moisture, temperature, pH, and gas concentration are usually monitored. Once the process of composting is finalized (after approximately 40 days), the windrows are disassembled and the compost is taken to the final disposal (Vishnoi and Dixit 2019).
Bioreactor	Commonly considered in clayed soils. The most common type of bioreactors for treatment contaminated soils are mud reactors or slurry reactors. In this technology, after excavation and screening, the contaminated soil is mixed with an aqueous phase (which may contain microorganisms and/or nutrients and/or surfactants). The "mud" generated contains more or fewer solids (from 10 to 40%) depending on soil type, stirring equipment and aeration system available. The treated sludge is usually dehydrated or alternatively may undergo solid-phase bioremediation. Another option in terms of bioreactor configuration is the reactor's solid phase, where it works with reduced soil moisture contents (approximately 15%) (Tripathi et al. 2017, Naeem and Qazi 2019).
Land farming	This technique does not require an extensive preliminary assessment of polluted sites prior to remediation. This makes the preliminary stage short, and the work simpler and cheaper (Azubuike et al. 2016). It is a method that consists of the biological degradation of contaminants into an upper layer of soil that is periodically turned over for aeration. The spreading of contaminating oily material on the soil and incorporation into the arable layer, also named reactive layer, may directly and differentially affect the microorganisms responsible for biodegradation. Microbial biodegradation, which is the primary mechanism for the elimination of organic pollutants from the environment, forms the basis of this treatment, and the maintenance of an active heterotrophic microbial community is very important (Bicki and Felsot 2018).

addition of organic amendments like biosolid, compost, and municipal solid waste, which is used as both nutrients and conditioners (Varjani et al. 2018).

2.2 The induction of the process of bioremediation

In terms of induction of the process of transformation by microorganisms, three approaches can be considered (Adams et al. 2015, Shishir et al. 2019): bio-augmentation, bio-stimulation and bio-enhancement (Table 3). Such approaches might be considered for work with indigenous microorganisms or by means of inoculation of exogenous microbial species that were previously isolated in culture media.

Table 3. Mechanisms of induction of bioremediation.

Name of the mechanism of induction	Description
Bio-augmentation	It is the inoculation at the contaminated site with microorganisms selected to augment the remediation process.
Bio-stimulation	It consists of the alteration of the environmental conditions to stimulate the degradation of contaminants. This service usually facilitates the work of microorganisms.
Bio-enhancement	When nutrients are provided to enhance the site for the native microorganisms. Furthermore, other environmental factors may be modified in order to improve the working conditions of microorganisms.

Source: Information compiled from Adams et al. (2015) and Shishir et al. (2019).

Table 4. Technologies embraced by phytoremediation. Explanation and comparison.

Approach	Description
Phytostabilization	Usually applied in organics and metals. Because the contaminant is treated *in situ* (retained), land cover (vegetation) is maintained.
Phytodegradation/Rhizodegradation	Mostly for organic products, which are attenuated *in situ*. Vegetation cover is preserved. Complementarily, rhizodegradation occurs in a region surrounding the plant roots. It is an integrative approach, where that the exudates from plants stimulate rhizosphere bacteria to improve biodegradation of pollutants.
Phytovolatilization	Mostly for organic products, which are removed *in situ*. Vegetation cover is preserved.
Phytoextraction	Used frequently for metals, which are removed *in situ*. Vegetation cover is harvested continually.

Source: Modified from Mahar et al. (2016).

Regarding phytoremediation, this kind of technology is accessible for various environments and categories of contaminants. However, this technology has limited application where the concentrations of contaminants are toxic to plants (Mahar et al. 2016). When we consider vascular plants, the literature points out four basic mechanisms (Gratão et al. 2005, Pilon-Smiths 2005, Mahar et al. 2016): phytostabilization, phytodegradation (some researchers consider rhizodegradation as part of this category), phytovolatilization, and phytoextraction (Table 4).

3. Groups of microorganisms

The use of living organisms, including microorganisms, algae, fungi, and higher plants, for the degradation or assimilation of xenobiotic composts, aiming to the environmental decontamination, is an approach known as bioremediation (Banerjee et al. 2018). Such organisms are capable of transforming pollutants and such transformation may be the degradation of organic molecules that hold toxic properties or complexation of composts that contain heavy metals. Among the organic molecules, the most common is perhaps the petroleum-related molecules, whereas heavy metals are also of high importance in terms of contamination. Three main groups of microorganisms are widely used in bioremediation: bacteria, fungi, and algae (Biswas et al. 2015).

3.1 Bacteria

The degradation of toxic substances by bacteria present in the soil depends on the presence of several metabolizing enzymes for their growth, thus being able to remedy the chemical compounds, and reduce the concentrations present in the environment, making such substances less toxic (Williams and Inweregbu 2019).

Table 5. A succinct list of species of bacteria and the correspondent contaminant focused to be metabolized.

Genera/species	Contaminants
Pseudomonas spp., specially: *Pseudomonas aeruginosa* and *Pseudomonas fluorescens*	The *Pseudomonas* spp. are mostly environmental saprotrophs. They usually degrade aliphatic and aromatic hydrocarbons. *P. fluorescens* can be used for bioremediation of crude oil contaminated soil; *P. aeruginosa* can be used to remove or detoxify the heavy metals in most contaminated sites (Williams and Inweregbu 2019).
Rhodococcus wratislaviensis	Egorova et al. (2017) found a good performance of this species of bacteria for remediating soils contaminated by organochlorine compounds (DDT). The obtained final concentrations were permissible according to legislation.
Sphingomonas paucimobilis	In general, most of the species of this genera are of special interest for degrading PAH (Polycyclic aromatic hydrocarbons) – contaminated soils. They are especially adapted to degrade PAHs with moderately high bioavailability, for example, phenanthrene (Zhou et al. 2016). The species *S. paucimobilis* was noticeably efficient for degrading the PAH naphthalene in waters when combined with surfactant products (San Miguel Arnanz et al. 2009).
Escherichia coli	In environmental water samples using a temperature-controlled, Wang et al. (2019) found dual-functionality (biodetection and bioremediation) of copper ions by the cells of such species.

Bacteria with the ability to biodegrade petroleum hydrocarbons can be found in polluted areas or even in areas that have not had prior contact with hydrocarbons (Zhou et al. 2016, Egorova et al. 2017). However, it is easier to find them in already impacted environments because the pollutants provide assistance in the selection of strains with the ability to degrade these compounds. In Table 5, we find a brief list with some bacteria species with the correspondent product that such species usually degrades.

3.2 Algae

Due to the feature to be photosynthetic organisms, algae are the primary producers of biomass which have been the nutritional foundation of a large variety of live forms reliant on algae as the primary source of food. Such organisms also play a crucial role in foraging minerals and molecules from the (contaminated) environment, which will favor humans and the ecosystems, frequently removing them from the contaminants by means of degradation (Vidyashankar and Ravishankar 2016).

Although algae can be efficient in remediating contaminated soils by pesticides (Biwas et al. 2015), this group of organisms is famous for its ability to remediate contaminated soils with some kinds of heavy metals, such as titanium, lead, magnesium, zinc, cadmium, strontium, copper, mercury, nickel and cobalt (Ibuot et al. 2019). Algal systems, especially the cyanobacteria, are not only valuable in treating the waste, but also making a variety of suitable products from the biomass that are generated by such organisms (Vijayakumar and Manoharan 2012).

The culture of photosynthetic microorganisms, including algae, is an option in biological processes that have been shown that microorganisms that inhabit saline or hypersaline environments (halophilic) have a high ability of bioremediation, since they may be used as catalyst in various processes where extreme situations are of vital importance to efficiently repair a polluted environment. In the process of bioremediation of water, algae use their photosynthetic abilities, permitting them to convert sunlight energy into biomass, and then absorb nutrients such as nitrogen and phosphorus which are responsible for the process of eutrophication (Ramirez et al. 2017, Ibuot et al. 2019).

Microalgae, when consorted with bacteria and fungi, also have the capability to remove metals such as iron, aluminum, manganese, magnesium, and zinc from wastewater (Ramirez et al. 2018). Mechanisms of absorption and adsorption are commonly used by algae species to eliminate nutrients, heavy metals (dependent on the species) and other mineralized products from wastewater (Bwapwa et al. 2017). When absorbed, a substance is soaked by the absorbent substance or organism and

Table 6. A succinct list of species of bacteria and the correspondent contaminant focused to be metabolized.

Genera/species	Contaminant
Chlorella vulgaris	By means of a microbial consortium of a bacteria species (*Pseudomona putida*) and *C. vulgaris* (a common single-cell microalgae (green colored) that endures a variety of heavy metals and metalloids), Awasthi et al. (2018) reported an improvement in the growth of such microorganisms and reduction of arsenic-induced oxidative stress in rice, as well as an improvement of the level of nutritious elements in rice.
Pinnularia obscura	In very acidic environments, they have the ability to constitute biofilms (i.e., biological communities with a high degree of organization, where bacteria form structured, coordinated and functional communities). Such biofilms may be utilized to degrade diverse pollutants in the degraded environments, as well as in engineered systems.
Spirulina platensis	This species of algae uses the absorbency mechanism to remove chromium (Fernández et al. 2018).
Westiellopsis prolifica	It can be used both for *in situ* and off site works of remediation. This species was successfully employed in experiments that removed color and some nutrients from effluents (Vijayakumar and Manoharan 2012).

when a substance is adsorpted, the substance is only retained on the adsorbing surface, without being incorporated into the volume or body of the other (Caumette et al. 2015). Table 6 illustrates some examples of algae and their potentialities.

3.3 Fungi

Fungi represent a very important promising group of agents for biodegradation. As example, some species are depicted in Table 7. The capacity of fungi, both yeasts and molds, to transform a wide variety of dangerous chemical materials have attracted the researchers and engineers to use them in bioremediation (Kumar 2017, Vishnoi and Dixit 2019).

Fungi have the capability to mineralize xenobiotic composites to CO_2 and H_2O by means of their non-specific ligninolytic and highly oxidative enzyme apparatus, which also influences the degradation and decolorization of a varied range of dyes (Biswas et al. 2015). Fungi show the capacity to absorb a substantial amount of metals in their cell wall, or also by extracellular polysaccharide slime (Das et al. 2009).

Table 7. A succinct list of species of bacteria and the correspondent contaminant focused to be metabolized.

Genera/species	Contaminant
Gloeophyllum trabeum	They are able to eliminate the pesticide DDT in contaminated soils and can be used directly for the degradation of DDT in soil without any other additional treatment (Purnomo et al. 2011). They are also able to excrete organic acids that may react with copper to render it soluble (Vishwakarma 2019).
Fomitopsis pinicola	They are able to eliminate the pesticide DDT in contaminated soils (Purnomo et al. 2011).
Daedalea dickinsii	They are able to eliminate the pesticide DDT in contaminated soils (Purnomo et al. 2011).
Candida viswanathii (yeast)	They have the capability to biodegrade petroleum hydrocarbons and diesel oil (Junior et al. 2009).
Armillaria mellea	They present the skills to accumulate heavy metals like mercury, lead, cadmium, and copper. Furthermore, as the concentration of mercury increases in the soil, *A. mellea* is shown to store higher concentrations of Hg^{2+} (Kumar 2017).

4. Chemical and biochemical pathways and processes of bioremediation

Some pollutants, such as heavy metals, cannot be degraded by means of any physical, chemical or biological processes that involve microorganisms. However, microorganisms can change the bioavailability of such pollutants and their potential biotoxicity by altering the valence state of specific elements by oxidation or reduction (Rockne and Reddy 2003, Adams et al. 2015, Roane et al. 2015).

In an oxidative environment, the pollutants are oxidized by external electron acceptors such as oxygen or sulfate (Roane et al. 2015). When the reaction is reductive, the electrophilic halogen or nitro groups on the pollutant are reduced by those microorganism groups that consume sugars, fatty acids, or even hydrogen (Adams et al. 2015). The halo- or nitro-group on the pollutant acts as the external electron acceptor (Figure 6).

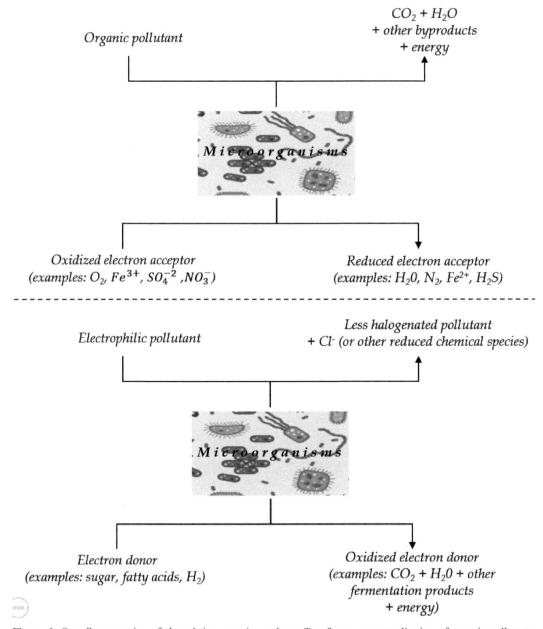

Figure 6. Overall progression of degradation organic products. Top figure: contextualization of organic pollutants; Bottom figure: contextualization of electrophilic pollutants. Source: modified from Rockne and Reddy (2003) and Adams et al. (2015).

Hence, it can be seen that the degradation of any organic molecule requires the production and efficient utilization of enzymes (Rockne and Reddy 2003).

Another important skill of some species of microorganisms is the capacity of making biofilms, being also useful in the process of bioremediation and considered a promising technology. Biofilms are the microbial communities that produce extracellular matrix. Biofilm-linked processes are intrinsically related to biotransformation of contaminants in groundwater and soil, as well as in engineered reactors. In most biofilms formation, the microorganisms represent minor mass (less than 10% of the dry mass), while the (mucilaginous) matrix is the major substance and may account for over 90% (Satpathy et al. 2016).

Bioemulsifiers are exopolymeric products that help the microorganism communities in the biofilm development. Emulsifiers are found in various natural resources and are synthesized mainly by bacteria, but also by fungi and yeast (Alizadeh-Sani et al. 2018). Such materials aid the cells in survival, as well as protect themselves from hostile and extreme situations, predators and especially from the loss of water from the cell. Bacterial adhesion occurs in mobile and stagnant phases. Biofilm formation is a complex process of surface attached community transition from numerous free-floating cells (Karlapudi et al. 2018). Bioemulsifiers have double lipophilic and hydrophilic properties and they are higher in molecular weight when compared to biosurfactants, since they are complex mixtures of products such as heteropolysaccharides, lipopolysaccharides, lipoproteins, and proteins (Alizadeh-Sani et al. 2018).

In turn, biosurfactants and bioemulsifiers are amphiphilic molecular chains and are produced as extracellular or as a part of the cell membrane by bacteria (an amphiphilic molecule is a molecule that has hydrophilic (polar) and hydrophobic (non-polar) characteristics). On the other hand, biofilms are bacterial communities that protect the bacterial cells from some hostile conditions (Karlapudi et al. 2018).

Biosurfactants emulsify the molecular chains, intensify the water solubility and make the molecular chains more reachable for the microorganisms (Elkhawaga 2018). Because of their potential advantages, such kind of product is largely used in several industrial segments such as chemistry, aliment production, agriculture, pharmaceutics, as well as cosmetics (Pacwa-Płociniczak et al. 2011).

On the other hand, the fungi, with their enzymes, have the ability to degrade a wide variety of environmentally persistent pollutants and transform industrial and agro-industrial wastes into products. Specifically, mushrooms can produce extracellular peroxidases, ligninase (lignin peroxidase, manganese-dependent peroxidase, and laccase), cellulases, pectinases, xylanases, and oxidases (Nyanhongo et al. 2007). These are able to oxidize recalcitrant pollutants *in vitro*. These enzymes are typically induced by their substrates. The uptake of pollutants/xenobiotics by mushrooms involves a combination of two processes: (i) bioaccumulation, i.e., active metabolism-dependent processes, which include both transport into the cell and partitioning into intracellular components, and (ii) biosorption, i.e., the binding of pollutants to the biomass without requiring metabolic energy.

5. Final remarks

The strategic approaches of bioremediation are ideally useful in circumstances when the level of pollution is moderately low or does not require fast restoration or where chemical treatment is not ideal (Mitra and Mukhopadhyay 2016).

Site specification and severity of contamination are crucial factors that determine the success of the application of bioremediation. Finally, we draw attention to the fact that insufficient or inadequate information about the environmental parameters on the extent and rate of biodegradation generate a source of ambiguity (Naeem and Qazi 2019).

References

Abatenh, E., Gizaw, B., Tsegaye, Z. and Wassie, M. 2017. Application of microorganisms in bioremediation—review. Open J Environ Biol 1(1): 38–46.

Adams, G.O., Fufeyin, P.T., Okoro, S.E. and Ehinomen, I. 2015. Bioremediation, biostimulation, and bioaugmention: a review. Int. J. of Environmental Bioremediation & Biodegradation 3: 28–39.

Alizadeh-Sani, M., Hamishehkar, H., Khezerlou, A., Azizi-Lalabadi, M., Azadi, Y., Nattagh-Eshtivani, E., Fasihi, M., Ghavami, A., Aynehchi, A. and Ehsani, A. 2018. Bioemulsifiers derived from microorganisms: applications in the drug and food industry. Adv. Pharma. Bulletin. 8(2): 191–199.

Aronson, J., Blatt, C. and Aronson, T. 2016. Restoring ecosystem health to improve human health and well-being: physicians and restoration ecologists unite in a common cause. Ecol. Soc. 21: 39.

Arponen, A. 2019. Restoration where it pays off. Nature Eco. & Evolution 3(1): 16.

Awasthi, S., Chauhan, R., Dwivedi, S., Srivastava, S., Srivastava, S. and Tripathi, R.D. 2018. A consortium of alga (*Chlorella vulgaris*) and bacterium (*Pseudomonas putida*) for amelioration of arsenic toxicity in rice: A promising and feasible approach. Environ. Exp. Bot. 150: 115–126.

Azubuike, C.C., Chikere, C.B. and Okpokwasili, G.C. 2016. Bioremediation techniques–classification based on site of application: principles, advantages, limitations, and prospects. World J. Microb. Biot. 32: 180.

Banerjee, A., Jhariya, M.K., Yadav, D.K. and Raj, A. 2018. Micro-remediation of metals: a new frontier in bioremediation. Handbook of Environmental Materials Manage. 1–36.

Bharagava, R.N. and Saxena, G. 2020. Progresses in bioremediation technologies for industrial waste treatment and management: Challenges and future prospects. pp. 531–538. *In*: Saxena, G. and Bharagava, R.N. (eds.). Bioremediation of Industrial Waste for Environmental Safety. Springer, Singapore.

Best, J. 2019. Anthropogenic stresses on the world's big rivers. Nat. Geosci. 12: 7–21.

Bicki, T.J. and Felsot, A.S. 2018. Remediation of pesticide contaminated soil at agrichemical facilities. pp. 81–100. *In*: Honeycutt, R. (ed.). Mechanisms of Pesticide Movement into Ground Water. CRC Press.

Biswas, K., Paul, D. and Sinha, S.N. 2015. Biological agents of bioremediation: A concise review. Front. in Environ. Microbiol. 1(3): 39–43.

Burges, A., Alkorta, I., Epelde, L. and Garbisu, C. 2018. From phytoremediation of soil contaminants to phytomanagement of ecosystem services in metal contaminated sites. Int. J. of Phytoremed. 20(4): 384–397.

Bwapwa, J.K., Jaiyeola, A.T. and Chetty, R. 2017. Bioremediation of acid mine drainage using algae strains: A review. S. African J. of Chemical Eng. 24: 62–70.

Caumette, P., Lebaron, P., Matheron, R., Normand, P. and Sime-Ngando, T. 2015. Environmental microbiology: fundamentals and applications. pp. 993. Bertrand, J.C. (ed.). Dordrecht: Springer.

Das, B.K., Roy, A., Singh, S. and Bhattacharya, J. 2009. Eukaryotes in acidic mine drainage environments: potential applications in bioremediation. Rev. Environ. Sci. Bio. 8(3): 257–274.

Diehl, P. 2018. Environmental Conflict: An Anthology. Routledge.

Egorova, D.O., Farafonova, V.V., Shestakova, E.A., Andreyev, D.N., Maksimov, A.S., Vasyanin, A.N., Buzmakov, S.A. and Plotnikova, E.G. 2017. Bioremediation of soil contaminated by dichlorodiphenyltrichloroethane with the use of aerobic strain *Rhodococcus wratislaviensis* Ch628. Eurasian Soil Sci. 50(10): 1217–1224.

Elkhawaga, M.A. 2018. Optimization and characterization of biosurfactant from *Streptomyces griseoplanus* NRRL-ISP 5009 (MS 1). J. Appl. Microbiol. 124(3): 691–707.

EPA (Environmental Protection Agency). 1999. Use of monitored natural attenuation at Superfund, RCRA Corrective Action, and underground storage tank sites. Directive 9200.4-17P, 32 pp., EPA, Office of Solid Waste and Emergency Response, Washington, D.C. Link: https://www.epa.gov/sites/production/files/2014-02/documents/d9200.4-17.pdf.

Fernández, P.M., Vinarta, S.C., Bernal, A.R., Cruz, E.L. and Figueroa, L.I. 2018. Bioremediation strategies for chromium removal: current research, scale-up approach and future perspectives. Chemosphere 208: 139–148.

Gomes, H.I., Dias-Ferreira, C. and Ribeiro, A.B. 2013. Overview of *in situ* and *ex situ* remediation technologies for PCB-contaminated soils and sediments and obstacles for full-scale application. Sci. Total Environ. 445: 237–260.

Gratão, P.L., Prasad, M.N.V., Cardoso, P.F., Lea, P.J. and Azevedo, R.A. 2005. Phytoremediation: green technology for the cleanup of toxic metals in the environment. Brazilian J. of Plant Phys. 17(1): 53–64.

Ibuot, A.A., Gupta, S.K., Ansolia, P. and Bajhaiya, A.K. 2019. Heavy metal bioremediation by microalgae. Microbial Biodegradation of Xenobiotic Compounds 57.

IPCC (Inter Panel Governmental on Climate Change). 2019. Climate change and Land: An IPCC special report on climate change, desertification, land degradation, sustainable land management, food security, and greenhouse gas fluxes in terrestrial ecosystems. Access in August 2019. Link: https://www.ipcc.ch/report/srccl/.

Junior, J.S., Mariano, A.P. and de Angelis, D. 2009. Biodegradation of biodiesel/diesel blends by Candida viswanathii. Afr. J. Biotechnol. 8(12): 2774–2778.

Karlapudi, A.P., Venkateswarulu, T.C., Tammineedi, J., Kanumuri, L., Ravuru, B.K., ramu Dirisala, V. and Kodali, V.P. 2018. Role of biosurfactants in bioremediation of oil pollution—A review. Petroleum. 4(3): 241–249.

Khan, F.I., Husain, T. and Hejazi, R. 2004. An overview and analysis of site remediation technologies. J. Environ. Manage. 71(2): 95–122.

Kumar, V.V. 2017. Mycoremediation: A step toward cleaner environment (Chapter 10). p. 240. *In*: Prasad, R. (ed.). Mycoremediation and Environmental Sustainability. Cham: Springer.

Lal, R. 2015. Restoring soil quality to mitigate soil degradation. Sustainability 7: 5875–5895.

Lopes, J.A., Silva, G., Marques, M. and Correa, S.M. 2014. Bioremediation of clayey soil contaminated with crude oil: comparison of dynamic and static biopiles in lab-scale. Proceedings from Linnaeus Eco-Tech'14. Kalmar, Sweden. https://doi.org/10.15626/Eco-Tech.2014.

Mahar, A., Wang, P., Ali, A., Awasthi, M.K., Lahori, A.H., Wang, Q., Li, R. and Zhang, Z. 2016. Challenges and opportunities in the phytoremediation of heavy metals contaminated soils: a review. Ecotox. Environ. Safe. 126: 111–121.

Mitra, A. and Mukhopadhyay, S. 2016. Biofilm mediated decontamination of pollutants from the environment. AIMS Bioeng. 3(1): 44–59.

Maitra, S. 2018. *In situ* bioremediation—an overview. Research J. of Life Sciences, Bioinformatics, Pharmaceutical and Chem. Sci. 4(6): 576–598.

Naeem, U. and Qazi, M.A. 2019. Leading edges in bioremediation technologies for removal of petroleum hydrocarbons. Environ. Sci. Pollut. R. 1–13.

Pacwa-Płociniczak, M., Płaza, G.A., Piotrowska-Seget, Z. and Cameotra, S.S. 2011. Environmental applications of biosurfactants: recent advances. Int. J. Mol. Sci. 12(1): 633–654.

Paliwal, V., Puranik, S. and Purohit, H.J. 2012. Integrated perspective for effective bioremediation. Appl. Biochem. Biotech. 166: 903–924.

Pilon-Smiths, E. 2005. Phytoremediation. Annu Rev Plant Biol. 56: 15–39.

Purnomo, A.S., Mori, T., Takagi, K. and Kondo, R. 2011. Bioremediation of DDT contaminated soil using brown-rot fungi. Int. Biodeter. Biodegr. 65(5): 691–695.

Ramírez, M.E., Vélez, Y.H., Rendón, L. and Alzate, E. 2018. Potential of microalgae in the bioremediation of water with chloride content. Braz. J. Biol. 78(3): 472–476.

Rydgren, K., Halvorsen, R., Töpper, J.P., Auestad, I., Hamre, L.N., Jongejans, E. and Sulavik, J. 2019. Advancing restoration ecology: A new approach to predict time to recovery. J. Appl. Ecol. 56(1): 225–234.

Roane, T.M., Pepper, I.L. and Gentry, T.J. 2015. Microorganisms and metal pollutants. pp. 415–439. *In*: Environmental Microbiol. Academic Press.

Rockne, K. and Reddy, K. 2003. Bioremediation of Contaminated Sites. University of Illinois at Chicago. http://tigger.uic.edu/~krockne/proceeding9.pdf#search=%22bioremediation%20of%20pesticides%20and%20herbicides%22.

San Miguel Arnanz, V., Peinado, C., Catalina, F. and Abrusci Bernal, C. 2009. Bioremediation of naphthalene in water by Sphingomonas paucimobilis using new biodegradable surfactants based on poly (ε-caprolactone). Int. Biodeter. Biodegr. 63(2): 217–223.

Satpathy, S., Sen, S.K., Pattanaik, S. and Raut, S. 2016. Review on bacterial biofilm: A universal cause of contamination. Biocatalysis and Agri. Biotech. 7: 56–66.

Sayler, G.S., Layton, A., Lajoie, C., Bowman, J., Tschantz, M. and Fleming, J.T. 1995. Molecular site assessment and process monitoring in bioremediation and natural attenuation. Appl. Biochem. Biotech. 54(1-3): 277–290.

Scow, K.M. and Hicks, K.A. 2005. Natural attenuation and enhanced bioremediation of organic contaminants in groundwater. Curr. Opin. Biotech. 16(3): 246–253.

SER (Society for Ecological Restoration–Australasia Chapter). 2017. National standards for the practice of ecological restoration in Australia, 2nd edition. Accessed in December 2019. Link: http://www.seraustralasia.com/standards/appendix1.html.

Shishir, T.A. and Mahbub, N. 2019. Review on bioremediation: A tool to resurrect polluted rivers. Pollution 5(3): 555–568.

Silva, A.M. and Rodgers, J. 2018. Deforestation across the world: Causes and alternatives for mitigating. Int. J. Environ. Sci. Dev. 9: 67–73.

Silva, R.A., Oliveira Afonso, A.A., Francescony, W. and Silva, A.M. 2019. Technical assessment and decision making for the environmental recovery of waterways and their banks: a science-based protocol. Int. J. Environ. Sci. Technol. 16: 2083–2090.

Tripathi, V., Edrisi, S.A., Chen, B., Gupta, V.K., Vilu, R., Gathergood, N. and Abhilash, P.C. 2017. Biotechnological advances for restoring degraded land for sustainable development. Trends Biotechnol. 35: 847–859.

Trujillo-Miranda, A.L., Toledo-Aceves, T., López-Barrera, F. and Gerez-Fernández, P. 2018. Active versus passive restoration: Recovery of cloud forest structure, diversity and soil condition in abandoned pastures. Ecolog. Eng. 117: 50–61.

Varjani, S.J., Agarwal, A.K., Gnansounou, E. and Gurunathan, B. (eds.). 2018. Bioremediation: Applications for Environmental Protection and Manage (ISSN 2522-8366). New York, NY: Springer.

Varshney, K. 2019. Bioremediation of pesticide waste at contaminated sites. J. of Emerging Tech. and Innovative Research 6(5): 128–134.

Velázquez-Fernández, J.B. and Muñiz-Hernández, S. 2014. Bioremediation: Processes, Challenges and Future Prospects. Nova Science Publishers, Inc.

Vijayakumar, S. and Manoharan, C. 2012. Treatment of dye industry effluent using free and immobilized cyanobacteria. J. of Bioremediation and Biodegradation 3(10): 1–6.

Vidyashankar, S. and Ravishankar, G.A. 2016. Algae-based bioremediation: bioproducts and biofuels for biobusiness. pp. 457–493. *In*: Labuto, G. and Carrilho, E.N.V.M. (eds.). Bioremediation and Bioeconomy. Elsevier.

Vishnoi, N. and Dixit, S. 2019. Bioremediation: new prospects for environmental cleaning by fungal enzymes. pp. 17–52. *In*: Yadav, A.N., Mishra, S., Singh, S. and Gupta, A. (eds.). Recent Advancement in White Biotech. Through Fungi. Springer, Cham.

Vishwakarma, P. 2019. Role of macrofungi in bioremediation of pollutants. pp. 285–304. *In*: Arora, P.K. (ed.). Microbial Metabolism of Xenobiotic Compounds. Springer, Singapore.

Wadgaonkar, S.L., Ferraro, A., Nancharaiah, Y.V., Dhillon, K.S., Fabbricino, M., Esposito, G. and Lens, P.N. 2019. *In situ* and *ex situ* bioremediation of seleniferous soils from northwestern India. J. Soils Sediments 19(2): 762–773.

Wang, W., Jiang, F., Wu, F., Li, J., Ge, R., Li, J., Tan, G., Pang, Y., Xiaofeng, Z., Ren, X., Fan, B. and Lyu, J. 2019. Biodetection and bioremediation of copper ions in environmental water samples using a temperature-controlled, dual-functional *Escherichia coli* cell. Appl. Microbiol. Biot. 1–11.

Williams, J.O. and Inweregbu, O.A. 2019. Comparative bioremediation potentials of *Pseudomonas aeruginosa* KX828570 and *Bacillus megaterium* KY085976 on polluted terrestrial soil treated with oil spill dispersant. S. Asian J. of Res. in Microbiol. 4(2): 1–17.

Zhou, L., Li, H., Zhang, Y., Han, S. and Xu, H. 2016. *Sphingomonas* from petroleum-contaminated soils in Shenfu, China and their PAHs degradation abilities. Braz. J. Microbiol. 47(2): 271–278.

2

Bioremediation

Current Status, Prospects and Challenges

Ruby Patel,[1] *Anandkumar Naorem,*[2] *Kaushik Batabyal*[3,*] *and Sidhu Murmu*[4]

1. Introduction

The growing industrialization also brings environmental pollution which has become a big concern in today's world. The pollution not only contaminates the environment but also makes the living organisms seriously ill. There are several types of pollutants such as solid waste, liquid waste, toxic gases, radioactive waste and even heavy metals that are highly persistent in soil. A novel technology should be constructed in order to assure the safety of the living organisms and to maintain the sustainability of the environment from the harmful reactions of environmental pollution, and bioremediation is one of those methods. Bioremediation is the utilization of microorganisms to clean up contaminated soils, aquifers, sludge, residues, and is considered an ecofriendly, economical, efficient and sustainable technology to improve the polluted sites.

Human health and natural ecosystem have gone under threat because of the huge load of contaminants such as polycyclic aromatic hydrocarbons (PAHs), petroleum and related products, pesticides, chlorophenols, and heavy metals that enter the soil and water. Chemical and physical technologies for soil remediation are either inefficient or too costly and bioremediation is believed to be one of the most cost-effective methods for soil remediation. Metals and metalloids like Cr, Pb, Hg, U, Se, Zn, As, Cd, Ag and Ni contaminate the water as they are hazardous to human health and environment. Conventional physicochemical methods such as electrochemical treatment, ion exchange, precipitation, osmosis, evaporation and sorption are too costly and not environmentally friendly. However, bioremediation processes show efficient removal of metals, while present in very low concentrations where physicochemical removal methods fail to operate (Mani and Kumar 2014).

---

[1] Department of Agricultural Chemistry and Soil Science, Bidhan Chandra Krishi Viswavidyalaya, Mohanpur, Nadia, West Bengal, India.
  Email: rubypatelssac@gmail.com
[2] ICAR-Central Arid Zone Research Institute, Jodhpur-342003 (Rajasthan), India.
  Email: naoremanand@gmail.com
[3] Department of Agricultural Chemistry and Soil Science, Bidhan Chandra Krishi Viswavidyalaya, Mohanpur, Nadia, West Bengal, India.
[4] Department of Agricultural Chemistry and Soil Science, Bidhan Chandra Krishi Viswavidyalaya, Mohanpur, Nadia, West Bengal, India.
  Email: sidhumurmu@gmail.com
* Corresponding author: kbatabyal@rediffmail.com

Bioremediation technologies have been developed in the last few decades and are used to alleviate environmental accidents and systematic contaminations. Bioremediation works through degradation, eradication, immobilization, or detoxification of different chemical wastes and hazardous materials through the microbial activities with remediating power from the surrounding sites. The main principle of biodegradation is degrading and transforming pollutants such as hydrocarbons, oil, heavy metal, pesticides and dyes, etc. into a less toxic form. Therefore, bioremediation is referred often as an ecofriendly biological mechanism of recycling wastes into another form that can be less toxic to the environment.

2. Bioremediation of solid waste

Numerous studies suggest that exposure to chemicals and other substances or gases like suspended particulate matter, carbon dioxide (CO_2), carbon monoxide (CO), nitrogen dioxide (N_2O), sulphur dioxide (SO_2) and hydrogen sulfide (H_2S) emitted from burning of solid waste in the dumping sites is highly dangerous, causing a significant risk to human health (Rim-Rukeh 2014). Compostable organic matter (fruit and vegetable peels, food waste), recyclables (paper, plastic, glass, metals, etc.), toxic substances (paints, pesticides, used batteries, medicines) and soiled waste (blood-stained cotton, sanitary napkins and disposable syringes) are few examples of municipal solid waste (Kausal et al. 2012). Depending on factors such as water content, and temperature, harmful odorous gases are often produced due to the decomposition of piled up solid wastes and particulate solid wastes can be dispersed into the atmosphere and contaminate the surrounding atmospheric environment (Tian et al. 2013). Municipal solid waste (MSW) open dump site operation is an important component of waste management throughout the country. Dump site fires are common practice. Both direct and indirect greenhouse gases have been emitted from the MSW management process such as landfill, composting and incineration. Where land is scarce, incineration has been the most successful method but the costs of landfills are too high and environmental benefits of incineration are still in question (Kinnaman 2017). In the future, better technology with regard to municipal solid waste open dump site operation and emission control should be developed.

Bioremediation involving bacteria and fungi or yeast is a comparatively far less expensive and more ecofriendly technology to remediate solid waste than other techniques. Awasthi et al. (2017) reported that some fungal strains of *Trichoderma* sp., *Aspergillus niger* and *Aspergillus flavus* are potential strains which are capable to exclude metal from the leachate via biosorption process, among which the *Trichoderma* sp. was found to be an excellent bioremediating agent for Cd^{2+} absorption. Bacterial isolates such as *Serratiaproteamaculans* S1BD1, *Alcaligenes* sp. OPKDS2 and *Rhodococcus erythropolis* OSDS1 strains with high bioremediating efficiency could potentially not only be used for the bioaugmentation of the solid waste management site but also possess a high tolerance level to a wide range of salinity and pH (Xia et al. 2017).

Apart from microbial bioremediation, another form of bioremediation, phytoremediation uses green plants with higher biomass to uptake pollutants in their tissues and reduce the availability of the pollutants especially heavy metals in the soil. There are two different approaches such as contaminant extraction/degradation and stabilization that can be used for phytoremediation of landfill sites. In the first approach, plants grown in heavy metal-contaminated sites can accumulate heavy metals through their uptake mechanism into different plant tissues like leaves, stem, grain, and roots, which can be harvested and disposed from the contaminated sites. The second approach involves establishing the plant vegetation near landfill site of solid waste that can help in the stabilization and prevention of runoff to surrounding areas from the contaminated sites through acting as a barrier. Moreover, plants can prevent inorganic contaminants by stabilizing them near the root zone from leaching to groundwater. Trees such as *Hibiscus tiliaceus, Acacia mangium, Dendrocalamus maroochy, Casuarina* species and *Eucalyptus grandis* growing on landfill caps can also reduce methane emissions from landfills by providing favorable environments for methane-oxidizing bacteria (Venkatraman and Ashwath 2007).

Phytostabilization is a useful technique to rectify toxicity of municipal solid waste containing heavy metals such as Cd, Cu, As, Zn and Cr. Tree species such as *Populus nigra* L. or *Agrostis capillaris* L. is often used for phytostabilization of Cu-polluted soil and municipal solid waste compost. A few endophytes like *Enterobacter* sp. and *Pseudomonas putida* isolated from poplar (*Populus deltoides*) were found to be able to degrade volatile organic compounds such as trichloroethylene (TCE) (Touceda-Gonzalez et al. 2017).

3. Bioremediation of liquid waste

Wastewater, fats, oils or grease (FOG), used oil, liquids, solids, gases, or sludge and hazardous household liquids are known as liquid waste. These liquids are hazardous or potentially toxic to human health or the environment. They can also be the discarded commercial products classified as "Liquid Industrial Waste" including cleaning fluids or pesticides, or the by-products of manufacturing processes. Liquid waste can be generated by society sector including sewage and wastewater from industrial processes such as food and agricultural processing, and manufacturing. Essential nutrients (phosphate, nitrates, micronutrients, etc.) and organic dyes (potential carbon source) are found in industrial wastewater like textile wastewater for algae cultivation. The application of microalgae in the treatment of industrial wastewater (dyes and nutrients removal) before disposal is one of the study areas. The integration of bioremediation-biodiesel production process can potentially improve biodiesel production as well as wastewater treatment (Fazal et al. 2018).

Petroleum production process causes a load of waste, effluents, atmospheric emissions and risks to the transport of petroleum products (Szklo and Uller 2008). Oily sludge, the most abundant oily wastes generated in petroleum refineries, is a semisolid material contaminated by oil, the water produced and the chemicals used in petroleum processing (Heidarzadeh et al. 2010).

The conventional techniques of remediation follow digging up of contaminated soil and transfer it to a landfill that calls for certain issues such as risks in excavation, handling, and transport of hazardous material. Moreover, the liquid waste from the landfill during monsoon contaminates the surrounding area and may lead to secondary pollution. Bioremediation is an option to destroy various contaminants using natural biological activity. Because of the difficulty in achieving sufficient oil removal by physical washing and collection, especially for oil that had moved into the subsurface, bioremediation became an efficient method for continuing treatment of the shoreline.

In cold regions, the biological-based remediation process is implemented during the warm season when the temperature is favorable for the microbial activities that might be one of the limitations of bioremediation in cold areas. Therefore, in order to overcome the problem, utilization of psychrophiles in oil-polluted sites can be effective in low temperatures. There have been several studies that address the oil-degrading potential of microbes at low temperatures (Yang et al. 2016, Wang et al. 2016).

The construction of contaminated soil mixed with oily sludge in cells or piles to stimulate internal aerobic microbial activity by aeration is known as biopile technology. Biostimulation process is employed to improve the microbial activity through the inclusion of humidity or nutrients (nitrogen and phosphorus). Bacteria degrade hydrocarbons adsorbed in the soil particles, ultimately reducing their concentration in the soil. Mostly, biopiles are constructed on an impermeable base to reduce leachate probability to the subsurface environment (Kriipsalu and Nammari 2010).

Different types of bioreactors have been developed, mainly for the treatment of liquid effluents. Slurry bioreactors are one of the most important bioreactors for treatment of oily sludge. They might be suitable for the degradation of highly resistant compounds present in the liquid effluents (Rizzo et al. 2010). This bioprocess was inoculated with hydrocarbonoclastic microorganisms such as *Acinetobacter, Alcaligenes, Ochrabactrum, Pseudomonas/Flavimonas, Rhodococcus* and *Stenotrophnomonas* with which 50 percent reduction of oils and greases was achieved after 50–100 days of inoculation in bioreactor (Ward and Singh 2003).

Numerous methods have been investigated such as incineration, chlorination, UV oxidation and solvent extraction for hydrocarbon treatment. However, most of these approaches are associated with several drawbacks such as less removal efficiency and adverse ecological impact. Therefore, biological technologies like phytoremediation and bioremediation are more suitable than other methods for the cleanup of contaminated sites.

The hydrocarbonoclastic bacteria like *Alcanivorax, Marinobacter, Thallassolituus, Cycloclasticus* and *Oleispira* are the most efficient genera that can degrade the hydrocarbons. In the hydrocarbon-polluted marine environment, obligate hydrocarbonoclastic bacteria like *Alcanivorax borkumensis* is ubiquitous and this strain was able to metabolize linear and branched alkanes, with the exception of aromatic hydrocarbons (Yakimov 1998). *Thallassolituus oleivorans, Oleiphilus* and *Oleispira* are highly specific in breaking of aliphatic hydrocarbons from C7 to C20 carbons (Yakimov 2004). *Cycloclasticus* is addressed to mineralize various PAHs such as naphthalene, phenanthrene, anthracene, pyrene and fluorene (Kasai 2002) and other non-obligate hydrocarbonoclastic bacteria have been described to degrade various classes of hydrocarbons. Moreover, *Acinetobacter* and *Sphingomonas* are well reported for the degradation of aliphatic compounds (Harayama 2004). However, biostimulation treatment through nutrient amendment to hydrocarbon-contaminated sediments enhanced the biodegradation activity of hydrocarbonoclastic bacteria such as *Alcanivorax, Marinobacter,* and *Cycloclasticus*. Among fungi, *Amorphoteca, Neosartorya, Talaromyces, Aspergillus* and *Graphium* are able to degrade hydrocarbon.

4. Bioremediation of toxic gases

Toxic gases are such as hydrogen sulfide or sewer gas (H_2S), carbon monoxide (CO), methane gas (CH_4), carbon dioxide (CO_2), nitrous oxide (N_2O) and solvents such as kerosene, gasoline, paint strippers, and degreasers cause damage to living tissues and severe illness or in extreme cases, death of living organisms. Carbon monoxide inhibits the body's ability to transport oxygen to all parts of the body. H_2S is a toxic, flammable and corrosive gas which must be reduced to acceptable levels (less than 10 ppmv in air) (Amirfakhri 2006) and H_2S in the gas stream can cause serious problems such as corrosion and environmental pollutions after combustion. Various methods were introduced in gas sweetening process. Biodesulfurization (BDS) is an economical method that can remove H_2S using sulfur-oxidizing bacteria. In this biochemical reaction, H_2S is utilized as electron donor and oxygen and nitrate act as an electron acceptor in aerobic and anaerobic conditions, respectively (Irani et al. 2011). Partial oxidation of hydrogen sulfide under denitrification conditions leads to the formation of sulfur and nitrogen which are less toxic to the environment (Fernandez 2014). Global warming induced by increasing concentrations of greenhouse gases in the atmosphere is of great concern. Carbon dioxide (CO_2) is the principal greenhouse gas and its concentrations have increased rapidly in atmosphere from 280 ppm at the initiation of the industrial revolution to 400 ppm today due to an increase in industrialization (IPCC 2013). A highly efficient biological system is microalgae incorporated photo-bioreactor used for transforming CO_2 into biomass. The air streams containing a high concentration of CO_2 (2–15%) may be introduced directly into a high-density culture of *Chlorella* sp. in a semicontinuous photo-bioreactor, which results in higher algae growth, biomass productivity, lipid productivity with higher amount of CO_2 removed from the air streams (Chiu 2008). Wastewater discharges into water bodies cause serious health and environmental concerns: emission of toxic (hydrogen sulfide, H_2S) or noxious (methane, CH_4) gases from sewer systems.

Biofiltration is the use of microorganisms to break down pollutants present in an air current in which microbes are locked to a porous medium. The microorganisms grow in a biofilm on the surface of a medium surrounding the medium particles. As the air passes through the filter bed, the contaminants in the air contaminants sorb into the biofilm and are biodegraded onto the filter medium. The filter-bed medium consists of inert substances such as compost and peat which supply additional nutrient and ensure large surface areas. Mostly, the air stream is humidified before entering the biofilter reactor; if humidification proves inadequate, direct irrigation of the bed may be required (Devinny et al. 2017).

5. Bioremediation of heavy metal

Soil, groundwater and air contamination by heavy metals is a big problem concerning the safety of environments and human health. Heavy metal might be rashly introduced to soils via anthropogenic activities such as mining, smelting, warfare and military training, electronic industries, fossil fuel consumption, waste disposal, agrochemical use, and irrigation or by geogenic processes. The fossil fuel coal contains different heavy metals including Hg, Pb, Cd, Cr, Cu, Co, Zn and Ni with a concentration range of 0.1 to 18 mg kg^{-1}; these heavy metals are discharged into the environment in various forms including vapor, flue gas particulate matter, fly ash, and bottom ash upon coal combustion (Nalbandian 2012). Mine spoils, industrial waste, and construction waste when discharged into soil results in heavy metal contamination and also land application of fertilizers, pesticides, biosolids, and animal manure and crop irrigation with sewage water and poorly-treated industrial wastewater are major pathways for heavy metals contamination to land.

Numerous remediation techniques such as *in situ* and *ex situ* including surface capping, encapsulation, landfilling, soil flushing, soil washing, electrokinetic extraction, stabilization, solidification, vitrification, phytoremediation, and bioremediation methods have been developed to remediate the heavy metal-contaminated sites (Liu 2018). Among all the available technology, bioremediation is productive and effective to rectify the heavy metal contaminated environment. Heavy metal toxicity can also be reduced by microorganisms via valence transformation (e.g., Cr (VI) to Cr (III), SeO_4^{2-} to Se), biosorption (to cell surface), extracellular chemical precipitation (e.g., by S^{2-} from sulfur-reducing bacteria), and volatilization (e.g., dimethylselenide, trimethylarsine and Hg vapor) (Garbisu and Alkorta 2003).

Plant roots secrete metal-mobilizing substances in the rhizosphere called phytosiderophores (Lone et al. 2008). Metal dissolution is accomplished by secretion of H$^+$ ions by roots which can acidify the rhizosphere and H$^+$ ions can displace heavy metal cations adsorbed to soil particles (Alford et al. 2010). Poplar (*Populus* sp.) and willow (*Salix* sp.) can be used on the edges of wetlands as phytosiderophores to remediate heavy metal contaminated soil (Pilon-Smits 2005). The aquatic plants such as water lily (*Nymphaea spontanea*) accumulate chromium metals by their roots (Choo et al. 2006).

Thlaspi sp., *Arabidopsis* sp., *Sedum alfredii* species have been extensively used as hyperaccumulator of heavy metal. *Thlaspi* sp. are well known to accumulate more than one metal, i.e., *T. caerulescens* for Cd, Ni, Pb and Zn, *T. goesingense* for Ni and Zn, *T. ochroleucum* for Ni and Zn, and *T. rotundifolium* for Ni, Pb and Zn. Due to specific rooting strategy and a high uptake rate, *T. caerulescens* has higher uptake of Cd resulting from the Cd-specific transport channels or carriers in the root membrane (Prasad and Freitas 2003).

Heavy metal ions trapped by the cellular structure of a microorganism and subsequently their sorbtion onto the binding sites of the cell surface is known as biosorption (Malik 2004). Various heavy metals like Cu, Cr, Hg, Pb and Ni have been tested on bacterial species such as *Pseudomonas, Enterobacter, Bacillus,* and *Micrococcus* species to remediate the heavy metal contaminated site via biosorption process. Their high surface-to-volume ratios and active chemosorption sites (teichoic acid on the cell wall) result in excellent sorption capacity of heavy metal (Mosa et al. 2016). *Bacillus cereus* and *Micrococcus luteus* of bacterial sp. are used to remove Hg and Pb, respectively and also *Desulfovibrio desulfuricans* has been used to remove Cu, Cr, and Ni from contaminated seawater.

According to Akar et al. (2005), lead (II) ions were removed by *Botrytis cinerea* in a batch reactor and Pb extracellularly accumulated on the cell surface and the rate of accumulation was affected by the pH, contact time, and initial metal concentration. Fu et al. (2012) recently reported the biosorption of Cu (II) ions by mycelial pellets of *Rhizopus oryzae. Paecilomyces lilacinus* and *Aspergillus* sp. were also reported for the chromium removal from tannery effluent. Algae (red, green, and brown algae) have also been used as biosorbents for heavy metal removal in marine and fresh water environments (Srivastava 2015). The sorption capacity of six different algae such as *Asparagopsis armata, Codium vermilara, Cystoseira barbata, Lessonia nigrescens, Sargassum muticum, Spirogyra* sp. was evaluated for the recovery of Cd, Cu, Ni, Pb, and Zn from an aqueous solution by Romera et al. (2007). Among these algae, the maximum sorption of Cd, Pb, and Zn were obtained for *Asparagopsis armata*, while the maximum Cd, Pb, and Zn uptake is found in *Codium vermilara*.

Minerals synthesized by organisms are known as biomineralisation and it may be biologically induced mineralization (BIM) or biologically controlled mineralization (BCM). In BIM, microbes result in extra-cellular mineral growth by altering the geochemical reactions in the close environment through its uncontrolled metabolic activity. However, biologically controlled mineralization process is completely regulated and minerals are formed either intracellularly or epicellularly. Microorganisms use a specific metabolic process and synthesize nearly sixty-four types of minerals of phosphorites, carbonates, silicates, iron and manganese oxides, sulfide minerals and amorphous silica (Knoll 2003). Among these minerals precipitated by microorganisms, microbially induced carbonate precipitation (MICP) is the main focus of BIM. Bacterium *Sporosarcina ginsengisoli* has potential for calcite production and remediation of As-contaminated soil (Achal et al. 2012). However, growth of *Sporosarcina ginsengisoli* was reduced by arsenic; huge urease production further causes urea degradation rate and calcite production and removal of 96.3% of exchangeable arsenic fraction from liquid solution occurs.

To enhance solubilization of heavy metals prior to extraction, bioremediation is employed together with other techniques such as soil flushing and phytoextraction to clean up heavy metal-contaminated soils. For example, siderophores produced by *Alcaligenes eutrophus* form heavy metal complexes; inclusion of the bacteria significantly increased the water extraction of Cd, Zn, and Pb from sandy soil (Diels et al. 1999). The presence of *Desulfuromonas palmitatis* (iron reducing bacterium) significantly enhanced the release of As in a calcareous soil (Vaxevanidou et al. 2008). Many bacteria such as *Bacillus subtilis, Torulopsisbombicola* can release biosurfactants like surfactin, rhamnolipids, sophorolipids, aescin, and saponin to solubilize metals in soils (Acikel 2011). Some rhizospheric microbes enhance the plants' tolerance to heavy metals and also enhance their growth in contaminated soils. Thus, genetically engineered microorganisms are efficient in hyperaccumulating heavy metals.

Bacterial species like *bacillus, Escherichia* and *Mycobacterium* and fungi (*Acremonium, Pleurotus* and *Fusarium*) are used for bioremediation of polycyclic aromatic hydrocarbons (PAHs) and heavy metals. They can break down PAHs such as anthracene, naphthalene, phenanthrene, pyrene and benzopyrene in the companionship of heavy metals, and can alleviate the suppression carried by heavy metals such as Cd, Cu, Cr and Pb along with PAHs (Li et al. 2016). Immobilization (Huang et al. 2015) and compost (Poulsen and Bester 2010, Tang et al. 2008) are other techniques that can either protect the microbes from heavy metals and PAHs or enhance the microbial activity, ultimately enhancing the remediation effect.

Chromium is one of the heaviest and most hazardous metals. Chromium occurs in various valency states like Cr (II) to Cr (VI); among these, hexavalent chromium Cr (VI) is more toxic. The maximum permissible limit of Cr (VI) in natural water should be 0.05 mg l^{-1}. Vala et al. (2007) reported that two seaweed-associated fungi such as *Aspergillus flavus* and *Aspergillus niger* showed remarkable hexavalent chromium tolerance, and Cr content was found to be 22.26 (mg g^{-1} wt) dry and 18.1 (mg g^{-1} dry wt) in *A. flavus* and *A. niger*, respectively. Khambhaty et al. (2009) investigated hexavalent chromium removal potential of three marine aspergilli, namely, *Aspergillus niger, Aspergillus wentii*, and *Aspergillus terreus*, isolated from Gujarat coast and reported *A. niger* to be the most promising among the three. Gomathi et al. (2012) used three species of thraustochytrids, viz. *Aplanochytrium* sp., *Thraustochytrium* sp. and *Schizochytrium* sp. for chromium accumulation efficiency and reported 69.4 percent chromium removal obtained by *Aplanochytrium* spp. Further, Kumari et al. (2014) have evaluated the effect of microbially induced calcite precipitation (MICP) bioremediation technique for Cr (VI) contaminated soil by knowing Cr phytoaccumulation. The technique involves the role of ureolytic bacteria *Bacillus cereus*, which binds the exchangeable Cr

(VI) as a stable carbonate bio-mineral and therefore, *Pisum sativum* seeds were planted in the treated soil and uptake of chromium was measured.

Recently, the concept of "nanobioremediation"—use of nanoparticles (e.g., nanoiron, nanosillicates, and nanousnic acid) produced by particular plants, bacterial, algae, and fungi and bacteria under controlled conditions for remediation of heavy metals and organic contaminants from wastewater and soil—has been widely studied (Yadav et al. 2017). In nanobioremediation methods, the contaminated soil is inoculated with specific microorganisms by spray irrigation or infiltration galleries and injection wells can be used if the contaminants are deep in soil. To stimulate microbial activity and enhance the bioremediation process, nutrients, oxygen, and other amendments are utilized. This technique has been used to cleanup organic pollutants (e.g., PAHs, non-halogenated volatile organic compounds, and petroleum) in soils and ground water (FRTR 2012).

6. Bioremediation of radioactive waste

The radioactive waste (U, Np, Am, Sr, Pt and Tc) is released from sources such as nuclear energy generation programs, testing of nuclear weapons and accidental release and it has become a nuisance to environment and living organisms (Lloyd and Gadd 2011). Numerous researches have been initiated since the last two decades to develop an environmentally safe and publicly acceptable method through the use of microbes in bioremediation processes to treat radionuclide-contaminated land and water (Kimber et al. 2011). Microorganisms, including bacterial genera such as *Geobacter, Deinococcus, Shewanella, Serratia, Kineococcus radiotolerans,* and *Hymenobacter metalli* can be used to treat the radioactive waste by affecting their solubility, mobility, and bioavailability (Lloyd and Gadd 2011, Chung 2010).

Uranium as U (VI) and also technetium as Tc (VII) are susceptible to enzymatic reduction by microbes. The oxidized forms of U and Tc are toxic and highly mobile in groundwater because of their high solubility in the aqueous medium, whereas the reduced forms are less toxic, insoluble and precipitable (Istok 2004). Lovley and Phillips (1992) first reported that the Fe (III)-reducing bacteria like *Geobacter metallireducens* and *Shewanella oneidensis* can conserve energy for anaerobic growth via the U (VI) reduction (Lovley et al. 1991). Other organisms, including a *Clostridium* sp. (Francis 1994) and *Desulfovibrio desulfuricans* (Lovley and Phillips 1992) and *Desulfovibrio vulgaris* (Lovley and Phillips 1992), also reduce uranium but are inefficient to conserve energy for their growth. It means that these bacteria convert the toxic form of U and Tc into less toxic via reduction process.

Neptunium, an alpha-emitting transuranic radionuclide, is of great concern because of its long half-life (2.14×10^6 years), high radio-toxicity, and relatively high solubility as Np (V) under toxic conditions. In comparison, Np (IV) species dominate under-reducing conditions and can be removed from the solution by hydrolysis and a reaction with the surfaces (Dozol and Hagemann 1993, Kaszuba and Runde 1999). Low Np uptakes (10 mg g^{-1} dry weight) were reported with *Pseudomonas aeruginosa, Streptomyces viridochromogenes, Scenedesmus obliquus,* and *Micrococcus luteus* (Strandberg and Arnold 1988). Songkasiri and coworkers suggested that *Pseudomonas fluorescens* can biosorb significant quantities of Np, removing 85 percent Np (V) from the solution at pH 7 (Songkasiri 2002). Neptunyl species (NpO^{2+}) can also be biologically reduced to insoluble Np (IV) under anaerobic conditions (Lloyd et al. 2000, Rittmann et al. 2002). *Shewenella putrefaciens* reduced Np (V) to Np (IV), which was then precipitated from solution as Np (IV) phosphate in the presence of a *Citrobacter* species with high phosphatase activity. *Desulfovibrio desulfuricans* have also been reported to reduce Np (V) into Np (IV).

The fission product Strontium-90 (90Sr) is generated in nuclear explosions and Strontium-90 undergoes beta decay, with a half-life of about 29 years. *Microccous luteus* is capable of Sr-binding on the cell envelope and is sensitive to pretreatment. Bound Sr can be displaced by chelating agents like divalent cations or H$^+$. Sr binding in *M. luteus* is reversible, via both ion exchange, mediated by acidic cell surface components and intracellular uptake may be involved (Faison et al. 1990).

Plutonium is comparatively far more toxic and challenging than for the other radioactive element (U, Th, Am and Tc) because of its high radiotoxicity and complex redox chemistry. The oxidation state of plutonium is Pu (IV), Pu (III), Pu (V), and Pu (VI); among these, Pu (IV) is most stable under environmental conditions. Pu (VI) and Pu (V) are reduced into Pu (IV) via direct enzymatic reduction by bacterial cell suspension of *Shewanella putrefaciens* CN32, *Shewanella oneidensis* MR1, and *Geobacter metallireducens* GS15 reported by Neu et al. (2005). Fe (II), Mn (II), and sulfide are common inorganic reductants that are produced by bacteria and reduce plutonium (Newton 2002).

Americium can stand in oxidation states ranging from II to VII and the environmentally important oxidation state of Am is trivalent oxidation state. *E. coli*, a marine bacterium, *Rhizopusarrhizus*, and *Candida utilis* can play a pivotal role in changing Am (III) solubility and efficient biosorption (Watson and Ellwood 1994, Fisher et al. 1983, Wurtz et al. 1986, Dhami et al. 1998). Marumo et al. demonstrated the new bacterial strains such as *Flavobacterium* spp., *Pseudomonas gladioli*, *Chryseobacterium indologenes*, and *Ochrobactrum anthropi*, which are radionuclides-tolerant and also responsible for the degradation processes of organic waste (De Padua Ferreira et al. 2011, Marumo et al. 2008).

Technetium-99 is a long-lived (half-life, 2.13×10^5 years), beta-emitting radionuclide and is an important component of radioactive wastes. Under oxic conditions, technetium is present as pertechnetate ion (Tc (VII); TcO^{4-}), which is one of the most mobile radionuclide species and less sorbed (Bondietti and Francis 1979). Bioreduction, a novel phosphor imaging technique, was used to show a reduction of the radionuclide by *Shewanella putrefaciens* and *Geobacter metallireducens*, with similar activities of *Rhodobacter sphaeroides*, *Paracoccus denitrificans*, some *Pseudomonas* species (Lloyd and Renshaw 2005), *Escherichia coli* (Lloyd et al. 1997) and a range of sulfate-reducing bacteria (*Thiobacillus* sp.) (Dhami 1998).

7. Current status and prospects of bioremediation

7.1 Introduction

Recently, emphasis on bioremediation has been increasing in the field of hazardous-waste management such as solid waste, liquid waste, toxic gases, heavy metal, and radioactive waste. However, bioremediation is still an immature technology. Although microbes are the primary stimulant in the bioremediation of contaminated environment and play an essential role in biogeochemical cycles, it is difficult to assess the impact of bioremediation on the ecosystem due to limited understanding of the changes in microbial communities during bioremediation.

7.2 Bioremediation technologies

Bioremediation of toxic wastes including solid waste, liquid waste, toxic gases, heavy metal and radioactive waste can be categorized as *in situ* and *ex situ* bioremediation. The main objective of bioremediation is to degrade and transform organic pollutants into a less toxic form. The different strategies of *in situ* and *ex situ* bioremediation and phytoremediation technologies which remediate the contaminants from soil, water and air are presented in Table 1. In *in situ* techniques, the soil and groundwater are remediated in place without excavation, while in *ex situ* applications it is excavated prior to treatment. Selection of appropriate technology among the different bioremediation strategies to treat contaminants depends upon three basic principles, i.e., biochemistry, bioavailability and the bioactivity (Shukla et al. 2010).

7.3 New bioremediation methodologies for waste management

Phycoremediation

Phycoremediation involves algae (micro and macro) for the remediation of contaminants in a water body. At minimal cost, algae remove excess nutrients effectively and fix carbon-dioxide by

Table 1. Technologies available for bioremediation of waste.

Bioremediation methods		Applications	Advantage	References
Ex situ	Land farming	Surface application, aerobic process, application of organic materials to natural soils followed by irrigation and tilling.	Cost-effective, Simple and self-heating.	Silva-Castro et al. (2012)
	Biopiling	Above-ground piling of excavated polluted soil, followed by nutrient amendment. Limited volatilization of low molecular weight pollutants.	Treat large volume of polluted soil in a limited space. Biopile setup can easily be scaled up to a pilot system.	Dias et al. (2015) Chemlal et al. (2013)
	Windrow	Periodic turning of polluted soil with addition of water. Uniform distribution of pollutants, and microbial degradative activities.	Higher efficiency towards hydrocarbon removal.	Coulon et al. (2010)
	Composting	Addition of nutrients, watering, tilling, addition of suitable microflora and bulking agents were considered an alternative option to improve the bioremediation of oil sludge.	Economical and effective way to treat oil sludge. Waste stabilization.	Prakash et al. (2015)
	Bioreactors (Slurry reactors and Aqueous reactors)	Treat soil or water polluted with volatile organic compounds (VOCs) including benzene, toluene, ethylbenzene and xylenes (BTEX). Toxic concentrations of contaminants.	Excellent control of bioprocess. Increased pollutant bioavailability, and mass transfer parameters. Effective use of inoculants and surfactant.	Chikere et al. (2016)
In situ	Biosparging	Used in treating aquifers contaminated with petroleum products, especially diesel and kerosene and BTEX contaminated ground water.	Most efficient. Non invasive.	Kao et al. (2008)
	Bioventing	Restoring sites polluted with lightly spilled petroleum products. Biodegradability of pollutants.	It can be used in anaerobic condition.	Hohener and Ponsin (2014)
	Bioaugmentation	Hazardous waste remediation as well as aerobic waste treatment.	Naturally attenuated process, treats soil and water.	Tale et al. (2015)

Table 1 Contd. ...

...Table 1 Contd.

Bioremediation methods		Applications	Advantage	References
	Bioslurping	Remediating capillary, unsaturated and saturated zones. Remediate soils contaminated with volatile and semi-volatile organic compounds.	Cost-effective.	Kim et al. (2014) Philp and Atlas (2005)
Phytoremediation				
	Phytodegradation	Plants and associated microorganisms degrade organic pollutants like petroleum hydrocarbon.	Cost-effective.	Al-Baldawi et al. (2015)
	Phytoextraction	Remove metal pollutants, petroleum, hydrocarbons and radionuclides and accumulate in plants. Remove organics from soil by concentrating them in plant parts.	Used in both soil and ground water. Contaminants permanently removed from soil.	Zhuang et al. (2007)
	Phytotransformation	Plant uptake and degradation of organic compounds such as xenobiotic substances.	Both economically and environmental friendly.	Al-Baldawi et al. (2018)
	Rhizofiltration	Roots absorb mainly heavy metals such as Zn, Pb, Cd, and As from water and aqueous waste streams.	*In situ* practice resulting in no disturbance. Contaminants do not have to be translocated into shoots.	Benavides et al. (2018)
	Phytostabilization	Plants reduced the bioavailability of heavy metal such as Sb, Cd, Cr, Ni, Pb, and As via precipitation.	Cost-effective. Capable of remediating heavy metal contaminated soil without impairing the soil quality.	Sylvain et al. (2016)
	Phytovolatilization	Organic pollutants are absorbed through plant roots and transported through the plant.	The contaminant like Hg may transform into less toxic form.	Limmer and Burken (2016)

photosynthesis. By algal metabolism, xenobiotics and heavy metals are detoxified or transformed to less toxic form or volatilized. Numerous recent studies show accumulation and degradation of polycyclic aromatic hydrocarbons and heavy metal by fresh-water algae like *Scenedesmus quadricauda, Chlorella vulgaris, Selenastrum capricornutum,* and *Scenedesmus platydiscus*. Ajayan et al. (2015) reported that heavy metals such as Cr, Cu, Pb, and Zn were found to be removed very effectively by *Scenedesmus quadricauda* microalgae with removal rates ranging from 60 percent to 100 percent.

Cyanophyceae, Chlorophyceae, Euglenophyceae, Bacillariophyceae, and *Desmidiaceae* were used in wastewater treatment plant (WWTP) at Shimoga Town, Karnataka State, India, recorded by Shanthala et al. (2009). The highest Cu, Zn and Co removal of 60, 42.9 and 29.6 percent, respectively, was observed with *Oscillatoria quadripunctutata*, while highest Pb removal of 34.6 percent was found with *Scenedesmus bijuga* in sewage wastewater (Ajayan et al. 2011). El-Sheekh et al. (2005) also observed the removal of heavy metals from paper production Verta Company, sewage wastewater and salt and soda company wastewater by mixed culture of *Nostoc muscorum* and *Anabaena subcylindrica* microalgae.

Chlorella pyrenoidosa was investigated by Pathak et al. (2015) as phycoremediation of dye removal from textile wastewater (TWW) in batch cultures and he observed that alga potentially grows up to 75 percent concentrated textile wastewater and reduces phosphate, nitrate, and BOD by 87 percent, 82 percent, and 63 percent, respectively. Removal of methylene blue dye (MB) was also observed by using dry and wet algal biomass in which dry algal biomass (DAB) was a more efficient biosorbent of MB dye as compared to wet algal biomass (WAB) because of large surface area and high binding affinity for MB dye.

Long term utilization of pesticides creates dangerous effects on human health and environment via biomagnification and eutrophication process. Among all the pesticides, organophosphate pesticides are widely used pesticides in the world and bioremediation is the best method of removing the pesticide that contaminates the land and water. Numerous studies show algae sp. *Spirulinaplatensis* and *Spirogyra* were used for the biodegradation of pesticide chlorpyrifos and biosorption of heavy metal chromium at various concentrations (Samuel Reinhard et al. 2019). Heavy metals are the most hazardous pollutants such as chromium (Cr) compounds which are highly toxic to plants and retard their growth and development. The utilization of algae to remediate toxicants or pollutants from the contaminated environment is named as phycoremediation. Phycoremediation is a novel technique in bioremediation methods to degrade the pesticide using the algal spp. Zainith et al. (2019) reported that microalgae *Scenedesmus rubescens* KACC 2 remove nitrogen, phosphorus, and heavy metals more efficiently from industrial and domestic effluents.

Mycoremediation

The use of fungi to degrade or remediate contaminants is known as mycoremediation. Yeast and fungi are unique in metal biosorption, and this process is known as mycosorption. *Trichoderma viride*, for example, can be used to remove Cd and Pb from aqueous media (Sahu, Asha et al. 2012). *Trichoderma asperellum* and *Trichoderma viride* can remove arsenic from liquid media through biovolatilization in laboratory conditions (Srivastava et al. 2011). *Trichoderma asperellum, T. harzianum,* and *T. tomentosum* were studied for removing Cd under different pH conditions. *Saccharomyces cerevisiae* removed 65–79 percent heavy metals like lead and cadmium from contaminated soil by biosorption process within 30 days (Damodaran et al. 2011). *Mucor hiemalis*, an effective fungus, was shown for bioremediation of pharmaceutical xenobiotics, e.g., acetaminophen (Esterhuizen-Londt et al. 2016). During the inoculation of mycelia from oyster mushrooms (*Pleurotus ostreatus*) in soil contaminated with diesel oil, it was found that after 4 weeks, 95 percent PAHs had been converted to nontoxic compounds and finally to CO_2 and H_2O (Rhodes 2014).

The marine fungi *Corollospora lacera* and *Monodictyspelagica* have been observed to accumulate lead and cadmium extracellularly in mycelia, respectively (Taboski et al. 2005). Fan et al. (2008) reported efficiency of *P. simplicissimum* to remove Cd (II), Zn (II) and Pb (II) from aqueous solutions. Fluorine degradation by *P. italicum* in the presence of several cyclodextrins was reported by Garon et al. (2004). Efficient degradation of a potentially toxic synthetic dye like Remazol brilliant blue R, by use of marine-derived fungus *Tinctoporellus* sp. CBMAI 1061, was revealed by Bonugli-Santos et al. (2012).

Hexavalent chromium removal efficiency of three marine aspergilla such as *Aspergillus niger, Aspergillus wentii,* and *Aspergillus terreus*, isolated from Gujarat coast, have been investigated by Khambhaty et al. (2009) and reported *Aspergillus niger* as the best among the three. Bioaccumulation of arsenic was also observed in both fungi, viz. *Aspergillus flavus* and *Rhizopus* sp. and results showed *Rhizopus* sp. to be a better accumulator (Vala and Sutariya 2012). *Aspergillus sydowii* has been considered as a potential candidate for arsenic bioremediation because of volatilization of

15.75 percent of supplied As (III) via., biovolatilization process (Vala 2017). Marine-derived fungi have been observed as efficient microbes for bioremediation of various pollutants.

Rhizoremediation (Plant and microbe interaction)

The rhizospheric breakdown of soil contaminants such as heavy metals, pesticides, petroleum products, fly ash, herbicides, etc., through plant roots in association with microbes is known as rhizoremediation. The root exudates stimulate micro-organisms' population in the soil, which subsequently degrade pollutants (Kuiper et al. 2004). The success of rhizoremediation is dependent on plant and microbe interaction.

Some plants like *Vetiveria zizanioides, Phragmites, Sacharum*, etc. have extensive root systems which make plants most efficient for rhizoremediation. *Vetiver* has been observed to be efficient in remediating Pb and Zn (Antiochia et al. 2007) and 2,4,6 trinitrotoluene from soil amended with urea (Das et al. 2010) as well as phenolics (Phenrat et al. 2017). *Phragmitesaustralis* in association with bacteria was capable of remediating 4-n-butylphenol (Toyama et al. 2011). Abou-Shanab et al. (2003) investigated the effect of *Microbacterium liquefaciens, Sphingomonas macrogoltabidus* and *Microbacterium arabinogalactanolyticum* on the rhizospheric zone of *Alyssum murale* and found 17, 24, and 32.4% magnified uptake of nickel, respectively, into the shoot and leaves as compared to control. The different plants and associated microbes which remediate the heavy metal contaminants from soil are presented in Table 2.

Table 2. List of plants and associated microbes to remediate contaminants.

Plant	Microbes	Contaminants	References
Brassica juncea	*Sinorhizobium* sp.	Lead	Di Gregorio et al. (2006)
Ricinus communis	*Pseudomonas jessenii*	Zinc, nickel, copper	Rajkumar and Freitas (2008)
Lycopersicum esculentum	*Burkholderia* sp.	Lead	Jiang et al. (2008)
Amaranthus hypochondriacus, Amaranthus Mangostanus, and *Solanum nigrum*	*Rahnella* sp.	Cadmium	Yuan et al. (2014)
Oryza sativa L.	*Brevundimonas diminuta*	Arsenic	Singh et al. (2016)
Sedum plumbizincicola	*Bacillus* sp. SC2b	Lead, zinc	Ma et al. (2015)
Oryza sativa L. seedling	*Kocuria flava* AB402 and *Bacillus vietnamensis* AB403	Arsenic	Mallick et al. (2018)
Brassica nigra	*Microbacterium* sp. CE3R2, *Microbacterium* sp. NE1R5, *Curtobacterium* sp. NM1R1, and *Microbacterium* sp. NM3E9	Arsenic, Zinc	Roman-Ponce et al. (2017)
Brassica napus L.	*Pantoea agglomerans*	Zn, Cu and Cd Biodiesel	Zhang et al. (2011), Lacalle et al. (2018)

7.4 Emerging technologies in bioremediation

Genoremediation

Genetic modifications in plants for the increased expression of metal chelators, metal transporters, metallothioneins, and phytochelatins are known as genoremediation. The main classes of metal chelators are phytochelatins (PCs) and metallothioneins (MT), between which only metallothioneins are direct products of gene expression. Model plants of tobacco (*Nicotiana glauca* and *Nicotiana tabacum*) were either showing greater accumulation of Cd and Pb, or were tolerant to Cd depending on the origin of the transgene (Huang et al. 2012, Chen et al. 2015). Phytochelatins gene from *Populus tomentosa* reduced cadmium translocation to the aerial parts of tobacco (Chen et al. 2015), while phytochelatins gene from aquatic macrophyte ceratophyllum demersum enhanced cadmium translocation (Shukla et al. 2013). To overcome this issue, co-transformation could be an efficient way to produce potential hyperaccumulator. Guo et al. (2012) transformed *A. thaliana* with two genes such as PCS (responsible for phytochelatin synthesis) and YCF1, ABC metal transporter and plants showing co-tolerance to Cd and As with increased bioaccumulation in vacuoles. Zhao et al. (2014) proved that co-transformation with Phytochelatins (PCS) and glutamyl cysteine synthetase (GCS) genes was beneficial than transformation with PCS gene alone because of overexpression of GCS enhanced PCS activity, resulting in higher production of phytochelatins.

Recently, transgenic plants developed for PCBs phytoremediation has been evaluated by several researchers. Extracellular oxidative enzymes like lignin peroxidase (Lip), manganese-dependent peroxidase (Mnp) and laccase (Lac) produced by *Basidiomycetes* (white rot fungus) were found to be important in degradation of PCBs. Genes produced by *Phaenerochete chrysoporium* responsible for generation of Lip, Mnp and Lac enzymes have been introduced into the DNA of *Arabidopsis thaliana* to make a transgenic species that potentially degrade PCBs (Sonoki et al. 2007). Francova et al. (2003) developed transgenic tobacco plants (*Nicotiana tabacum*) by inserting a gene which is responsible for 2,3-dihydroxybiphenyl ring cleavage, *bph*C, from the PCB degrader *Comamonas testosteroni*.

Electro-assisted bioremediation

The bioconversion of a wide range of groundwater contaminants such as hydrocarbons, aromatic organics into inorganic substances like nitrate, sulfate and heavy metals is promoted by electrochemical approaches. The combination of bioremediation and electrochemical stimulation (electrokinetic enhancement) has significantly enhanced the degradation of contaminant. An accelerated bioreduction of perchloroethene (PCE) and further cathodic oxidation of the intermediates can be achieved by the use of *in situ* generated hydrogen and oxygen from water electrolysis, respectively, and complete mineralization of PCE in groundwater occurs (Lohner and Tiehm 2009). An energy efficiency improvement can be achieved by establishing a direct delivery of electrons between the electrodes and degrading bacteria (Leitao et al. 2015). The incorporation of EDTA enhancement and bioaugmentation improved electrokinetics process, resulting in 92.7 percent lead (Pb), 64.3 percent copper (Cu) and 45.9 percent zinc (Zn) removal simultaneously with low power consumptions, as reported by Lee and Kim (2010). Similar strategies have also been implied to remediate arsenite (As) from the soil by combining anaerobic bioleaching and electrokinetics, yielding 66.5 percent As removal within 16 days compared to 17.3 percent removal in sole bioleaching system (Lee et al. 2009).

These electrochemical reduction and precipitation methods could be applied to release most heavy metal species including copper, chromium, cadmium, mercury, uranium, and vanadium (Huang et al. 2011, Williams et al. 2010). Many electro-active bacteria such as *Geobacter* species could efficiently catalyze metal reduction that are mostly present in groundwater (Williams et al. 2010). Hao et al. (2015) reported remediation of vanadium significantly with a removal efficiency of 93.6 percent within 12 hours under accelerated microbial vanadium reduction in groundwater via bioelectrochemical simulation. Biosorption and precipitation of metals such as Cd (II) and Zn (II), with *in situ* formed sulfide at the alkaline cathode, are also essential mechanisms for their removal (Abourached et al. 2014, Varia et al. 2014). The bio-electrochemical process can also be used to improve nitrate removal efficiency from groundwater.

8. Challenges of bioremediation

8.1 Limitations of bioremediation

Bioremediation also has its own limitations like other methods. Only biodegradable compounds are remediated by bioremediation, and not all compounds are susceptible to rapid and complete degradation such as chlorinated organic or high aromatic hydrocarbons that are resistant to microbial attack. Sometimes, the products of biodegradation may be more persistent or toxic than the parent compound. Bioremediation may take more time than other treatment options, such as excavation and removal of soil or incineration to remediate the contaminants (Zeyaullah et al. 2009).

The field release of genetically engineered microorganisms (GEMs) has some drawbacks such as decreased levels of fitness and the extra energy demands required by the presence of foreign genetic material in the cells (Singh et al. 2011). There are numerous limitations with electro-bioremediation technology that need to be overcome such as solubility of the pollutant and its desorption from the soil matrix, the availability of potential microorganisms at the site of contamination, the ratio between target and non-target ion concentrations, and toxic electrode effects on microbial metabolism (Virkutyte et al. 2002).

However, some of the more recalcitrant and toxic xenobiotic compounds involving highly nitrated and halogenated aromatic compounds and explosives are generally stable, chemically inert under natural conditions, and not common to remediate efficiently by many microorganisms. Some compounds are co-metabolized by microorganisms only in the presence of an alternative carbon source; these compounds cause problems for biodegradation and bioremediation for the reasons of the toxicity of these organic pollutants to the existing microbial populations. Pollutants, such as petroleum hydrocarbons, polycyclic aromatic hydrocarbons and some chlorinated pesticides among others, are not easily available to the microorganisms because of their hydrophobicity and persistence in soil.

The composting efficiency depends on the type of contaminants, temperature, and soil/waste amendment ratio for bioremediation (Antizar-Ladislao et al. 2005). The spent mushroom waste from *Pleurotus ostreatus* was found to degrade and mineralize DDT in soil (Purnomo et al. 2010). But, Alvey and Crowley (1995) observed that additions of compost suppressed soil mineralization of atrazine were relative to rates in un-amended soils.

The use of windrow treatment may not be the best option to adopt in remediating soil polluted with toxic volatiles due to periodic turning and production of CH_4, i.e., greenhouse gas because of development of anaerobic zone within piled polluted soil.

Bioreactor-based bioremediation is not a popular full-scale practice as it requires more manpower and is cost ineffective (Philp and Atlas 2005). Limitations of land farming bioremediation techniques are large operating space, reduction in microbial activities due to unfavorable environmental conditions, additional cost due to excavation, and reduced efficacy in inorganic pollutant removal (Khan et al. 2004, Maila and Colete 2004). In hot (tropical) climate regions, it is not suitable for treating soil polluted with toxic volatiles due to its design and mechanism of pollutant removal (volatilization).

In bioslurping technique, excessive soil moisture limits air permeability and decreases oxygen transfer rate, in turn reducing microbial activities and these are the major concerns of this particular *in situ* technique. The major limitation of biosparing is predicting the direction of airflow. Longer remediation time, pollutant concentration, toxicity, and bioavailability to plant and slow plant growth rate are limiting factors which check the application of phytoremediation (Kuiper et al. 2004, Vangronsveld et al. 2009, Ali et al. 2013). There is a possibility that bioaccumulated toxic contaminants may be transferred along the food chain.

8.2 Toxic intermediates produced during the bioremediation process

HCH isomers show resistance towards biodegradation due to their different chemical configuration and complex nature. The complete degradation of small concentrations of HCH was successful, while partial/incomplete reductions resulted when the concentration of contaminant was increased in many of the bioremediation experiments. Incomplete degradation also results in formation of toxic intermediates such as pentachlorocyclohexanols, pentachlorocyclohexenes, and tetrachlorocyclohexane-diols as some of the metabolites of α and γ HCH degradation has a higher solubility in water than their parental counterpart (Raina et al. 2008).

The growth of the inoculated microorganisms may be affected by the high concentration of pollutant and the presence of the autochthonous species (Cycon et al. 2017, Chen et al. 2014). The degradative abilities of the inoculated microorganisms may also be attributed to the production of toxic intermediates that are formed during the degradation of lindane and HCH which might inhibit their further growth (Megharaj et al. 2011). The associated challenges of the bioaugmentation technology include both abiotic and biotic factors and environmental conditions like temperature, pH, moisture and organic content of the soil, initial pesticide concentration and additional carbon sources. Besides this, there are many factors that affect the bioaugmentation of the polluted soils and have been discussed in detail by Cycon et al. (2017).

Microorganisms produce biosurfactants under specific growth conditions, which may be anionic or nonionic (Zhang and Miller 1995). Biosurfactant (rhamnolipid) was found to enhance the solubility and the subsequent degradation of phenanthrene by *Sphingomonas* sp. (Pei et al. 2010). Biosurfactants can be toxic or even utilized preferentially by the pollutant-degrading microorganisms. In a recent report, Hua et al. (2010) demonstrated that a salt-tolerant *Enterobacter cloacae* mutant could be used as an agent for bioaugmentation of petroleum- and salt contaminated soil due to increased K^+ accumulation inside and exopolysaccharide level outside the cell membrane.

Won et al. (1974) used *Pseudomonos* sp. strain to degrade TNT, which resulted into 2-Amino-4, 6-dinitrotoluene (2-ADNT), 4-Amino-2,6-dinitrotoluene (4-ADNT), 2,6-dinitro-4-hydroxylaminotoluene (2,6-DHAT), and the corresponding nitrodiaminotoluenes, 2,2',6,6'-tetranitro-4,4'-azoxytoluene and 2,2',4,4'-tetranitro-6,6'-azoxytoluene. Lachance et al. (2004) reported that TNT metabolites (4-amino-2,6-dinitrotoluene, 2-amino-4,6-dinitrotoluene) are as toxic as TNT itself. In addition, 2,4-DANT and 2,4-DNT were also observed to be more toxic than TNT to Hyalella azteca and Rana catesbeiana, respectively (Sims and Steevens 2008, Paden et al. 2011). The intermediate metabolites of TNT such as 2-Amino-4, 6-dinitrotoluene (2-ADNT) and 4-Amino-2,6-dinitrotoluene (4-ADNT) are more toxic than parental TNT on the basis of their reproductive toxicities (Karnjanapiboonwong et al. 2009).

Aromatic amines are released after cleavage of azo bonds by microflora containing azoreductases (Prasad and Aikat 2014). Aromatic amines are more persistent in the environment than dyes (Chen et al. 2009). Nitroanilines are produced during the biodegradation of azo dyes under anaerobic conditions, which are toxic and affect the ability and efficiency of the dye decolorizing bacteria (Khalid et al. 2009). According to Tsuboy et al. (2007), the acetoxy group ($COCH_3$) located on the benzene ring can be metabolized by the P450 enzyme and other hepatic enzymes generating radical mutagenic intermediates. Biochemical activation through N-hydroxylation, followed by sulfation, esterification or acetylation reactions, generates reactive intermediates that are able to bind to DNA and largely account for the carcinogenicity of arylamines (Pinheiro et al. 2004).

8.3 Failure of bioremediation process

Due to lesser efficiency, competitiveness, and adaptability, relative to the indigenous members of natural communities, bioaugmentation efforts have failed to remediate the contaminated soils. For example, the bacteria capable of degrading polychlorinated biphenyls (PCBs) in laboratory culture media survived poorly in natural soils, and when these strains were inoculated to remediate PCB-contaminated soils, the result was the failure of bioaugmentation (Blasco et al. 1995). Further investigations revealed that formation of an antibiotic compound, protoanemonin, from 4-chlorocatechol via the classical 3-oxoadipate pathway by the native microorganisms was the reason for poor survival of the introduced specialist PCB-degrading strains (Blasco et al. 1997).

9. Conclusion

Future studies and research are required to develop environmental friendly and socially acceptable potent bioremediation technologies to ensure the elimination of emerging contaminants as these contaminants coupled with toxic intermediate metabolites continue to cause new and serious challenges to water, air, soil, natural resources, ecosystems, and human health. The possibility of isolating, culturing, and studying natural microorganisms with novel metabolic pathways and its survival under contaminated environment is of great concern for bioremediation purposes.

Biological processes are often highly specific. Presence of metabolically capable microbial populations, suitable environmental conditions, and sufficient levels of nutrients and contaminants are important site factors required for the success of bioremediation technology. It is challenging to extrapolate from the lab and to field operations. Research is needed to engineer bioremediation technologies that are more efficient and suitable with complex mixtures of contaminants which are randomly spread in the environment.

Considering the current situation, it is clear that several factors must be taken into account such as contaminant concentration, characteristics and category, scale and level of contamination, the risk intensity generated for health or the environment and available resources for the success of bioremediation process. Moreover, applying multidisciplinary techniques to remediate pollutants from any given environment would be made more predictable. It is also necessary to find future challenges for reducing the environmental impacts from emerging contaminants, with greater emphasis on innovative and advanced technologies for monitoring, prevention, and mitigation of environmental and health risks.

References

Abbas, S.H., Ismail, I.M., Mostafa, T.M. and Sulaymon, A.H. 2014. Biosorption of heavy metals: A review. J. Chem. Sci. Technol. 3: 74–102.

Abourached, C., Catal, T. and Liu, H. 2014. Efficacy of single-chamber microbial fuel cells for removal of cadmium and zinc with simultaneous electricity production. Water Res. 51: 228–233.

Abou-Shanab, R.A., Angle, J.S., Delorme, T.A., Chaney, R.L., Van Berkum, P., Moawad, H., Ghanem, K. and Ghozlan, H.A. 2003. Rhizobacterial effects on nickel extraction from soil and uptake by *Alyssum murale*. New Phytol. 158: 219–224.

Achal, V., Pan, X., Zhang, D. and Fu, Q. 2012. Bioremediation of Pb-contaminated soil based on microbially induced calcite precipitation. J. of Microbio. and Biotech. 2: 244–247.

Acikel, Y.S. 2011. Use of biosurfactants in the removal of heavy metal ions from soils. pp. 183–223. *In*: Khan, M.S., Zaidi, A., Goel, R. and Musarrat, J. (eds.). Biomanagement of Metal Contaminated Soils. Springer, Heidelberg, Germany.

Ahmad, M.M., Azoddein, A.A.M. and Jami, M.S. 2018. Eco-friendly Industrial wastewater treatment: Potential of mesophilic bacterium, *Pseudomonas putida* (ATCC 49128) for hydrogen sulfide oxidation. J. of Appli. Bio. & Biotech. 6: 53–57.

Ajayan, K.V., Selvaraju, M. and Thirugnamoorthy, K. 2011. Growth and heavy metals accumulation potential of microalgae grown in sewage wastewater and petrochemical effluent. Pak. J. Biol. Sci. 14: 805–811.

Ajayan, K.V., Selvaraju, M., Unnikannan, P. and Sruthi, P. 2015. Phycoremediation of tannery wastewater using microalgae *Scenedesmus* species. Inter. Jour. of Phytore. 17: 907–916.

Akar, T., Tunali, S. and Kiran, I. 2005. *Botrytis cinerea* as a new fungal biosorbent for removal of Pb (II) from aqueous solutions. Bioche. Engi. J. 25: 227–235.

Al-Baldawi, I.A., Abdullah, S.R.S., Anuar, N., Suja, F. and Mushrifah, I. 2015. Phytodegradation of total petroleum hydrocarbon (TPH) in diesel-contaminated water using *Scirpus grossus*. Ecolo. Engi. 74: 463–473.

Al-Baldawi, I.A., Abdullah, S.R.S., Anuar, N. and Hasan, H.A. 2018. Phytotransformation of methylene blue from water using aquatic plant (*Azolla pinnata*). Enviro. Tech. & Inno. 11: 15–22.

Alford, E.R., Pilon-Smits, E.A. and Paschke, M.W. 2010. Metallophytes—a view from the rhizosphere. Plant and Soil. 337: 33–50.

Ali, H., Khan, E. and Sajad, M.A. 2013. Phytoremediation of heavy metals—concepts and applications. Chemosphere 91: 869–881.

Alvey, S. and Crowley, D.E. 1995. Influence of organic amendments on biodegradation of atrazine as a nitrogen source. J. Environ. Qual. 24: 1156–62.

Amirfakhri, J., Vossoughi, M. and Soltanieh, M. 2006. Assessment of desulfurization of natural gas by chemoautotrophic bacteria in an anaerobic baffled reactor. Chem. Eng. Proc. 45: 232–237.

Antizar-Ladislao, B., Lopez-Real, J. and Beck, A.J. 2005. In-vessel composting-bioremediation of aged coal tar soil: effect of temperature and soil/green waste amendment ratio. Environ. Int. 31: 173–8.

Antizar-Ladislao, B., Beck, A.J., Spanova, K., Lopez-Real, J. and Russell, N.J. 2007. The influence of different temperature programmes on the bioremediation of polycyclic aromatic hydrocarbons (PAHs) in a coal-tar contaminated soil by in-vessel composting. J. of Hazar. Materi. 14: 340–347.

Awasthi, A.K., Pandey, A.K. and Khan, J. 2017. Biosorption an innovative tool for bioremediation of metal-contaminated municipal solid waste leachate: optimization and mechanisms exploration. Inter. J. of Environ. Sci. and Tech. 14: 729–742.

Benavides, L.C.L., Pinilla, L.A.C., Serrezuela, R.R. and Serrezuela, W.F.R. 2018. Extraction in laboratory of heavy metals through rhizofiltration using the plant *Zea mays* (maize). Inter. J. of Appli. Environ. Sci. 13: 9–26.

Blasco, R., Wittich, R.M., Megharaj, M., Timmis, K.N. and Pieper, D.H. 1995. From xenobiotic to antibiotic: formation of protoanemonin from 4-chlorocatechol by enzymes of the 3-oxoadipate pathway. J. Biol. Chem. 270: 29–35.

Blasco, R., Megharaj, M., Wittich, R.M., Timmis, K.N. and Pieper, D.H. 1997. Evidence that formation of protoanemonin from metabolites of 4-chlorobiphenyl degradation negatively affects the survival of 4-chlorobiphenyl-cometabolizing microorganisms. Appl. Environ. Microbiol. 63: 27–34.

Bondietti, E.A. and Francis, C.W. 1979. Geologic migration potentials of technetium-99 and neptunium-237. Science 203: 1337–1340.

Bonugli-Santos, R.C., Durrant, L.R. and Sette, L.D. 2012. The production of ligninolytic enzymes by marine-derived basidiomycetes and their biotechnological potential in the biodegradation of recalcitrant pollutants and the treatment of textile effluents. Water Air Soil Pollut. 223: 2333–2345.

Chemlal, R., Abdi, N., Lounici, H., Drouiche, N., Pauss, A. and Mameri, N. 2013. Modeling and qualitative study of diesel biodegradation using biopile process in sandy soil. Int. Biodeterior. Biodegradation 78: 43–48.

Chen, Y., Liu, Y., Ding, Y., Wang, X. and Xu, J. 2015. Overexpression of Pt PCS enhances cadmium tolerance and cadmium accumulation in tobacco. Plant Cell Tissue Organ Cult. 121: 389–396.

Chen, B.Y., Yen, C.Y., Chen, W.M., Chang, C.T., Wang, C.T. and Hu, Y.C. 2009. Exploring threshold operation criteria of biostimulation for azo dye decolorization using immobilized cell systems. Bioresour. Technol. 100: 5763–70.

Chikere, C.B., Okoye, A.U. and Okpokwasili, G.C. 2016. Microbial community profiling of active oleophilic bacteria involved in bioreactor based crude-oil polluted sediment treatment. J. Appl. Environ. Microbiol. 4: 1–20.

Chiu, S.Y., Kao, C.Y., Chen, C.H., Kuan, T.C., Ong, S.C. and Lin, C.S. 2008. Reduction of CO_2 by a high-density culture of *Chlorella* sp. in a semicontinuous photobioreactor. Bioreso. Tech. 99: 3389–3396.

Choo, T.P., Lee, C.K., Low, K.S. and Hishamuddin, O. 2006. Accumulation of chromium (VI) from aqueous solutions using water lilies (*Nymphaea spontanea*). Chemosphere 62: 961–967.

Chung, A.P., Lopes, A., Nobre, M.F. and Morais, P.V. 2010. *Hymenobacter perfusus* sp. nov., *Hymenobacter flocculans* sp. nov. and *Hymenobacter metalli* sp. nov. three new species isolated from an uranium mine waste water treatment system. Syst. Appl. Microbiol. 33: 436–443.

Coulon, F., Al Awadi, M., Cowie, W., Mardlin, D., Pollard, S., Cunningham, C., Risdon, G., Arthur, P., Semple, K.T. and Paton, G.I. 2010. When is a soil remediated? Comparison of biopiled and windrowed soils contaminated with bunker-fuel in a full-scale trial. Environ. Pollut. 158: 3032–3040.

Damodaran, D., Suresh, G. and Mohan, R. 2011. Bioremediation of soil by removing heavy metals using *Saccharomyces cerevisiae*. pp. 22–35. *In*: 2nd International Conference on Environmental Science and Technology, IPCBEE.

Dash, H.R. and Das, S. 2015. Bioremediation of inorganic mercury through volatilization and biosorption by transgenic *Bacillus cereus* BW-03. Int. Biodeterior. Biodegrad. 103: 179–185.

De Padua Ferreira, R.V., Sakata, S.K., Isiki, V.L.K., Miyamoto, H., Bellini, M.H., de Lima, L.F.C. and Marumo, J.T. 2011. Influence of americium-241 on the microbial population and biodegradation of organic waste. Environm. Chem. Letters. 9: 209–216.

Devinny, J.S., Deshusses, M.A. and Webster, T.S. 2017. Biofiltration for Air Pollution Control. CRC Press. Pub. Location Boca Raton. Imprint CRC Press. Taylor and Francis Group. https://doi.org/10.1201/9781315138275.

Dhami, P.S., Gopalakrishnan, V., Kannan, R., Ramanujam, A., Salvi, N. and Udupa, S.R. 1998. Biosorption of radionuclides by *Rhizopus arrhizus*. Biotech. Letters 20: 225–228.

Di Gregorio, S., Barbafieri, M., Lampis, S., Sanangelantoni, A.M., Tassi, E. and Vallini, G. 2006. Combined application of Triton X-100 and *Sinorhizobium* sp. Pb002 inoculum for the improvement of lead phytoextraction by *Brassica juncea* in EDTA amended soil. Chemosphere 63: 293–299.

Dias, R.L., Ruberto, L., Calabro', A., Balbo, A.L., Del Panno, M.T. and Mac Cormack, W.P. 2015. Hydrocarbon removal and bacterial community structure in on-site biostimulated biopile systems designed for bioremediation of diesel-contaminated Antarctic soil. Polar Biol. 38: 677–687.

Diels, L., De Smet, M., Hooyberghs, L. and Corbisier, P. 1999. Heavy metals bioremediation of soil. Mol. Biotechnol. 12: 149–158.

Dozol, M. and Hagemann, R. 1993. Radionuclide migration in groundwaters: review of the behaviour of actinides (technical report). Pure and Appl. Chem. 65: 1081–1102.

El-Sheekh, M.M., El-Shouny, W.A., Osman, M.E.H. and El-Gammal, E.W.E. 2005. Growth and heavy metals removal efficiency of *Nostoc muscorum* and *Anabaena subcylindrica* in sewage and industrial wastewater effluents. Environ. Toxicol. Pharmacol. 19: 357–365.

Esterhuizen-Londt, M., Schwartz, K. and Pflugmacher, S. 2016. Using aquatic fungi for pharmaceutical bioremediation: uptake of acetaminophen by *Mucor hiemalis* does not result in an enzymatic oxidative stress response. Fungal Biol. 120: 1249–1257.

Faison, B.D., Cancel, C.A., Lewis, S.N. and Adler, H.I. 1990. Binding of dissolved strontium by *Micrococcus luteus*. Appl. Environ. Microbiol. 56: 3649–3656.

Fan, T., Liu, Y., Feng, B., Zeng, G., Yang, C., Zhou, M., Zhou, H., Tan, Z. and Wang, X. 2008. Biosorption of cadmium(II), zinc(II) and lead(II) by *Penicillium simplicissimum*: Isotherms, kinetics and thermodynamics. J. of Hazard. Mater. 160: 655–661.

Fazal, T., Mushtaq, A., Rehman, F., Khan, A.U., Rashid, N., Farooq, W. and Xu, J. 2018. Bioremediation of textile wastewater and successive biodiesel production using microalgae. Renew. and Sustain. Ener. Revie. 82: 3107–3126.

Fernandez, M., Ramirez, M., Gomez, J.M. and Cantero, D. 2014. Biogas biodesulfurization in an anoxic biotrickling filter packed with open-pore polyurethane foam. J. Hazard. Mater. 264: 529–535.

Fisher, N.S., Bjerregaard, P., Huynh-Ngoc, L. and Harvey, G.R. 1983. Interactions of marine plankton with transuranic elements. II. Influence of dissolved organic compounds on americium and plutonium accumulation in a diatom. Marine Chem. 13: 45–56.

Francis, A.J. 1994. Microbial transformations of radioactive wastes and environmental restoration through bioremediation. J. of Alloys and Comp. 213: 226–231.

Francova, K., Sura, M., Macek, T., Szekeres, M., Bancos, S., Demnerova, K., Sylvestre, M. and Mackova, M. 2003. Preparation of plants containing bacterial enzyme for degradation of polychlorinated biphenyls. Fresenius Environ. Bull. 12: 309–313.

Fu, Y.Q., Li, S., Zhu, H.Y., Jiang, R. and Yin, L.F. 2012. Biosorption of copper (II) from aqueous solution by mycelial pellets of *Rhizopus oryzae*. African J. of Biotech. 11: 1403–1411.

Garbisu, C. and Alkorta, I. 2003. Basic concepts on heavy metal soil bioremediation. Eur. J. Miner. Process. Environ. Prot. 3: 58–66.

Garon, D., Sage, L., Wouessidjewe, D. and Seigle-Murandi, F. 2004. Enhanced degradation of fluorene in soil slurry by *Absidia cylindrospora* and *maltosyl-cyclodextrin*. Chemosphere 56: 159–166.

Guo, J., Xu, W. and Ma, M. 2012. The assembly of metals chelation by thiols and vacuolar compartmentalization conferred increased tolerance and accumulation of cadmium and arsenic in transgenic *Arabidopsis thaliana*. J. Hazard Mater. 199–200: 309–313.

Hao, L., Zhang, B., Tian, C., Liu, Y., Shi, C., Cheng, M. and Feng, C. 2015. Enhanced microbial reduction of vanadium (V) in groundwater with bioelectricity from microbial fuel cells. J. Power Sources 287: 43–49.

Harayama, S., Kasai, Y. and Hara, A. 2004. Microbial communities in oil-contaminated seawater. Curr. Opin. Biotechnol. 15: 205–214.

Heidarzadeh, N., Gitipour, S. and Abdoli, M.A. 2010. Characterization of oily sludge from a Tehran oil refinery. Waste Manage. & Research. 28: 921–927.

Hohener, P. and Ponsin, V. 2014. *In situ* vadose zone bioremediation. Curr. Opin. in Biotechnol. 27: 1–7.

Huang, J., Zhang, Y., Peng, J.S., Zhong, C., Yi, H.Y., Ow, D.W. and Gong, J.M. 2012. Fission yeast HMT1 lowers seed cadmium through phytochelatin-dependent vacuolar sequestration in *Arabidopsis*. Plant Physiol. 158: 1779–1788.

Huang, L., Chai, X., Chen, G. and Logan, B.E. 2011. Effect of set potential on hexavalent chromium reduction and electricity generation from bio-cathode microbial fuel cells. Environ. Sci. Technol. 45: 5025–5031.

Huang, R., Tian, W. and Yu, H. 2015. Enhanced biodegradation of HMW-PAHs using immobilized microorganisms in an estuarine reed wetlands simulator. pp. 1151–1154. *In*: Lee, M.S. and Wu, H.T. (eds.). Proceedings of the 2015 Inter. Power, Electronics and Mater. Eng. Conference.

Irani, Z.A., Mehrnia, M.R., Yazdian, F., Soheily, M., Mohebali, G. and Rasekh, B. 2011. Analysis of petroleum biodesulfurization in an airlift bioreactor using response surface methodology. Bioresour. Technol. 102: 10585–10591.

Istok, J.D., Senko, J.M., Krumholz, L.R., Watson, D., Bogle, M.A., Peacock, A. and White, D.C. 2004. *In situ* bioreduction of technetium and uranium in a nitrate-contaminated aquifer. Environ. Sci. & Technol. 38: 468–475.

Jiang, C.Y., Sheng, X.F., Qian, M. and Wang, Q.Y. 2008. Isolation and characterization of a heavy metal-resistant *Burkholderia* sp. from heavy metal-contaminated paddy field soil and its potential in promoting plant growth and heavy metal accumulation in metal-polluted soil. Chemosphere 72: 157–164.

Kao, C.M., Chen, C.Y., Chen, S.C., Chien, H.Y. and Chen, Y.L. 2008. Application of *in situ* biosparging to remediate a petroleum hydrocarbon spill site: field and microbial evaluation. Chemosphere 70: 1492–1499.

Karnjanapiboonwong, A., Zhang, B., Freitag, C.M., Dobrovolny, M., Salice, C.J., Smith, P.N., Kendall, R.J. and Anderson, T.A. 2009. Reproductive toxicity of nitroaromatics to the cricket, *Acheta domesticus*. Sci. Tot. Environ. 407: 5046–5049.

Kasai, Y., Kishira, H. and Harayama, S. 2002. Bacteria belonging to the genus *Cycloclasticus* play a primary role in the degradation of aromatic hydrocarbons released in a marine environment. Appl. Environ. Microbiol. 68: 5625–5633.

Kaszuba, J.P. and Runde, W.H. 1999. The aqueous geochemistry of neptunium: Dynamic control of soluble concentrations with applications to nuclear waste disposal. Environ. Sci. & Technol. 33: 4427–4433.

Kausal, R.K., Varghese, G.K. and Chabukdhara, M. 2012. Municipal solid waste management in India—current state and future challenges: a review. Int. J. Eng. Sci. Technol. 4: 1473–1489.

Keith-Roach, M.J. and Livens, F.R. 2002. Elsevier, Amsterdam. M. Vidali, Pure Appl. Chem. 73: 1163.

Khalid, A., Arshad, M. and Crowley, D.E. 2009. Biodegradation potential of pure and mixed bacterial cultures for removal of 4-nitroaniline from textile dye wastewater. Water Res. 43: 1110–16.

Khambhaty, Y., Mody, K., Basha, S. and Jha, B. 2009. Biosorption of Cr(VI) onto marine *Aspergillus niger*: experimental studies and pseudo-second order kinetics. World J. Microbiol. Biotechnol. 25: 1413–1421.

Khan, F.I., Husain, T. and Hejazi, R. 2004. An overview and analysis of site remediation technologies. J. Environ. Manag. 71: 95–122.

Kim, S., Krajmalnik-Brown, R., Kim, J.-O. and Chung, J. 2014. Remediation of petroleum hydrocarbon-contaminated sites by DNA diagnosis-based bioslurping technology. Sci. Tot. Environ. 497: 250–259.

Kimber, R., Livens, F.R. and Lloyd, J.R. 2011. Management of land contaminated by the Nuclear Legacy. Nuclear Power and the Environ. 82.

Kinnaman, T.C. 2017. The Economics of Residential Solid Waste Management. Routledge.

Knoll, A. 2003. Biomineralization and evolutionary history. pp. 329–356. Reviews in Miner. and Geochem.

Kriipsalu, M. and Nammari, D. 2010. Monitoring of biopile composting of oily sludge. Waste Manag. & Rese. 28: 395–403.

Kuiper, I., Lagendijk, E.L., Bloemberg, G.V. and Lugtenberg, B.J.J. 2004. Rhizoremediation: a beneficial plant-microbe interaction. Mol. Plant Microbe. Interact. 7: 6–15.

Kumari, D., Li, M., Pan, X. and Xin-Yi, Q. 2014. Effect of bacterial treatment on Cr(VI) remediation from soil and subsequent plantation of *Pisum sativum*. pp. 404–408. Ecologi. Eng.

Lacalle, R.G., Gomez-Sagasti, M.T., Artetxe, U., Garbisu, C. and Becerril, J.M. 2018. *Brassica napus* has a key role in the recovery of the health of soils contaminated with metals and diesel by rhizoremediation. Sci. of the Tot. Environ. 618: 347–356.

Lachance, B., Renoux, A.Y., Sarrazin, M., Hawari, J. and Sunahara, G.I. 2004. Toxicity and bioaccumulation of reduced TNT metabolites in the earthworm *Eisenia andrei* exposed to amended forest soil. Chemosphere 55: 1339–1348.

Lee, K.Y., Yoon, I.H., Lee, B.T., Kim, S.O. and Kim, K.W. 2009. A novel combination of anaerobic bioleaching and electrokinetics for arsenic removal from mine tailing soil. Environ. Sci. & Technol. 43: 9354–9360.

Lee, K.Y., K.W. Kim and S.O. Kim. 2010. Geochemical and microbial effects on the mobilization of arsenic in mine tailing soils. Environ. Geochem. and Health. 32: 31–44.

Leitao, P., Rossetti, S., Nouws, H.P.A., Danko, A.S., Majone, M. and Aulenta, F. 2015. Bioelectrochemically-assisted reductive dechlorination of 1,2-dichloroethane by a Dehalococcoides-enriched microbial culture. Bioresour. Technol. 195: 78–82.

Li, H., Zhang, X., Liu, X., Hu, X., Wang, Q., Hou, Y., Chen, X. and Chen, X. 2016. Effect of rhizodeposition on alterations of soil structure and microbial community in pyrene–lead co-contaminated soils. Environ. Earth Sci. 75: 1–8.

Limmer, M. and Burken, J. 2016. Phytovolatilization of organic contaminants. Environ. Sci. & Technol. 50: 6632–6643.

Liu, L., Li, W., Song, W. and Guo, M. 2018. Remediation techniques for heavy metal-contaminated soils: principles and applicability. Sci. of the Tot. Environ. 633: 206–219.

Lloyd, J.R., Cole, J.A. and Macaskie, L.E. 1997. Reduction and removal of heptavalent technetium from solution by *Escherichia coli*. J. of Bacteriol. 179: 2014–2021.

Lloyd, J.R., Yong, P. and Macaskie, L.E. 2000. Biological reduction and removal of Np (V) by two microorganisms. Environ. Sci. & Technol. 34: 1297–1301.

Lloyd, J.R. and Renshaw, J.C. 2005. Bioremediation of radioactive waste: radionuclide–microbe interactions in laboratory and field-scale studies. Curr. Opi. in Biotechnol. 16: 254–260.

Lloyd, J.R. and Gadd, G.M. 2011. The geomicrobiology of radionuclides. Geomicrobiol. J. 28: 383–386.

Lohner, S.T. and Tiehm, A. 2009. Application of electrolysis to stimulate microbial reductive PCE dechlorination and oxidative VC biodegradation. Environ. Sci. & Technol. 43: 7098–7104.

Lone, M.I., He, Z.L., Stoffella, P.J. and Yang, X.E. 2008. Phytoremediation of heavy metal polluted soils and water: progresses and perspectives. J. of Zhejiang University Sci. B. 9: 210–220.

Lovley, D.R., Phillips, E.J., Gorby, Y.A. and Landa, E.R. 1991. Microbial reduction of uranium. Nature 350: 413.

Lovley, D.R. and Phillips, E.J. 1992. Reduction of uranium by *Desulfovibrio desulfuricans*. Appl. Environ. Microbiol. 58: 850–856.

Lovley, D.R. and Phillips, E.J. 1994. Reduction of chromate by *Desulfovibrio vulgaris* and its C3 cytochrome. Appl. Environ. Microbiol. 60: 726–728.

Ma, Y., Oliveira, R.S., Wu, L., Luo, Y., Rajkumar, M., Rocha, I. and Freitas, H. 2015. Inoculation with metal-mobilizing plant-growth-promoting rhizobacterium *Bacillus* sp. SC2b and its role in rhizoremediation. J. of Toxicol. and Environ. Health. 78: 931–944.

Maila, M.P. and Colete, T.E. 2004. Bioremediation of petroleum hydrocarbons through land farming: are simplicity and cost-effectiveness the only advantages? Rev. Environ. Sci. Bio./Technol. 3: 349–360.

Malik, A. 2004. Metal bioremediation through growing cells. Environ. Int. 30: 261–278.

Mallick, I., Bhattacharyya, C., Mukherji, S., Dey, D., Sarkar, S.C., Mukhopadhyay, U.K. and Ghosh, A. 2018. Effective rhizoinoculation and biofilm formation by arsenic immobilizing halophilic plant growth promoting bacteria (PGPB) isolated from mangrove rhizosphere: a step towards arsenic rhizoremediation. Sci. of the Tot. Environ. 610: 1239–1250.

Mani, D. and Kumar, C. 2014. Biotechnological advances in bioremediation of heavy metals contaminated ecosystems: an overview with special reference to phytoremediation. Inter. J. of Environ. Sci. and Technol. 11: 843–872.

Marumo, J.T., Isiki, V.L., Miyamoto, H., Ferreira, R.V., Bellini, M.H. and de Lima, L.F. 2008. Investigation of the radiation risk due to environmental contamination by 241 Am from lightning rods disposed at uncontrolled garbage dumps. Radia. and Environ. Biophy. 47: 131–137.

Megharaj, M., Ramakrishnan, B., Venkateswarlu, K., Sethunathan, N. and Naidu, R. 2011. Bioremediation approaches for organic pollutants: a critical perspective. Environ. Inter. 37: 1362–1375.

Mohebali, G., Ball, A.S., Rasekh, B. and Kaytash, A. 2007. Biodesulfurization potential of a newly isolated bacterium, *Gordonia alkanivorans* RIPI90A. Enzyme and Microbial. Technol. 4: 578–584.
Mosa, K.A., Saadoun, I., Kumar, K., Helmy, M. and Dhankher, O.P. 2016. Potential biotechnological strategies for the cleanup of heavy metals and metalloids. Front. in Plant Sci. 7: 303.
Neu, M.P., Icopini, G.A. and Boukhalfa, H. 2005. Plutonium speciation affected by environmental bacteria. Radiochimica Acta. 93: 705–714.
Nalbandian, H. 2012. Trace Element Emissions From Coal. IEA Clean Coal Center, London, UK.
Newton, T.W. 2002. Redox reactions of plutonium ions in aqueous solutions. Adva. in Plutonium Chem. 1967–2000: 24–60.
Paden, N.E., Smith, E.E., Maul, J.D. and Kendall, R.J. 2011. Effects of chronic 2,4,6,-trinitrotoluene, 2,4-dinitrotoluene, and 2,6-dinitrotoluene exposure on developing bullfrog (Rana *catesbeiana*) tadpoles. Ecotoxicol. Environ. Safe. 74: 924–928.
Pathak, V.V., Kothari, R., Chopra, A.K. and Singh, D.P. 2015. Experimental and kinetic studies for phycoremediation and dye removal by *Chlorella pyrenoidosa* from textile wastewater. J. of Environ. Manag. 163: 270–277.
Phillips, R.W., Wiegel, J., Berry, C.J., Fliermans, C., Peacock, A.D., White, D.C. and Shimkets, L.J. 2002. *Kineococcus radiotolerans* sp. nov., a radiation-resistant, gram-positive bacterium. Inter. J. of Syst. and Evolutionary Microbiol. 52: 933–938.
Philp, J.C. and Atlas, R.M. 2005. Bioremediation of contaminated soils and aquifers. pp. 139–236. *In*: Atlas, R.M. and Philp, J.C. (eds.). Bioremediation: applied microbial solutions for real-world environmental cleanup. Ameri. Soc. for Microbiol. (ASM) Press. Washington.
Pilon-Smits, E. 2005. Phytoremediation. Annu. Rev. Plant Biol. 56: 15–39.
Pinheiro, H.M., Touraud, E. and Thomas, O. 2004. Aromatic amines from azo dye reduction: status review with emphasis on direct UV spectrophotometric detection in textile industry wastewaters. Dyes Pigments 61: 121–39.
Poulsen, T.G. and Bester, K. 2010. Organic micropollutant degradation in sewage sludge during composting under thermophilic conditions. Environ. Sci. Technol. 44: 5086–5091.
Prakash, V., Saxena, S., Sharma, A., Singh, S. and Singh, S.K. 2015. Treatment of oil sludge contamination by composting. J. of Bioreme. & Biodegra. 6: 1.
Prasad, M.N.V. and Freitas, H.M.D. 2003. Metal hyperaccumulation in plants—Biodiversity prospecting for phytoremediation technology. Electron. J. Biotechnol. 93: 285–321.
Prasad, S.S. and Aikat, K. 2014. Study of bio-degradation and biodecolourization of azo dye by Enterobacter sp. SXCR. Environ. Technol. 35: 956–65.
Purnomo, A.S., Mori, T., Kamei, I., Nishii, T. and Kondo, R. 2010. Application of mushroom waste medium from Pleurotus ostreatus for bioremediation of DDT-contaminated soil. Int. Biodeterior. Biodegrad. 64: 397–402.
Raina, V., Suar, M., Singh, A., Prakash, O., Dadhwal, M., Gupta, S.K., Dogra, C., Lawlor, K., Lal, S., van der Meer, J.R. and Holliger, C. 2008. Enhanced biodegradation of hexachlorocyclohexane (HCH) in contaminated soils via inoculation with *Sphingobium indicum* B90A. Biodegradation 19: 27–40.
Rajkumar, M. and Freitas, H. 2008. Influence of metal resistant-plant growth-promoting bacteria on the growth of *Ricinus communis* in soil contaminated with heavy metals. Chemosphere 71: 834–842.
Ramirez, M., Fernandez, M., Granada, C., Le borgne, S., Gomez, J.M. and Cantero, D. 2011. Biofiltration of reduced sulphur compounds and community analysis of sulphur-oxidizing bacteria. Bioresour. Technol. 102: 4047–4053.
Rashtchi, M., Mohebali, G., Akbarnejad, M., Towfighi, J., Rasekh, B. and Keytash, A. 2006. Analysis of biodesulfurization of model oil system by the bacterium, strain RIPI-22. Biochem. Eng. J. 29: 169–173.
Rhodes, C.J. 2014. Mycoremediation (bioremediation with fungi)—growing mushrooms to clean the earth. Chem. Spec. Bioavailab. 26: 196–198.
Rim-Rukeh, A. 2014. An assessment of the contribution of municipal solid waste dump sites fire to atmospheric pollution. Open J. of Air Pollu. 3: 53.
Rittmann, B.E., Banaszak, J.E. and Reed, D.T. 2002. Reduction of Np (V) and precipitation of Np (IV) by an anaerobic microbial consortium. Biodegradation 13: 329–342.
Rizzo, A.C.D.L., dos Santos, R.D.M., dos Santos, R.L., Soriano, A.U., da Cunha, C.D., Rosado, A.S., Sobral, L.G.D.S. and Leite, S.G. 2010. Petroleum-contaminated soil remediation in a new solid phase bioreactor. J. of Chemi. Technol. and Biotechnol. 85: 1260–1267.
Roh, C., Kang, C. and Lloyd, J.R. 2015. Microbial bioremediation processes for radioactive waste. Korean J. of Chemi. Engi. 32: 1720–1726.
Roman-Ponce, B. 2017. Plant growth-promoting traits in rhizobacteria of heavy metal-resistant plants and their effects on *Brassica nigra* seed germination. Pedosphere. 27: 511–526.
Romera, E., González, F., Ballester, A., Blázquez, M.L. and Munoz, J.A. 2007. Comparative study of biosorption of heavy metals using different types of algae. Biores. Technol. 98: 3344–3353.
Saher, N.U. and Siddiqui, A.S. 2019. Occurrence of heavy metals in sediment and their bioaccumulation in sentinel crab (*Macrophthalmus depressus*) from highly impacted coastal zone. Chemosphere 221: 89–98.
Sahu, A., Mandal, A., Thakur, J., Manna, M.C. and Rao, A.S. 2012. Exploring bioaccumulation efficacy of *Trichoderma viride*: an alternative bioremediation of cadmium and lead. Natl. Acad. Sci. Lett. 35: 299–302.

Schwartz, C., Echevarria, G. and Morel, J.L. 2003. Phytoextraction of cadmium with *Thlaspi caerulescens*. Plant and Soil. 249: 27–35.

Shanthala, M., Hosmani, S.P. and Hosetti, B.B. 2009. Diversity of phytoplanktons in a waste stabilization pond at Shimoga town, Karnataka State, India. Environ. Monit. Assess. 151: 437–44.

Shukla, D., Kesari, R., Tiwari, M., Dwivedi, S., Tripathi, R.D., Nath, P. and Trivedi, P.K. 2013. Expression of *Ceratophyllum demersum* phytochelatin synthase, CdPCS1, in *Escherichia coli* and *Arabidopsis* enhances heavy metalloids accumulation. Protoplasma. 250: 1263–1272.

Shukla, K.P., Singh, N.K. and Sharma, S. 2010. Bioremediation: Developments, current practices and perspectives. Genetic Engi. & Biotechnol. J. 3: 1–20.

Silva-Castro, G.A., Uad, I., Gonzalez-Lopez, J., Fandino, C.G., Toledo, F.L. and Calvo, C. 2012. Application of selected microbial consortia combined with inorganic and oleophilic fertilizers to recuperate oil-polluted soil using land farming technology. Clean Technol. Environ. Policy. 14: 719–726.

Sims, J.G. and Steevens, J.A. 2008. The role of metabolism in the toxicity of 2,4,6-trinitrotoluene and its degradation products to the aquatic amphipod Hyalella Azteca. Ecotoxicol. Environ. Saf. 70: 38–46.

Singh, J.S., Abhilash, P.C., Singh, H.B., Singh, R.P. and Singh, D.P. 2011. Genetically engineered bacteriam emerging tool for environmental remediation and future research perspectives. Gene. 480: 1–9.

Singh, N., Marwa, N., Mishra, J., Verma, P.C., Rathaur, S. and Singh, N. 2016. *Brevundimonas diminuta* mediated alleviation of arsenic toxicity and plant growth promotion in *Oryza sativa* L. Ecotoxicol. Environ. Saf. 125: 25–34.

Songkasiri, W., Reed, D.T. and Rittmann, B.E. 2002. Bio-sorption of neptunium (V) by *Pseudomonas fluorescens*. Radiochimica Acta. 90: 785–789.

Sonoki, S., Fujihiro, S. and Hisamatsu, S. 2007. Genetic engineering of plants for phytoremediation of polychlorinated biphenyls. pp. 3–13. *In*: Willey, N. (ed.). Phytoremediation: Methods and Reviews, Humana Press, New York.

Srivastava, P.K., Vaish, A., Dwivedi, S., Chakrabarty, D., Singh, N. and Tripathi, R.D. 2011. Biological removal of arsenic pollution by soil fungi. Sci. Total Environ. 409: 2430–2442.

Srivastava, S., Agrawal, S.B. and Mondal, M.K. 2015. A review on progress of heavy metal removal using adsorbents of microbial and plant origin. Environ. Sci. Pollut. Res. 22: 15386–15415.

Stocker, T.F., Qin, D., Plattner, G.K., Tignor, M., Allen, S.K., Boschung, J., Nauels, A., Xia, Y., Bex, V. and Midgley, P.M. 2013. Climate Change 2013: The physical science basis contribution of working group I to the fifth assessment report of IPCC the intergovernmental panel on climate change.

Strandberg, G.W. and Arnold, W.D. 1988. Microbial accumulation of neptunium. J. of Indus. Microbiol. 3: 329–331.

Sylvain, B. 2016. Phytostabilization of As, Sb and Pb by two willow species (*S. viminalis* and *S. purpurea*) on former mine technosols. Catena. 136: 44–52.

Szklo, A.S. and Uller, V.C. 2008. Fundamentals of Oil Refining: Technology and Economics. Interciencia. Rio de Janeiro, Brazil.

Taboski, M., Rand, T. and Piorko, A. 2005. Lead and cadmium uptake in the marine fungi *Corollospora lacera* and *Monodictys pelagic*. FEMS Microbiol. Ecol. 53: 445–453.

Tale, V.P., Maki, J.S. and Zitomer, D.H. 2015. Bioaugmentation of overloaded anaerobic digesters restores function and archaeal community. Water Research 70: 138–147.

Tandukar, M., Huber, S.J., Onodera, T. and Pavlostathis, S.G. 2009. Biological chromium(VI) reduction in the cathode of a microbial fuel cell. Environ. Sci. Technol. 43: 8159–8165.

Tang, L., Zeng, G.M., Shen, G.L., Li, Y.P., Zhang, Y. and Huang, D.L. 2008. Rapid detection of picloram in agricultural field samples using a disposable immune membrane based electrochemical sensor. Environ. Sci. Technol. 42: 1207–1212.

Tian, H., Gao, J., Hao, J., Lu, L., Zhu and Qiu, C.P. 2013. Atmospheric pollution problems and control proposals associated with solid waste management in China: a review. J. of Hazardous Materials 252: 142–154.

Touceda-González, M., Álvarez-López, V., Prieto-Fernández, Á., Rodríguez-Garrido, B., Trasar-Cepeda, C., Mench, M., Puschenreiter, M., Quintela-Sabarís, C., Macías-García, F. and Kidd, P.S. 2017. Aided phytostabilisation reduces metal toxicity, improves soil fertility and enhances microbial activity in Cu-rich mine tailings. J. of Environ. Manag. 186: 301–313.

Tsuboy, M.S., Angeli, J.P.F., Mantovani, M.S., Knasmüller, S., Umbuzeiro, G.A. and Ribeiro, L.R. 2007. Genotoxic, mutagenic and cytotoxic effects of the commercial dye CI disperse blue 291 in the human hepatic cell line HepG2. Toxicol. *In Vitro* 21: 1650–1655.

Vala, A.K. and Sutariya, V. 2012. Trivalent arsenic tolerance and accumulation in two facultative marine fungi. Jundishapur J. Microbiol. 5: 542–545.

Vala, A.K. and Dave, B.P. 2017. Marine-derived fungi: prospective candidates for bioremediation. pp. 17–37. *In*: Ram Prasad (ed.). Mycoremediation and Environmental Sustainability. Springer, Cham.

Vangronsveld, J., Herzig, R., Weyens, N., Boulet, J., Adriaensen, K., Ruttens, A., Thewys, T., Vassilev, A., Meers, E., Nehnevajova, E. and van der Lelie, D. 2009. Phytoremediation of contaminated soils and groundwater: lessons from the field. Environ. Sci. Pollut. Res. 16: 765–794.

Vara Prasad, M.N. and de Oliveira Freitas, H.M. 2003. Metal hyperaccumulation in plants: biodiversity prospecting for phytoremediation technology. Electronic J. of Biotechnol. 6: 285–321.

Vaxevanidou, K., Papassiopi, N. and Paspaliaris, I. 2008. Removal of heavy metals and arsenic from contaminated soils using bioremediation and chelant extraction techniques. Chemosphere 70: 1329–1337.

Venkatraman, K. and Ashwath, N. 2007. Phytocapping: an alternative technique to reduce leachate and methane generation from municipal landfills. The Environmentalist 27: 155–164.

Virkutyte, J., Sillanpaa, M. and Latostenmaa, P. 2002. Electrokinetic soil remediation—critical review. Sci. Tot. Environ. 289: 97–121.

Ward, O.P. and Singh, A. 2003. Biodegradation of oil sludge. U.S. Patent # 6,652,752.

Watson, J.H.P. and Ellwood, D.C. 1994. Biomagnetic separation and extraction process for heavy metals from solution. Minerals Engi. 7: 1017–1028.

Whelan, M.J. 2015. Fate and transport of petroleum hydrocarbons in engineered biopiles in polar regions. Chemosphere 131: 232–240.

Williams, K.H., Nevin, K.P., Franks, A., Englert, A., Long, P.E. and Lovley, D.R. 2010. Electrode-based approach for monitoring *in situ* microbial activity during subsurface bioremediation. Environ. Sci. Technol. 44: 47–54.

Williams, K.H. 2011. Acetate availability and its influence on sustainable bioremediation of Uranium-contaminated groundwater. Geomicrobiol J. 28: 519–539.

Won, W.D., Heckly, R.J., Glover, D.J. and Hoffsommer, J.C. 1974. Metabolic disposition of 2,4,6-trinitrotoluene. Appl. Microhiol. 27: 513–516.

Wurtz, E.A., Sibley, T.H. and Schell, W.R. 1986. Interactions of *Escherichia coli* and marine bacteria with 241Am in laboratory cultures. Health Physics 50: 79–88.

Xia, M., Liu, Y., Taylor, A.A., Fu, D., Khan, A.R. and Terry, N. 2017. Crude oil depletion by bacterial strains isolated from a petroleum hydrocarbon impacted solid waste management site in California. Inter Biodeter. & Biodegr. 123: 70–77.

Yadav, K.K., Singh, J.K., Gupta, N. and Kumar, V. 2017. A review of nanobioremediation technologies for environmental cleanup: a novel biological approach. J. Mater. Environ. Sci. 8: 740–757.

Yakimov, M.M., Golyshin, P.N., Lang, S., Moore, E.R., Abraham, W.R., Lünsdorf, H. and Timmis, K.N. 1998. *Alcanivorax borkumensis* gen. nov., sp. nov., a new, hydrocarbon-degrading and surfactant-producing marine bacterium. Inter. J. Syst. Bacteriol. 48: 339–348

Yakimov, M.M., Giuliano, L., Denaro, R., Crisafi, E., Chernikova, T.N., Abraham, W.R., Luensdorf, H., Timmis, K.N. and Golyshin, P.N. 2004. *Thalassolituus oleivorans* gen. nov., sp. nov., a novel marine bacterium that obligately utilizes hydrocarbons. Int. J. Syst. Evol. Microbiol. 54: 141–148.

Yuan, M., He, H., Xiao, L., Zhong, T., Liu, H., Li, S., Deng, P., Ye, Z. and Jing, Y. 2014. Enhancement of Cd phytoextraction by two *Amaranthus* species with *endophytic Rahnella* sp. JN27.Chemosphere 103: 99–104.

Zeyaullah, M.D., Atif, M., Islam, B., Abdelkafe, A.S., Sultan, P., ElSaady, M.A. and Ali, A. 2009. Bioremediation: A tool for environmental cleaning. Afr. J. Microbio. Res. 3: 310–314.

Zhang, Y.F., He, L.Y., Chen, Z.J., Wang, Q.Y., Qian, M. and Sheng, X.F. 2011. Characterization of ACC deaminase producing endophytic bacteria isolated from copper tolerant plants and their potential in promoting the growth and copper accumulation of *Brassica napus*. Chemosphere 83: 57–62.

Zhang, Y. and Miller, R.M. 1995. Effect of rhamnolipid (biosurfactant) structure on solubilization and biodegradation of n-alkanes. Appl. Environ. Microbiol. 61: 2247–2251.

Zhuang, P., Yang, Q.W., Wang, H.B. and Shu, W.S. 2007. Phytoextraction of heavy metals by eight plant species in the field. Water, Air and Soil Pollu. 184: 235–242.

3

Integrative Approaches for Understanding and Designing Strategies of Bioremediation

Shiv Prasad,[1,]* *Sudha Kannojiya,*[1] *Sandeep Kumar,*[1] *Krishna Kumar Yadav,*[2] *Monika Kundu*[3] *and Amitava Rakshit*[4]

1. Introduction

Quality of life on Earth is solely associated with healthy environmental conditions and an abundance of natural resources. Natural resources play a vital role in life on Earth. They form the basis of everything humans need to survive. Currently, most of the world's resources are depleting or losing their quality due to pollution, thereby affecting humans, wildlife health, and the whole ecosystem (Vidali et al. 2001, Sharma and Reddy 2004). This demonstrates our carelessness and negligence regarding the environment and natural resources. Land and water are the two fundamental pillars of natural resources on which the sustainability of agriculture and the continued existence of civilization rely. Unfortunately, both have been drastically degraded through toxic chemicals due to some natural and anthropogenic activities (Sharma and Reddy 2004, Chitra et al. 2011). These activities generate wastes and pose risks to public health, natural habitat, and all or part of the ecosystem (Yadav et al. 2018).

Currently, the contamination of water bodies and soil pollution due to toxic substances is a primary environmental concern that affects large areas worldwide. Industrial waste, domestic effluents, and agricultural practices have been the prime source of heavy metals in the soil, such as lead, chromium, arsenic, zinc, cadmium, copper, and nickel. For example, the one-sixth part of the agricultural land is affected by heavy metals in China. It is reported that 42 rivers in India are polluted with at least two toxic heavy metals (Central Water Commission 2014). The contaminated water bodies and polluted soil sites are a dormant warning to human health and the environment.

On the other hand, India is facing a considerable waste management crisis. As a developing country, this is a big setback to India. Poor waste management leads to a decline in quality of life

---

[1] Centre for Environment Science and Climate Resilient Agriculture, Indian Agricultural Research Institute, New Delhi 110012, India.
[2] Institute of Environment and Development Studies, Bundelkhand University, Kanpur Road, Jhansi, 284128, India.
[3] Division of Agricultural Physics, Indian Agricultural Research Institute, New Delhi 110012, India.
[4] Department of Soil Science & Agricultural Chemistry, Institute of Agricultural Science, Banaras Hindu University, Varanasi-221005, UP, India.
* Corresponding author: shiv_drprasad@yahoo.co.in

in the aspects of health and hygiene (Yadav et al. 2018). Additionally, it is tough and increasingly expensive to find new landfill sites for the final disposal of the material. The cap and contain method are only an interim solution since the contamination remains on-site, requiring monitoring and maintenance. A healthier approach than these traditional methods is to destroy the pollutants if possible, or at least to transform them into innocuous substances (Sharma 2012). Some techniques, such as incineration, are used for decomposition of types of chemicals (e.g., base-catalyzed dechlorination). They can be very operational at reducing levels of a range of contaminants but have many drawbacks, principally their technological complexity and the lack of public acceptance (Sharma 2012, Yadav et al. 2018). Hence, considering all other methods, bioremediation is an option that offers the possibility to destroy or reduce various harmless contaminants using natural biological activity (Vidali et al. 2001, Naik et al. 2012).

2. Bioremediation: characterization and concepts

Bioremediation is a process that principally uses microorganisms, plants, or microbial or plant enzymes to detoxify contaminants in the soil and other polluted sites (Shukla et al. 2014, Sharma et al. 2018), and known efficient and eco-friendly strategies for sustainable development (Varol 2011, Sharma 2012, Rizwan et al. 2014). Bioremediation concept includes biodegradation, which refers to the partial, and sometimes total, transformation, or detoxification of contaminants by microorganisms and plants (Shukla et al. 2014, Yadav et al. 2018). In bioremediation, microorganisms must enzymatically attack the pollutants and convert them into harmless products (McCutcheon et al. 2004). As bioremediation can be practical only where environmental conditions permit microbial growth and activity, its use often involves the manipulation of environmental factors to allow microbial growth and degradation to keep at a faster rate (Sharma 2012). Bioremediation methods are naturally more cost-effective than traditional methods such as incineration, and some pollutants can be treated on-site, thus decreasing exposure risks for clean-up personnel or potentially more extensive exposure as a result of transport accidents. Since bioremediation is created based on natural attenuation, the community considers it more acceptable than other methods. Most bioremediation systems are operated under aerobic conditions. However, running a bioremediation system under anaerobic conditions may allow microbial organisms to degrade otherwise recalcitrant molecules (Colberg and Young 1995).

2.1 Contaminants' characteristics, behavior, and their fate

Due to their toxicity, persistence, and non-biodegradable nature, trace metals containing mercury, arsenic, lead, cadmium, copper, zinc, nickel, chromium, and organic pollutants in soil and water have drawn considerable attention worldwide (Yang et al. 2009). These contaminants in the environment may accumulate in aquatic flora, fauna, and microorganisms, which may enter the food chain and consequently appear in humans (Varol 2011, Shukla et al. 2012). These contaminants may be anthropogenic as well as natural in origin. To prevent environmental pollution, especially soil and water, and to improve the ecosystem's health, it is essential to thoroughly understand the contamination characteristics of pollutants in ecological segments and target their potential sources (Sakan et al. 2009, Shukla et al. 2014).

Petroleum and other hydrocarbons are a complicated mixture composed mainly of carbon and hydrogen, in which most of the alkanes are determined to be narcotic and irritant, and most of PAHs have substantial toxicity, carcinogenicity, teratogenicity, and mutagenicity (Kao et al. 2008, Shukla et al. 2014). Many oil pollutants in the water cause severe pollution to the water ecosystems and are detrimental to the health of living creatures and human bodies (Zhang et al. 2003).

Once contaminants are in soils, where they go and how quickly they travel depends on several factors. Few organic pollutants can undergo chemical modifications or degrade into products that may be toxic than the original compound (Garrison et al. 2000, Newman and Reynolds 2004). The chemical elements which cannot break down, but their characteristics may change, can be quickly taken up by plants or animals. Many contaminants vary in their tendency to end up in water held in the soil or in the underlying groundwater (by leaching through the land), volatilize (evaporate) into the air, and bind tightly to the soil (Shayler et al. 2009).

Biological and physico-chemical processes usually bound the presence of organic contaminants in soils. It includes sorption–desorption, volatilization, and bio-chemical degradation, uptake by plants, run-off, and leaching (Mamy et al. 2005). The pattern (horizontally and vertically) of organic pollutants in soils depends on their action and degradation. Movement and degradation of organic contaminants, in turn, depend on three common factors: (i) Chemical and biological properties of the pollutant, (ii) hydraulic properties of the soil, and (iii) weather conditions. Movement of the organic pollutants can be due to diffusion and mass flow (Mamy et al. 2005, Todorovic 2009).

Bioremediation also requires minimal effort and can often be carried out on-site, usually without causing a significant separation of normal activities. That also reduces the need to transport loads of waste off-site and potential risks to human health and environment that can arise during transportation (NAP 1993, Sharma 2012). It is also a cost-effective process as it costs less than the other conventional methods that are used for cleaning up of hazardous waste. It also helps in the destruction of the pollutants; many of the dangerous compounds can be transformed into harmless products. It does not use any hazardous chemicals (NAP 1993).

Bioremediation is excellent for biodegrading organic pollutants. Common examples of organic contaminants that respond to bioremediation include petroleum hydrocarbons (PAHs), non-chlorinated chemicals (e.g., acetone), wood treatment chemicals (e.g., creosote and pentachlorophenol (PCP), certain chlorinated aromatic compounds (e.g., chlorobenzenes and biphenyls) and certain chlorinated aliphatic compounds (e.g., trichloroethene (TCE)) (USEPA 2001a). Inorganic contaminants such as heavy metals from acid mine drains (AMD) are often treated using bioremediation techniques (NAP 1993, Sharma and Reddy 2004). Table 1 below shows the behavior of pollutants.

Table 1. Behavior of contaminants.

S. No.	Behavior	Examples
(i)	Bio-degradable	a) Petroleum products (gas, diesel, fuel oil) b) Crude oil compounds (benzene, toluene, xylene, naphthalene) c) Pesticides (malathion), some industrial solvents d) Coal compounds (phenols, cyanide in coal tars, and coke waste)
(ii)	Partially degradable/ persistent	e) TCE (trichloroethane): threat to groundwater f) PCE (perchloroethylene): dry cleaning solvent g) PCBs: have been degraded in labs, but not in fieldwork h) Arsenic, Chromium, Selenium
(iii)	Not degradable/ Recalcitrant	i) Uranium j) Mercury k) DDT

2.2 Principles of bioremediation

"Remediate" means to solve a problem, and "bioremediation" means to use biological organisms to solve an environmental issue such as contaminated soil or water. The term bioremediation is used for a natural remedy approach to decrease or clean up contamination. According to the EPA 2006, bioremediation is a treatment that uses naturally occurring microorganisms to break down dangerous substances into less toxic or non-toxic materials (Gonzalez et al. 2019).

Bioremediation is used to treat sites contaminated with organic materials (USEPA 2001a). However, it can also be applied to immobilize inorganic contaminants such as heavy metals, although this is a developing area (Sharma and Reddy 2004). The most common organic pollutants typically include polycyclic aromatic hydrocarbons (PAHs), benzene, toluene, ethylbenzene and xylene (BTEX), polychlorinated biphenyls (PCBs), pesticides and herbicides (Sharma and Reddy 2004, USEPA 2001a).

Microorganisms are ideally accommodated to the task of contaminant elimination because they have enzymes that enable them to use contaminants as food or energy. Without the activity of microbes, the Earth would be buried in wastes, and the nutrients required for the survival of life would be locked up in detritus. Microbes can successfully degrade and diminish human-made contaminants in the soil. However, it depends on three factors: the presence of types of organisms, contaminants nature, and the geological and chemical conditions at the contaminated sites (NAP 1993).

In bioremediation, microorganisms with biological activity, including algae, bacteria, fungi, and yeast, are used to digest toxic contaminants (USEPA 2001b, Sharma and Reddy 2004). The process of biodegradation occurs in the presence of oxygen or without oxygen, known as aerobic and anaerobic digestion, respectively. An example of hazardous oil spill digestion and degradation and its conversion into the water and harmless gases by microbes is shown diagrammatically in Figure 1. Another example of microbial degradation of organic contaminants into water and CO_2 is shown by the simplified equation (1).

$$\text{Organic contaminants} + O_2 \rightarrow H_2O + CO_2 + \text{cell material} + \text{energy} \tag{1}$$

Similarly, sulfate-reducing bacteria use ferric iron or SO_4^- as an electron acceptor. They reduce it to ferrous iron and H_2S. Denitrifying bacteria such as *Thiobacillus denitrificans, Micrococcus denitrificans* use NO_3^- as an electron acceptor and emit NO_2^-, N_2O, and N_2 as reduced gases. Likewise, methanogens use CO_2 as an electron acceptor to produce CH_4.

Under the aerobic conditions, microorganisms can transform several organic contaminants to carbon dioxide, water, and microbial cell mass. Aerobic bioremediation utilizes oxygen as the electron acceptor (Parsons Corporation 2004). Aerobic metabolism is generally more utilized and can be useful for hydrocarbons, and other organic compounds, such as petroleum hydrocarbons and methyl tertiary-butyl ether. Several organisms can degrade hydrocarbons (HC) using oxygen as the electron acceptor and HC as carbon and energy sources. Aerobic technologies may also alter the ionic form of metals (EPA 2006).

Anaerobic bioremediation includes microbial reactions happening in the absence of oxygen and encompasses many processes, including fermentation, methanogenesis, reductive dechlorination, and sulfate- and nitrate-reducing conditions (EPA 2000, Tomei and Daugulis 2013). Depending on the contaminant, a subset of these activities may be cultivated. In anaerobic metabolism, nitrate, sulfate, carbon dioxide, oxidized materials, or organic compounds may replace oxygen as the electron acceptor. Heavier petroleum products such as lubricating oils commonly take a longer time to biodegrade than the lighter products. However, it is also not practical to use to enhance aerobic bioremediation to address petroleum contamination in low permeability, especially in clay soil (EPA 2004). A plume moving with groundwater flow typically produces distinct redox zones—once an electron acceptor is depleted, a new redox reaction utilizing a new electron acceptor occurs. The

Figure 1. Diagram representing the basic concept of bioremediation (USEPA 2001b).

electron acceptor that would lead to the next largest production of energy during the reaction will dominate (EPA 2000). The main reason for the lack of knowledge and skepticism of bioremediation is that the technology needs experience not only of such fields as environmental engineering but also hydrology. The multidisciplinary characteristics of bioremediation present problems not only for clients and regulators but also for the vendors of environmental clean-up services (NAP 1993).

3. Factors affecting bioremediation

Various factors affect bioremediation, such as contaminant concentration, availability of the microbes, and environmental factors such as temperature, pH, the presence of oxygen or other electron acceptors, soil type, and nutrients status (Vidali et al. 2001, Parsons Corporation 2004, ICSCS 2006). Although the microbes exist in contaminated soil, they cannot be there in the numbers needed for bioremediation of the site. Their growth and activity must be stimulated, which usually requires the addition of nutrients and oxygen to help indigenous microorganisms (ESTCP 2005, Naik et al. 2012).

Nutrients are the vital building blocks of the microbe's life and enable them to create the necessary enzymes to break down the contaminants (Sharma et al. 2018). Carbon is an essential element of living forms of life and is required in higher amounts than other elements. In addition to hydrogen, oxygen, and nitrogen, carbon constitutes about 95% of the weight of cells. Phosphorous and sulfur contribute 70% of the remainder (Boopathy 2000). The nutritional demand for carbon to nitrogen ratio is 10:1, and carbon to phosphorous is 30:1 (Vidali et al. 2001). Several factors' optimum environmental conditions for the degradation of contaminants are given below in Table 2. Some of the additional information about important factors that affect bioremediation have also been explained below.

Table 2. Optimum environmental conditions for the degradation of contaminants.

Parameter	Condition required for microbial activity	Optimum value for oil degradation
Soil moisture	25–28% of water holding capacity	30–90%
Soil pH	5.5–8.8	6.5–8.0
Oxygen content	Aerobic, minimum air-filled pore space of 10%	10–40%
Nutrient content	N and P for microbial growth	C:N:P = 100:10:1
Temperature (°C)	15–45°C	20–30°C
Contaminants	Not too toxic	Hydrocarbon 5–10%
Heavy metals	Total content 2000 ppm	
Type of soil	Low clay or silt content	

Source: Vidali et al. (2001).

3.1 Contaminant concentrations

Pollutants instantly influence the microbial activity in the soil. When levels are too high, then they may have toxic effects on the existing bacteria. In contrast, low contaminant concentration may prevent the induction of bacterial degradation enzymes (ESTCP 2005, Sharma et al. 2018).

3.2 Contaminant bioavailability

The contaminant bioavailability depends on the degree to which they sorb to solids by molecules in contaminated areas and other factors such as whether contaminants are present in non-aqueous phase liquid (NAPL) form. Bioavailability for microbial reactions is lower for pollutants that are more actively sorbed to solids, embedded in molecules' matrices of contaminated media, likewise widely diffused in macropores of soil and sediments, or are present in NAPL form (ICSCS 2006).

3.3 Site characteristics

The location of the contaminated area has a significant impact on the effectiveness of any bioremediation approach. Site environmental conditions essential to consider for bioremediation applications include pH, temperature, water content, nutrient availability, and redox potential.

3.3.1 pH

pH influences the solubility and biological availability of nutrients, metals, and other constituents for optimal bacterial growth (ESTCP 2005).

3.3.2 Redox potential and oxygen content

Both the factors typify oxidizing or reducing conditions. Redox potential is influenced by the appearance of electron acceptors such as nitrate, manganese oxides, iron oxides, and sulfate (ICSCS 2006).

3.3.3 Nutrients

They are needed for microbial cell growth and division (ESTCP 2005). Suitable amounts of minimum nutrients for microbial growth are generally present, but nutrients can be supplied in a useable form or via an organic substrate supplement (Parsons Corporation 2004), which also serves as an electron donor, to stimulate bioremediation.

3.3.4 Temperature

It directly influences the rate of microbial metabolism and, consequently, microbial activity in the environment. The biodegradation rate, to an extent, increases with rising temperature and decreases with reducing temperature (ESTCP 2005).

4. Role of plant roots and associated microbes to clean contaminated sites

The root-associated microorganisms establish a synergism with plant roots and can help the plant to absorb nutrients improving plant performance and, consequently, the quality of soils (Barea et al. 2002, Yang et al. 2009, Coelho et al. 2015). Plants growing in metal contaminated soils harbor a diverse group of microorganisms that can tolerate high concentration of heavy metals and provide some benefits to both the soil and the plant. Several microorganisms are used in heavy metal remediation of contaminated sites (Barea et al. 2002, Verma et al. 2006, Dong et al. 2013). Among the microorganisms involved in heavy metal phytoremediation, the rhizosphere bacteria deserve attention because they can directly improve the phytoremediation process by changing the metal bioavailability through altering soil pH, the release of chelators (e.g., organic acids, siderophores) and oxidation/reduction reactions (Yang et al. 2009, Rajkumar et al. 2012).

Various metabolites (e.g., 1-aminocyclopropane-1-carboxylic acid deaminase, indole-3-acetic acid, siderophores, organic acids, etc.) produced by plant root-associated microbes (e.g., plant growth-promoting bacteria, mycorrhizae) have been proposed to be involved in many biogeochemical processes operating in the plant root or rhizosphere (Barea et al. 2002). The salient functions include nutrient acquisition, cell elongation, metal detoxification, and alleviation of biotic/abiotic stress in plants. Plant root rhizosphere microbes accelerate metal mobility or immobilization. Plants and associated microbes release inorganic and organic compounds possessing acidifying, chelating, and reductive power (Meagher 2000, Kuiper et al. 2004, Rajkumar et al. 2012).

These functions are implicated to play an essential role in plant metal uptake. Overall, the plant root-associated beneficial microbes enhance the efficiency of the phytoremediation process directly by altering the metal accumulation in plant tissues and indirectly by promoting the shoot and root biomass production (Rajkumar et al. 2012). Similarly, the metal-tolerant mycorrhizal

Table 3. Microbial strains characterized by their potential to mobilize metals to alter the plant metal uptake.

S. No.	Microbial metabolites/reactions	Microorganisms	Microbial effects on metals and its uptake by plants
1.	Siderophores (Pyoverdine, pyochelin and alcaligin E)	*Pseudomonas aeruginosa, Pseudomonas fluorescens*	Enhanced Cr and Pb uptake by plants by facilitating their mobilization
2.	Organic acids Oxalic acid, Tartaric acid, Formic acid Acetic acid	*Burkholderia cepacia Pseudomonas aeruginosa*	Solubilized ZnO, $ZnCO_3$ and $CdCO_3$ Solubilized ZnO and $Zn_3(PO_4)_2$
3.	Biosurfactants Di-rhamnolipid	*Pseudomonas aeruginosa* BS2	Mobilized Cd and Pb
4.	Polymeric substances and glycoprotein	*Azotobacter* spp.	Immobilized Cd and Cr and decreased their uptake by *Triticum aestivum*
5.	Oxidation and reduction reaction	A consortium of oxidizing sulfur bacteria *Stenotrophomonas maltophilia*	Increased bioavailability of Cu Reduced soluble and harmful Se (IV) to insoluble and unavailable Se (0) and thereby decreased the plant Se uptake
6.	Bioaccumulation	*Brevibacillus* sp. B-I *Serratia* sp. MSMC541	Decreased the concentration of Zn in shoot tissues of *Trifolium repens* Reduced translocation of As, Cd, and Cu from roots to shoots in *Lupinus luteus*

Source: Rajkumar et al. (2012).

fungi have also been frequently reported in hyperaccumulators growing in metal-polluted soils, indicating that these fungi have evolved a heavy metal tolerance and that they may play a significant role in the phytoremediation. Table 3 given above summarizes the published studies on plant roots and associated microbes that clean contaminated soil through metabolites/actions on heavy metal mobilization/immobilization and their uptake by plants.

5. Bioremediation management strategies and application

Bioremediation uses living organisms that interact directly or indirectly with the environment to neutralize or remove contaminants (Vidali 2001). Bioremediation can be applied both *in situ* (i.e., field conditions) and *ex situ* (i.e., mesocosm/controlled conditions). Many small-scale examples exist of both approaches using plants, fungi, and bacteria as bio-remediators of organic contaminants, although results have been varied (Vidali 2001, Harms et al. 2011). However, applications of important bio-remediators at a large scale has been reported by Boopathy 2000, Adriaens et al. 2006, Gonzalez et al. 2019. The classification of bioremediation processes is given in Figure 2.

5.1 In situ bioremediation

In situ bioremediation means that you allow bioremediation to take place while leaving the soil or water in its original location. *In situ* bioremediation can be designed with or without plant species. Plant species have been used because they take up vast amounts of water that assists in controlling contaminated water, such as a groundwater contaminant plume, in the soil. Since *in situ* bioremediation, mostly applications, is based on aerobic biodegradation, anaerobic-based processes have also become of interest. That can be essential to improve the aerobic degradation of organic compounds (Azubuike et al. 2016). Some of the *in situ* bioremediation methods have been addressed below.

5.1.1 Natural bioremediation

Natural bioremediation has been happening for millions of years. Naturally, existing bacteria and fungi or plants themselves degrade or detoxify substances hazardous to human health and the

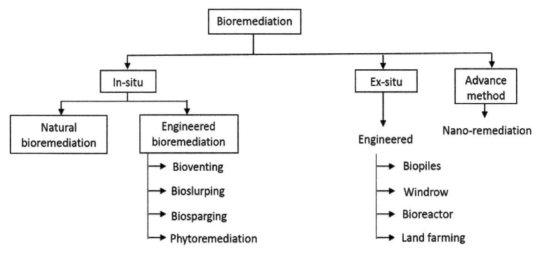

Figure 2. Classification of bioremediation processes. Source: (Oyetibo et al. 2017).

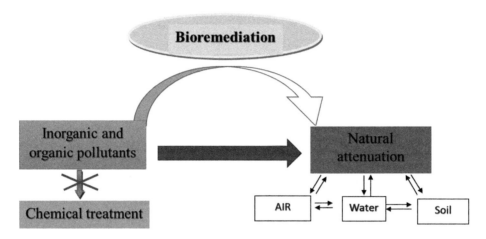

Figure 3. Natural attenuation and bioremediation are widely accepted environment clean-up procedures. Source: (Shukla et al. 2014).

environment. Biodegradation of dead animals or decay of vegetation is a kind of natural remediation. Bioremediation is a natural part of the carbon, nitrogen, and sulfur cycles. Chemical energy present in waste matters is used by microbes to grow while they transform organic carbon and hydrogen to water and CO_2 (Figure 3).

5.1.2 Managed or engineered bioremediation

A type of remediation that enhances the growth and degradative activity of microorganisms by using engineered systems that supply nutrients (e.g., nitrogen and phosphorus), electron acceptors, and other growth-stimulating materials (Fiedler 2000). An example of managed bioremediation is land farming, which refers to the regulated organic compounds degradation that are distributed onto the soil surface and then tilled. Since around 1980, prepared bed practices are used for bioremediation. In this procedure, contaminated soil is excavated and deposited with suitable fertilizers into a shallow layer over an impermeable base. Conditions are managed to achieve degradation of the contaminants of concern. Essential techniques of engineered bioremediation are described below with the results of some of the case studies.

5.1.2.1 Bioventing

This technique has gained popularity among other *in situ* bioremediation techniques, especially in restoring sites polluted with light spilled petroleum products (Hohener and Ponsin 2014). This technique includes controlled stimulation of airflow by carrying oxygen to the unsaturated zone to increase bioremediation, by enhancing movements of indigenous microbes. In bioventing, amendments are made by supplementing nutrients and moisture to improve bioremediation, with the final goal being to produce a microbial transformation of pollutants to a safe state (Philp and Atlas 2005).

A study by Hong and Xingang (2011) showed the effect of air injection rate on volatilization, biodegradation, and biotransformation of toluene-contaminated location by bioventing. It was seen that at two separate air injection rates (81.504 and 407.52 m^3/d), no significant difference in contaminant elimination was seen (200 days). Though, at the earlier stage of the study (day 100), it was observed that high air injection rates resulted in enhanced toluene removal by volatilization compared to low air injection rates. In other words, a high airflow rate does not bring about an increase in the biodegradation rate, nor makes pollutant biotransformation more efficient. That is due to the initial saturation of air in the subsurface for oxygen requirement during biodegradation. However, a low air injection rate resulted in a significant improvement in biodegradation. It, therefore, shows that in bioventing, the air injection rate is among the essential parameters for pollutant dispersal, redistribution, and surface loss.

The bioremediation method, particularly in treating the vadose zone polluted with chlorinated compounds, is generally recalcitrant under aerobic conditions. In this process, in pure oxygen, a mixture of nitrogen together with low concentrations of CO_2 and H_2 can also be injected to bring about loss of chlorinated vapor, with H_2 acting as the electron donor (Mihopoulos et al. 2000, Shah et al. 2001). In soil with low-permeability, injection of pure oxygen might lead to higher oxygen concentration compared to air injection. Moreover, ozonation might be beneficial for the incomplete oxidation of recalcitrant compounds to accelerate biodegradation (Philp and Atlas 2005). Unlike bioventing that relies on improving the microbial biodegradation process at the vadose-zone by controlled air injection, soil vapor extraction (SVE) maximizes volatile organic compound (VOCs) volatilization via vapor extraction (Magalhaes et al. 2009).

Soil vapor extraction (SVE) may be regarded as a physical method of remediation due to its mechanism of pollutant removal. During on-site field trials, obtaining similar results achieved during lab investigations is not always achievable due to additional environmental factors and various characteristics of the unsaturated zone to which air is injected. As a result, bioventing treatment time may be extended. High airflow rate begins to transfer volatile organic compounds (VOCs) to the soil vapor phase, which needs off-gas treatment of the resulting gases before releasing into the atmosphere (Burgess et al. 2001).

5.1.2.2 Bioslurping

This technique connects vacuum-enhanced pumping, soil vapor extraction, and bioventing to achieve soil groundwater remediation by the indirect provision of oxygen and stimulation of contaminant biodegradation (Gidarakos and Aivalioti 2007). The method is intended for free product recovery, such as light non-aqueous phase liquids (LNAPLs), thus remediating capillary, unsaturated, and saturated zones. It can also be applied to remediate soils contaminated with volatile and semi-volatile organic compounds. The system uses a "slurp" that continues into the free product layer, which draws up liquids from this layer in a way comparable to that of how a straw draws fluid from any vessel. The pumping mechanism produces the upward movement of LNAPLs to the surface, where it becomes isolated from water and air.

In this technique, excessive soil moisture restricts air permeability and reduces oxygen transfer rate, in turn reducing microbial activities. Although the method is not fit for soil bioremediation with low permeability, it saves costs due to less volume of groundwater resulting from the operation. Thus, it minimizes storage, treatment, and waste disposal costs (Philp and Atlas 2005). Building a vacuum on a deep high permeable site and fluctuating water table, which could produce saturated soil lenses that are hard to aerate, is amongst the major concerns of this *in situ* technique (Azubuike et al. 2016).

5.1.2.3 Biosparging

This technique is very comparable to bioventing in that air is injected into soil subsurface to excite microbial activities to promote pollutant removal from polluted sites. Unlike bioventing, the air is introduced at the saturated zone, which can cause upward mobility of VOCs to the unsaturated zone to improve biodegradation. The effectiveness of biosparging depends on two significant factors, namely, soil permeability, which describes pollutant bioavailability to microbes, and pollutant degradability (Philp and Atlas 2005).

As with bioventing and soil vapor extraction (SVE), biosparing is comparable in operation with an almost similar technique known as *in situ* air sparging (IAS), which depends on high airflow rates to realize pollutant volatilization, whereas biosparging increases biodegradation. Moreover, both mechanisms of pollutant removal are not mutually exclusive for both techniques. Biosparging has been extensively applied in managing aquifers contaminated with petroleum products, especially diesel and kerosene. Kao et al. (2008) reported that biosparging of benzene, toluene, ethylbenzene, and xylene (BTEX) contaminated aquifer plume resulted in a transfer from anaerobic to aerobic conditions (Azubuike et al. 2016).

A case study conducted in the Damodar Valley in Eastern India revealed that biosparging is effective at removing 75% of contaminants present within a year (Kruger et al. 1997). The initial results achieved in the field trials were also explained using computer modeling programs. The results from the study were used to set the optimum conditions for bioremediation, including proper moisture content, pH, temperature, nutrients, and carbon sources. The field tests used six separate test sites. Different parameters were tested in each location to investigate the optimum condition.

5.1.2.4 Phytoremediation

This technique relies on the application of plant interactions (i.e., physical, biochemical, biological, chemical, and microbiological interactions) in contaminated sites to mitigate the toxic effects of pollutants. Depending on the pollutant type (elemental or organic), there are several mechanisms (accumulation or extraction, degradation, filtration, stabilization, and volatilization) involved in phytoremediation (Kruger et al. 1997, McCutcheon et al. 2004, Subramaniam et al. 2006, Yadav et al. 2018). Toxic heavy metals and radionuclides pollutants are mostly removed by extraction, transformation, and sequestration (Ojuederie et al. 2017). On the other hand, organic pollutants like hydrocarbons and chlorinated compounds are dominantly reduced by degradation, rhizoremediation, stabilization, and volatilization (Table 4) by several types of plant species (Meagher 2000, Kuiper et al. 2004).

Some critical factors must be considered when picking a plant species for phytoremediation. This includes-plant root system, i.e., fibrous or tap depending on the depth of pollutant, above-ground biomass, which should not be accessible for animal eating, toxicity of pollutant to plant, plant survival and its adaptability to existing environmental conditions, plant growth rate, site monitoring and above all, time required to achieve the desired level of cleanliness. Besides, the plant should be resistant to diseases and pests (Lee 2013). It has been reported that in some contaminated environments, the process of contaminant removal by plant involves: uptake, which is mostly by passive means, translocation from roots to shoots, which is carried out by xylem flow, and accumulation in the shoot (San et al. 2013).

Purakayastha et al. (2009) conducted a phytoremediation study on crop species of *Brassica juncea, Brassica campestris, Brassica carinata*, and *Brassica napus*. They concluded *Brassica carinata* cv. DLSC17 to be a suitable species for heavy metals remediation from the soil, which was able to reduce the metals load by 15% for Zn, 12% Pb, and 11% for Ni from a naturally contaminated soil from peri-urban Delhi. While Unterbrunner et al. (2007) conducted a phytoremediation study

Table 4. Process of phytoremediation with a suitable example of pollutants and their potential plants.

Process	Function	Pollutant	Medium	Plant	References
Phytoextraction	Remove metals pollutants that accumulate in plants. Remove organics from the soil by concentrating them in plants	Cd, Pb, Zn, As, Petroleum, and Radionuclides	Soil and groundwater	Viola baoshanensis, Helianthus annus, Alfalfa, Poplar, Indian mustard cabbage	(Macek et al. 2000, Zhuang et al. 2007)
Phyto-transformation	Plant uptake and degradation of organic compounds	Xenobiotic substances	Soil	Cannes	(Subramanian et al. 2006)
Phyto-degradation	Plant and associated microorganism degrade organic pollutants	DDT, Explosives, waste, and Nitrates	Groundwater	Elodea Canadensis, Duckweed, Hybrid poplar	(Garrison et al. 2000, Newman and Reynolds 2004)
Rhizofiltration	Roots absorb and adsorb pollutants, primarily metals, from water and aqueous waste streams	Zn, Pb, Cd, As Also, Radionuclei	Groundwater	Brassica juncea, Helianthus annus	(Dushenkov et al. 1995, Verma et al. 2006)
Phyto-stabilization	Use of plant species to reduce the bioavailability of pollutants in the environment	Cu, Cd, Cr, Ni, Pb, Zn	Soil	Anthyllisvallesiana, Lupinusalbus Hybrid poplar grasses	(Vazquez et al. 2006)
Phyto-volatilization	Use of plants to volatilize pollutants	Se, CCl4, EDB, TCE		Zea mays, Brassica sp.	(Ayotamuno and Cogbara 2007)

on tree species and observed a considerable amount of Cd and Zn accumulation in various willow, poplar, and birch tree species with up to 116 mg Cd kg^{-1} and 4680 mg Zn kg^{-1} in leaves of Salix caprea tree, metal concentrations in leaves were not related to total (aqua regia), or labile (1 M NH$_4$NO$_3$ extract), concentrations in soil but the accumulation factors (leaf concentration: soil concentration) for Cd and Zn, which followed an inverse log type function.

That the application of plant growth-promoting rhizobacteria (PGPR) might play an essential role in phytoremediation is also reported by many researchers, as PGPR promotes biomass production and tolerance of plants to heavy metals and other unfavorable soil (edaphic) conditions (Barea et al. 2002, Yancheshmeh et al. 2011, De-Bashan et al. 2012, Rajkumar et al. 2012). *Brachiaria mutica* and maize (corn) have also been listed as potent phytoremediators of heavy metal-polluted soil (Tiecher et al. 2016).

5.2 *Ex-situ* bioremediation

Ex-situ bioremediation techniques entail the removal of the contaminated object to be treated elsewhere (Philp and Atlas 2005, Tomei and Daugulis 2013). For instance, in the case of material polluted by kerosene, phenols, cresols, polycyclic aromatic hydrocarbons and semi-volatile organic compounds, these techniques involve excavating pollutants from contaminated zones and finally transporting them to another site for treatment. *Ex-situ* bioremediation techniques are usually considered based on the cost of treatment, depth of pollution, type of pollutant, degree of contamination, geographic location, and geology of the polluted site (EPA 2006). Performance criteria also determine the selection of *ex-situ* bioremediation techniques (Philp and Atlas 2005, Tomei and Daugulis 2013). Some of the *ex-situ* bioremediation methods have been addressed below.

5.2.1 Biopile-mediated bioremediation

A biopile-mediated method is the most effective method in remediating pollutants such as BTEX, phenols, PAHs, and explosives such as TNT and RDX. This is a full-scale technology which involves above-ground piling of excavated polluted soil on a treatment area, followed by nutrient supplement, and sometimes forced aeration is used to enhance bioremediation by basically increasing microbial activities (Chemlal et al. 2013, Dias et al. 2015). The components of the biopile technique are aeration system, irrigation, nutrient and leachate collection systems, and a treatment bed. The use of this *ex situ* technique is increasingly being considered due to its useful features, including cost-effectiveness, which enables effective biodegradation on the condition that nutrients, temperature, oxygen, moisture, and pH are well controlled (Whelan et al. 2015). The utilization of biopile by polluted sites can help limit the volatilization of low molecular weight (LMW) contaminants. It can also be used efficiently to remediate extremely polluted environments such as the frigid regions (Gomez and Sartaj 2014, Whelan et al. 2015).

A study was conducted by Gomez and Sartaj (2014) to investigate the effects of the different applications of microbial consortia rates, i.e., 3 and 6 ml/m^3, with compost at the rate of 5 and 10% on total petroleum hydrocarbon (TPH) reduction in field-scale biopiles at low-temperature conditions. The result showed that at the end of 94 days, 90.7% TPH reduction in the bioaugmented and biostimulated setups was achieved over the control setup. The high TPH removal was attributed to the synergistic interaction between bioaugmentation and biostimulation, thus demonstrating the flexibility of biopiles for bioremediation. Biopile-mediated bioremediation setup can readily be scaled up to a pilot system to get similar performance obtained during lab experiments (Chemlal et al. 2013).

Essential to the efficiency of biopiling is the sieving and aeration of contaminated soil before processing (Delille et al. 2008). Bulking agents such as straw, sawdust, bark, or wood chips, and other organic materials have been added to enhance the remediation process in a biopile construct (Rodríguez-Rodríguez et al. 2010). Although biopile-mediated bioremediation systems conserve space compared to other *ex situ* bioremediation methods, including land farming, robust engineering, cost of maintenance and operations, lack of power supply particularly at remote places, which would facilitate uniform dispersion of air in contaminated gathered soil through air pump, are some of the limitations of this system. More so, extreme heating of air can lead to drying of soil undergoing bioremediation, which results in inhibition of microbial activities and increase in volatilization rather than biodegradation (Delille et al. 2008, Rodríguez-Rodríguez et al. 2010).

5.2.2 Windrows

Windrow system is just like composting, another form of *ex-situ* solid-phase bioremediation. Windrows rely on the periodic turning of piled polluted soil to enhance aeration, which increases degradation activities of indigenous and transient hydrocarbon clastic bacteria present in contaminated soil (McCutcheon et al. 2004). The regular turning of polluted soil, together with adding water, brings the uniform distribution of pollutants, nutrients, and microbial degradative enzymatic activities, thus speeding up bioassimilation, biotransformation, and mineralization rate to achieve efficient bioremediation (Barr et al. 2002).

Windrow-mediated treatment, when compared with biopile-mediated treatment, showed a greater hydrocarbon removal. However, the high efficiency of the windrow towards hydrocarbon removal is a result of the soil type, which is stated to be more friable (Coulon et al. 2010). Nevertheless, due to the requirement of periodic turning in windrow treatment, especially in the case of soil polluted with toxic volatiles, it is not considered as the better option to adopt in bioremediation. The use of windrow-mediated treatment has been associated with CH_4 (greenhouse gas) release due to the creation of anaerobic conditions within piled polluted soil (Hobson et al. 2005).

5.2.3 Bioreactor-mediated bioremediation

Bioreactors, which can be applied in bioremediation strategies, are tanks in which pollutants are converted, by living organisms, into a specific product(s) following a series of biological reactions. In bioreactors-mediated bioremediation, various operating modes are used like batch, fed-batch, sequencing batch, continuous, and multistage, etc. (Castaldi and Ford 1992, Wang and Vipulanandam 2001, Adriaens et al. 2006). The selection of operating mode often depends on financial and capital expenditure. Conditions in a bioreactor support the natural process of cells by mimicking and maintaining their natural environment to provide optimum growing conditions. Polluted samples are fed into a bioreactor either as dry solid or slurry to treat contaminated soil. It has many benefits as compared to other *ex-situ* bioremediation procedures. In bioremediation, bioreactor provision of controlling all bioprocess parameters (*i.e.*, pH, agitation and aeration rates, temperature, substrate loading, and inoculum concentrations) makes it more meaningful than the others (Azubuike et al. 2016).

The provision to control and manipulate process parameters in a bioreactor helps in speeding up biological reactions to reduce bioremediation time effectively. Most importantly, in controlled bioaugmentation practice, where cultured microbes and nutrients are added into the subsurface for biodegrading specific contaminants, increased pollutant bioavailability, and mass transfer (contact between pollutants and microbes), can conclusively be established in a bioreactor, thus making bioreactor-based bioremediation more efficient (Wang and Vipulanandam 2001, Adriaens et al. 2006). It can be applied to efficiently treat soil or water polluted with volatile organic compounds (VOCs), including ethylbenzene, xylenes (BTEX), benzene, and toluene.

Castaldi and Ford (1992) treated petroleum waste sludges containing 680 mg kg^{-1} of 2–3 rings PAHs and 38 mg kg^{-1} of 4–6 rings PAHs in a continuous flow multistage A-SB that operated at relatively short residence times with the minimal loss of volatile constituents. The higher molecular weight PAHs, including the four-ring derivatives of pyrene, benzo[a]pyrene, and chrysene, were removed with efficiencies higher than 90%. In another study, Wang and Vipulanandam (2001) observed the effect of naphthalene concentration on the A-SB remediation potential. Naphthalene was generally found to rapidly desorb from the spiked soil by undergoing rapid and extensive biodegradation. 500 mg kg^{-1} of naphthalene was reduced to 20 mg kg^{-1} after 65 h days of treatment, whereas 5000 mg kg^{-1} of the same pollutants were reduced to 40 mg kg^{-1} after 100 h.

5.2.4 Landfarming

Landfarming is a full-scale bioremediation technology, which generally combines liners and additional methods to control leaching of pollutants, which needs excavation and placement of polluted soils, sediments, and sludges. Contaminated and polluted media is applied to lined beds and periodically turned over to enhance aeration, which improves bioremediation efficiency. Landfarming often needs controlled and optimized soil conditions to increase the contaminant degradation rate. Controlled conditions are typically maintained by irrigation or spraying (for moisture content), tilling the soil (for aeration), liming (for pH), and nutrient supplement (nitrogen, phosphorus, and potassium). In most cases, it is regarded as *ex situ* bioremediation, while, in some cases, it is viewed as an *in-situ* bioremediation technique (Kao et al. 2008).

In agriculture, especially in conventional practice, polluted soils are usually excavated and tilled, but the site of treatment determines the type of bioremediation. When excavated contaminated soil is treated on-site, it can be regarded as *in-situ*; otherwise, it is *ex-situ* as it has more in common with other *ex-situ* bioremediation techniques (EPA 2000). It has been stated that when a pollutant lies 1 m below ground surface, bioremediation continued without excavation contaminated media, while pollutant lying 1.7 m below needs transportation on the ground surface for practical bioremediation (Nikolopoulou et al. 2013). Generally, excavated polluted soils are carefully applied to fixed layer support above the ground surface to allow aerobic biodegradation of pollutants by microorganisms (Philp and Atlas 2005, Silva-Castro et al. 2015).

Moreover, landfarming is not fit for soil polluted with toxic volatiles treatment because of its design and pollutant removal mechanism (volatilization), particularly in hot (tropical) climate regions. These limitations make landfarming-based bioremediation time consuming and less efficient compared to other *ex-situ* bioremediation techniques. One of the notable advantages of *ex-situ* bioremediation techniques is that they do not require an extensive preliminary assessment of polluted sites before remediation; this makes the initial stage short, less complicated, and less expensive.

6. Advanced technique: Nano-bioremediation

Despite *in-situ* as well as *ex-situ* bioremediation techniques, some advanced technologies have come to the fore in recent years, but their application is still limited (Harms et al. 2011, Amin et al. 2014). Nanotechnology is one of them and is characterized by microscopic manufactured particles (< 100 nm), called nanoparticles (NPS) or ultrafine particles (Zhang et al. 2003, Dong et al. 2013, Hong et al. 2017). NSP are atomic or molecular aggregates that can drastically modify their physicochemical properties compared with the bulk material. The technique where nanoparticles/nanomaterial are formed by plant, fungi, and bacteria with the help of nanotechnology, which are used to remove environmental contaminants (such as heavy metals, organic and inorganic pollutants) from contaminated sites, is called nano-bioremediation (Zhang et al. 2003, Singh and Walker 2006, Stafiej and Pyrzynska 2007, Yadav et al. 2017). Table 5 below shows some plant nanoparticles used in the bioremediation of heavy metals of contaminated sites.

Nano-bioremediation is the emerging technique for the removal of pollutants for environmental clean-up. Prevailing technologies for contaminated-site remediation are chemical and physical remediation and incineration, including bioremediation (Varol et al. 2011). With recent advances, bioremediation offers an environmentally friendly and economically feasible option to remove contaminants from the environment (Mueller and Nowack 2010). Three main approaches are used in nano-bioremediation; it includes the use of microbes, plants, and enzymatic activities. Nanotechnology increases phytoremediation efficiency and can also be used for the eco-friendly restoration of soils, and water contaminated with heavy metals, organic and inorganic pollutants (Wang 2007, Rizwan et al. 2014, Yadav et al. 2017).

A new concept of single enzyme nanoparticles (SENs), related to nanoparticle, has been developed in which each enzyme molecule is contained by a hybrid organic/inorganic polymer network (Dong et al. 2013). Each enzyme molecule is modified with a porous organic/inorganic structure (or armor) of less than a few nanometers thick. This approach represents a new way to change and stabilize the enzymes, which form a new type of nanostructure as well (Chang et al. 2005, 2007, Hong et al. 2017). These nanoparticles have the potential to bind with the xenobiotic compounds and degrade them completely or transform in less harmful derivatives, which further help in cleaning the environment.

Recent studies have shown that organic contaminants such as atrazine, molinate, and chlorpyrifos can be degraded with nano-sized zero-valent ions (Zhang et al. 2003, Chang et al. 2005, 2007, Ghormade et al. 2011). Nanoparticles in enzyme-based bioremediation can also be used in combination with phytoremediation (Nowack 2008, Singh et al. 2009, 2010, Sharma et al. 2018). For example, several complex organic compounds, such as long-chain hydrocarbons and

Table 5. Some plant nanoparticles used in bioremediation of heavy metals.

Plant species	Heavy metals
Noaea mucronata	Pb (98%), Zn (79.03%), Cu (73.38%), Cd (72.04%) and Ni (33.61%)
Euphorbia macroclada	Pb (92%), Zn (76.05%), Cu (74.66), Cd (69.08%) and Ni (31.50%)

Source: Mohsenzadeh et al. (2012).

Table 6. Examples of the use of nanoparticles in remediation.

Process exploited	Target compounds	Nanomaterials used	References
Adsorption	Heavy metals, organic compounds, arsenic, phosphate, Cr (IV), mercury, PAHs, DDT, Dioxin	Iron oxides, carbon-based nanomaterials such as dendrimers and polymers, carbon nanotubes (CNTs)	(Bhaumik et al. 2012, Pan et al. 2010, Mueller and Nowack 2009, 2010, Rickerby and Morrison 2007, Stafiej and Pyrzynska 2007, Chang et al. 2005, 2007)
Photocatalysis	Organic pollutants, NOX, VOCs Azo dye, Congo red dye, 4-chlorophenol and Orange II, PAHs	TiO_2, ZnO, species of iron oxides (Fe III, Fe_2O_3, FeO_4)	(Khedr et al. 2009, Wang 2007, Bandara et al. 2007, Banhnemann 2004, Kim et al. 2001, Feng et al. 2000)
Redox reactions	Halogenated organic compounds, metals, nitrate, arsenate, oil, PAH, PCB	Nanoscale zero-valent iron (nZVI), nanoscale calcium peroxide	(Zhang et al. 2003) (Tratnyek and Johnson 2006) (Nowack 2008, Klimkova et al. 2008, Varansi et al. 2007)
Disinfection	Diamines, phenols formaldehyde, hydrogen peroxide, silver ions, halogens, glutaraldehyde, acridines	Nano silver/titanium dioxide (Ag/TiO_2) and CNTs	(Amin et al. 2014, Mcdonnell et al. 1999)
Membranes	Chlorinated compounds, polyethylene, 1,2-dichlorobenzene, organic and inorganic solutes, halogenated organic solvents	Nano Ag/TiO_2/Zeolites/Magnetite and CNTs	(Amin et al. 2014, Mcdonnell et al. 1999)

organochlorines, are particularly resistant to microbial and plant degradation. A combined approach involving nanotechnology and biotechnology could overcome this limitation: complex organic compounds would be degraded into simpler compounds by nano-encapsulated enzymes, which in turn would be rapidly degraded by the joint activities of microbes and plants. Table 6 below shows examples of the use of nanoparticles in the remediation of contaminated sites (Yadav et al. 2017).

7. Advantages and limitations of bioremediation

7.1 Advantages of bioremediation

The most significant benefits of using bioremediation are its contribution to the environment because it is a natural process and accepted by society as the best process for decontaminating soil and water. After contaminant treatment and degradation, no offensive products such as carbon dioxide, wastewater, and cell biomass are released and thus making it a sustainable technology. Some of its benefits are pointed out below (Sharma and Reddy 2004, Vivaldi 2001).

- It is possible to completely breakdown toxic organic contaminants into other non-toxic chemicals.
- Equipment demands are minimum compared to other remediation technologies.
- It can be executed as an *in situ* or *ex situ* method depending on conditions.
- It is a cost-effective technology per unit volume of soil or groundwater compared to other remediation methods.
- Low-technology equipment is needed, i.e., readily available equipment, e.g., pumps, well-drilling equipment, etc.

- Bioremediation is perceived arbitrarily by society because it is a natural process.
- A complete breakdown of pollutants into non-toxic compounds is possible because the process does not involve transferring contaminants to another environmental medium.

7.2 Limitations of bioremediation

There are some limitations and challenges associated with bioremediation. It is limited to compounds that are biodegradable and susceptible to rapid and complete degradation. Sometimes, by-products of biodegradation may be more persistent or toxic than the parent compound. Some of its limitations and challenges are pointed out below (Sharma and Reddy 2004, Vivaldi 2001).

- Most significant limitations have been countered, and if the bioremediation process is not controlled, there is a possibility that organic contaminants may not be broken down completely, resulting in more toxic by-products that may be more mobile than initial contamination.
- The process is sensitive to toxicity level and prevailing conditions on the ground, and the terms such as temperature, pH, must be favorable to microbial activity.
- Area monitoring to track the biodegradation rate of contaminants is advised.
- If an *ex-situ* technique is practiced, managing volatile organic compounds (VOCs) may be challenging.
- Treatment time is typically higher than that of other remediation methods.
- The range of contaminants that can be effectively treated is limited to biodegradable compounds.
- Left residual levels can be too high (not meeting regulative specifications), persistent, and toxic.
- Performance evaluations are challenging because a defined level of a clean site is not there, and therefore, performance criteria regulations are uncertain.

8. Conclusion and future perspectives

The introduction of toxic contaminants into the environment causes adverse effects on humans, wildlife, and natural habitats. In this context, decontamination and restoration of polluted sites have become a priority in any part of the world. Modern society is also concerned about the wide range of contaminants and pollutants. Scientists are working continuously to provide an effective solution for increased pollution. They are trying to develop relevant site-specific strategies and cost-effective technologies through the identification and characterization of novel plant species, microbial strains, and their enzymatic activities, and nanoparticle-based materials found in nature. Bioremediation has enough capability and immense potential to overcome a wide range of contaminants through its biodegradation.

Therefore, it is necessary to study the complex behavior and identify the metabolites and their degradation pathways to find a more suitable solution for reducing environmental pollution, which is a significant thrust area. On the other hand, a systems biology approach involving omics tools such as genomics, proteomics, transcriptomics, phenomics, lipidomics, and metabolomics could play an essential role in the study of this complex behavior of microbes (Singh and Shukla 2015). Recently, nanoparticle-based materials have been attracting considerable interest for their unique properties and the immense application potential in diverse areas. These nanoparticles have the potential to bind with the xenobiotic compounds and degrade them completely or transform in less harmful derivatives, which further help in efficient and eco-friendly environment cleaning.

Acknowledgments

The authors are grateful to the Director, ICAR (Indian Agricultural Research Institute), New Delhi, and Head, CESCRA for providing necessary facilities and support service for this work.

References

Adriaens, P., Li, M.Y. and Michalak, A.M. 2006. Scaling methods of sediment bioremediation processes and applications. Eng. Life. Sci. 6(3): 217–227.

Amin, M.T., Alazba, A.A. and Manzoor, U. 2014. A review of removal of pollutants from water/wastewater using different types of nanomaterials. Adva. Mater. Sci. Eng. 825910. http://dx.doi.org/10.1155/2014/825910.

Ayotamuno, J.M. and Kogbara, R.B. 2007. Determining the tolerance level of Zea mays (maize) to a crude oil polluted agricultural soil. Afr. J. Biotechnol. 6(11).

Azubuike, C.C., Chikere, C.B. and Okpokwasili, G.C. 2016. Bioremediation techniques—classification based on site of application: principles, advantages, limitations, and prospects. World J. Microbiol. 32(11): 180.

Bahnemann, D. 2004. Photocatalytic water treatment: solar energy applications. Solar Energy 77(5): 445–459.

Bandara, J., Klehm, U. and Kiwi, J. 2007. Raschig rings-Fe_2O_3 composite photocatalyst activate in the degradation of 4-chlorophenol and Orange II under daylight irradiation. Appl. Catal. B. 76(1-2): 73–81.

Barea, J.M., Azcón, R. and Azcón-Aguilar, C. 2002. Mycorrhizosphere interactions to improve plant fitness and soil quality. Anton Leeuw 81: 343–51.

Barr, D., Finnamore, J.R., Bardos, R.P., Weeks, J.M. and Nathanail, C.P. 2002. Biological methods for assessment and remediation of contaminated land: case studies. CIRIA.

Bhaumik, M., Maity, A., Srinivasu, V.V. and Onyango, M.S. 2012. Removal of hexavalent chromium from aqueous solution using polypyrrole-polyaniline nanofibers. Chem. Eng. J. 181: 323–333.

Boopathy, R. 2000. Factors limiting bioremediation technologies. Bioresour. Technol. 74(1): 63–67.

Burgess, J.E., Parsons, S.A. and Stuetz, R.M. 2001. Developments in odour control and waste gas treatment biotechnology: a review. Biotechnol. Adv. 19(1): 35–63.

Chang, M.C., Shu, H.Y., Hsieh, W.P. and Wang, M.C. 2005. Using nanoscale zero-valent iron for the remediation of polycyclic aromatic hydrocarbons contaminated soil. J. Air Waste Manag. Assoc. 55(8): 1200–1207.

Chemlal, R., Abdi, N., Lounici, H., Drouiche, N., Pauss, A. and Mameri, N. 2013. Modeling and qualitative study of diesel biodegradation using biopile process in sandy soil. Int. Biodeter. Biodegr. 78: 43–48.

Chitra, K., Sharavanan, S. and Vijayaragavan, M. 2011. Tobacco, corn, and wheat for phytoremediation of cadmium polluted soil. Recent Res. Sci. Technol. 3(2): 148–151.

Coelho, L.M., Rezende, H.C., Coelho, L.M., de Sousa, P.A.R., Melo, D.F.O. and Coelho, N.M.M. 2015. Bioremediation of polluted waters using microorganisms. pp. 1–22. *In*: Shiomi, N (ed.). Advances in Bioremediation of Wastewater and Polluted Soil.

Colberg, P.J. and Young, L.Y. 1995. Anaerobic degradation of nonhalogenated homocyclic aromatic compounds coupled with nitrate, iron, or sulfate reduction. pp. 307–330. *In*: Young, L.Y. and Cerniglia, C.E. (eds.). Microbial Transformation and Degradation of Toxic Organic Chemicals. New York, N.Y: Wiley-Liss.

Coulon, F., Awadi, M.A.I., Cowie, W., Mardlin, D., Pollard, S., Cunningham, C., Risdon, G., Arthur, P., Semple, K.T. and Paton, G.I. 2010. When is a soil remediated? Comparison of biopiled and windrowed soils contaminated with bunker-fuel in a full-scale trial. Environ. Pollut. 158(10): 3032–3040.

De-Bashan, L.E., Hernandez, J.P. and Bashan, Y. 2012. The potential contribution of plant growth-promoting bacteria to reduce environmental degradation—A comprehensive evaluation. Appl. Soil Ecol. 61: 171–189.

Delille, D., Duval, A. and Pelletier, E. 2008. Highly efficient pilot biopiles for on-site fertilization treatment of diesel oil-contaminated sub-Antarctic soil. Cold Reg. Sci. Technol. 54(1): 7–18.

Dias, R.L., Ruberto, L., Calabro, A., Balbo, A.L., Del Panno, M.T. and Mac Cormack, W.P. 2015. Hydrocarbon removal and bacterial community structure in on-site biostimulated biopile systems designed for bioremediation of diesel-contaminated Antarctic soil. Polar Bio. 38(5): 677–687.

Dong, G., Wang, Y., Gong, L., Wang, M., Wang, H., He, N., Zheng, Y. and Li, Q. 2013. Formation of soluble Cr (III) end-products and nanoparticles during Cr (VI) reduction by *Bacillus cereus* strain XMCr-6. Biochem. Eng. J. 70: 166–172.

Dushenkov, V., Kumar, P.N., Motto, H. and Raskin, I. 1995. Rhizofiltration: the use of plants to remove heavy metals from aqueous streams. Environ. Sci. Technol. 29(5): 1239–1245.

ESTCP. 2005. Environmental Security Technology Certification Program. Bioaugmentation for Remediation of Chlorinated Solvents: Tech. Development Status and Res. Needs.

EPA. 2000. Engineered Approaches to *In Situ* Bioremediation of Chlorinated Solvents: Fundamentals and Field Applications.

EPA. 2004. How to evaluate alternative cleanup technologies for underground storage tank sites: a guide for corrective action plan reviewers. US Environmental Protection Agency (EPA 510-R-04-002).

EPA. 2006. Engineering Issue: *In Situ* and *Ex Situ* Biodegradation Technologies for Remediation of Contaminated Sites. EPA-625-R-06-015.

Farhan, S.N. and Khadom, A.A. 2015. Biosorption of heavy metals from aqueous solutions by *Saccharomyces cerevisiae*. Int. J. Ind. Chem. 6(2): 119–130.

Feng, W., Nansheng, D. and Helin, H. 2000. Degradation mechanism of azo dye CI reactive red 2 by iron powder reduction and photooxidation in aqueous solutions. Chemosphere 41(8): 1233–1238.

Fiedler, L. 2000. Engineered approaches to *in situ* bioremediation of chlorinated solvents: Fundamentals and field applications. EPA Washington DC Office of Solid Waste.

Garrison, A.W., Nzengung, V.A., Avants, J.K., Ellington, J.J., Jones, W.J., Rennels, D. and Wolfe, N.L. 2000. Phytodegradation of p, p '-DDT and the enantiomers of o, p '-DDT. Environmental Science and Tech. 34(9): 1663–1670.

Ghormade, V., Deshpande, M.V. and Paknikar, K.M. 2011. Perspectives for nano-biotechnology enabled protection and nutrition of plants. Biotech. Adv. 29(6): 792–803.

Gidarakos, E. and Aivalioti, M. 2007. Large scale and long term application of bioslurping: the case of a Greek petroleum refinery site. J. Hazard. Mater. 149(3): 574–581.

Gomez, F. and Sartaj, M. 2014. Optimization of field scale biopiles for bioremediation of petroleum hydrocarbon contaminated soil at low temperature conditions by response surface methodology (RSM). Int. Biodeter. Biodegr. 89: 103–109.

Gonzalez, S.V., Johnston, E., Gribben, P.E. and Dafforn, K. 2019. The application of bioturbators for aquatic bioremediation: Review and meta-analysis. Environ. Pollut. 250(2019): 426–436.

Harms, H., Schlosser, D. and Wick, L.Y. 2011. Untapped potential: exploiting fungi in bioremediation of hazardous chemicals. Nat. Rev. Microbiol. 9(3): 177.

Hobson, A.M., Frederickson, J. and Dise, N.B. 2005. CH_4 and N_2O from mechanically turned windrow and vermicomposting systems following in-vessel pre-treatment. Waste Manag. 25(4): 345–352.

Hohener, P. and Ponsin, V. 2014. *In situ* vadose zone bioremediation. Current Opinion in Biotech. 27: 1–7.

Hong, S.U.I. and Xingang, L.I. 2011. Modeling for volatilization and bioremediation of toluene-contaminated soil by bioventing. Chin. J. Chem. Eng. 19(2): 340–348.

Hong, S.G., Kim, B.C., Na, H.B., Lee, J., Youn, J., Chung, S.W., Lee, C.W., Lee, B., Kim, H.S., Hsiao, E. and Kim, S.H. 2017. Single enzyme nanoparticles armored by a thin silicate network: Single enzyme caged nanoparticles. Chem. Eng. J. 322: 510–515.

ICSCS. 2006. International Centre for Soil and Contaminated Sites 2006. Manual for biological remediation techniques. 81 pp.

Jan, A.T., Azam, M., Ali, A. and Haq, Q.M.R. 2014. Prospects for exploiting bacteria for bioremediation of metal pollution. Crit. Rev. Environ. Sci. Technol. 44(5): 519–560.

Kao, C.M., Chen, C.Y., Chen, S.C., Chien, H.Y. and Chen, Y.L. 2008. Application of *in situ* biosparging to remediate a petroleum-hydrocarbon spill site: Field and microbial evaluation. Chemosphere 70(8): 1492–1499.

Khedr, M.H., Halim, K.S.A. and Soliman, N.K. 2009. Synthesis and photocatalytic activity of nano-sized iron oxides. Mater. Lett. 63(6-7): 598–601.

Kim, S., Krajmalnik-Brown, R., Kim, J.O. and Chung, J. 2014. Remediation of petroleum hydrocarbon-contaminated sites by DNA diagnosis-based bioslurping technology. Sci. Total Environ. 497: 250–259.

Kim, Y.C., Sasaki, S., Yano, K., Ikebukuro, K., Hashimoto, K. and Karube, I. 2001. Photocatalytic sensor for the determination of chemical oxygen demand using flow injection analysis. Anal. Chim. Acta 432(1): 59–66.

Klimkova, S., Cernik, M., Lacinova, L. and Nosek, J. 2008. Application of nanoscale zero-valent iron for groundwater remediation: laboratory and pilot experiments. Nano 3(04): 287–289.

Kuiper, I., Lagendijk, E.L., Bloemberg, G.V. and Lugtenberg, B.J.J. 2004. Rhizoremediation: a beneficial plant-microbe interaction. Mol. Plant Microbe Interact. 17(1): 6–15.

Lee, J.H. 2013. An overview of phytoremediation as a potentially promising technology for environmental pollution control. Biotechnol. Bioprocess Eng. 18(3): 431–439.

Macek, T., Mackova, M. and Kas, J. 2000. Exploitation of plants for the removal of organics in environmental remediation. Biotechnol. Adv. 18(1): 23–34.

Magalhaes, S.M.C., Jorge, R.M.F. and Castro, P.M.L. 2009. Investigations into the application of a combination of bioventing and biotrickling filter technologies for soil decontamination processes-a transition regime between bioventing and soil vapour extraction. J. Hazard. Mater. 170(2-3): 711–715.

Mamy, L., Barriuso, E. and Gabrielle, B. 2005. Environmental fate of herbicides trifluralin, metazachlor, metamitron and sulcotrione compared with that of glyphosate, a substitute broad spectrum herbicide for different glyphosate, a resistant crops. Pest Manag. Sci. 61(9): 905–916.

McCutcheon, S.C. and Schnoor, J.L. 2004. Phytoremediation: Transformation and Control of Contaminants. Vol. 121: John Wiley & Sons.

McDonnell, G. and Russell, A.D. 2001. Antiseptics and disinfectants: activity, action, and resistance. Clin. Microbiol. Rev. 14(1): 227.

Meagher, R.B. 2000. Phytoremediation of toxic elemental and organic pollutants. Current Opinion in Plant Bio. 3(2): 153–162.

Mihopoulos, P.G., Sayles, G.D., Suidan, M.T., Shah, J. and Bishop, D.F. 2000. Vapor phase treatment of PCE in a soil column by lab-scale anaerobic bioventing. Water Res. 34(12): 3231–3237.

Mohsenzadeh, F. and Rad, C.A. 2012. Bioremediation of heavy metal pollution by nano-particles of Noaea mucronata. Int. J. Biosci. Biochem. Bioinformatics 2: 85–89.

Mueller, N.C. and Nowack, B. 2009. Nanotechnology Developments for the Environment Sector. Report of the Observatory NANO.

Mueller, N.C. and Nowack, B. 2010. Nanoparticles for remediation: solving big problems with little particles. Elements 6(6): 395–400.

Naik, M.G. and Duraphe, M.D. 2012. Review paper on-parameters affecting bioremediation. Int. J. Life Sci. Pharma. Res. 2(3): L77–L80.

NAP. 1993. *In situ* bioremediation: When does it work? National Research Council. National Academies Press.

Newman, L.A. and Reynolds, C.M. 2004. Phytodegradation of organic compounds. Curr. Opin. Biotechnol. 15(3): 225–230.

Nikolopoulou, M., Pasadakis, N., Norf, H. and Kalogerakis, N. 2013. Enhanced *ex-situ* bioremediation of crude oil contaminated beach sand by supplementation with nutrients and rhamnolipids. Marine Poll. Bulletin 77(1-2): 37–44.

Nowack, B. 2008. Nanotechnology (Ed: Krug H). Wiley-VCS Verlag GmbH & Co, Weinheim, 1–15.

Ojuederie, O.B. and Babalola, O.O. 2017. Microbial and plant-assisted bioremediation of heavy metal polluted environments: a review. Int. J. Environ. Res. Public Health 14(12): 1504.

Oyetibo, G.O., Miyauchi, K., Huang, Y., Chien, M.F., Ilori, M.O., Amund, O.O. and Endo, G. 2017. Biotechnological remedies for the estuarine environment polluted with heavy metals and persistent organic pollutants. Int. Biodeterior. Biodegradation 119: 614–625.

Pan, G., Li, L., Zhao, D. and Chen, H. 2010. Immobilization of non-point phosphorus using stabilized magnetite nanoparticles with enhanced transportability and reactivity in soils. Environ. Pollut. 158(1): 35–40.

Parsons Corporation. 2004. Principles and Practices of Enhanced Anaerobic Bioremediation of Chlorinated Solvents. AFCEE, NFEC, ESTCP, https://apps.dtic.mil/sti/pdfs/ADA487293.pdf

Philp, J.C. and Atlas, R.M. 2005. Bioremediation of contaminated soils and aquifers. pp. 139–236. *In*: Atlas, R.M. and Philp, J.C. (eds.). Bioremediation: Applied Microbial Solutions for Real-World Environmental Cleanup. American Society for Microbiology (ASM) Press, Washington.

Purakayastha, T.J., Bhadraray, S. and Chhonkar, P.K. 2009. Screening of brassica for hyper-accumulation of zinc, copper, lead, nickel and cadmium. Indian J. Plant Physiol. 14: 344–352.

Rajkumar, M., Sandhya, S., Prasad, M.N.V. and Freitas, H. 2012. Perspectives of plant-associated microbes in heavy metal phytoremediation. Biotechnol. Adv. 30(6): 1562–1574.

Rickerby, D. and Morrison, M. 2007. Report from the workshop on nanotechnologies for environmental remediation. JRC Ispra 200.

Rizwan, M., Singh, M., Mitra, C.K. and Morve, R.K. 2014. Ecofriendly application of nanomaterials: nano-bioremediation. J. Nanopart, https://doi.org/10.1155/2014/431787.

Rodríguez-Rodríguez, C.E., Marco-Urrea, E. and Caminal, G. 2010. Degradation of naproxen and carbamazepine in spiked sludge by slurry and solid-phase Trametes Versicolor systems. Bioresour. Technol. 101(7): 2259–2266.

Sakan, S.M., Djordjevic, D.S., Manojlovic, D.D. and Polic, P.S. 2009. Assessment of heavy metal pollutants accumulation in the Tisza river sediments. J. Environ. Manag. 90: 3382–3390.

San, A.M., Ravanel, P. and Raveton, M. 2013. A comparative study on the uptake and translocation of organochlorines by Phragmites australis. J. Hazard. Mater. 244: 60–69.

Shah, J.K., Sayles, G.D., Suidan, M.T., Mihopoulos, P. and Kaskassian, S. 2001. Anaerobic bioventing of unsaturated zone contaminated with DDT and DNT. Water Sci. Technol. 43(2): 35–42.

Sharma, B., Dangi, A.K. and Shukla, P. 2018. Contemporary enzyme-based technologies for bioremediation: a review. J. Environ. Manage. 210: 10–22.

Sharma, H.D. and Reddy, K.R. 2004. Geoenvironmental Engineering: Site Remediation, Waste Containment, and Emerging Waste Management Technologies. John Wiley & Sons.

Sharma, S. 2012. Bioremediation: features, strategies, and applications. Asian J. of Pharmacy and Life Sci. ISSN 2231: 4423.

Shayler, H., McBride, M. and Harrison, E. 2009. Sources and impacts of contaminants in soils. http://cwmi.css.cornell.edu/sourcesandimpacts.pdf.

Shukla, K.P., Singh, N.K. and Sharma, S. 2012. Bioremediation: developments, current practices, and perspectives. Genet. Eng. Biotechnol. J. 3: 1–20.

Shukla, S.K., Mangwani, N., Rao, T.S. and Das, S. 2014. Biofilm-mediated bioremediation of polycyclic aromatic hydrocarbons. pp. 203–232. *In*: Surajeet Das (ed.). Microbial Biodegradation and Bioremediation. DOI:: http://dx.doi.org/10.1016/B978-0-12-800021-2.00008-X.

Silva-Castro, G.A., Uad, I., Gónzalez-López, J., Fandiño, C.G., Toledo, F.L. and Calvo, C. 2012. Application of selected microbial consortia combined with inorganic and oleophilic fertilizers to recuperate oil-polluted soil using land farming technology. Clean Technol. Environ. Policy 14(4): 719–726.

Singh, B.K. and Walker, A. 2006. Microbial degradation of organophosphorus compounds. FEMS Microbiol. Rev. 30: 428–471.

Singh, B.K. 2009. Organophosphorus-degrading bacteria: ecology and industrial applications. Nat. Rev. Microbiol. 7: 156–164.

Singh, B.K. 2010. Exploring microbial diversity for biotechnology: the way forward. Trend Biotechnol. 28: 111–116.

Singh, P.K. and Shukla, P. 2015. Systems biology as an approach for deciphering microbial interactions. Briefings Funct. Genom. Oxf. J. 14(2): 166–168. https://doi.org/10.1093/bfgp/elu023.

Stafiej, A. and Pyrzynska, K. 2007. Adsorption of heavy metal ions with carbon nanotubes. Sep. Purif. Technol. 58(1): 49–52.

Subramanian, M., Oliver, D.J. and Shanks, J.V. 2006. TNT phytotransformation pathway characteristics in Arabidopsis: role of aromatic hydroxylamines. Biotech. Progress 22(1): 208–216.

Tiecher, T.L., Ceretta, C.A., Ferreira, P.A., Lourenzi, C.R., Tiecher, T., Girotto, E. and Cesco, S. 2016. The potential of Zea mays L. in remediating copper and zinc contaminated soils for grapevine production. Geoderma. 262: 52–61.

Todorovic, G.R. 2009. Behavior of organic pollutants in the soil environment. Special focus on glyphosate and AMPA. Air, Water, and Soil Quality 99–113.

Tomei, M.C. and Daugulis, A.J. 2013. *Ex-situ* bioremediation of contaminated soils: an overview of conventional and innovative technologies. Crit. Rev. Environ. Sci. Technol. 43(20): 2107–2139.

Tratnyek, P.G. and Johnson, R.L. 2006. Nanotechnologies for environmental cleanup. Nano Today 1(2): 44–48.

USEPA. 2001a. United States Environmental Protection Agency. Use of Bioremediation at Superfund Sites. EPA 542-R-01-019.

USEPA. 2001b. United States Environmental Protection Agency. A Citizen's Guide to Bioremediation. EPA 524-F-01-001.

Varanasi, P., Fullana, A. and Sidhu, S. 2007. Remediation of PCB contaminated soils using iron nanoparticles. Chemosphere 66(6): 1031–1038.

Varol, M. 2011. Assessment of heavy metal contamination in sediments of the Tigris River (Turkey) using pollution indices and multivariate statistical techniques. J. Hazard. Mater 195: 355–364.

Vázquez, S., Agha, R., Granado, A., Sarro, M.J., Esteban, E., Penalosa, J.M. and Carpena, R.O. 2006. Use of white lupin plant for phytostabilization of Cd and As polluted acid soil. Water Air Soil Pollut. 177(1-4): 349–365.

Verma, P., George, K.V., Singh, H.V., Singh, S.K., Juwarkar, A. and Singh, R.N. 2006. Modeling rhizofiltration: heavy-metal uptake by plant roots. Environ. Model. Assess. 11(4): 387–394.

Vidali, M. 2001. Bioremediation. An overview. Pure Appl. Chem. 73(7): 1163–1172.

Wang, C.T. 2007. Photocatalytic activity of nanoparticle gold/iron oxide aerogels for azo dye degradation. J. Non Cryst. Solids 353(11-12): 1126–1133.

Wang, S.Y. and Vipulanandan, C. 2001. Biodegradation of naphthalene-contaminated soils in slurry bioreactors. J. Environ. Eng. 127(8): 748–754.

Whelan, M.J., Coulon, F., Hince, G., Rayner, J., McWatters, R., Spedding, T. and Snape, I. 2015. Fate and transport of petroleum hydrocarbons in engineered biopiles in polar regions. Chemosphere 131: 232–240.

Yadav, K.K., Singh, J.K., Gupta, N. and Kumar, V. 2017. A review of nano-bioremediation technologies for environmental cleanup: a novel biological approach. J. Mater. Environ. Sci. 8(2): 740–757.

Yadav, K.K., Gupta, N., Kumar, A., Reece, L.M., Singh, N., Rezania, S. and Khan, S.A. 2018. Mechanistic understanding and holistic approach of phytoremediation: a review on application and future prospects. Ecol. Eng. 120: 274–298.

Yancheshmeh, J.B., Pazira, E. and Solhi, M. 2011. Evaluation of inoculation of plant growth-promoting rhizobacteria on cadmium and lead uptake by canola and barley. Afr. J. Microbiol. Res. 5(14): 1747–1754.

Yang, Z.F., Wang, Y., Shen, Z.Y., Niu, J.F. and Tang, Z.W. 2009. Distribution and speciation of heavy metals in sediments from the mainstream, tributaries, and lakes of the Yangtze River catchment of Wuhan, China. J. Hazard. Mater. 166(2–3): 1186–1194.

Zhang, X.J., Ji, W. and Kang, Z.J. 2009. Harmfulness of petroleum pollutants in water and its treating techniques. Petrochem. Tech. Application 27: 181–186.

Zhang, W.X. 2003. Nanoscale iron particles for environmental remediation: an overview. J. Nanopart. Res. 5(3-4): 323–332.

Zhuang, P., Yang, Q.W., Wang, H.B. and Shu, W.S. 2007. Phytoextraction of heavy metals by eight plant species in the field. Water Air Soil Pollut. 184(1-4): 235–242.

4
Ecological Tools for Remediation of Soil Pollutants

Nayan Moni Gogoi, Bhaswatee Baroowa and *Nirmali Gogoi**

1. Introduction

Soil is an extremely complex medium responsible for vital functions and ecosystem services linked with productivity and sustainability, environmental quality, biodiversity, and human wellbeing (Halvorson 2008, Cachada et al. 2018, Durães et al. 2018). Various anthropogenic activities have resulted in the intentional or unintentional release of both organic and inorganic pollutants leading to decline in soil quality, which in turn reduces its capacity to perform ecosystem functions and services. Factors such as soil type, topography, geology, and the erosive processes influence the concentration, distribution and bioavailability of pollutants in the environment. Remediation of contaminated sites has been carried out by several strategies which involve biological, physicochemical, and thermal processes (Rubilar et al. 2011). Incineration, excavation, landfilling and storage are some of the expensive decontamination techniques and in most cases it is difficult to execute and involve generation of toxic byproducts (Bustamante et al. 2012). However, the selection of soil remediation approaches depends on soil type, soil composition physical properties, nature of contaminants, handling intensity, cost, etc. Bioremediation is a widely recognized technique of treating polluted soil because of its usability, eco-friendly and cost-effective nature. The rationale of using bioremediation method in the recent years has been justified with non-toxic inputs and byproducts throughout the processes of remediation. Hence, bioremediation has been modified or amalgamated with other methods to obtain better results in a lesser time frame. Cost-benefit analysis highlights the effectiveness of combined bioremediation technologies and their role in world environmental markets (Arora 2018).

Remediation of contaminated sites has been carried out by several biological methods which involve biological, physicochemical, and thermal processes (Rubilar et al. 2011). Methods such as incineration, excavation, landfilling and storage are expensive, sometimes difficult to execute, inefficient, and often exchange one problem for another (Bollag and Bollag 1995). On the other hand, biological processes offer several advantages over conventional technologies, because they are often more environmentally friendly, economic and versatile, and they can reduce the concentration and toxicity of a large number of contaminants (Bustamante et al. 2012).

Department of Environmental Science, Tezpur University, Napaam-784028, Assam, India.
* Corresponding author: nirmalievs@gmail.com

2. Types of soil pollutants and their characteristics

There are several chemicals that may pollute soils, ranging from simple inorganic ions to complex organic molecules. The two major types of soil pollutants are organic pollutants (OPs) and the inorganic pollutants (IPs). Both IPs and OPs can have natural or anthropogenic origins and the majority of polluted soils in the world contain complex mixtures of each or from both.

2.1 Organic pollutants (OPs)

Several groups of compounds containing carbon in their structure (with or without functional groups) such as pesticides, hydrocarbons, PAHs, PCBs, polychlorinated dibenzo-p-dioxins (PCDDs) are counted in OPs. These OPs may occur naturally as a result of volcanic emission, forest fires, or related to fossil fuels. But, the greatest source is associated with anthropogenic production of a huge number of chemical compounds. These compounds vary with a wide range of properties like polarity, solubility, volatility, etc. even within the same group, resulting in different behaviors in the environment and toxicity to organisms (Durães et al. 2018). Many organic chemicals are readily bio-transformed or degraded, while others are very resistant to both chemical and biochemical transformation and have long half-lives (e.g., polyhalogenated compounds). The most important factors of OPs with regard to its toxicity include persistence, solubility (water/organic solvents) and volatility (Walker et al. 2001). Persistent organic pollutants (POPs) are one of the most serious groups of organic contaminants and are considered as a global environmental issue due to their potential for bioaccumulation, carcinogenicity and mutagenicity. Their massive use or continued emission, persistence, and mobility throughout the environment make the situation more serious (Pacyna 2011). Sources of pollution from POPs include the incorrect use and/or disposal of agrochemicals and industrial chemicals and improper burning. There are eight initial POPs pesticides such as aldrin, chlordane, DDT, dieldrin, endrin, heptachlor, hexachlorobenzene, mirex, and toxaphene and five new POP chemicals which may be categorized as pesticides, viz. alpha hexachlorocyclohexane, beta hexachlorocyclohexane, chlordecone, lindane, and pentachlorobenzene (Stojić et al. 2018). These chemical compounds have long half-lives in environment due to their resistance to biological, chemical, and photolytic degradation. Their great affinity for lipids allows accumulation in food chains by storage in fatty tissues of organisms. The semi-volatile character of these pollutants also allows them to move long distances in the atmosphere causing a high spatial distribution away from the source (Jones and Voogt 1999).

Despite ban on several organic compounds, there are many others still being manufactured and widely in use (e.g., PBDEs, fluorinated compounds, synthetic musk fragrances, organophosphate esters). Undoubtedly, the number of chemicals introduced by many different applications like pesticides, pharmaceuticals, personal care products, detergents, flame-retardants, di-electric fluids and combustion by-products has been drastically increasing in the environment (Cachada et al. 2018). PAHs are important soil pollutants due to the potential risks to human health as some of them are probable human carcinogens (having natural or anthropogenic origin). Due to their hydrophobicity, these pollutants are readily adsorbed by soil particles, and have difficulty being degraded (Wang et al. 2000), whereas the polar nature of phenolic compounds makes it easier to transport in soils and can exist as dissolved in soil solution, sorbed to soil particles, or polymerized in humic compounds. For this reason, they are easily degraded under aerobic conditions (Min et al. 2000). The organic based pollutants have the potential to disrupt hormonal systems and modify the natural growth of humans and animals and to significantly alter the diversity of organisms in the soil system. However, the impact of many of the OPs in soil quality is not well recognized and hence, monitoring and identifying their fate in the soil environment is much more difficult. For example, perfluorinated compounds, used since 1960 as surfactants or fire retardants, are widely distributed but little is known about their toxicity and behavior in the environment (Thomaidi et al. 2016).

2.2 Inorganic pollutants (IPs)

Inorganic pollutants (IPs) are released into the environment due to anthropogenic activities like mining, smelting, electroplating, usage in agriculture of pesticides, phosphatic fertilizers and biosolids, sludge dumping, and emission from industry (Fangueiro et al. 2018). Natural causes such as weathering of minerals, erosion, and volcanic activity release inorganic pollutants to the environment. Environmental risks associated with inorganic pollutants vary widely due to several complex interactions at both intracellular and extracellular levels (Saha et al. 2017). Over the last decades, heavy metals have been recognized as the group of metal(loid)s widely associated with contamination or toxicity processes in the environment. Heavy metals include elements with an atomic density greater than 4 g cm^{-3}. Therefore, concerning the high toxicological relevance of this group of metals and metalloids, they are also referred as Potential Toxic Elements (PTEs) and include elements such as arsenic (As), cadmium (Cd), chromium (Cr), cobalt (Co), copper (Cu), mercury (Hg), nickel (Ni), lead (Pb), and zinc (Zn) (Patinha et al. 2018).

PTEs are the best examples of noxious inorganic pollutants due to their non-degradability and persistence in environment for longer periods allowing their transfer from the contamination sources (e.g., mining or industrial areas) to other locations where direct exposure to living organisms may be more favored. Soil is a major reservoir for PTEs and can occur in various forms in association with the solid fractions (Durães et al. 2018). Ionic, molecular, chelated and colloidal forms of PTEs in soil confirm their high mobility. On the basis of availability, PTEs forms are exchangeable ions in mineral or organic particles > complexed or chelated to organic colloids > sorbed to inorganic constituents > incorporated in supergenic phases as (oxy) hydroxides, clay minerals, or insoluble salts, and > fixed in crystal lattice of the minerals (Romic 2012). The higher content of available PTEs in soil environment can compromise the ecosystem services and the growth of plants. Free ionic forms of PTEs present in the soil solution may be toxic to the soil microflora as it leads to reduction in biota activity, population size, biodiversity and inactivation of some extracellular enzymes responsible for the cycling of many nutrients (Patinha et al. 2018). As a result, the cycling of organic matter will be disturbed due to its limited biodegradation leading to reduction in soil nutrients and plant productivity. Plants are also sensitive to higher PTEs content as it limits their metabolic activities (Reeve and Baker 2000). Higher amounts of PTEs absorbed by plants from soils may enter in human and other animals by direct (inhalation or ingestion of soil particles) or indirect (plant consumption) routes, causing remarkable physiological or metabolic disorders (Patinha et al. 2018).

Both PTEs and OPs are present in majority of polluted soils in the world as a complex mixture of one or both types of pollutants. However, these two types of pollutants highly differ in their behavior in soils due to the non-degradable nature of PTEs over OPs, which can be decomposed by living organisms (Domene 2016). Thus, PTEs may experience increased or decreased availability over time, depending on their form when deposited in soil and changes in physicochemical conditions during their accumulation in soil (Wuana et al. 2014). On the other hand, the decreased availability of OPs over time is closely related to their susceptibility to degradation for producing simple units or functional groups of lowest toxicity (Wuana et al. 2014).

3. Ecological tools used in remediation of soil pollutants

The development of environmentally sound, low-cost, low-input technologies with economic benefits is very much essential to tackle remediation of polluted soils. The uses of microbes, plants and organic materials to clean up contaminated soils are cost-effective, environment friendly and equally protective of human health and the environment. Advantages and disadvantages of bioremediation are given in Table 1.

Table 1. Advantages and disadvantages of bioremediation.

Sl. No.	Advantages	Disadvantages
1	Bioremediation is a natural, eco-friendly process	Bioremediation is limited to those compounds that are biodegradable
2	Less expensive than other waste removal technologies	The process is slow. Time required is in day to months
3	Many hazardous compounds can be transformed to harmless products	For *in situ* bioremediation, site must have soil with high permeability
4	Low capital expenditure. Less manual supervision	It does not remove all quantities of contaminants
5	Less energy is required as compared to other technologies	Heavy metals are not removed
6	Complete destruction of target pollutants is possible	A stronger scientific base is required for rational designing of process and success

3.1 Microbes

Soil microorganisms are involved in many biochemical processes in soil that enhance revegetation, thereby increasing the stability of contaminated ecosystems. Microorganisms are the key components of bioremediation enabling both aerobic as well as anaerobic degradation of soil contaminants (Abatenh et al. 2017). Microorganisms survive and sustain in a wide range of environmental conditions, this property of microorganisms makes them most suitable for utilization in treatment of contaminated soils through various techniques. Microorganisms survive and undergo changes by mutation and formation of spores to sustain in unfavorable conditions. This property of microorganisms enables them to survive and sustain in a wide range of environmental conditions and makes them most suitable for utilization in treatment of contaminated soils through various techniques. They are also known as biocatalyst because they can sustain and function in extreme environmental conditions (Tomei and Daugulis 2013). Microorganisms such as bacteria, fungi, algae and actinomycetes can effectively degrade soil contaminants. Microorganisms are an inherent part of the Earth system and play a major role in flow of nutrients as carbon, nitrogen, phosphorus and sulfur through biogeochemical cycles. Flow of nutrients is mediated by biogeochemical processes as mineralization, immobilization and redox reactions. These processes are foundational basis of microbial metabolism for utilization of energy source from various substrates. In microbial remediation, microorganisms obtain energy and nutrition from biodegradation or biotransformation of soil pollutants. Polluted soils contain compounds as aliphatic and aromatic hydrocarbons, inorganics such as heavy metals, radioactive elements, biomolecules, and antibiotics mostly in polymeric forms. These compounds primarily contain carbon, nitrogen, phosphorus and sulfur, and trace elements essential for growth and development of microorganisms.

3.1.1 Microbial metabolism and soil pollutants

Microorganisms derive nutrition and primary energy source, i.e., carbon, from the organic and inorganic pollutants present in the soil. Microbial physiology or mode of action determines the transformation rate of toxic soil contaminants into non-toxic and simpler forms and utilize these products of degradation as source of nutrition. The metabolic pathways in all microorganisms are almost similar, involving a series of enzymes functioning in aerobic or anaerobic conditions. According to Suthersan et al. (1999), biotransformation or biodegradation by microorganisms may be categorized depending on requirement for carbon as energy source. In the first category, the requirement of carbon is inherently growth-related, microorganisms bio-transform toxic soil contaminants to derive carbon for growth and development, indicating metabolic transformation. Secondly, microorganisms degrade soil contaminants and utilize carbon for sustaining respiration and maintaining cell viability mostly when concentration of carbon is low, such biotransformation is known as cometabolic transformation. Metabolism and cometabolism may coexist in both aerobic and anaerobic degradation of soil contaminants by microorganisms (Suthersan and McDonough 1999). For example, aerobic degradation and anaerobic degradation of PAHs and pesticides are accomplished through metabolic as well as cometabolic pathways (Tomei and Dauglulis 2013, Krishna and Philip 2011, Lei et al. 2005, Quintero et al. 2006). Apart from carbon source, nitrogen, sulfur, phosphorus and trace elements as potassium, iron, molybdenum, etc. are equally important for microbial metabolism.

Most common bacteria used in bioremediation of contaminated soils are *Pseudomonas, Aeromonas, Flavobacteria, Chlorobacteria, Corynebacteria, Acinetobacter, Mycobacteria, Streptomyces, Bacillus, Arthrobacter, Aeromonas, Cyanobacteria, Citrobacter, Burkholderiaxenovorans, Desulfovibrio vulgaris, Amycolatopsistucumanensis*, etc. (Francis and Nancharaiah 2015, Tomei and Dauglulis 2013, Mani and Kumar 2014, Amoroso and Abate 2010). Fungal strains include *Cladosporium resinae, Rhizopus arrhizus, Aspergillus niger, Phanarochaetechrysosporium, Bierkanderaadusta, Phlebia* sp., *Fusarium* sp. and *Hypocrea* sp. (Francis and Nancharaiah 2015, Tomei and Dauglulis 2013, Quintero et al. 2009, Ingle et al. 2014). Among fungal species, most commonly utilized yeast in bioremediation are *Candida lipolytica, C. tropicalis, Rhodoturularubra*, and *Aureobasidion (Trichosporon) pullulans, Rhodotorulaaurantiaca* and *C. ernobii* (Miranda et al. 2007). Algal species that can bioremediate polluted soils are *Ascophyllum nodosum, Chlorella vulgaris, Cladophora fascicularis, Fucusvesiculosus, Oscillatoria quadripunctulata, Scenedesmus acutus, Sargassum filipendula*, etc. (Yadav et al. 2017).

3.1.2 Microbial remediation

Microorganisms are equally important in *in situ* as well as *ex situ* bioremediation. Indigenous microorganisms naturally degrade soil pollutants in NMA, intrinsic bioremediation and amended bioremediation techniques (Brown et al. 2017). Genetically modified microorganisms are popular means of obtaining targeted specific treatment of soil pollutants. Recent development in microbial biotechnology highlights DNA fingerprinting to understand or monitor the changes in performance of the microorganisms corresponding to different soil contaminants (Alvarez et al. 2011). Microorganisms are successfully utilized in treatment of contaminants in the subsurface soils and vadose zones under aerobic or anaerobic conditions. In bioventing remediation technique, microbial degradation is enhanced by adding oxygen in the vadose zone. Similarly, nutrients and aeration are provided for better performance of the microorganisms in the subsurface zone during biosparging. In bioaugmentation, microorganisms are utilized in the form of natural consortium. Indigenous or allochthonous or genetically modified microbes are used (Mrozik and Piotrowska-Seget 2010). In land farming and biopiling, role of microbes is very crucial to ensure degradation of soil pollutants in aerobic condition. Windrow method facilitates proper aeration for increased degradation activities by the microorganisms. Microbial consortium is also utilized in slurry bioreactors in a liquid phase system.

Microbial enzymes are responsible for degradation of toxic oil contaminants, xenobiotics and heavy metals (Fan and Krishnamurthy 1995). These enzymes are mostly microbial oxygenases, microbial lacases, microbial peroxidases, and hydrolytic enzymes like amylases, cellulases, proteases, lipases, DNases, pullulanases, and xylanases (Karigar and Rao 2011). Microorganisms also contain metal binding proteins known as metallothioneins (MT) and other constituents as bacterial siderophores known as catecholates and fungal siderophores called as hydroxamate (Ojuederie and Babalola 2017). These components are responsible for removal of toxic metals from polluted soils. In soil remediation, microbial machinery functions by modification in their metabolic pathways to uptake metals, enzymatic transformation of toxic metals into metal chelate complexes and minimizing metal concentrations using efficient microbial efflux systems (Ojuederie and Babalola 2017, Jan et al. 2014).

Most of the toxic metals are absorbed by the cell walls of microorganisms known as biosorption. Pollutants such as thorium are absorbed by bacterial strain *Pseudomonas* and fungal strains such as *Rhizopus arrhizus* and *Aspergillus niger* in their cell walls (Francis and Nancharaiah

2015, Kazy et al. 2009). *Citrobacter* sp. can precipitate uranium and other toxic metals through enzymatic transformation of uranium containing soil contaminants (Francis and Nancharaiah 2015). *Brevibacterium* sp., *Aeromicrobium* sp., *Dietzia* sp., *Burkholderia* sp., *Mycobacterium* sp., *Gordoniadesulfuricans, Pseudomonas* sp., and *Ralstoniapickettii* are mostly used in biopiles for treatment of hydrocarbons (Chaillan et al. 2004, Lin 2010). Some other bacterial species that biodegrade petroleum wastes are *Aeromonas, Moraxella, Beijerinckia, Flavobacteria, Chrobacteria, Nocardia, Corynebacteria, Atinetobacter, Mycobactena, Modococci, Streptomyces, Bacilli, Arthrobacter, Aeromonas,* and *Cyanobacteria* (Ingle et al. 2014, Robles-González et al. 2008). *Fusarium* and *Hypocrea* are known to degrade pyrene, a highly carcinogenic PAH, along with uptake of trace metals as copper and zinc (Ingle et al. 2014). Fungal strains such as *Amorphoteca* sp., *Neosartorya* sp., *Talaromyces* sp., and *Graphium* sp. and yeasts such as *Candida* sp., *Yarrowia* sp., and *Pichia* sp. degrade petroleum wastes (Chaillan et al. 2004, Das and Chandran 2010). *Cladosporium resinae,* a fungus, has reportedly degraded polyurethane from polluted soils. Algal species such as *Ascophyllum nodosum, Chlorella vulgaris, Cladophora fascicularis, Fucusvesiculosus, Oscillatoria quadripunctulata, Scenedesmus acutus, Sargassum filipendula,* etc. are known to remediate heavy metals from polluted soils. Microorganisms also produce biomolecules that enhance or facilitate biodegradation processes like biosurfactants to facilitate bioavailability of soil contaminants in liquid phase treatment. Some bacterial species that release biosurfactants are *Pseudomonas aeruginosa, Corynebacterium, Nocardia,* and *Rhodococcus* spp. and *Bacillus subtilis* (Tomei and Dauglulis 2013).

3.2 Plants

Biodegradation or removal of toxic contaminants from soil with the help of plants is known as phytoremediation. This method is cost-effective, less labor intensive and the most common and traditional method of treating polluted soils since past centuries as compared to other bioremediation methods. In phytoremediation, plants either completely extract or remove toxic contaminants or completely transform or stabilize them into non-toxic forms. Plant species remove soil pollutants in a synergistic combination with soil microbiota inhabiting the rhizosphere zone (Ojuederie and Babalola 2017, Dixit et al. 2015). Phytoremediation is mostly reported as a sub-category of *in situ* bioremediation because polluted soil is treated at the site of pollution (Azibuke et al. 2016). Phytoremediation is known to improve physicochemical and biological properties of soil due to hyperaccumulating properties of the plants utilized (Jutsz and Gnida 2015, Mahmood et al. 2015). Plants remove or treat toxic soil contaminants by complete extraction, accumulation in plant parts, stabilization, enhancement of degradation by rhizosphere microflora, volatilization or complete degradation of contaminations, and filtration through roots; therefore phytoremediation is broadly categorized into the following subcategories (Ojuederie and Babalola 2017).

3.2.1 Phytoextraction or phytoaccumulation

Phytoextraction refers to the hyperaccumulating property of plants that help in rapid extraction of contaminants from polluted soils. Heavy metals such as Mn, Zn, As, Co, Cr, Cu, Ni, Pb, Sb, Se, Ti, Cd are extensively removed by phytoextraction method; these metals are extracted from soil and transported to various parts of the hyperaccumulating plants (Verbruggen et al. 2009). These plants have unique physiology and morphology that enable sequestration and detoxification of large amounts of pollutants from the contaminated soils. Some of the hyperaccumulating plants are *Berkheyacoddii, Helianthus annuus, Minuartiaverna, Euphorbia cheiradenia, Astragalus racemosus, Pteris vittata, Viola boashanensis, Arabidopsis halleri, Thlaspigoesingense, Sedum alfredi, Thlaspicaerulescens, Nicotiana tabacum* and *Thymus praecox* (Ojuederie and Babalola 2017, Liu et al. 2018, Jaffré et al. 2013, Yang et al. 2017).

3.2.2 Phytostabilization

This method involves immobilization or absorption of toxic contaminants by plant roots in the rhizosphere zone itself, followed by separation and stabilization without affecting the surrounding environment. Phytostabilization prevents translocation of toxic metals to the above ground parts of the plants and the food web. Stabilization of toxic metals can be enhanced by application of soil amendments involving optimization of soil pH. Addition of organic matter, biochar and compost documented positive response (Ojuederie and Babalola 2017, Radziemska et al. 2017). A grass species, *Festuca rubra*, also known as red fescue, shows good phytostabilization of copper (Radziemska et al. 2017).

3.2.3 Phytofiltration

This process involves active absorption of toxic metals by various plant parts such as roots (rhizofiltration), excised shoots (caulofiltration) and seedlings (blastofiltration) (Ojuederie and Babalola 2017, Dixit et al. 2015, Mesjasz-Przybyłowicz et al. 2004). These plant parts have been identified as sites for faster absorption of metals from soil through chelation, exchange of ions, and adsorption. Metal absorption process depends on plant physiology, its efficiency in intracellular uptake of metals, vacuolar deposition of metals and translocation of metals to above ground parts of the plant (Salt et al. 1995).

3.2.4 Phytostimulation

In this method, microbial activity is stimulated in the rhizosphere zone by root exudates to increase degradation of toxic soil contaminants (Ojuederie and Babalola 2017). *Mucor* sp. has been reported as endophytic phytostimulant fungus in the rhizosphere of *Brassica campestris* degrading heavy metals such as Mn, Co, Cu and Zn (Zahoor et al. 2017). Another study showed good proliferation of cyanobacteria in the presence of phytohormones that enhanced its nitrogen fixing activity (Hussain and Hasnain 2011).

3.2.5 Phytovolatilization

Phytovolatilization mediates volatilization of toxic soil contaminants and releases them into the atmosphere (Ali et al. 2013). This may occur in two ways—direct or indirect. In the direct method, the contaminant vapors are transcribed from the plant body through vascular pathway and escape through cuticle and stomata of leaves and shoots. In indirect method, vapors escape from the rhizophere zone due to root activities, without being transcribed through the plant body (Limmer and Burken 2016). Methane gas released from wetlands is an example of direct volatilization of organic compounds from plants such as *Phragmites australis* and *Scirpuslacustris* in wetland soils (Van Der Nat et al. 1998). *Nicotiana tabacum* has the ability to volatilize highly toxic soil contaminant such as methyl mercury from Hg into elemental Hg vapors that escapes through cuticle and stomata of the leaves (Rayu et al. 2012, Mukhopadhyay and Maiti 2010). It is reported that phytovolatilization is a result of combined activity of plant and microbial metabolism in the rhizosphere zone (Ojuederie and Babalola 2017, Tak et al. 2013).

3.2.6 Phytodegradation

Phytodegradation involves enzymatic degradation of organic soil pollutants; phytoenzymes such as nitroreductases and dehalogenases are known to transform toxic organic pollutants into non-toxic forms (Ojuederie and Babalola 2017, Ali et al. 2013). Phytodegradation activity may be enhanced by microorganisms in the rhizosphere, known as rhizodegradation (Newman and Reynolds 2004, Ojuederie and Babalola 2017). Microbes in the rhizosphere receive ample amount of nutrients from the root exudates that help in additional degradation of organic pollutants (Ojuederie and Babalola 2017, Babalola 2010).

3.3 Organic materials

Role of organic matter during remediation of polluted soil has been recognized by many researchers (Rada et al. 2019, Urzelai et al. 2000). Improvement of physical, chemical and biological health of soil under application of organics was recognized long back. Microbes responsible for biodegradation can obtain an interrupted supply of carbon with addition of organic matter in polluted soil. Lee et al. (2003) have documented that carbon in the form of pyruvate enhance the growth of PAH degrading microbes. Composting is a method of improving organic carbon in soil and has been found to be helpful during biostimulation. In this process, contaminated soil is mixed thoroughly with the primary ingredients of compost (Semple et al. 2001). Efficacy in removal of organopollutants such as PAH has been achieved with the addition of both mushroom compost and spent mushroom compost (SMC) in the polluted sites apart from enhanced microbial growth, enzymatic activity and nutrient contents in the mixture (Lau et al. 2003). Addition of compost increases the functional diversity of native microbial community. Some other organic wastes such as banana skin, melon shell and brewery spent grain were found useful during bioremediation of oil contaminated soils (Abioye et al. 2012). Use of poultry manure as organic fertilizer with alternate carbon substrate is found to be effective in biodegradation (Okolo et al. 2005).

Soil remediation of heavy metals is successful with the addition of vermicompost, humic substances and biochar (Wang et al. 2018a, Burlakovs et al. 2013). Application of vermicompost improves soil nutrient content and supports plant growth and development, which is crucial for phytoremediation process. Earlier studies documented positive role of biochar and humic substances in retention of heavy metals in contaminated soil (Wang et al. 2018b, Piccolo 1989, Fonseca et al. 2013). More research is needed on organic materials as cheap sources of nutrients, which can support microbial growth and plant activities during bioremediation. Generation of organic wastes arising from economic and demographic growth can be a potent resource for nutrients. Organic amendments derived from off field sources such as solid wastes and animal manures are effective in revitalization of soil nutrient status (Reeve et al. 2016). Immobilization of heavy metals promote the diversity of Arbuscular mycorrhiza in metal polluted soil (Montiel-Rozas et al. 2016). Uses of animal manure and biosolids have side effects causing elevated content of both amonical and nitrate nitrogen, and emission of greenhouse gases such as methane and carbon dioxide (Sun et al. 2008, Petersen 2018, Alvarenga et al. 2015).

Higher concentration of heavy metals like Cu and Zn at 10–20 cm topsoil layer after continuous application of swine compost might be basically due to higher content of these metals in the compost (Zhao et al. 2005). This higher content of metal also leads to reduced activities of some important soil enzymes like dehydrogenase and urease (Yang et al. 2006). Use of technology for lowering metal concentration before composting of these materials can overcome this issue. Soil factors such as soil pH, cation exchange capacity, temperature, and moisture status of soil along with content of humic substances play vital role in binding of metals in soil. Production methods (aerobic/anaerobic) used for preparation of compost are important as they determine the degree of heavy metal complexion with the produced organics. Aerobic composting is found superior over anaerobic composting (Smith 2009). With long time application, heavy metals were found to be detached from the complex, leading to contamination of soil profiles. Organic matter dynamics is reported to be responsible for this metal movement rather than the mineralization of applied heavy metal contaminated organics (Sukkariyah et al. 2005). Therefore, constant monitoring of heavy metal fractions and their contributions to total metal concentration is important.

4. Role of biotechnology in remediation of soil pollutants

Biotechnology is being applied in recent years for decontamination of soils polluted with both organic and inorganic contaminants. Bioremediation technology (also known as environmental biotechnology) is an ecologically sound emerging tool and can be defined as the elimination, attenuation or transformation of polluting or contaminating substances by the use of biological processes, i.e., the use of natural strains of bacteria (Pal et al. 2010). During bioremediation, microbes utilize chemical contaminants in the soil as an energy source and, through oxidation-reduction reactions, metabolize the target contaminant into useable energy for microbial growth. By-products (metabolites) released back into the environment are typically in a less toxic form than the parent contaminants. For example, petroleum hydrocarbons can be degraded by microorganisms in the presence of oxygen through aerobic respiration. The hydrocarbon loses electrons and is oxidized, while oxygen gains electrons and is reduced forming carbon dioxide and water (Nester et al. 2001). Because of their appearance in nature, the population of the strains explodes which drives the process of breakdown of hazardous wastes or pushes the bioremediation process forward. These bacteria increase in number when a food source, i.e., the waste is present. When the contaminant is degraded, the microbial population naturally declines with the production of less harmless products (Sen and Chakrabarti 2009). Bioremediation of contaminated soil is carried out at the place of contamination or in a specially prepared place. Bioremediation techniques that are applied to soil at the site with minimal disturbance are referred to as *in situ*, whereas *ex situ* techniques are applied at the site which has been removed via excavation soil (Dzionek et al. 2016).

4.1 In situ bioremediation

This method facilitates treatment of polluted soils at the site of contamination, thereby minimizing the expenses for excavation, and many effects on the integrity of soil contaminants before treatment (Azubuike et al. 2016). It is cost-effective and less labor intensive as compared to other methods of bioremediation. *In situ* bioremediation is a promising eco-friendly solution for complete transformation of hazardous soil pollutants into non-hazardous forms without affecting ecological communities (Perelo 2010).

This method is based on strategies that decide the best-fit bioremediation techniques for polluted soils, considering environmental conditions at the polluted site, bioavailability of the soil contaminants, and the limiting factors during the remediation processes (Brown et al. 2017). *In situ* bioremediation techniques are grouped as 'intrinsic bioremediation' and 'engineered bioremediation', the former takes place without any artificial manipulation or stimulation during the treatment and the latter occurs in a manipulated or stimulated treatment environment. In intrinsic method, natural monitored attenuation (NMA) plays a significant role in keeping track of degradation activities without human intervention in a natural and undisturbed treatment environment (Brown et al. 2017). These amendments are foundational basis of various bioengineering techniques applied during *in situ* bioremediation. NMA is one such strategy that ensures active degradation of soil pollutants naturally in the polluted sites. It involves activities like allowing natural sedimentation, and biochemical transformation of soil contaminants at site itself, thereby reducing their bioavailability (Perelo 2010). However, extensive monitoring is required in NMA to obtain results within a pre-defined time period. Other strategies include incorporation of various amendments to the polluted soils, installation of injection well for subsurface soil contaminants, spray irrigation for shallow soil contaminants and recirculation of treated ground water from polluted sites (Brown et al. 2017). Both nutrients and mediator compounds that enhance degradation of soil contaminants are used as amendments. Nutrient amendments also instigate effective degradation of soil contaminants by microorganisms (Brown et al. 2017). In engineered bioremediation approach, the polluted topsoil is capped or encapsulated with moisture proof material to prevent spreading and diffusion of contaminants (Lee and Lee 1997, Liu et al. 2018). Modified conventional *in situ* bioremediation technologies can effectively treat the polluted soils with minimal effort. It includes treatment of contaminated soils at pollution site in combination with interdisciplinary understanding of scientific and engineering techniques. These techniques encompass a wide array of multidisciplinary applications such as biotechnological, biophysical, biochemical and bioengineering. Some of the *in situ* bioremediation techniques are broadly discussed below.

4.1.1 Biosparging

In this technique, subsurface soil (saturated zone) or ground water table is injected with oxygen (air sparging) and occasionally nutrients to improve natural degradation of soil contaminants (Johnson et al. 2010). Biosparging is used to treat petroleum waste or hydrocarbons in contaminated soil and ground water (Kao et al. 2008). Air sparging promotes volatilization of organic compounds and nutrients to obtain enhanced microbial degradation of the organic compounds in the subsurface soils. Biosparging enables vertical movement of volatile organic compounds from saturated to unsaturated zone in the soil for effective volatilization as well as microbial degradation (Azubuike et al. 2016). This technique does not much affect the treatment site and requires a time period of 6 months to 24 months for complete treatment of the contaminated soils under favorable conditions (EPA 2004). The efficacy of biosparging depends on soil texture and structure, soil biophysicochemical properties and density of microbial population in the subsurface soil (EPA 2004).

4.1.2 Bioventing

This technique may be considered as a part of or modified biosparging technique that adds or injects oxygen (air sparging) under a controlled flow rate. Oxygen is injected in the unsaturated zone, also known as vadose zone, to improve and sustain microbial degradation of contaminants (Azubuike et al. 2016). Bioventing involves air injection with a maintained flow rate vis-a-vis high rate or low rate that helps in obtaining uniform distribution of air in the vadose zone. Therefore, bioventing may be applied to treat recalcitrant compounds like chlorinated hydrocarbons through partial oxidation under aerobic conditions (Azubuike et al. 2016, Philp and Atlas 2005). Bioventing is used to treat soils contaminated with petroleum waste, hydrocarbons, pesticides and organic chemicals, mostly organic compounds that could be degraded under aerobic conditions (Hoeppel et al. 1991, Hellekson 1999, EPA 2004). Bioventing depends on soil physicochemical factors such as moisture, permeability, stratification, temperature, etc. (Brown et al. 2017).

4.1.3 Bioaugmentation

This technique involves introduction or addition of natural microbial consortium or genetically modified microorganisms into the contaminated soils (Mrozik and Piotrowska-Seget 2010). Bioaugmentation is useful in degrading a wide range of inorganic and mostly organic compounds such as aromatic hydrocarbons, chlorinated compounds, and petroleum hydrocarbons, and is also most effective in degrading recalcitrant compounds that are highly resistant to degradation (Mrozik and Piotrowska-Seget 2010, Nzila et al. 2016). It has some major challenges which include (i) survival of microorganisms or acclimatization of the microorganism in the new environment after injection, and (ii) uniform distribution of microorganisms, availability of oxygen and nutrients or interference by unwanted growth near the injection site (da Silva and Alvarez 2010). Newer approaches to bioaugmentation have been developed by combined bioengineering techniques which include encapsulation of microbial consortium in alginate, inoculation of microbial strains in the rhizosphere of plants or genetically modified plants incorporating microbial genes (Gentry et al. 2004). Overall bioaugmentation is a feasible *in situ* technique with minimal complexity that leads to faster degradation of soil contaminants. It depends on the nature of microbial strains, soil physicochemical properties and bioavailability of the contaminants (da Silva and Alvarez 2010, Brown et al. 2017, Federici et al. 2012).

4.1.4 Bioslurping

This technique involves bioventing and vacuum extraction of free products of ground water and subsurface soils (Brown and Ulrich 2014). Basically, it is a vapor extraction process that also removes slugs and liquid droplets with the help of a slurp tube from vadose zone. Bioslurping is accompanied by bioventing to provide aeration for degradation of various soil contaminants (Miller 1996). It is very effective in removing petroleum hydrocarbons, primarily light non-aqueous phase liquids (LNAPL), that are found on the top of ground water table and capillary fringe, i.e., the region immediately above the saturated zone holding water in soil capillaries (Place et al. 2003). Combined bioslurping and bioengineering methods have utilized DNA microarrays to identify and investigate genes that are responsible for degradation of aliphatic and aromatic hydrocarbons. Based on this genetic information, bioslurping and bioventing are performed (Kim et al. 2014). Bioslurping depends on soil physicochemical properties, height of ground water table and soil porosity. Therefore, it has certain limitations like inadequate soil moisture and soil permeability tend to dry the soil, slow down biodegradation of the contaminants and reduce effectiveness of bioventing. Biolsurping focuses on removal of toxic products and most of the extraction products are removed or discharged from vadose zone without treatment. It is less effective in aerobic degradation of recalcitrant chlorinated compounds in absence of a co-metabolite. However, bioslurping is considered as a cost-effective technique as it pumps lower quantity of ground water, minimizing additional costs of storage, treatment of vapor and free product and disposal of other extraction products (Philp and Atlas 2005, Held and Dörr 2000, Azubuike et al. 2016).

4.1.5 Phytoremediation

In this method, removal and degradation of soil contaminants are carried out by plant uptake at the pollution site. One of the following actions take place during this process: (i) toxic soil pollutants are extracted by the plants and translocated to different parts (ii) degraded into less toxic forms (iii) volatilized as vapors (iv) stabilized in the rhizosphere zone degraded by phytoenzymes (Ojuederie and Babalola 2017). Phytoremediating plants have excellent morphology and physiology for hyperaccumulating heavy metals and degrading toxic organic pollutants. Some of these plants are—*Aeolanthusbioformifollus, Lemna minor, Vigna radiata, Larreatridan-tata, Bacopa monnieri, Brassica juncea, Alyxiarubricaulis, Macademianeurophylla, Thlaspicaerulescens, Lemna minor, Pistia stratiotes, Eichhorniacrassipes, Hydrilla verticillata, Phyllanthus serpentines, Lemna minor, Salvinia molesta, Spirodelapolyrhiza, Haumaniastrumrobertii,* etc. (Ingle et al. 2014). Phytoremediation is comprehensively discussed in Section 3.2.

4.2 Ex situ bioremediation

This bioremediation method involves transfer or excavation of pollutants from the site of contamination and treated safely away from the site of origin. *Ex situ* bioremediation has no complex strategies for selecting the most suitable technique for treatment of polluted soils because the soils are transported from contaminated sites to new sites for treatment. However, safe transport and outspread of soil contaminants are taken care of following the nature and stability of pollutants, environmental health and biogeochemical conditions of the treatment sites (Philp and Atlas 2005, Frutos et al. 2010, Azubuike et al. 2016). Excavation and treatment costs, nature of contaminated soils, active transport system required to transport the contaminated soils, environmental policies and their social impact determine the preference of active treatment methods for *ex situ* bioremediation (Azubuike et al. 2016). *Ex situ* bioremediation is often considered as a traditional method of treatment of polluted soils amending conventional sewage and solid waste treatment methods. Depending on the substrate used for bioremediation, *ex situ* method is categorized into two systems known as solid phase system and slurry phase system. The former involves treatment of contaminated soil through land farming, developing piles of organic wastes (biopiles) and periodical turning of the biopiles, and the latter includes treatment of contaminated soils in the form of slurry or liquid suspension in bioreactors. Techniques used in *ex situ* bioremediation are discussed below.

4.2.1 Landfarming

Landfarming is performed by applying excavated polluted soils over the land chosen as a treatment site and allowing degradation of contaminants by microorganisms growing on it. It is a common practice for treating petroleum contaminated soils—the volatile fraction evaporates as gases and the remaining fractions are biodegraded through microbial activity (Al-Awadhi et al. 1996,

Cerqueira et al. 2014, Khan et al. 2014, Nikolopoulou and Kalogerakis 2016). This technique is also considered as advantageous over other methods because its requirements are cost-effective and fulfilled through minimal energy consumption, ensuring low risk of contaminant outspread, complying with government rules and regulations, following environmental policies and assuring benefits to the society and geographical location (Maila and Cloete 2004, Besaltatpour et al. 2011, Nikolopoulou and Kalogerakis 2016, Azibuke et al. 2016, Alfke et al. 1999). Landfarming depends on physicochemical properties of soil, pollutant characteristics and climatic conditions (Morgan and Watkinson 1989, Zhu et al. 2004, Khan et al. 2004, Paudyn et al. 2008, Besaltatpour et al. 2011). This technique performs well with addition of autochthonous microorganisms, nutrient amendment, proper supply of water through irrigation and oxygen through tillage (Philp and Atlas 2005, Nikolopoulou and Kalogerakis 2016, Silva-Castro et al. 2015). These activities provide essential conditions such as soil moisture, growth factors and aeration to enhance contaminant degrading ability of microorganisms under favorable climatic conditions. In extreme climatic conditions like in desert areas, amendment of nutrients and water may lead to soil compaction (Philp and Atlas 2005, Hejazi and Husain 2004). Landfarming has few limitations: (i) requirement of huge land areas (ii) management of highly volatile organic compounds in highly cold and highly hot and humid climate (iii) sustaining useful microorganisms in unfavorable environmental conditions (Philp and Atlas 2005, Khan et al. 2004, Azibuke et al. 2016, Maila and Colete 2004).

4.2.2 Biopiling

This technique involves piling or heaping of excavated soils in a defined area for treatment with microorganisms. Biopiling is often defined as a modified landfarming bioremediation limiting irrigation and tillage of treated soils, promoting growth of aerobic and anaerobic microorganism in the heap of contaminated soils (Mani and Kumar 2013). Microbial degradation in biopiling is enhanced with nutrient, moisture and oxygen amendment in a well-organized treatment bed. Contaminant leaching from treatment bed is prevented by applying an impermeable layer or lining that impedes or limits the flow of leachate into ground water. Biopiling treatment beds are also equipped with leachate drainage and collection unit that collects seeping leachates for further treatment. Sometimes bulking agents like straw, bark, wood chips and similar biomass may be added to the biopliles to improve bioavailability of soil contaminants and enhance microbial remediation (Azibuke et al. 2016, Rodríguez-Rodríguez et al. 2010). This method is effortlessly utilized in treating low molecular weight (LMW) organic pollutants. Biopiling is less costly as it requires low capital input and easily controllable parameters ensuring effective degradation of soil contaminants in favorable as well as extreme climatic conditions. It depends on soil physicochemical characteristics, nutrient concentration and constituent characteristics of soil contaminants (Whelan et al. 2015, Gomez and Sartaj 2014). In biopiling, aeration is induced by air pumps, thereby continuous heating and drying of biopiles affect the microbial flora and their degrading capabilities (Sanscartier et al. 2009). Other limitations include (i) lack of uniform contaminant mixing with soils due to poor aeration, and (ii) low efficiency for high molecular weight organic pollutants, heavy metals inhibiting microbial growth and rapid volatilization of untreated lower fractions of organic pollutants (USEPA 1995).

4.2.3 Windrow

Windrow is a method of turning piles of contaminated soils at regular intervals to provide aeration for enhanced degradation of soil contaminants. It ensures proper mixing of contaminated soils with nutrients, and uniform distribution of oxygen and water for better results. Windrow has a very good option in removal of greenhouse gases like methane produced during anaerobic conditions of degradation, and periodic turning of soils which causes release or oozing of methane gas from the anaerobic zone of biopiles (Andersen et al. 2010). Nutrient, oxygen and water amendments in biopiling can be immediately followed by windrow treatment; this will lead to better integration of the amendment components and improve microbial degradation activities (Shi et al. 1999). Windrow is effectively utilized in treating crude oil contaminated soils with the help of specialized machinery

(Al-Daher et al. 1998, Balba et al. 1998). This technique depends on soil physicochemical properties and degree of bioavailability and biodegradability of soil contaminants (Coulon et al. 2010, Azibuke et al. 2016, Semple et al. 2001). Drawbacks of windrow treatment include (i) rapid removal of toxic untreated volatile gases challenging the environmental health, and (ii) requirement of ample of amount of time and labor (Azibuke et al. 2016).

4.2.4 Bioreactor

Treatment of polluted soils in bioreactors is carried out in liquid phase system known as slurry bioreactors. Contaminated soils are fed into bioreactors in the form of slurry with optimized bioprocess parameters such as pH, temperature, aeration, microbial population, and nutrient concentration (Azibuke et al. 2016). Slurry bioreactors are supplied with nutrients for growth of microorganisms and surfactants for increasing bioavailability of soil contaminants. These are easily manipulated and controlled, allowing enhanced and faster degradation of soil contaminants in an optimized operation mode and design of the bioreactor system (Robles-González et al. 2008, Mueller et al. 1991). This technique is very effective in treating polluted soils with high concentration of toxic and recalcitrant contaminants (Robles-González et al. 2008, Mueller et al. 1991, Christodoulatos et al. 1998). Slurry bioreactors are artificially instigated with microbes and their growth factors following bioaugmentation, i.e., introduction of autochthonous microorganisms and biostimulation, i.e., input of nutrients for better growth and performance of microorganisms (Robles-González et al. 2008). Advanced studies on slurry bioreactor system and bioprocessing showed that DNA (molecular) fingerprinting may be used for identifying changes in the pattern of microbial communities corresponding to varied experimental conditions. This may serve as useful information in optimizing the bioreactor parameters for better performance of the microorganisms (Rizzo et al. 2010). These bioreactors depend on bioavailability and biodegradability of soil contaminants, microorganisms and bioprocess parameters such as pH, temperature, nutrient concentration and exchange of gases. However, slurry bioreactors are limited to small scale operations where space and manpower constraints prevail (Robles-González et al. 2008). Another drawback of slurry bioreactor is that the contaminated soils require pre-treatment like crushing and sieving before incorporation into the bioreactor unit (Woo and Park 1999). Moreover, cost of the bioreactor system narrows down its applicability and most of the times it is inconvenient to conduct slurry bioreactor treatment in absence of proper optimization of bioreactor operation unit, operation design and bioprocess parameters (Philp and Atlas 2005). Different *in situ* and *ex situ* methods are given in Figure 1.

5. Importance of biosurfactants for ecological remediation of soil pollutants

Bioremediation processes are negatively affected by the lower aqueous solubility of some contaminants. This necessitates the use of biosurfactants to enhance the solubility of contaminants for their effective remediation. Biosurfactants are surface-active biomolecules produced by microorganisms with relative ease of preparation and have unique properties like specificity and low toxicity. These biomolecules play three major functions in bioremediation through the following activities: (i) increase the surface area of hydrophobic substrates (ii) increase the bioavailability of hydrophobic substrates through solubilization/desorption, and (iii) regulate the attachment and removal of microorganisms from the surfaces (Vijayakumar and Saravanan 2015). Biosurfactants contain both water soluble and water insoluble portions. The water solubility of the surfactants is due to the hydrophilic portion (polar group), while the hydrophobic portion (nonpolar chain) tends to concentrate at the air-water interfaces or in the center of micelles, reducing the surface tension of the solution. Their diverse structures along with superior properties qualify them as potential candidates for application in several industrial sectors such as organic chemicals, petroleum, petrochemicals, mining, metallurgy (mainly bioleaching), agrochemicals, fertilizers, foods, beverages, cosmetics, pharmaceuticals and many others (Rangarajan and Narayanan 2018). Mechanism of metal removal by biosurfactants is given in Figure 2.

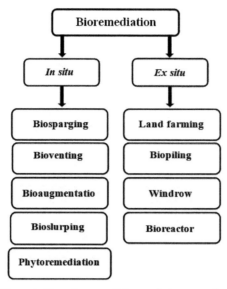

Figure 1. Flow diagram of bioremediation strategies.

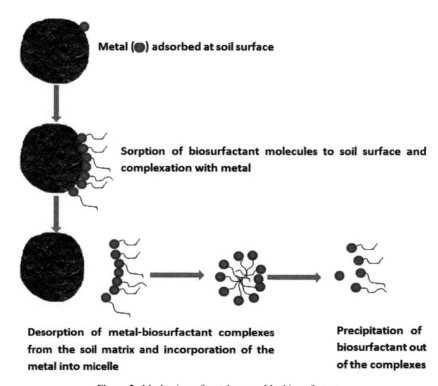

Figure 2. Mechanism of metal removal by biosurfactants.

Based on their composition, biosurfactants can be of high or low molecular weight and are categorized into glycolipids, phospholipids, lipopeptides, or a mixture of amphiphilic polysaccharides, proteins, lipoproteins, or lipopolysaccharides. Biosurfactants produced by microorganisms are the combination of many chemicals that are referred as polymeric microbial surfactants (Sáenz-Marta et al. 2015). Biosurfactants can be produced extracellularly or as part of the cell membrane by a wide variety of microorganisms such as bacteria, fungi, and yeast. Biodegradability, biocompatibility and digestibility of biosurfactants promote their application in environmental cleanup of industrial effluents and bioremediation of contaminated soils (Vijayakumar and Saravanan 2015). *Bacillus salmalaya, Candida lipolytica, Pseudomonas aeruginosa*, and *Saccharomyces lipolytica* are some examples of microorganisms with the ability to produce biosurfactant. However, the best known biosurfactants are produced by bacteria, and there are many studies on them, especially on *Pseudomonas* spp., strains that produce rhamnolipids (Usman et al. 2016).

Studies reveal that biosurfactants influence cells' surface properties, causing some significant alterations in cell surface hydrophobicity, electrokinetic potential, biomorphology, and surface functional groups (Kaczorek et al. 2018). The efficacy of bioremediation can be significantly enhanced by the addition of biosurfactants through the following mechanisms. First mechanism includes increasing substrate bioavailability for microorganisms, while the second one is connected with cell modification, which involves interaction with the cell surface, which increases their hydrophobicity, allowing hydrophobic substrates to associate more easily with bacterial cells (Kaczorek et al. 2018). Recent research findings have confirmed that biosurfactants can effectively augment the rate of hydrocarbon biodegradation by increasing microbial accessibility to insoluble substrates, enhancing their water solubility and increasing the displacement of oily substances from soil particles (Vijayakumar and Saravanan 2015). The affinity of microbial cell to adhere to hydrophobic substrates, such as hydrocarbons, is referred to as microbial cell hydrophobicity that promotes those cells to perform better degradation of hydrocarbons. The increase of microbial adhesion to hydrocarbons is directly related to the ability of such microorganisms to grow in the medium where hydrocarbons or other hydrophobic substrates are present. Binding of the biosurfactant compound to the microbial cell wall enhances hydrophobicity. Microorganisms can use their biosurfactants to regulate their cell-surface properties—to attach or detach from surfaces according to their needs (Ron and Rogenberg 2002).

Biosurfactants possess numerous advantages beyond chemically synthesized surfactants. For example, biosurfactants have high biodegradability as they can be easily degraded by bacteria and other microbes, thus having lesser harm on the environment. They also have lower toxicity compared to chemically synthesized surfactants, and productions from extremophiles have high efficiency at critical pH and temperature values. Biocompatibility and digestibility of biosurfactants guarantee their usages in remediating cosmetics, pharmaceuticals, and oil. In addition, biosurfactants can also be employed in managing industrial emulsions, oil spillage, detoxification and biodegradation of industrial effluents and polluted soils (Usman et al. 2016).

The goal of the use of biosurfactants for both organics and metals is almost similar; only there are some key differences between metal-contaminated and organic-contaminated soils. The most obvious difference between organic and metal contaminants is that unlike organic contaminants, metals cannot be biodegraded. In some cases, metals may be transformed but transformation often only increases metal toxicity (e.g., $Hg^{++} \rightarrow CH_3 - Hg^+$). Secondly, organics of the most concern are neutral molecules, while metals are most often found as cationic species (Miller 1995). As contaminant sorption depends on the chemical properties of both the soil and the contaminant, the choice of surfactant used for contaminant complexation is very much important in remediation of contaminated soil.

6. Case study

Arsenic resisting and metabolizing microorganisms isolated from various ecosystems are employed to remediate arsenic polluted soil and water. Reports from India carried out by Mondal et al. in 2014 documented both *ex situ* and *in situ* decontamination of petroleum hydrocarbons from contaminated soils using microbes. Rate of bioremediation was found to be depending on initial oil content of the sample, and geo-climatic parameters. Oil degrading microbial consortia were produced in a bioreactor and mixed in the soil of both the sites. For *ex situ* experiment, high density polyethylene lined were used for *ex situ* application, where oily wastes collected from oil fields were dumped at the site. Aeration and moisture were maintained during the experimentations. The used microbial consortia was prepared from the microbial straining previously collected from the oil-contaminated sites and screened based on their ability to degrade TPH components of oily waste. After the treatment, oily waste and ground water near the experimental sites were tested at regular interval for TPH, pH and heavy metals. Results showed a decrease of TPH content of 57.50–62.70 g/kg to 0.50 to 57.10 g/kg waste within a period of 2 to 12 months. The role of biodegradation was found to be 0.07–1.93 kg TPH/day/m^2 area. Heavy metals present in oily waste have no impact on the process of bioremediation.

Successful bioremediation using switchgrass (*Panicum virgutam*) and plant associated microbes has been made in Pb and Cd polluted soils by Arora et al. (2016) from India. Both AM fungi and Azospirillum were taken as plant associated microbes. Experiment was conducted in pot using different concentrations of Pb and Cd. Results showed a shift in soil pH towards neutral with AMF and Azospirillum inoculations. An increased root length, branches surface area, along with both root and shoot biomass of switchgrass, were also noted. The calculated bioconcentration factor (BCF) for Pb (12 mg kg^{-1}) and Cd (10 mg kg^{-1}) was found to be 0.25 and 0.23, respectively, whereas translocation index (TI) obtained was 17.8 and 16.7, respectively, which was approximately 45% higher than control. The lower values of BCF and TI even at the highest concentration of Pb and Cd revealed the capacity of switchgrass for accumulating high concentration of Pb and Cd in roots, while preventing the translocation of Pb and Cd to aerial biomass.

7. Conclusion

Different ecological tools are found effective in remediating the polluted soil. Significant efforts are needed to apply these ecological tools at pilot scale.

8. Future prospective

Mechanism of heavy metal retention, availability and subsequent assimilation of retained heavy metal is essential by materials like biochar and humic substances not only for development of cheaper ecological tools but also for future safety of the planet. Simultaneously, emphasis should be made on molecular biology to optimize the screened bioremediation process. Social aspect such as employment generation should be emphasized during planning of any bioremediation method which will not only attract attention from all sections of the society but also help in sustainable use of resources for conducting bioremediation activities without affecting the economy of a region. Thus, future strategy of bioremediation needs a broader vision for treating polluted soils along with sustaining soil health for benefit of the humanity.

References

Abatenh, E., Gizaw, B., Tsegaye, Z. and Wassie, M. 2017. Application of microorganisms in bioremediation—review. J. Environ. Microbiol. 1(1).
Abioye, O.P., Agamuthu, P. and Abdul Aziz, A.R. 2012. Biodegradation of used motor oil in soil using organic waste amendments. Biotechnol. Res. Int. doi:10.1155/2012/587041.
Al-Awadhi, N., Al-Daher, R., ElNawawy, A. and Balba, M.T. 1996. Bioremediation of oil-contaminated soil in Kuwait. I. Land farming to remediate oil-contaminated soil. Soil Sediment Contam. 5: 243–260.
Al-Daher, R., Al-Awadhi, N. and El-Nawawy, A. 1998. Bioremediation of damaged desert environment using the windrow soil pile system in Kuwait. Environ. Int. 24: 175–180.
Alfke, G., Bunch, G., Crociani, G., Dando, D., Fontaine, M., Goodsell, P., Green, A., Hafker, W., Isaak, G., Marvillet, J., Poot, B., Sutherland, H., van der Rest, A., van Oudenhoven, J., Walden, T., Martin, E. and Schipper, H. 1999. Best Available Techniques to Reduce Emissions from Refineries—Introduction. CONCAWE 99/01–1. CONCAWE, Brussels, Belgium.
Ali, H., Khan, E. and Sajad, M.A. 2013. Phytoremediation of heavy metals—Concepts and applications. Chemosphere 91: 869–881.

Alvarenga, P., Mourinha, C., Farto, M., Santos, T., Palma, P. and Sengo, J. 2015. Sewage sludge, compost and other representative organic wastes as agricultural soil amendments: benefits versus limiting factors. Waste Manag. 40: 44–52. doi: 10.1016/j.wasman.2015.01.027.

Alvarez, V.M., Marques, J.M., Korenblum, E. and Seldin, L. 2011. Comparative bioremediation of crude oil-amended tropical soil microcosms by natural attenuation, bioaugmentation, or bioenrichment. Appl. Environ. Soil Sci. 2011: 1–10.

Amoroso, M.J. and Abate, C.M. 2012. Bioremediation of copper, chromium and cadmium by actinomycetes from contaminated soils. pp. 349–364. *In:* Kothe, E. and Varma A. (eds.). Bio-Geo Interactions in Metal Contaminated Soils. Vol. 31.

Andersen, J.K., Boldrin, A., Samuelsson, J., Christensen, T.H. and Scheutz, C. 2010. Quantification of greenhouse gas emissions from windrow composting of garden waste. J. of Environ. Quality 39(2): 713–724.

Arora, K., Sharma, S. and Monti, A. 2016. Bioremediation of Pb and Cd polluted sites by switchgrass: A case study in India. Int. J. Phytoremediation 18(7): 704–709.

Arora, N.K. 2018. Bioremediation: a green approach for restoration of polluted ecosystems. Environ. Sustain. 1: 305–307.

Azubuike, C.C., Chikere, C.B. and Okpokwasili, G.C. 2016. Bioremediation techniques—classification based on site of application: principles, advantages, limitations and prospects. World J. Microbiol. Biotechnol. 32: 1–18.

Balba, M.T., Al-Daher, R., Al-Awadhi, N., Chino, H. and Tsuji, H. 1998. Bioremediation of oil-contaminated desert soil: The Kuwaiti experience. Environ. Int. 24: 163–173.

Banat, I.M., Franzetti, A., Gangolfi, I., Bestetti, G., Martinotti, M.G., Fracchia, L., Smyth, T.J. and Marchant, R. 2010. Microbial biosurfactants production, applications and future potential. Appl. Microbiol. Biotechnol. 87: 427–444.

Besalatpour, A., Hajabbasi, M.A., Khoshgoftarmanesh, A.H. and Dorostkar, V. 2011. Landfarming process effects on biochemical properties of petroleum-contaminated soils. Soil Sediment Contam. 20: 234–248.

Bollag, J.M. and Bollag, W.B. 1995. Soil contamination and the feasibility of biological remediation. Bioremediation: Science and Applications, SSSA Special Publication. 43: 1–10.

Brown, D.M., Okoro, S., van Gils, J., van Spanning, R., Bonte, M., Hutchings, T., Linden, O., Egbuche, U., Bruun, K.B. and Smith, J.W.N. 2017. Comparison of landfarming amendments to improve bioremediation of petroleum hydrocarbons in Niger Delta soils. Sci. Total Environ. 596–597: 284–292.

Brown, L.D. and Ulrich, A.C. 2014. Bioremediation of oil spills on land. Handbook of Oil Spill Science and Technology 395–406.

Burlakovs, J., Kļaviņš, M., Osinska, L. and Purmalis, O. 2013. The impact of humic substances as remediation agents to the speciation forms of metals in soil. APCBEE Procedia. 5: 192–196.

Bustamante, M., Duran, N. and Diez, M.C. 2012. Biosurfactants are useful tools for the bioremediation of contaminated soil: a review. J. Soil Sci. Plant Nut. 12(4): 667–687.

Cachada, A., Rocha-Santos, T. and Duarte, A.C. 2018. Soil and pollution: an introduction to the main issues. pp. 1–28. *In*: Duarte, A., Cachada, A. and Rocha-Santos, T. (eds.). Soil Pollution, Academic Press.

Cerqueira, V.S., do, M., Peralba, C.R., Camargo, F.A.O. and Bento, F.M. 2014. Comparison of bioremediation strategies for soil impacted with petrochemical oily sludge. Int. Biodeterior. Biodegrad. 95: 338–345.

Chaillan, F., Le Flèche, A., Bury, E., Phantavong, Y.H., Grimont, P., Saliot, A. and Oudot, J. 2004. Identification and biodegradation potential of tropical aerobic hydrocarbon-degrading microorganisms. Res. Microbiol. 155: 587–595.

Chibuogwu, O., Amadi, E.N. and Odu, C.T.I. 2005. Effects of soil treatments containing poultry manure on crude oil degradation in a sandy loam soil. Appl. Ecol. Environ. Res. 3(1): 47–53.

Christodoulatos, C. and Koutsospyros, A. 1998. Bioslurry reactors. pp. 69–103. *In*: Lewandowsky, G.A., DeFilippi, L.J. (eds.). Biological Treatment of Hazardous Wastes. New York: John Wiley & Sons. Inc.

Concetta Tomei, M. and Daugulis, A.J. 2013. *Ex situ* bioremediation of contaminated soils: An overview of conventional and innovative technologies. Crit. Rev. Environ. Sci. Technol. 43: 2107–2139.

Coulon, F., Al Awadi, M., Cowie, W., Mardlin, D., Pollard, S., Cunningham, C., Risdon, G., Arthur, P., Semple, K.T. and Paton, G.I. 2010. When is a soil remediated? Comparison of biopiled and windrowed soils contaminated with bunker-fuel in a full-scale trial. Environ. Pollut. 158(10): 3032–3040.

Da Silva, M.L.B. and Alvarez, P.J.J. 2010. Bioaugmentation. Handbook of Hydrocarbon and Lipid Microbiology 4531–4544.

Das, N. and Chandran, P. 2011. Microbial degradation of petroleum hydrocarbon contaminants: an overview. Biotechnol. Res. Int. 2011: 1–13.

Dixit, R., Wasiullah, Malaviya, D., Pandiyan, K., Singh, U.B., Sahu, A., Shukla, R., Singh, B.P., Rai, J.P., Sharma, P.K., Lade, H. and Paul, D. 2015. Bioremediation of heavy metals from soil and aquatic environment: An overview of principles and criteria of fundamental processes. Sustain. 7: 2189–2212.

Domene, X. 2016. A critical analysis of meso- and macrofauna effects following biochar supplementation. pp. 268–292. *In*: Komang Ralebitso-Senior, T. and Caroline H. Orr (eds.). Biochar Application, Elsevier.

Durães, N., Novo, L.A., Candeias, C. and Silva, E.F. 2018. Distribution, transport and fate of pollutants. pp. 29–57. *In*: Duarte, A.C., Cachada, A. and Rocha-Santos, T.A. (eds.). Soil Pollution, Academic Press.

Dzionek, A., Wojcieszyńska, D. and Guzik, U. 2016. Natural carriers in bioremediation: A review. Elect. J. Biotechnol. 19(5): 28–36.

E.P.A. 2004. How to evaluate alternative cleanup technologies for underground storage tank sites: a guide for corrective action plan reviewers. US Environmental Protection Agency.

Fan, C.Y. and Krishnamurthy, S. 1995. Enzymes for enhancing bioremediation of petroleum-contaminated soils: A brief review. J. Air Waste Manag. Assoc. 45: 453–460.

Fangueiro, D., Kidd, P.S., Alvarenga, P., Beesley, L. and de Varennes, A. 2018. Strategies for soil protection and remediation. pp. 251–281. In: Armando, C., Duarte, Anabela, C. and Teresa, R.-S. (eds.). Soil Pollution. Academic Press.

Federici, E., Giubilei, M., Santi, G., Zanaroli, G., Negroni, A., Fava, F., Petruccioli, M. and D'Annibale, A. 2012. Bioaugmentation of a historically contaminated soil by polychlorinated biphenyls with Lentinustigrinus. Microb. Cell Fact. 11: 1–14.

Fonseca, E.M., Baptista Neto, J.A., Mcalister, J., Smith, B., Fernandez, M.A. and Balieiro, F.C. 2013. The role of the humic substances in the fractioning of heavy metals in Rodrigo de Freitas Lagoon, Rio de Janeiro-Brazil. Anais da Academia Brasileira de Ciências 85(4): 1289–1301.

Francis, A.J. and Nancharaiah, Y.V. 2015. In situ and ex situ bioremediation of radionuclide-contaminated soils at nuclear and norm sites. Page Environmental Remediation and Restoration of Contaminated Nuclear and Norm Sites. Elsevier Ltd.

Frutos, F.J.G., Escolano, O., García, S., Babín, M. and Fernández, M.D. 2010. Bioventing remediation and ecotoxicity evaluation of phenanthrene-contaminated soil. J. Hazard. Mater. 183: 806–813.

Frutos, F.J.G., Pérez, R., Escolano, O., Rubio, A., Gimeno, A., Fernandez, M.D., Carbonell, G., Perucha, C. and Laguna, J. 2012. Remediation trials for hydrocarbon-contaminated sludge from a soil washing process: Evaluation of bioremediation technologies. J. Hazard. Mater. 199–200: 262–271.

Gentry, T.J., Rensing, C. and Pepper, I.L. 2004. New approaches for bioaugmentation as a remediation technology. Crit. Rev. Environ. Sci. Technol. 34: 447–494.

Golodyaev, G.P., Kostenkov, N.M. and Oznobikhin, V.I. 2009. Bioremediation of oil-contaminated soils by composting. Eurasian Soil Sci. 42: 926–35.

Gomez, F. and Sartaj, M. 2014. Optimization of field scale biopiles for bioremediation of petroleum hydrocarbon contaminated soil at low temperature conditions by response surface methodology (RSM). Int. Biodeterior. Biodegrad. 89: 103–109.

Halvorson, A.D. 2008. Soil in the environment. Soil Science Society of America Journal 72(4): 1185.

Hazen, T.C. 2010. In situ: groundwater bioremediation. pp. 2583–2594. In: Timmis, K.N. (ed.). Handbook of Hydrocarbon and Lipid Microbiology. Springer, Berlin.

Hejazi, R.F. and Husain, T. 2004. Landfarm performance under arid conditions. 2. Evaluation of parameters. Environ. Sci. Technol. 38: 2457–2469.

Held, T. and Dörr, H. 2000. In situ remediation. Biotechnology 11(b): 350–370.

Hellekson, D. 1999. Bioventing principles, applications and potential. Restor. Reclam. Rev. 5: 1–9.

Hoeppel, R.E., Hinchee, R.E. and Arthur, M.F. 1991. Bioventing soils contaminated with petroleum hydrocarbons. J. Ind. Microbiol. 8: 141–146.

Hussain, A. and Hasnain, S. 2011. Phytostimulation and biofertilization in wheat by cyanobacteria. J. Ind. Microbiol. Biotechnol. 38: 85–92.

Ingle, A.P., Seabra, A.B., Duran, N. and Rai, M. 2014. Nanoremediation: a new and emerging technology for the removal of toxic contaminant from environment. pp. 233–250. In: Das, S.(ed.). Microbial Biodegradation and Bioremediation. Elsevier.

Jaffre, T., Brooks, R.R., Lee, J. and Reeves, R.D. 1976. Sebertiaacuminata-Hyperaccumulator of nickel from New-Caledonia. Science 193: 579–580.

Jain, P.K., Gupta, V.K., Gaur, R.K., Lowry, M., Jaroli, D.P. and Chauhan, U.K. 2011. Bioremediation of petroleum oil contaminated soil and water. Res. J. Environ. Toxicol. 5(1): 1–26.

Jan, A.T., Azam, M., Ali, A. and Haq, Q.M.R. 2014. Prospects for exploiting bacteria for bioremediation of metal pollution. Crit. Rev. Environ. Sci. Technol. 44: 519–560.

Jones, K.C. and de Voogt, P. 1999. Persistent organic pollutants (POPs): state of the science. Environ. Pollut. 100: 209–221.

Jutsz, A.M. and Gnida, A. 2015. Mechanisms of stress avoidance and tolerance by plants used in phytoremediation of heavy metals. Arch. Environ. Prot. 41: 104–114.

Kaczorek, E., Pacholak, A., Zdarta, A. and Smułek, W. 2018. The impact of biosurfactants on microbial cell properties leading to hydrocarbon bioavailability increase. Colloids and Interfaces 2(3): 35.

Kao, C.M., Chen, C.Y., Chen, S.C., Chien, H.Y. and Chen, Y.L. 2008. Application of in situ biosparging to remediate a petroleum-hydrocarbon spill site: Field and microbial evaluation. Chemosphere 70: 1492–1499.

Karigar, C.S. and Rao, S.S. 2011. Role of microbial enzymes in the bioremediation of pollutants: A review. Enzyme Res. 2011: 805187.

Kazy, S.K., D'Souza, S.F. and Sar, P. 2009. Uranium and thorium sequestration by a Pseudomonas sp.: Mechanism and chemical characterization. J. Hazard. Mater. 163: 65–72.

Khan, F.I., Husain, T. and Hejazi, R. 2004. An overview and analysis of site remediation technologies. J. Environ. Manage. 71: 95–122.

Kim, S., Krajmalnik-Brown, R., Kim, J.O. and Chung, J. 2014. Remediation of petroleum hydrocarbon-contaminated sites by DNA diagnosis-based bioslurping technology. Sci. Total Environ. 497: 250–259.

Krishna, K.R. and Philip, L. 2011. Bioremediation of single and mixture of pesticide contaminated soils by mixed pesticide-enriched cultures. Appl. Biochem. Biotechnol. 164: 1257.

Lau, K.L., Tsang, Y.Y. and Chiu, S.W. 2003. Use of spent mushroom compost to bioremediate PAH-contaminated samples. Chemosphere 52(9): 1539–1546.

Lee, G.F. and Jones-Lee, A. 1997. Hazardous chemical site remediation through capping: Problems with long-term protection. Remediation 7: 51–57.

Lee, K., Park, J.W. and Ahn, I.S. 2003. Effect of additional carbon source on naphthalene biodegradation by *Pseudomonas putida* G7. J. Hazard. Mater. B105: 157–167.

Lei, L., Khodadoust, A.P., Suidan, M.T. and Tabak, H.H. 2005. Biodegradation of sediment-bound PAHs in field-contaminated sediment. Water Res. 39: 349–361.

Limmer, M. and Burken, J. 2016. Phytovolatilization of organic contaminants. Environ. Sci. Technol. 50: 6632–6643.

Lin, T.C., Pan, P.T. and Cheng, S.S. 2010. *Ex situ* bioremediation of oil-contaminated soil. J. Hazard. Mater. 176: 27–34.

Liu, L., Li, W., Song, W. and Guo, M. 2018. Remediation techniques for heavy metal-contaminated soils: Principles and applicability. Sci. Total Environ. 633: 206–219.

Mahmood, Q., Mirza, N. and Shaheen, S. 2015. Phytoremediation using algae and macrophytes: I. pp. 265–289. *In*: Ansari, A., Gill, S., Gill, R., Lanza, G. and Newman, L. (eds.). Phytoremediation. Springer. Cham.

Maila, M.P. and Cloete, T.E. 2004. Bioremediation of petroleum hydrocarbons through landfarming: Are simplicity and cost-effectiveness the only advantages? Rev. Environ. Sci. Biotechnol. 3: 349–360.

Mani, D. and Kumar, C. 2014. Biotechnological advances in bioremediation of heavy metals contaminated ecosystems: An overview with special reference to phytoremediation. Int. J. Environ. Sci. Technol. 11: 843–872.

Mesjasz-Przybylowicz, J., Nakonieczny, M., Migula, P., Augustyniak, M., Tarnawska, M., Reimold, W.U., Koeberl, C., Przybylowicz, W. and Glowacka, E. 2004. Uptake of cadmium, lead nickel and zinc from soil and water solutions by the nickel hyperaccumulator Berkheyacoddii. Acta Biol. Cracoviensia Ser. Bot. 46: 75–85.

Miller, R.R. 1996. Bioslurping: Technology Overview Report. Analysis.

Miller, R.M. 1995. Biosurfactant-facilitated remediation of metal-contaminated soils. Environ. Health Persp. 103(suppl 1): 59–62.

Min, K., Freeman, C., Kang, H. and Choi, S. 2015. The regulation by phenolic compounds of soil organic matter dynamics under a changing environment. BioMed. Res. Intern. Article ID 825098.

Miranda, R.D.C., De Souza, C.S., Gomes, E.D.B., Lovaglio, R.B., Lopes, C.E. and Sousa, M.D.F.V.D.Q. 2007. Biodegradation of diesel oil by yeasts isolated from the vicinity of Suape Port in the State of Pernambuco-Brazil. Brazilian Arch. Biol. Technol. 50: 147–152.

Mohan, S.V., Purushotham Reddy, B. and Sarma, P.N. 2009. *Ex situ* slurry phase bioremediation of chrysene contaminated soil with the function of metabolic function: Process evaluation by data enveloping analysis (DEA) and Taguchi design of experimental methodology (DOE). Bioresour. Technol. 100: 164–172.

Mondal, A.J., Jana, A., Dutta, A., Priyangshu, M., Sarma, B.L. and Dutta, J. 2014. Monitoring ground water quality and heavy metals in soil during large scale bioremediation of petroleum hydrocarbon contaminated waste in India: case study. J. Natu. Res. Dev. 4: 65–74.

Montiel-Rozas, M.M., López-García, Á., Kjøller, R., Madejón, E. and Rosendahl, S. 2016. Organic amendments increase phylogenetic diversity of arbuscular mycorrhizal fungi in acid soil contaminated by trace elements. Mycorrhiza 26: 575–585. doi: 10.1007/s00572-016-0694-3.

Morgan, P. and Atlas, R.M. 1989. Hydrocarbon degradation in soils and methods for soil biotreatment. Crit. Rev. Biotechnol. 8: 305–333.

Mrozik, A. and Piotrowska-Seget, Z. 2010. Bioaugmentation as a strategy for cleaning up of soils contaminated with aromatic compounds. Microbiol. Res. 165: 363–375.

Mukhopadhyay, S. and Maiti, S.K. 2010. Phytoremediation of metal mine waste. Appl. Ecol. Environ. Res. 8: 207–222.

Nester, E.W., Denise, G., Anderson, C., Roberts Jr., E., Pearsall, N.N. and Nester, M.T. 2001. Microbiology: A Human Perspective. 3rd ed. New York: McGraw-Hill.

Newman, L.A. and Reynolds, C.M. 2004. Phytodegradation of organic compounds. Curr. Opin. Biotechnol. 15(3): 225–230.

Nikolopoulou, M. and Kalogerakis, N. 2016. *Ex situ* bioremediation treatment (Landfarming). Hydrocarb. Lipid Microbiol. Protoc. 195–220.

Nzila, A., Razzak, S.A. and Zhu, J. 2016. Bioaugmentation: An emerging strategy of industrial wastewater treatment for reuse and discharge. Int. J. Environ. Res. Public Health 13.

Ojuederie, O.B. and Babalola, O.O. 2017. Microbial and plant-assisted bioremediation of heavy metal polluted environments: A review. Int. J. Environ. Res. Public Health 14.

Pacwa-Płociniczak, M., Płaza, G.A., Piotrowska-Seget, Z. and Cameotra, S.S. 2011. Environmental applications of biosurfactants: recent advances. Int. J. Mol. Sci. 12(1): 633–654.

Pacyna, J. 2011. Environmental emissions of selected persistent organic pollutants. pp. 49–56. *In*: Quante, M., Ebinghaus, R. and Flöser, G. (eds.). Persistent Pollution–Past, Present and Future, Springer.

Pal, S., Patra, A.K., Reza, S.K., Wildi, W. and Pote-Wembonyama, J. 2010. Use of bio-resources for remediation of soil pollution. Nat. Res. 1(2): 110–125.

Parween, T., Bhandari, P., Sharma, R., Jan, S., Siddiqui, Z.H. and Patanjali, P.K. 2018. Bioremediation: a sustainable tool to prevent pesticide pollution. pp. 215–227. *In*: Oves, M., Zain Khan, M. and. Ismail, M.I. (eds.). Modern Age Environmental Problems and their Remediation. Springer, Cham.

Patinha, C., Armienta, A., Argyraki, A. and Durães, N. 2018. Inorganic pollutants in soils. pp. 127–159. *In*: Armando, C.D., Anabela, C. and Teresa, R.-S. (eds.). Soil Pollution. Academic Press.

Paudyn, K., Rutter, A., Kerry Rowe, R. and Poland, J.S. 2008. Remediation of hydrocarbon contaminated soils in the Canadian Arctic by landfarming. Cold Reg. Sci. Technol. 53: 102–114.

Perelo, L.W. 2010. *In situ* and bioremediation of organic pollutants in aquatic sediments. J. Hazard. Mater. 177(1-3): 81–89.

Petersen, S.O. 2018. Greenhouse gas emissions from liquid dairy manure: prediction and mitigation. J. dairy Sci. 101(7): 6642–6654.

Philp, J.C. and Atlas, R.M. 2005. Bioremediation of contaminated soils and aquifers. pp. 139–236. *In*: Atlas, R. and Philip, J. (eds.). Bioremediation. American Society of Microbiology.

Piccolo, A. 1989. Reactivity of added humic substances towards plant available heavy metals in soils. Sci. Tot. Env. 81: 607–614.

Place, M., Hoeppel, R., Chaudhry, T., McCall, S. and Williamson, T. 2003. Application Guide for Bioslurping, Principles and Practices of Bioslurping, Addendum: Use of Pre-Pump Separation for Improved Bioslurper System Operation. White Pap. 1–20.

Quintero, J.C., Moreira, M.T., Lema, J.M. and Feijoo, G. 2006. An anaerobic bioreactor allows the efficient degradation of HCH isomers in soil slurry. Chemosphere 63: 1005–1013.

Rada, E.C., Andreottola, G., Istrate, I.A., Viotti, P., Conti, F. and Magaril, E.R. 2019. Remediation of soil polluted by organic compounds through chemical oxidation and phytoremediation combined with DCT. Int. J. Env. Res. Pub. He. 16(17): 3179.

Radziemska, M., Gusiatin, Z.M. and Bilgin, A. 2017. Potential of using immobilizing agents in aided phytostabilization on simulated contamination of soil with lead. Ecol. Eng. 102: 490–500.

Rangarajan, V. and Narayanan, M. 2018. Biosurfactants in soil bioremediation. pp. 193–204. *In*: Adhya, T., Lal, B., Mohapatra, B., Paul, D. and Das, S. (eds.). Advances in Soil Microbiology: Recent Trends and Future Prospects. Springer, Singapore.

Rayu, S., Karpouzas, D.G. and Singh, B.K. 2012. Emerging technologies in bioremediation: Constraints and opportunities. Biodegradation 23: 917–926.

Reeve, J.R., Hoagland, L.A., Villalba, J.J., Carr, P.M., Atucha, A., Cambardella, C. and Delate, K. 2016. Organic farming, soil health, and food quality: considering possible links. pp. 319–367. *In*: Donald L. Sparks (ed.). Adv. Agron. 137: Academic Press.

Reeves, R.D. and Baker, A.J.M. 2000. Phytoremediation of toxic metals: using plants to clean up the environment. Metal-accumulating Plants. pp. 193–229.

Rizzo, A.C.D.L., Dos Santos, R.D.M., Dos Santos, R.L.C., Soriano, A.U., Da Cunha, C.D., Rosado, A.S., Sobral, L.G.D.S. and Leite, S.G.F. 2010. Petroleum-contaminated soil remediation in a new solid phase bioreactor. J. Chem. Technol. Biotechnol. 85: 1260–1267.

Robles-González, I.V., Fava, F. and Poggi-Varaldo, H.M. 2008. A review on slurry bioreactors for bioremediation of soils and sediments. Microb. Cell Fact. 7: 1–16.

Rodrigues, S.M. and Römkens, P.F. 2018. Human health risks and soil pollution. pp. 217–250. *In*: Armando C. Duarte, Anabela Cachada and Teresa Rocha-Santos (eds.). Soil Pollution. Academic Press.

Rodríguez-Rodríguez, C.E., Marco-Urrea, E. and Caminal, G. 2010. Degradation of naproxen and carbamazepine in spiked sludge by slurry and solid-phase Trametes versicolor systems. Bioresour. Technol. 101: 2259–2266.

Romero-González, M., Nwaobi, B.C., Hufton, J.M. and Gilmour, D.J. 2016. *Ex-situ* bioremediation of U (VI) from contaminated mine water using acidithiobacillusferrooxidans strains. Front. Environ. Sci. 4: 1–11.

Romic, M. 2012. Bioavailability of trace metals in terrestrial environment: methodological issues. Eur. Chem. Bull. 1(11): 489–493.

Ron, E.Z. and Rosenberg, E. 2002. Biosurfactants and oil bioremediation. Current Opinion in Biotechnology 13: 249–252.

Rubilar, O., Tortella, G., Cea, M., Acevedo, F., Bustamante, M., Gianfreda, L. and Diez, M.C. 2011. Bioremediation of a Chilean Andisol contaminated with pentachlorophenol (PCP) by solid substrate cultures of white rot fungi. Biodegradation 22: 31–41.

Sáenz-Marta, C.I., de Lourdes Ballinas-Casarrubias, M., Rivera-Chavira, B.E. and Nevárez-Moorillón, G.V. 2015. Biosurfactants as useful tools in bioremediation. Advances in Bioremediation of Wastewater and Polluted Soil. pp. 94–109.

Saha, J.K., Selladurai, R., Coumar, M.V., Dotaniya, M.L., Kundu, S. and Patra, A.K. 2017. Major inorganic pollutants affecting soil and crop quality. pp. 75–104. *In*: Eric, L., Agroé, C., Dijon, Jan, S., Didier, R. and Saint, A. (eds.). Soil Pollution—An Emerging Threat to Agriculture. Springer, Singapore.

Salt, D.E., Blaylock, M., Kumar, N.P., Dushenkov, V., Ensley, B.D., Chet, I. and Raskin, I. 1995. Phytoremediation: A novel strategy for the removal of toxic elements from the environment using plants. Biotechnology 13: 468–474.

Sanscartier, D., Zeeb, B., Koch, I. and Reimer, K. 2009. Bioremediation of diesel-contaminated soil by heated and humidified biopile system in cold climates. Cold Reg. Sci. Technol. 55: 167–173.

Semple, K.T., Reid, B.J. and Fermor, T.R. 2001. Impact of composting strategies on the treatment of soils contaminated with organic pollutants. Environ. Pollut. 112: 269–283.

Sen, R. and Chakrabarti, S. 2009. Biotechnology—applications to environmental remediation in resource exploitation. Curr. Sci. pp. 768–775.

Seple, K.T., Reid, B.J. and Fermor, T.R. 2001. Impact of composting strategies on the treatment of soils contaminated with organic pollutants. Env. Poll. 112(2): 269–283.

Shi, W., Norton, J.M., Miller, B.E. and Pace, M.G. 1999. Effects of aeration and moisture during windrow composting on the nitrogen fertilizer values of dairy waste composts. Appl. Soil Ecol. 11: 17–28.

Silva-Castro, G.A., Uad, I., Rodríguez-Calvo, A., González-López, J. and Calvo, C. 2015. Response of autochthonous microbiota of diesel polluted soils to land-farming treatments. Environ. Res. 137: 49–58.

Simarro, R., González, N., Bautista, L.F. and Molina, M.C. 2013. Assessment of the efficiency of in situ bioremediation techniques in a creosote polluted soil: Change in bacterial community. J. Hazard. Mater. 262: 158–67.

Smith, E., Thavamani, P., Ramadass, K., Naidu, R., Srivastava, P. and Megharaj, M. 2015. Remediation trials for hydrocarbon-contaminated soils in arid environments: Evaluation of bioslurry and biopiling techniques. Int. Biodeterior. Biodegrad. 101: 56–65.

Smith, S.R. 2009. A critical review of the bioavailability and impacts and heavy metals in municipal solid waste compost compared to sewage sludge. Environ. Int. 35: 142–156.

Stojić, N., Štrbac, S. and Prokić, D. 2018. Soil pollution and remediation. pp. 1–34. In: Hussain, C. (ed.). Handbook of Environmental Materials Management.

Sukkariyah, B.F., Evanylo, G., Zelazny, L. and Chaney, R.L. 2005 Cadmium, copper, nickel, and zinc availability in a biosolids-amended piedmont soil years after application, J. Environ. Qual. 34: 2255–2262.

Sun, H., Trabue, S.L., Scoggin, K., Jackson, W.A., Pan, Y. and Zhao, Y. 2008. Alcohol, volatile fatty acid, phenol, and methane emissions from dairy cows and fresh manure. J. Environ. Qual. 37: 615–622.

Suthersan, S.S. and Payne, F.C. 2004. *In situ* Remediation Engineering, 1st edition, CRC Press 1–487.

Tahri, N., Bahafid, W., Sayel, H. and El Ghachtouli, N. 2013. Biodegradation: involved microorganisms and genetically engineered microorganisms. Biodegrad. - Life Sci. 56194.

Tak, H.I., Ahmad, F. and Babalola, O.O. 2013. Advances in the application of plant growth-promoting rhizobacteria in phytoremediation of heavy metals. Rev. Environ. Contam. Toxicol. 223: 33–52.

Thomaidi, V.S., Stasinakis, A.S., Borova, V.S. and Thomaidis, N.S. 2016. Assessing the risk associated with the presence of emerging organic contaminants in sludge-amended soil: a country-level analysis. Sci. Total Environ. 548-549: 280–288.

Tomei, M.C. and Daugulis, A.J. 2013. *Ex situ* bioremediation of contaminated soils: An overview of conventional an innovative technologies. Crit. Rev. Environ. Sci. Technol. 43: 2107–39.

Urzelai, A., Vega, M. and Angulo, E. 2000. Deriving ecological risk-based soil quality values in the Basque Country. Sci. Tot. Env. 247: 279–284.

US Environmental Protection Agency (USEPA). 1995. How to evaluate alternative cleanup technologies for underground storage tank sites.

Usman, M.M., Dadrasnia, A., Lim, K.T., Mahmud, A.F. and Ismail, S. 2016. Application of biosurfactants in environmental biotechnology; remediation of oil and heavy metal. AIMS Bioengineering 3(3): 289–304.

Van Der Nat, F.J.W.a. and Middelburg, J.J. 1998. Seasonal variation in methane oxidation by the rhizosphere of Phragmites australis and Scirpuslacustris. Aquat. Bot. 61: 95–110.

Verbruggen, N., Hermans, C. and Schat, H. 2009. Molecular mechanisms of metal hyperaccumulation in plants. New Phytologist. 181(4): 759–776.

Vijayakumar, S. and Saravanan, V. 2015. Biosurfactants—types, sources and applications. Res. J. Microbiol. 10(5): 181–92.

Walker, C.H., Hoplin, Sibley, S.P. and Peakall, D.B. 2001. Principles of Ecotoxicology. Second ed., Taylor & Francis, London.

Wang, S., Xu, Y., Norbu, N. and Wang, Z. 2018a. Remediation of biochar on heavy metal polluted soils. In IOP Conference Series: Earth and Env. Sci. 108(4): 042113.

Wang, W., Simonich, S.L.M., Xue, M., Zhao, J., Zhang, N. and Wang, R. 2000. Concentrations, sources and spatial distribution of polycyclic aromatic hydrocarbons in soils from Beijing, Tianjin and surrounding mareas, North China. Environ. Pollut. 158: 1245–1251.

Wang, Y., Xu, Y., Li, D., Tang, B., Man, S., Jia, Y. and Xu, H. 2018b. Vermicompost and biochar as bio-conditioners to immobilize heavy metal and improve soil fertility on cadmium contaminated soil under acid rain stress. Sci. Tot. Env. 621: 1057–1065.

Whelan, M.J., Coulon, F., Hince, G., Rayner, J., McWatters, R., Spedding, T. and Snape, I. 2015. Fate and transport of petroleum hydrocarbons in engineered biopiles in polar regions. Chemosphere 131: 232–240.

Williams, J. 2002. Bioremediation of Contaminated Soils: A comparison of *in situ* and *ex situ* techniques. Available at: http://home.eng.iastate.edu/~tge/ce421-521/jera.pdf.

Woo, S.H. and Park, J.M. 1999. Evaluation of drum bioreactor performance used for decontamination of soil polluted with polycyclic aromatic hydrocarbons. J. Chem. Technol. Biotechnol. 74: 937–944.

Wuana, R.A., Okieimen, F.E. and Vesuwe, R.N. 2014. Mixed pollutant interactions in soil: implications for bioavailability, risk assessment and remediation, Afr. J. Environ. Sci. Technol. 8(12): 691–706.

Yadav, K.K., Gupta, N., Kumar, V. and Singh, J.K. 2017. Bioremediation of heavy metals from contaminated sites using potential species: A review. Indian J. Environ. Prot. 37: 65–84.

Yang, L.Y., Wang, L.T., Ma, J.H., Ma, E.D., Li, J.Y. and Gong, M. 2017. Effects of light quality on growth and development, photosynthetic characteristics and content of carbohydrates in tobacco (*Nicotiana tabacum* L.) plants. Photosynthetica 55: 1–11.

Yang, Z. X., S.Q. Liu, D.W. Zheng, and S.D. Feng. 2006. Effects of cadium, zinc and lead on soil enzyme activities. J Env. Sci. 18(6): pp.1135-1141.

Zahoor, M., Irshad, M., Rahman, H., Qasim, M., Afridi, S.G., Qadir, M. and Hussain, A. 2017. Alleviation of heavy metal toxicity and phytostimulation of *Brassica campestris* L. by endophytic *Mucor* sp. MHR-7. Ecotoxicol. Environ. Saf. 142: 139–149.

Zhou, D.M., Hao, X.Z., Wang, Y.J., Dong, Y.H. and Cang, L. 2005. Copper and Zn uptake by radish and pakchoi as affected by application of livestock and poultry manures. Chemosphere 59: 167–175.

Zhu, X., Venosa, A.D., Suidan, M.T. and Lee, K. 2004. Guidelines for the Bioremediation of Oil-Contaminated Salt Marshes. National Risk Management Research Laboratory Office of Research and Development, U.S. Environmental Protection Agency, Cincinnati, Ohio.

5

Phytoremediation

A Green Approach for the Restoration of Heavy Metal Contaminated Soils

Sivakoti Ramana,* Vassanda Coumar Mounissamy and *Jayant Kumar Saha*

1. Introduction

Soil is the uppermost layer of Earth's crust on which every person's life, well-being and fulfilment depend. It takes over thousands of years or even longer to develop one inch of soil. Agriculture productivity and production of good quality food is dependent on good quality soil, necessitating protection of fertile land area. In recent years, soil degradation is a major global concern as a result of increasing demand on the land for food production and waste disposal. Every year in every country, soil resources are impaired and in some cases lost for productive use because of misuse, application of toxic materials, or poor land management systems. In India, about 60% of the geographical area is occupied by agricultural land, most of which is facing one or more kind degradation stresses (NAAS 2010).

Among the various soil degradation types, soil pollution is one of the most sensitive environmental issues faced today across the globe. Soil pollution can occur either by natural processes or man-made (anthropogenic) by the introduction of chemicals to the natural soil environment through industrial activity (industrial sector), chemicals used in agriculture (agricultural sector), or improper disposal of waste (urban sector). It brings undesirable change in the physical, chemical or biological properties of the soil, which adversely affects crop production, soil quality, human nutrition, surrounding environment and thereby causes huge disturbance in the ecological balance (Abrahams 2002).

2. Soil contaminants' types and status

Based on the nature of the soil contaminants, they can be broadly classified as organic (those that contain carbon) and inorganic (metals, inorganic ions and salts) contaminants. Both organic and inorganic contaminants cause potential hazard to soil, plants, animals and human beings. The most prominent organic contaminants are petroleum hydrocarbons (benzene, xylene, toluene, ethylene, and alkanes), polynuclear aromatic hydrocarbons (PAHs), polychlorinated biphenyls (PCBs),

---

ICAR-Indian Institute of Soil Science, Nabibagh, Berasia Road, Bhopal (M.P.).
Emails: vassanda.coumar@gmail.com; jk_saha12000@yahoo.com
* Corresponding author: ramana.sivakoti@gmail.com

trichloroethylene (TCE), chlorinated aromatic compounds, detergents, pesticides (including their chemical impurities such as dioxins), dyes and antibiotics. Inorganic contaminants in soil includes nitrates, phosphates, and heavy metals such as lead, cadmium, chromium, cobalt, nickel, arsenic, zinc, iron, copper and mercury, inorganic acids, and radionuclides (radioactive substances).

Among different kinds of contaminants, the problem of heavy metal pollution has emerged as a matter of concern at local, regional and global scales. Heavy metals are often referred to as trace or toxic elements because of their toxic effect on living things (Jarup 2003). The level of toxicity and the bioavailability to the plant species depends on several factors, i.e., chemical form (species) and the amount that is presented to the plant in the environment, as well as certain external factors like soil pH, oxidation-reduction potential, presence of other cations and anions in the system and clay and organic matter content in soil (Loska and Wiechula 2000, Wyszkowska and Wyszkowski 2003). Because of the persistent and non-biodegradable nature of metals in soil, remediation of heavy metal polluted soil is a greater task (Gao 1986).

In India, heavy metal contamination in soil due to anthropogenic activity has been reported from different areas (Sachan 2007, Shanker et al. 2005, Deka and Bhattacharyya 2009). The Central Pollution Control Board (CPCB) in India has identified critically polluted industrial areas and clusters or potential impact zone based on its Comprehensive Environmental Pollution Index (CEPI) rating. Forty three critically polluted zones were reported in the 16 states which have CEPI rating more than 70. Among the 43 sites, 21 sites exist in only four states namely Gujarat, Uttar Pradesh, Maharashtra and Tamil Nadu. Further, the Central Pollution Control Board (CPCB) has identified 17 industries as highly polluting industries, the majority of which are manufacturing industries like aluminum, cement, chlor-alkali, copper, distillery, dyes and intermediates, fertilizers, iron and steel, oil refineries, pesticides, petrochemicals, pharmaceuticals, pulp and paper, sugar, tannery, power plant and zinc smelters.

3. Impact of soil contaminants on ecosystem and soil quality

A range of chemicals introduced to the soil system remain in the soil for many years and their negative impact on soil health are quite alarming because it can cause huge imbalance to the ecosystem and affects health of the living creatures on earth. Irrespective of their sources in the soil, accumulation of contaminants can degrade soil quality, reduce crop yield and the quality of agricultural products, and thus negatively impact the health of human, animals, and the ecosystem (Nagajyoti et al. 2010). Different soil contaminants interact differently with soil components and microorganisms. Based on the type of contaminants, the possible impact on various ecosystems also varies.

Plants that grow on polluted soil may have lower yields because the hazardous chemicals in the soil interfere with their growth. Besides, when plants take up the soil contaminants, they pass them up the food chain, endangering the health of animals and humans. In some parts of the world, heavily polluted soils with metals and chemicals such as lead, asbestos, and sulfur are considered unfavorable for crop production and cannot be used to grow crops. Further, soil pollution can also lead to scarcity of food if the soil is rendered infertile and crops fail to grow on it, thereby leading to acute food shortage or famines. The most prominent effect of soil pollution on soil chemical properties includes development of salinity, sodicity and acidity, changes in nutrient uptake pattern, decrease in nutrient use efficiency, etc. In addition, soil contaminants (inorganic salts, ions, organic contaminants, heavy metals, dyes and antibiotics) can have negative impact on the lives of the living organisms which can result in the gradual death of many organisms. Soil contaminants also affect several microbially mediated nutrient transformation processes like nitrogen fixation, nutrient mineralization, etc. (Bondarenko et al. 2010, Arora et al. 2010, Bianucci et al. 2011). Thus, the effects of pollution on soil are quite alarming and can cause huge disturbances in the ecological balance and health of living creatures on Earth.

4. Soil remediation techniques

In order to restore the natural ecosystem functions, to improve the quality of human and animal life dependent on the polluted land area, the search for effective and feasible remedial measures to address land pollution has become important. Location/site specific remediation management has been advocated by government organizations, environmentalists, policy makers and land owners for rational utilization of polluted soil. Remediation of large rural areas with marginally polluted soils, and agricultural fields should be approached differently than the remediation of heavily contaminated areas such as those around mining and smelting sites. In general, remediation technologies, whether in place or *ex situ* can be achieved either by removal of the heavy metals ("site decontamination or clean-up techniques") or by preventing their spread to surrounding soil and groundwater (isolation and/or immobilization), thereby reducing exposure risk.

Removal of contaminants and risk minimization are the major approaches for heavy metal polluted soil. Several technologies have been developed for their remediation based on clean-up, detoxification and risk minimization approaches. All of these technologies have both advantages and disadvantages in respect of the extent of applicability, side-effects on other components of environment, cost and ease of adoption, speed and effectiveness of remediation. Remediation of large areas of heavy metal contaminated soil by conventional methods (e.g., excavation, physico-chemical treatment) is too expensive for large sites (Willscher et al. 2013). The use of plants and associated microorganisms to stabilize inactivate, remove, or degrade toxic environmental pollutants, which is generally termed as phytoremediation, is gaining more and more attention. Phytoremediation represents an emerging and sustainable technology for the remediation of slightly to moderately contaminated sites.

4.1 Phytoremediation

The term phytoremediation is derived from the Greek *prefix* phyto, meaning "plant", and the Latin *suffix* remedium, "able to cure" or "restore". Phytoremediation technology uses metal accumulation and exclusion abilities of plants to clean up heavy metal polluted areas (Baker and Walker 1990, Schnoor 2002). It actually refers to a diverse collection of plant-based technologies that use either naturally occurring or genetically engineered plants for cleaning contaminated environments (Flathman and Lanza 1998). It is also called as "Green remediation" and "Botanical bioremediation" and is complementary to classical bioremediation techniques, which are based on the use of microorganisms. It can be applied to a wide range of organic (Schnoor et al. 1995) and inorganic contaminants. Phytoremediation is a general term which includes several processes, among which phytoextraction and phytostabilisation are the most reliable for heavy metals.

The research of phytoremediation has its origin in the nuclear disaster at Chernobyl in the year 1986 which caused severe radioactive contamination. The scientists were hopeful that plants could play a key role in cleansing some of the contamination. Three years after the explosion, the government requested the International Atomic Energy Agency (IAEA) to assess the radiological and health situation in the area surrounding the nuclear power plant. The study found radioactive emissions and toxic metals-including iodine, cesium-137, strontium, and plutonium-concentrated within the soil, plants, and animals which entered the food chain via grazers, such as cows and other livestock, which fed on plants grown in contaminated soils. These elements finally accumulated in the meat and milk products eventually consumed by humans. Therefore, a soil clean-up technique was employed using green plants to get rid of toxins from the soil. This technique is known as phytoremediation, a term coined by Ilya Raskin who was one of the members of the original task force sent by the IAEA to examine food safety at the Chernobyl site. According to him, using plants to alter the environment "has been around forever, since the time plants were used to drain swamps." What is new, he asserts, is the systematic and scientific investigation of how plants can be used to decontaminate soil and water. It takes advantage of the plant's ability to remove pollutants from the environment or to make them harmless or less dangerous.

The term phytoremediation, like the term itself, consists of two major processes that decontaminate the matrix (remove, extract, degrade, volatalise, etc.) and processes that stabilize the contaminant in the soil to reduce or prevent further environmental damage (sequester, solidify, precipiatate, etc.) (Cuuningham et al. 1996). The former process is known as phytodecontamination and the latter is called phytostabilization. The choice of which of these alternative techniques should be implemented at a site is not solely a matter of economics, for they have different constraints and applications and are sensitive to different site parameters such as concentration of the contaminant, soil chemistry, contamination depth or the time frame required for remediation. If an immediate reduction in risk is required, phytostabilisation would be chosen because of the length required for plants to remove the contaminant for extraction. However, as sites where decontamination is desired and feasible (phytodecontamination), phytoextraction is more appropriate technique despite the higher cost (Cuuningham and Berti 1999).

4.2 Phytodecontamination

Phytodecontamination is a subset of phytoremediation in which the concentration or the level of the contaminants of concern in the soil/water is reduced to an acceptable level through the action of plants, their associated microflora and agronomic practices. It consists of a collection of different plant-based technologies, each having a different mechanism of action for the remediation of metal-polluted soil, sediment or water. These include: phytoextraction, rhizofiltration, phytovolatilization and phytodegradation.

4.2.1 Phytoextraction (Phytoaccumulation)

Phytoextraction is the name given to the process in which plant roots take up the metal contaminants from the soil and translocate them to their above ground plant tissues. It is the most commonly recognized of all phytoremediation technologies. The technology of phytoextraction utilizes metal hyperaccumulating plants to extract heavy metal contaminants from the environment. Baker and Walker (1990) categorized plants into three groups based on their strategies for growing on metal-contaminated soils: metal excluders, indicators and accumulators or hyperaccumulators.

Metal excluders

Metal excluders are the plants which effectively limit the levels of heavy metal translocation within them and maintain relatively low levels in their shoot over a wide range of soil levels; however, they can still contain large amounts of metals in their roots. Boularbah et al. (2006) defined an excluder as a plant species that can have high levels of heavy metals in the roots, but always has a shoot/root quotient of less than one. Such plants have a low potential for metal extraction but may be efficient for phytostabilization purposes (Lasat 2002, Barcelo and Poschenrieder 2003).

Metal indicators

Metal indicators are the plants that accumulate metals in their above-ground tissues and the metal levels in the tissues of these plants generally reflect metal levels in the soil (Ghosh et al. 2005). The relationship between the concentration of metal in the plant and soil is generally linear. However, under continued uptake of heavy metals these plant species die-off. The indicator plants render biological and ecological functions in that they are possible indicators of pollution and useful in absorption of pollutants (Kvesitadze et al. 2006). Metal indicator plants have received attention since the late 1800s as metal accumulator plants (Anderson et al. 2005). A recent example demonstrates the potential use of Eucalyptus trees, which translocate Au from deep (> 10 m) mineral deposits to their aerial tissues in order to identify Au deposits (Lintern et al. 2013).

Metal accumulators

These plants accumulate high concentration of metals in plant tissue even at very low concentration of metal in the soil. However, at high external metal concentrations, these plants don't increase their uptake, probably due to competition between metal ions at the site of uptake. These plants have certain detoxification mechanisms within the tissue which allow the plant to accumulate such high amounts of metal.

Hyper accumulators

Yet another type of response of a plant to heavy metal stress is hyper accumulation of metals. Hyper accumulators are the plant species that concentrate metals in their above-ground tissues to levels far exceeding those present in the soil or in the non-accumulating species growing nearby. The most commonly postulated hypothesis regarding the reason or advantage of metal hyperaccumulation in plants is elemental defense against herbivores (by making leaves unpalatable or toxic) and pathogens (Meharg 2005, Prasad 2005, Dipu 2012). These plants grow on metalliferous soils and have developed the ability to accumulate massive amounts of the indigenous metals in their tissues without exhibiting symptoms of toxicity and therefore, they are widely used in phytoextraction.

The concept of metal hyperaccumulator plants was first described for nickel accumulating species *Sebertiaacuminata* by Jaffré et al. (1976). The Ni content of the latex (25.74% on a dry weight basis) was the highest nickel concentration ever found in living material. However, the term "hyperaccumulator" was first coined by Brooks et al. (1977) to define plants with Ni concentrations higher than 1000 mg kg^{-1} dry weigh (0.1%). Reeves (1992) defined Ni hyperaccumulation with greater precision as "a Ni hyperaccumulator" is a plant in which a Ni concentration of at least 1000 mg kg^{-1} has been recorded in the dry matter of any aboveground tissue in at least one specimen growing in its natural habitat. For establishing hyperaccumulator status, the aboveground tissue should be regarded as plant foliage only. However, the most frequently cited criteria for hyperaccumulation of metals is that of Baker and Brooks (1989), according to which "hyperaccumulators are plant species, which accumulate greater than 100 mg of Cd kg^{-1} dry weight, 1000 mg of Ni, Cu and Pb kg^{-1} dry weight, and 10000 mg of Zn and Mn kg^{-1} dry weight in their shoots when grown on metal rich soils". Since then, the definition of hyperaccumulation has been refined a number of times and was most recently updated by van der Ent et al. in 2013. According to them, the commonly used criteria for hyperaccumulation of some metals are unnecessarily conservative and therefore they proposed that criteria for hyperaccumulation of such metals be lowered. Accordingly, the following concentration criteria has been established for different metals and metalloids in dried foliage with plants growing in their natural habitats: 100 mg kg^{-1} for Cd, Se and Tl; 300 mg kg^{-1} for Co, Cu and Cr; 1000 mg kg^{-1} for Ni, Pb and As; 3000 mg kg^{-1} for Zn; 10000 mg kg^{-1} for Mn. If these criteria are adopted, more than 500 plant taxa have been cited in the literature to date as hyper accumulators for one or more elements (As, Cd, Co, Cu, Mn, Ni, Pb, Se, Tl, Zn), which still represents a very small proportion of the (approximately) 300,000 recognized vascular plant species. Generally, hyperaccumulators achieve 100-fold higher shoot metal concentration (without yield reduction) compared to crop plants or common nonaccumulator plants (Lasat 2002, Chaney et al. 2007). Hyperaccumulators achieve a shoot-to-root metal concentration ratio (called translocation factor, TF) > 1 and extraction coefficient, i.e., bio concentration factor (BCF) of > 1 (Tangahu et al. 2011, Badr et al. 2012). However, hyperaccumulators were later believed to have limited potential in this area because of their small size and slow growth, which limit the speed of metal removal (Cunningham et al. 1995, Ebbs 1997). Family Brassicaceae contains many metal accumulating species (Poniedziałek et al. 2010). Some classic examples of hyperaccumulators are *Thlaspicaerulescens* and *Alyssum bertolonii*. *Thlaspicaerulescens* (Alpine pennycress) is possibly the best-known metal hyperaccumulator (Lasat 2002) for Zn, Cd and Ni (Assuncao et al. 2003).

4.3 Phytodecontamination process

In practice, metal-accumulating or hyperaccumulator plants are sown or transplanted into metal polluted soil and are cultivated using standard agricultural practices. The roots of established plants absorb metal elements from the soil and translocate them to the aboveground shoots, where they accumulate. After sufficient plant growth and metal accumulation, the aboveground portions of the plant are harvested and removed, resulting in the permanent removal of metals from the site. This process is repeated several times to reduce contamination to acceptable levels. An important question is that what will be the fate of plants after being used for phytoextraction of heavy metals? Therefore, the use and safe disposal of biomass after phytoremediation has been addressed by many researchers. Koppolu et al. (2003a,b, 2004) extensively analyzed the use of heavy metal contaminated material for the production of carbon catalysts. Sas-Nowosielska et al. (2004) examined various strategies to effectively utilize the contaminated biomass like composting, compaction, incineration, ashing, pyrolysis, direct disposal and liquid extraction. According to him, the incineration (smelting) is the most feasible, economically acceptable and environmentally sound process. Further, the addition of biochar (the product of pyrolysis using contaminated biomass as raw material) to soil has been also suggested as a means to sequester carbon, thereby reducing the effects of human-induced climate change caused by CO_2 emissions (Lehmann 2007). According to Jadia and Fulekar (2008), the biomass can be either disposed as hazardous waste safely in specialized dumps or if economically feasible, processed for biorecovery of precious and semiprecious metals (a practice known as phytomining). It takes advantage of the plant's ability to remove pollutants from the environment or to make them harmless or less dangerous (Raskin 1996). The metals that have been successfully recovered from the phytoextracted biomass include zinc, copper, and nickel. Marion et al. (2013) reported that the biomass produced during phytoextraction may be usable for bioenergy production.

4.4 Quantification of efficiency of phytoextraction

The efficiency of phytoextraction can be quantified by calculating bioconcentration factor, translocation factor and phytoextraction rate or metal extraction rate.

Bioconcentration factor indicates the efficiency of a plant species in accumulating a metal into its tissues from the surrounding environment (Ladislas et al. 2012). It is calculated as follows (Zhuang et al. 2007):

Bioconcentration Factor = $C_{harvested\ tissue}/C_{soil}$

where $C_{harvested\ tissue}$ is the concentration of the target metal in the plant harvested tissue and C_{soil} is the concentration of the same metal in the soil (substrate).

Translocation factor or shoot to root ratio indicates the efficiency of the plant in translocating the accumulated metal from its roots to shoots. It is calculated as follows (Padmavathiamma and Li 2007):

Translocation Factor = C_{shoot}/C_{root}

where C_{shoot} is concentration of the metal in plant shoots and C_{root} is concentration of the metal in plant roots.

Translocation efficiency (TE %) was calculated as per formula proposed by Meers et al. (2004).

TE (%) = Metal content in the shoots ($\mu g\ g^{-1}$)/Metal content in the whole plant ($\mu g\ g^{-1}$) × 100

Phytoextraction rate or metal extraction ratio is used to determine the ability of the trees for the immobilization of heavy metals in their organism and to determine the suitability of a particular species of trees in the phytoremediation process of degraded soils (such as phytoextraction) (Zhao et al. 2003, Mertens et al. 2005).

Phytoextraction rate or metal extraction ratio is calculated as follows:

$(C_{plant} \times M_{plant}/C_{soil} \times M_{rooted\ zone}) \times 100$

where M_{plant} is the mass of the harvestable aboveground biomass produced in one harvest, C_{plant} is the metal concentration in the harvested component of the plant biomass, $M_{rooted\ zone}$ is the mass of the soil volume rooted by the species under study and C_{soil} is the metal concentration in the soil volume.

4.4.1 Rhizofiltration

Rhizofiltration refers to the approach of using hydroponically cultivated plant roots to remediate contaminated water through absorption, concentration, and precipitation of pollutants. Both phytoextraction and rhizofiltration follow the same basic path for remediation and are aimed more toward concentrating and precipitating heavy metals than organic contaminants. The major difference is that phytoextraction deals with soil remediation, while rhizofiltration is used for treatment in aquatic environment. Plants are hydroponically grown in clean water until a large root system develops. Once a large root system is in place, the water supply is substituted for a polluted water supply to acclimatise the plant. After the plants become acclimatised, they are planted in the polluted aquatic bodies where the roots uptake the polluted water and the contaminants along with it. As the roots become saturated, they are harvested and disposed of safely/processed further. Repeated treatments of the site can reduce pollution to suitable levels as was exemplified in Chernobyl where sunflowers were grown in radioactively contaminated pools. Aquatic higher plants like water hyacinth (*Eichhornia crassipes*) (Turnquist et al. 1990, Kay et al. 1984), pennywort (Hydrocotyleumbellata) (Dierberg et al. 1987), duckweed (*Lemna minor*) (Mo et al. 1989), and water velvet (*Azolla pinnata*) (Jain et al. 1989) have been utilized for water purification. However, because of their small size and small, slow-growing roots, the efficiency of metal removal by these plants is low. Further, the high water content of aquatic plants also complicates their drying, composting, and incineration. In contrast, terrestrial plants develop much longer, fibrous root systems covered with root hairs, which create an extremely high surface area (Dittmer 1937). These roots are easily dried in the open air. Terrestrial plants like sunflower, Indian mustard and various grasses effectively removed toxic metals such as Cu^{2+}, Cd^{2+}, Cr^{6+}, Ni^{2+}, Pb^{2+}, and Zn^{2+} from aqueous solutions. Roots of *B. juncea* concentrated these metals 131–563 fold (on a dry weight basis) above initial solution concentrations (Dushenkov et al. 1995). Dushenkov and Kapulnik (2000) describe the characteristics of the ideal plant for rhizofiltration. The characteristics they suggested were: the plants should be able to accumulate and tolerate significant amounts of the target metals; plants should produce significant amounts of root biomass or root surface area; handling should be easy; maintenance cost should be low and it should generate a minimum amount of secondary waste requiring disposal.

Rhizofiltration may be applicable for the treatment of surface water and ground water, domestic and industrial effluents, down washes from power lines, storm waters, runoff water from agriculture fields and radio nuclide contaminated solutions. The technique of rhizofiltration has some advantages, which is that both the terrestrial and aquatic plants can be used for *in situ* or *ex situ* treatment. Among them, terrestrial plants are preferred for rhizofiltration because these plants have fibrous and much longer root system which increases the surface area (Raskin and Ensley 2000). Another advantage is that since the pollutants need not be translocated to the shoot portion, the plant species other than hyperaccumulators (non hyperaccumulators) may be used after harvesting for conversion to biofuel briquette which acts as a substitute for fossil fuel. This process is relatively inexpensive with low capital costs.

4.4.2 Phytovolatilization

Phytovolatilization is defined as the technique by which the plants transform/volatilize the contaminants into volatile forms such as mercury- or arsenic-containing compounds and transpire them into the atmosphere (USEPA 2000). In this process, the plants take up toxic contaminants in water soluble form, which get translocated from the roots to the leaves along with water through plant's vascular system. The plant would then convert them to less toxic forms and release them into the atmosphere through plant transpiration mechanism along with water. This technique is relevant in the remediation of soils contaminated with metalloids like Se, Hg and As. Among these metalloids, phytovolatilization of Se has been given the most attention to date (Terry et al. 1992, McGrath 1998), as this element is a serious problem in many parts of the world where there are areas of Se-rich soil (Brooks 1998). The release of volatile Se compounds (Dimethyl selenide) from higher plants was first reported by Lewis et al. (1966). The members of Brassicaceae (Cruciferae) are capable of releasing up to 40 g Se ha^{-1} day^{-1} as various gaseous compounds (Terry et al. 1992) and some aquatic plants, such as cattail (*Typhalatifolia* L.), are also good for Se phytoremediation (Pilon-Smits et al. 1999). Volatile Se compounds, such as dimethyl selenide, are 1/600 to 1/500 as toxic as inorganic forms of Se found in the soil (De Souza et al. 2000). Similarly, the mercury ion is transformed into less toxic elemental mercury and lost to the atmosphere. There has been a considerable effort in recent years to insert bacterial Hg ion reductase genes into plants for the purpose of Hg phytovolatilization (Heaton et al. 1998, Rugh et al. 1998, Bizily et al. 1999). Unlike plants that are being used for Se volatilization, those which volatilize Hg are genetically modified organisms. *Arabidopsis thaliana* L. and tobacco (*Nicotianatabacum* L.) have been genetically modified with bacterial organomercuriallyase (*MerB*) and mercuric reductase (*MerA*) genes. These plants absorb elemental Hg(II) and methyl mercury (MeHg) from the soil and release volatile Hg(O) from the leaves into the atmosphere (Heaton et al. 1998). The phytovolatilization of Se and Hg into the atmosphere has several advantages. The volatilization of Se and Hg is a permanent solution for the contaminated site because the inorganic forms of these elements are removed and the gaseous species are also not likely to be redeposited at or near the site (Atkinson et al. 1990). Furthermore, sites that utilize this technology may not require much management after the original planting. Hence, this remediation method has the added benefits of minimal site disturbance, less erosion, and no need to dispose of contaminated plant material (Rugh et al. 2000). Unlike other remediation techniques, once contaminants have been removed via volatilization, there is a loss of control over their migration to other areas. However, its use is limited by the fact that it does not remove the pollutant completely; only it is transferred from one segment (soil) to another (atmosphere) from where it can be redeposited. Hence, phytovolatilization technique would not be a better option for remediation of contaminated sites near population centers or at places with unique meteorological conditions which promote the rapid deposition of volatile compounds (Heaton et al. 1998, Rugh et al. 2000). Therefore, this technique is the most controversial of all the phytoremediation technologies (Padmavathiamma and Li 2007). Despite the controversy surrounding phytovolatilization, this technique is a promising tool for the remediation of Se and Hg contaminated soils and also for the removal of organic contaminants.

4.4.3 Phytostabilization

Phytostabilization, also known as phytorestoration, is another subset of phytoremediation technology which aims at reducing the mobility of contaminants within the vadose zone through accumulation by roots or immobilization within the rhizosphere by establishing a plant cover on the surface of the contaminated sites (Bolan et al. 2011). It creates a vegetative cap for the long-term stabilization and containment of the contaminant. The plant canopy serves to reduce eolian dispersion, whereas plant roots prevent water erosion, immobilize metals by adsorption or accumulation, and provide a rhizosphere wherein metals precipitate and stabilize. The mobility of contaminants is reduced by the accumulation of contaminants by plant roots, absorption onto roots, or precipitation within the root zone. The immobilization of metals will be accomplished by decreasing wind-blown dirt, minimizing soil erosion, and reducing contaminant solubility or bioavailability to the food chain. It provides hydraulic control, which suppresses the vertical migration of contaminants into the groundwater, and physically and chemically immobilizes contaminants by root sorption and by chemical fixation with various soil amendments (Berti and Cunningham 2000, Flathman and Lanza 1998, Salt 1995, Schnoor 2000). This technique of phytostabilization is particularly attractive when other methods to clean areas are not feasible. Unlike phytoextraction, or hyperaccumulation of metals into shoot/root tissues (Ernst 2005), phytostabilization primarily focuses on sequestration of the metals within the rhizosphere but not in plant tissues. Consequently, metals become less bioavailable and livestock, wildlife, and human exposure is reduced (Cunningham et al. 1995, Wong 2003). However, since the contaminants are left in place, the site requires regular monitoring to ensure that the optimal stabilizing conditions are maintained.

Plants suitable for phytostabilization

Plants chosen for phytostabilization ideally should have the following characteristics: they must be metallophytes (metal-tolerant plants) but excluders, i.e., the metal tolerant plants that do not accumulate or limit metal accumulation to root tissues. The lack of appreciable metals in shoot tissue also eliminates the necessity of treating harvested shoot residue as hazardous waste (Flathman and Lanza 1998). Also, the plants should be native to the area, have high production of root biomass with the ability to immobilize contaminants and the ability to hold contaminants in the roots, should have the ability to tolerate soil conditions, their establishment and maintenance under field conditions should be easy and should have the ability to self-propagate, and they should have rapid growth to provide adequate ground coverage. Ideally, the plants that are suitable for phytostabilization should have bioconcentration factor (BCF) or accumulation factor (AF) (total element concentration in shoot tissue/total element concentration in mine tailings) and translocation factor (TF) or shoot:root (S:R) ratio (total element concentration in shoot tissue/total element concentration in the root tissue) should be < 1 (Brooks 1998). In addition to the metal accumulation ratios, several metal concentration guidelines, like soil plant toxicity levels, which can provide a guide in evaluating metal tolerance and the plant leaf tissue toxicity limits, which can help assess the long-term potential for plant establishment (Munshower 1994, Mulvey and Elliott 2000, Kataba-Pendias and Pendias 2001), can be used to evaluate metal toxicity issues that may arise during phytostabilization.

Phytostabilization is useful at sites with shallow contamination and where contamination is relatively low. The efficiency of phytostabilization depends on the plant and soil amendment used. Enzymes and proteins secreted by plant roots into adjacent soil results in immobilization and precipitation of the contaminants in soil or on root surface. Phytostabilization can be enhanced by using soil amendments that are effective in the immobilization of metal(loid)s, but may need to be periodically reapplied to maintain their effectiveness. Soil amendments used in phytostabilization techniques favors inactivate heavy metals, which prevents plant metal uptake (Marques et al. 2009). The best soil amendments are those that are easy to handle, safe to workers who apply them, easy to produce, inexpensive and most importantly are not toxic to plants (Marques et al. 2008). Most of the times, organic amendments are used because of their low cost and the other benefits they provide such as provision of nutrients for plant growth and improvement of soil physical properties (Marques et al. 2009). Marques et al. (2008) showed that Zn percolation through the soil reduced by 80% after application of manure or compost to polluted soils on which *Solanumnigrum* was grown. Other amendments that can be used for phytostabilization include phosphates, lime, biosolids, and litter (Adriano et al. 2004). Plants that accumulate heavy metals in the roots and in the root zone typically are effective at depths of up to 24 inches (Blaylock et al. 1995). It does not produce secondary waste that needs treatment and is best suited at sites having fine-textured soils with high organic matter content (Berti and Cunningham 2000). However, phytostabilization is not a permanent solution because the heavy metals remain in the soil; only their movement is limited. Actually, it is a management strategy for stabilizing (inactivating) potentially toxic contaminants (Vangronsveld et al. 2009). Though phytoremediation has shown promising results as an innovative clean-up technology, it is still at an infant stage. Intensive pilot scale research work is needed to manage post-harvest stages of this remediation technology.

5. Conclusion

Soil pollution is one of the most sensitive environmental issues today. According to numerous studies, pollution in agricultural soils has become a growing concern in most of the developed and developing nations due to enhanced industrialization and urbanization. The primary concern over soil pollution stems primarily from health risks, from direct contact with the contaminated soil, vapors from the contaminants, and from secondary contamination of water supplies within and underlying the soil. In the last decades, large scientific progress was achieved in the field of remediation of heavy metal contaminated sites. Conventional clean-up technologies are costly and feasible only for small but heavily polluted sites where rapid and complete decontamination is required. On the other hand, phytoremediation using plants and associated microorganisms is gaining more and more attention to stabilize, remove, or degrade contaminants and to rejuvenate slightly to moderately contaminated sites. However, the efficacy and economics depend on several factors, like concentration and type of the contaminant, plant type, soil chemistry, contamination depth or the time frame required for remediation. Though there has been ample evidence at national and international both at laboratory and field scale level on phytoremediation techniques' success to regenerate contaminated soil, still more fundamental research is needed to better exploit the metabolic diversity of the plants themselves, and also to better understand the complex interactions between contaminants, soil, plant roots, and microorganisms in the rhizosphere. Further, an integrated approach using chemical and biological remediation technologies are needed to remediate heavy metal contaminated soils so as to protect our precious natural resources for our future generations.

References

Abhilash, P. and Yunus, M. 2011. Can we use biomass produced from phytoremediation? Biomass Bioenergy 35: 1371–1372.
Abrahams, P.W. 2002. Soils: Their implications to human health. Sci. Total Environ. 291: 1–32.
Adriano, D.C., Wenzel, W.W., Vangronsveld, J. and Bolan, N.S. 2004. Role of assisted natural remediation in environmental cleanup. Geoderma 122: 121–142.
Anderson, C.W.N., Moreno, F. and Meech, J. 2005. A field demonstration of gold phytoextraction technology. Miner. Eng. 18: 385–392.
Arora, N.K., Khare, E., Singh, S. and Maheshwari, D.K. 2010. Effect of Al and heavy metals on enzymes of nitrogen metabolism of fast and slow growing rhizobia under explanta conditions. World J. Microbiol. Biotechnol. 26: 811–816.
Assuncao, A.G.L., Schat, H. and Aarts, M.G.M. 2003. Thlaspicaerulescens, an attractive model species to study heavy metal hyperaccumulation in plants. New Phytol. 159: 351–360.
Badr, N., Fawzy, M. and Al-Qahtani, K.M. 2012. Phytoremediation: an economical solution to heavy-metal-polluted soil and evaluation of plant removal ability. World Appl. Sci. J. 16: 1292–1301.
Baker, A.J.M. and Brooks, R.R. 1989. Terrestrial higher plants which hyperaccumulate metallic elements. A review of their distribution, ecology and phytochemistry. Biorecovery 1: 81–126.
Baker, A.J.M. and Walker, P.L. 1990. Ecophysiology of metal uptake by tolerant plants. pp. 155–177. In: Shaw, A.J. (ed.). Heavy Metal Tolerance in Plants: Evolutionary Aspects. CRC Press, Boca Raton.
Barceló, J. and Poschenrieder, C. 2003. Phytoremediation: Principles and perspectives. Contrib. Sci. 2: 333–344.
Beans, C. 2017. Core concept: Phytoremediation advances in the lab but lags in the field. Proc. Natl. Acad. Sci. USA 114(29): 7475–7477.
Berti, W.R. and Cunningham, S.D. 2000. Phytostabilization of metals. pp. 71–88. In: Raskin, I. and Ensley, B.D. (eds.). Phytoremediation of Toxic Metals: Using Plants to Clean-Up the Environment. New York, John Wiley & Sons, Inc.
Bianucci, E., Fabra, A. and Castro, S. 2011. Cadmium accumulation and tolerance in Bradyrhizobium spp. (Peanut microsymbionts). Curr. Microbiol. 62: 96–100.
Blaylock, M., Ensley, B., Salt, D., Kumar, N., Dushenkov, V. and Raskin, I. 1995. Phytoremediation: A novel strategy for the removal of toxic metals from the environment using plants. Biotechnol. 13(7): 468–474.
Bolan, N.S., Park, J.H., Robinson, B., Ravi Naidu and Huh, K.Y. 2011. Phytostabilization: A green approach to contaminant containment. Adv. Agron. 112: 145–204.
Bondarenko, O., Rahman, P.K.S.M., Rahman, T.J., Kahru, A. and Ivask, A. 2010. Effects of rhamnolipids from *Pseudomonas aeruginosa* DS10-129 on luminescent bacteria: toxicity and modulation of cadmium bioavailability. Microb. Ecol. 59: 588–600.
Boularbah, A., Schwartz, C., Bitton, G., Aboudrar, W., Ouhammou, A. and Morel, J.L. 2006. Heavy metal contamination from mining sites in South Morocco: 2. Assessment of metal accumulation and toxicity in plants. Chemosphere 63: 811–817.
Brooks, R.R., Lee, J., Reeves, R.D. and Jaffrre, T. 1977. Detection of nickeliferous rocks by analysis of herbarium specimens of indicator plants. J. Geochem. Explor. 7: 49–57.
Brooks, R.R. 1998. Plants that Hyperaccumulate Heavy Metals: Their Role in Phytoremediation, Microbiology, Archaeology, Mineral Exploration and Phytomining. Wallingford, UK: CAB International.

Chaney, R.L., Li, Y.-M., Angle, J.S., Baker, A.J.M., Reeves, R.D., Brown, S.L., Homer, F.A., Malik, M. and Chin, M. 1999. Improving metal-hyperaccumulators wild plants to develop commercial phytoextraction systems: Approaches and progress. pp. 129–158. *In*: Terry, N. and Bañuelos, G.S. (eds.). Phytoremediation of Contaminated Soil and Water. CRC Press, Boca Raton, FL.

Chaney, R.L., Angle, J.S., Broadhurst, C.L., Peters, C.A., Tappero, R.V. and Sparks, D.L. 2007. Improved understanding of hyperaccumulation yields commercial phytoextraction and phytomining technologies. J. Environ. Qual. 36: 1429–1443.

CPCB. 2000. Status of Municipal Solid Waste Generation, Collection Treatment, and Disposable in Class 1 Cities, Central Pollution Control Board, Ministry of Environmental and Forests, Governments of India, New Delhi.

Cunningham, S.D., Berti, W.R. and Huang, J.W. 1995. Phytoremediation of contaminated soils. Trends Biotechnol. 13: 393–397.

Daniel, H. and Perinaz, B.T. 2012. What a waste: a global review of solid waste management. Urban development series knowledge papers. World Bank 15: 1–98.

Deka, G. and Bhattacharyya, K.G. 2009. Enrichment of Cr, Mn, Ni and Zn in surface soil. Proceedings of International Conference on Energy and Environment, March 19–21, ISSN: 2070-3740. 301–303.

Desouza, M.P., Pilon-Smits, E.A.H. and Terry, N. 2000. The physiology and biochemistry of selenium volatilization by plants. pp. 171–190. *In*: Raskin, I. and Ensley, B.D. (eds.). Phytoremediation of Toxic Metals: Using Plants to Clean-Up the Environment. New York, John Wiley & Sons, Inc.

Dierberg, F.E., Débuts, T.A. and Goulet, J.R.N.A. 1987. Removal of copper and lead using a thin-film technique. pp. 497–504. *In*: Reddy, K.R. and Smith, W.H. (eds.). Aquatic Plants for Water Treatment and Resource Recovery. Magnolia.

Dipu, S., Kumar, A.A. and Thanga, S.G. 2012. Effect of chelating agents in phytoremediation of heavy metals. Remediation J. 22: 133–146.

Dushenkov, S. and Kapulnik, Y. 2000. Phytofilitration of metals. pp. 89–106. *In*: Raskin, I. and Ensley, B.D. (eds.). Phytoremediation of Toxic Metals—Using Plants to Clean-Up the Environment. New York, John Wiley & Sons, Inc.

Dushenkov, V., Kumar, P.B.A.N., Motto, H. and Raskin, I. 1995. Rhizofiltration: the use of plants to remove heavy metals from aqueous streams. Environ. Sci. Technol. 29: 1239–1245.

Ebbs, S., Lasat, M., Mitch, B.D., Cornish, J., Jay, G., Gordon, R. and Kochian, L. 1997. Phytoextraction of cadmium and zinc from a contaminated soil. J. Environ. Qual. 26: 1424–1430.

Ernst, W.H.O. 2005. Phytoextraction of mine wastes—options and impossibilities. ChemErde-Geochem. 65: 29–42.

Flathman, P.E. and Lanza, G.R. 1998. Phytoremediation: current views on an emerging green technology. J. Soil Contam. 7: 415–432.

Gao, Z.M. 1986. Ecological Research on the Pollution of Soil-Plant System. China Science and Technology Press, Beijing, China.

Ghosh, M. and Singh, S.P. 2005. A review on phytoremediation of heavy metals and utilization of its by-products. Appl. Ecol. Environ. Res. 3: 1–18.

Heaton, A.C.P., Rugh, C.L., Wang, N. and Meagher, R.B. 1998. Phytoremediation of mercury- and methylmercury-polluted soils using genetically engineered plants. J. Soil Cont. 7(4): 497–510.

Jadia, C.D. and Fulekar, M.H. 2008. Phytotoxicity and remediation of heavy metals by fibrous root grass (sorghum). J. Appl. Biosci. (10): 491–499.

Jaffre, T., Brooks, R.R., Lee, J. and Reeves, R.D. 1976. *Sebertiaacuminata*: A hyperaccumulator of nickel from New Caledonia. Science 193(4253): 579–580.

Jain, S.K., Vasudevan, P. and Jha, N.K. 1989. Removal of some heavy metals from polluted water by some aquatic plants: Studies on duck weed and water velvet. Biol. Wastes. 28: 115–126.

Järup, L. 2003. Hazards of heavy metal contamination. Br. Med. Bull. 68(1): 167–182.

Kataba-Pendias, A. and Pendias, H. 2001. Trace Elements in Soils and Plants. Boca Raton, FL: CRC Press.

Kay, S.H., Haller, W.T. and Garrard, L.A. 1984. Effect of heavy metals on water hyacinth [*Eichhorniacrassipes* (Mart.) Solms]. Aquat. Toxicol. 5: 117–128.

Koppolu, L. and Clements, L.D. 2003a. Pyrolysis as a technique for separating heavy metals from hyperaccumulators. Part I: Preparation of synthetic hyperaccumulator biomass. Biomass Bioenergy 24: 69–79 (doi:10.1016/S0961-9534(02)00074-0).

Koppolu, L., Agblevor, F.A. and Clements, L.D. 2003b. Pyrolysis as a technique for separating heavy metals from hyperaccumulators. Part II: Lab-scale pyrolysis of synthetic hyperaccumulator biomass. Biomass Bioenergy 25: 651–663.

Koppolu, L., Prasad, R. and Clements, L.D. 2004. Pyrolysis as a technique for separating heavy metals from hyperaccumulators. Part III: Pilot-scale pyrolysis of synthetic hyperaccumulator biomass. Biomass Bioenergy 26: 463–472.

Kvesitadze, G., Khatisashvili, G., Sadunishvili, T. and Ramsden, J. 2006. Biochemical Mechanisms of Detoxification in Higher Plants. Springer, Germany.

Ladislas, S., El-Mufleh, A., Gerente, C., Chazarenc, F., Andres, Y. and Bechet, B. 2012. Potential of aquatic macrophytes as bioindicators of heavy metal pollution in urban stormwater runoff. Water Air Soil Pollut. 223: 877–888.

Lasat, M.M. 2002. Phytoextraction of toxic metals: a review of biological mechanisms. J. Environ. Qual. 31: 109–120.

Lehmann, J. 2007. Bioenergy in the black. Front. Ecol. Environ. 5(7): 381–387.

Leiros, M.C., Trasar-Cepeda, C., Garcia-Fernandez, F. and Gil-Sotre, S.F. 1999. Defining the validity of a biochemical index of soil quality. Biol. Fertil. Soils 30(1-2): 140–146.

Lewis, B.G., Johnson, C.M. and Delwiche, C.C. 1966. Release of volatile selenium compounds by plants: collection procedures and preliminary observations. J. Agric. Food Chem. 14: 638–640.

Lintern, M., Anand, R., Ryan, C. and Paterson, D. 2013. Natural gold particles in Eucalyptus leaves and their relevance to exploration for buried gold deposits. Nature Communications.

Loska, K. and Wiechuła, D. 2000. Effects of pH and aeration on copper migration in above-sediment water. Pol. J. Environ. St. 9(5): 433–437.

Macek, T., Kotrba, P., Svatos, A., Novakova, M., Demnerova, K. and Mackova, M. 2008. Novel roles for genetically modified plants in environmental protection. Trends Biotechnol. 26: 146–52.

Marion, D., Serge, C., Florence, D.G., Benoit, S.R.G., Brett, R. and Valérie, B. 2013. The fate of metals and viability of the processes. Biomass Bioenergy 49: 160–170.

Marques, A.P.G.C., Oliveira, R.S., Rangel, A.O.S.S. and Castro, P.M.L. 2008. Application of manure and compost to contaminated soils and its effect on zinc accumulation by *Solanumnigrum* inoculated with arbuscularmycorrhizal fungi. Environ. Pollut. 151(3): 608–620.

Marques, A.P.G.C., Rangel, A.O.S.S. and Castro, P.M.L. 2009. Remediation of heavy metal contaminated soils: phytoremediation as a potentially promising clean-up technology. Crit. Rev. Environ. Sci. Technol. 39(8): 622–654.

McGrath, S.P. 1998. Phytoextraction for soil remediation. pp. 261–288. *In*: Brooks, R.R. (ed.). Plants that Hyperaccumulate Heavy Metals: Their Role in Phytoremediation, Microbiology, Archaeology, Mineral Exploration and Phytomining. New York, CAB International.

Meers, E., Hopgood, M., Lesage, E., Vervaeke, P., Tack, F.M.G. and Verloo, M. 2004. Enhanced phytoextraction. In Search for EDTA Alternatives. Int. J. Phytorem. 6(2): 95–109.

Meharg, A.A. 2005. Mechanisms of plant resistance to metal and metalloid ions and potential biotechnological applications. Plant Soil 274: 163–174.

Mertens, J., Luyssaert, S. and Verheyen, K. 2005. Use and abuse of trace metal concentrations in plants tissue for biomonitoring and phytoextraction. Environ. Pollut. 138: 1–4.

Mirck, J., Isebrands, J.G., Verwijst, T. and Ledin, S. 2005. Development of short-rotation willow coppice systems for environmental purposes in Sweden. Biomass Bioenergy 28: 219–228.

Mo, S.C., Choi, D.S. and Robinson, J.W. 1989. Uptake of mercury from aqueous solution by duckweed: the effects of pH, copper and humic acid. J. Environ. Sci. Health A24: 135–146.

Mulvey, P.J. and Elliott, G.L. 2000. Toxicities in soils. pp. 252–257. *In*: Charman, P.E.V. and Murphy, B.W. (eds.). Soils: Their Properties and Management. South Melbourne, Australia: Oxford University Press.

NAAS. 2010. Degraded and Wastelands of India—Status of Spatial Distribution. National Academy of Agricultural Sciences, New Delhi.

Nagajyoti, P.C., Lee, K.D. and Sreekanth, T.V.M. 2010. Heavy metals, occurrence and toxicity for plants: A review. Environ. Chem. Lett. 8(3): 199–216.

Padmavathiamma, P.K. and Li, L.Y. 2007. Phytoremediation technology: hyper-accumulation metals in plants. Water Air Soil Pollut. 184: 105–126.

Pilon-Smits, E.A.H., Desouza, M.P., Hong, G., Amini, A., Bravo, R.C., Paybyab, S.T. and Terry, N. 1999. Selenium volatilization and accumulation by twenty aquatic plant species. J. Environ. Qual. 28(3): 1011–1017.

Poniedziałek, M., Sękara, A., Jędrszczyk, E. and Ciura, J. 2010. Phytoremediation efficiency of crop plants in removing cadmium, lead and zinc from soil. Folia Hortic. Ann. 22: 25–31.

Prasad, M.N.V. 2005. Nickelophilous plants and their significance in phytotechnologies. Braz. J. Plant Physiol. 17: 113–128.

Raskin, I. 1996. Plant genetic engineering may help with environmental cleanup (commentary). Proc. Natl. Acad. Sci. USA 93: 3164–3166.

Raskin, I. and Ensley, B.D. 2000. Uptake of mercury from aqueous solution by duckweed: The effects of pH, copper and humic acid. p 352. *In*. Raskin, I. and Ensley, B.D. (eds.). Phytoremediation of Toxic Metals: Using Plants to Clean Up the Environment. New York, John Wiley and Sons. ISBN0-47-119254-6.

Reeves, R.D. 1992. Hyperaccumulation of nickel by serpentine plants. pp. 253–277. *In*: Baker, A.J.M., Proctor, J. and Reeves, R.D. (eds.). The Vegetation of Ultramafic (serpentine) Soils. Intercept, Andover UK.

Rugh, C.L., Gragson, G.M., Meaguer, R.B. and Merkle, S.A. 1998. Toxic mercury reduction and remediation using transgenic plants with a modified bacterial gene. Hortscience 33(4): 618–621.

Rugh, C.L., Bizily, S.P. and Meagher, R.B. 2000. Phytoreduction of environmental mercury pollution. pp. 151–170. *In*: Raskin, I. and Ensley, B.D. (eds.). Phytoremediation of Toxic Metals: Using Plants to Clean-Up the Environment. New York, John Wiley and Sons.

Sachan, S., Singh, S.K. and Srivastava, P.C. 2007. Buildup of heavy metals in soil water plant continuum as influenced by irrigation with contaminated effluent. J. Environ. Sci. Eng. 49: 293–296.

Sakakibara, M., Ohmori, Y., Ha, N.T.H., Sano, S. and Sera, K. 2011. Phytoremediation of heavy metal contaminated water and sediment by *Eleocharisacicularis*. Clean: Soil Air Water. 39: 735–741.

Salt, D.E., Blaylock, M., Kumar, N.B.P.A., Dushenkov, V., Ensley, B.D., Chet, I. and Raskin, I. 1995. Phytoremediation: a novel strategy for the removal of toxic metals from the environment using plants. Biotechnol. 13: 468–474.

Sas-Nowosielska, A., Kucharski,R., Małkowski, E., Pogrzeba, M., Kuperberg, J.M. and Kryński, K. 2004. Phytoextraction crop disposal—an unsolved problem. Environ. Pollut. 128: 373–379.

Schnoor, J.L, Licht, L.A., McCutcheon, S.C., Wolfe, N.L. and Carreira, LH. 1995. Phytoremediation of organic and nutrient contaminants. Environ. Sci. Technol. 29(7): 318A–23A.

Schnoor, J.L. 2000. Phytostabilization of metals using hybrid poplar trees. pp. 133–150. *In*: Raskin, I. and Ensley, B.D. (eds.). Phytoremediation of Toxic Metals: Using Plants to Clean-Up the Environment. New York: Wiley.

Schnoor, J.L. 2002. Technology Evaluation Report: Phytoremediation of Soil and Groundwater. GWRTAC Series TE-02-01.

Shanker, A.K., Cervantes, C., Loza-Tavera, H. and Avudainayagam, S. 2005. Chromium toxicity in plants. Environ. Int. 31: 739–753.

Tangahu, B.V., Abdullah, S.R.S., Basri, H., Idris, M., Anuar, N. and Mukhlisin, M. 2011. A review on heavy metals (As, Pb, and Hg) uptake by plants through phytoremediation. Int. J. Chem. Eng. vol. 2011, Article ID 939161, 31 pages, 2011. https://doi.org/10.1155/2011/939161.

Terry, N., Carlson, C., Raab, T.K. and Zayed, A. 1992. Rates of selenium volatilization among crop species. J. Environ. Qual. 21: 341–344.

Turnquist, T.D., Urig, B.M. and Hardy, J.K.J. 1990. Nickel uptake by the water hyacinth. Environ. Sci. Health A25: 897–912.

United States Protection Agency (USPA). 2000. Introduction to Phytoremediation. EPA 600/R99/107. U.S. Environmental Protection Agency, Office of Research and Development, Cincinnati, OH.

van der Ent, A., Baker, A.J.M., Reeves, R.D., Pollard, A.J. and Schat, H. 2013. Hyperaccumulators of metal and metalloid trace elements: facts and fiction. Plant Soil. 362: 319–334.

Vangronsveld, J., Herzig, R., Weyens, N., Boulet, J., Adriaensen, K., Ruttens, A., Thewys, A., Vassilev, A., Meers, E., Nehnevajova, E., Van der Lelie, D. and Mench, M. 2009. Phytoremediation of contaminated soils and groundwater: lessons from the field. Environ. Sci. Pollut. Res. 16: 765–794.

Willscher, S., Mirgorodsky, D., Jablonski, L., Ollivier, D., Merten, D., Büchel, G. and Werner, P. 2013. Field scale phytoremediation experiments on a heavy metal and uranium contaminated site, and further utilization of the plant residues. Hydrometallurgy 131–132: 46–53.

Wong, M.H. 2003. Ecological restoration of mine degraded soils, with emphasis on metal contaminated soils. Chemosphere 50: 775–780.

Wyszkowska, J. and Wyszkowski, M. 2003. Effect of cadmium and magnesium on enzymatic activity in soil. Pol. J. Environ. St. 12(4): 473–479.

Zhao, F.J., Lombi, E. and McGrath, S.P. 2003. Assessing the potential for Zn and Cd phytoremediation with hyperaccumulator *Thlaspicaerulescens*. Plant Soil 249: 37–43.

Zhuang, P., Yang, Q., Wang, H. and Shu, W. 2007. Phytoextraction of heavy metals by eight plant species in the field. Water Air Soil Pollut. 184: 235–242.

6
Soil Heavy Metal Pollution and its Bioremediation
An Overview

Swagata Mukhopadhyay, Swetha R.K. and *Somsubhra Chakraborty**

1. Introduction

The development of civilization gave birth to industrialization and now it is causing pollution all over the world. Continuous discharging of the effluents to the soil and water without any protection has led to the accumulation of the heavy metals in the soil and water especially in the industrially developed and populated areas. The soil minerals and organic matter tightly bind these heavy metals, which helps them to sustain in the environment for a long time, and their effects could be long-lasting. Among them, some could be toxic at a very low concentration, e.g., arsenic, cadmium, chromium, copper, lead, mercury, nickel, selenium, silver, zinc, etc. (Salem et al. 2000). These heavy metals are not only cytotoxic but also carcinogenic and mutagenic. Many of them are micronutrients and are essential for plant life but when the concentration of the heavy metal increases in the soil, it could be toxic for the plants and there could be a chance of them entering the food chains.

At some point, the environment needs rectification for human welfare. Some techniques are known to us including chemical precipitation, oxidation and reduction, filtration, ion-exchange, reverse osmosis, membrane technology, evaporation, and electrochemical treatments but they are not popular due to some reasons. First, these techniques are expensive and not easily applicable in the field condition. Another important reason is that these techniques become useless when the concentration of heavy metals goes below 100 mg/L (Ahluwalia and Goyal 2007). Salts of heavy metals are water-soluble and the physical separation of them is difficult (Hussein et al. 2004). In this condition, only the bioremediation process can show satisfactory outcomes. Bioremediation is a process where microorganisms are intentionally introduced to the environment to adsorb and breakdown the pollutants in order to clean the polluted sites and it is completely eco-friendly and economically feasible. The microorganisms mainly use bio-adsorption and bio-accumulation methods to remove the heavy metals from the contaminated site with high efficiency (Kapoor and Viraraghvan 1995). Though the exact mechanisms of these methods are still unknown, they are known to be the most sustainable techniques ever.

Agricultural and Food Engineering Department, IIT Kharagpur, Kharagpur-721302, West Bengal, India.
* Corresponding author: somsubhra@agfe.iitkgp.ac.in

Bioaccumulation is the process where microorganisms spend energy and for the supply of the energy, oxygen is badly required which automatically increases the biological or chemical oxygen demand in the polluted soils. So, the bioaccumulation process could be considered as the active process. On the other hand, bio-adsorption process is a passive process that does not require any metabolic energy, only the structure of the cell wall is important. The microbial cell wall consists of polysaccharides, proteins and lipids and several functional groups like phosphate and amino groups, carboxyl and hydroxyl groups. Another important thing is the maintenance of the healthy population of the microorganisms in the toxic environment. Some fungi and plant species are also capable of extracting heavy metals and their capabilities and potentials are going through different experiments. Nowadays, new approaches like biotechnology, designer plant approach, and rhizosphere modifications are extensively studied to improve the bioremediation technique.

2. Sources of heavy metals in the environment

Heavy metals are naturally present in the environment through the weathering of rocks, volcanic eruption, erosion of soils, etc. But the contamination depends on the intensity of anthropogenic activity. Different human activities like mining, disposal of by-products from different smelting, electroplating, fertilizers and pesticide industries as well as deposition of biosolids and atmospheric deposition intensify the contamination of the heavy metals in the soil environment (Wuana and Okieimen 2011, Sumner et al. 2000) (Table 1). These ever-increasing human activities are responsible for heavy metal pollution which occurs quickly than the geochemical cycle (D'Amore et al. 2005). A mass balance equation was predicted by (Lombi and Gerzabek 1998):

$$M_{total} = (Mp + Ma + Mf + Mag + Mow + Mip) - (Mcr + Ml + Mv + Mop) \tag{1}$$

M = heavy metals, p = parent materials, a = atmospheric deposition, f = fertilizer source, ag = agrochemical source, ow = organic waste, ip = inorganic pollutant, cr = crop removal, l = leaching, v = volatilization, op = other processes.

3. Status of heavy metal pollution

The metals whose atomic weight is more than iron (Fe 55.8 g mol^{-1}) or density is more than 5 g cm^{-3} are called heavy metals. Some metals with atomic weight lower than Fe (e.g., Cr) and some metalloids (As, Se) are also considered as heavy metals. Among them, some play an important role as micronutrients in the plant body (Fe, Cu, Mn, Mo, Zn, and Ni) and others may have dangerous effects on the ecosystem (Costa et al. 2006). Heavy metals are electrostatically attracted to the binding sites of micronutrients in different cellular structures, which is responsible for the distortion of the cellular structures, mutagenesis and genetic disorder (Perpetuo et al. 2011, Rajendran et al. 2003).

Not all the chemical forms of heavy metals are harmful to the environment. The various states of the chemical forms depend on the pH of the medium and the composition. Bioremediation techniques can be successfully used to biotransform heavy metals (Sutherland et al. 2000). The liquid phase has more adverse consequences than the solid phase as it can easily become a part of the nutrient cycle and can enter the plant body as well as the food chain causing bioaccumulation. After entering the food chain, the toxins pass from one trophic level to the next trophic level which increases the possibilities of biomagnification. Living tissues cannot metabolize the contaminants but these toxins can be stored in the tissues as they may be fat-soluble.

The term microorganism is mostly applied for bacteria, fungi, algae and yeast and sometimes also for other organisms. The relative proportion of the microorganisms in the soil depends upon types of soil, temperature and climate, pH of the soil, amount of water availability and some other reasons. But in general, bacteria and protozoa are at the top of the list regarding population followed by fungi. The proportion of algae and yeast varies.

Table 1. The sources of the heavy metals and their effect on human health.

Heavy metals	Sources		EPA regulatory limits (ppm)	Adverse health effects
	Natural	Anthropogenic		
Ag	Sulfides, in association with iron (pyrite), lead (galena), and tellurides, and with gold	Photographic industry	0.10	Skin problem, breathing problem, lung and throat irritation and stomach pain.
As	Arsenates, sulfides, sulfosalts, arsenites, arsenides, native elements, and metal alloys	Mining, metal smelting, burning of fossil, pesticides	0.01	Cancer in skin, bladders, lungs and kidney, skin pigmentation.
Ba	Sulfates, carbonates, oxides	From different industries	2.00	Breathing problems, swelling in brains, damage in kidney and nerve cells, changes in blood pressure, stomach irritation.
Cd	Mining, smelting, phosphate rocks	NiCd batteries, plating, pigments, plastics (ATSDR 1999)	5.00	Kidney, bone and lung diseases, muscles pain.
Cr	Chromium salts	Manufacturing chromic acid, chromium pigments, in leather tanning, anticorrosive products	0.10	Cancer, skin problems.
Cu	Copper salts	Fertilizers, mining, electroplating	1.3	Irritation in nose, mouth, and eyes, headaches, dizziness, nausea, stomach pain, diarrhea.
Hg	Volcanoes, forest fires, cannabar (ore) and fossil fuels such as coal and petroleum	Hydroelectric, mining, pulp, paper industries	2.00	Effects on nervous system, immune system, digestive system, kidney.
Ni	Igneous rocks	Agricultural chemicals, metal alloys, industrial waste, battery plants	0.20 (WHO permissible limit)	Chronic bronchitis, reduced lung function, and cancer of the lung, nasal sinus.
Pb	Galena	Industrial sources, deteriorated paint, and the combustion of leaded gasoline, aviation fuel	15.00	Anemia, weakness, and kidney and brain damage, affects baby's nervous system.
Se	Sedimentary rocks	Coal and oil combustion facilities, selenium refining factories, base metal smelting and refining factories, mining and milling operations, end-product manufacturers (e.g., some semiconductor manufacturers)	50.00	Dermatological effects.
Zn	Sulfides, oxides, silicates	Smelter slags and wastes, mine tailings, coal and bottom fly ash, fertilizers, wood preservatives	0.50	Nausea, vomiting, diarrhea, metallic taste, kidney and stomach damage, anemia.

4. Bioremediation

Bioremediation is possible if the biological species can be identified correctly. The organic toxic materials can be simplified into harmless chemicals by the process of immobilization, but inorganic toxins cannot be simplified. The only possible way is to convert them into a harmless form. Bioremediation is the process which completely depends upon the metabolic capability of the organisms and different organisms react differently to the heavy metals. So, the identification of the proper microbial species for the clearance of heavy metal pollution is an important task. Some of them act as an essential micronutrient for the growth and development of the microorganisms like iron. The mechanisms involved like adsorption, conversion of the redox state and formation of the insoluble forms depend upon the enzymatic activity, the population of the microorganisms and the geochemistry.

The exact cellular mechanisms about bioremediation is not understood properly, which limits its successful application. The native microorganisms should be explored for the detoxification of the environment (Garbisu et al. 2001). The capable microorganisms should be identified with their genes helping in bioremediation. The study of these genes could help us to develop genetically engineered microorganisms for bioremediation in the future.

5. Mechanisms of bioremediation

The things which are important for bioremediation are the enzymatic activity of the microorganisms and the resistance of the microorganisms towards heavy metal pollution. Many microbes can develop a defense mechanism from organic pollutants by forming hydrophobic outer cell membranes (Sikkema et al. 1995) or cells have the mechanisms of energy-driven ion/proton pump to release heavy metal cations. The enzymatic activity of the microbes is capable of dissolving or volatilizing the metals or transforming them into one redox state to another state. Some common mechanisms related to the bioremediation are discussed below.

5.1 Bioadsorption

Microbes adsorb different heavy metals in their extracellular structures without any expenditure of energy. So, this process can be called a passive process. The cell walls have extracellular polymeric substances (EPS) having effects on acid-base properties (Guiné et al. 2006) and can bind heavy metals. These substances bind metals through some mechanisms like proton exchange or micro-precipitation of the metals or electrostatic attraction (Comte et al. 2008, Fang et al. 2010) (Figure 1). *Saccharomyces cerevisiae* and *Cunnighamellaelegans* were identified as bioadsorbent and can remove Zn, Cd and other heavy metals through ion-exchange mechanisms. In order to find out the mechanisms of the EPS, this activity had been studied using the bacterial cells with EPS and without EPS (Fang et al. 2011) but the exact mechanisms in the genetic level were not understood properly. Because of that, the metabolic pathway of these heavy metals and their kinetics in the bacterial cells are not clear. This inability restricts their application in the fields and indicates that further developments are needed in scientific studies to predict their behavior (Gan et al. 2009, Haritash et al. 2009, Onwubuya et al. 2009, Carter et al. 2006, Kinya and Kimberly 1996).

5.2 Bioaccumulation

The biosorption process does not require any energy as the cell wall has a high affinity towards the heavy metals and the process continues until the equilibrium is reached between sorbate and sorbent (Das et al. 2008). But other bioabsorbtion processes require metabolic energy to accumulate the heavy metals within the cells of the microbes. The bioaccumulation process includes both processes, i.e., active process and passive process of removing heavy metals. The potential of fungi as a biocatalyst is more than the other microbes as they are eukaryotes and they can efficiently transform a more toxic compound to less toxic compound (Pinedo-Rivilla et al. 2009). The fungi

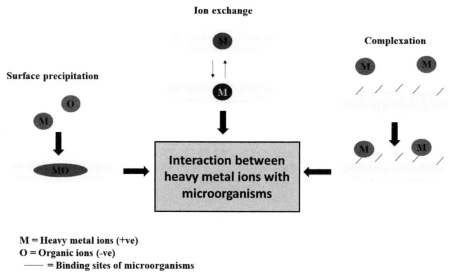

Figure 1. Mechanisms of bioadsorption.

like *Klebsiellaoxytoca*, *Allescheriella* sp., *Stachybotrys* sp., *Phlebia* sp., *Pleurotuspulmonarius*, *Botryosphaeriarhodina* have the potential to bind heavy metals. The contaminated soils with Pb (II) can be transformed to a remediated soil with the help of the biosorption process of *Aspergillus parasitica* and *Cephalosporiumaphidicola* (Tunali et al. 2006, Akar et al. 2007). *Hymenoscyphusericae* and *Neocosmosporavasinfecta* have the mechanism to biotransform Hg (II) state to a less toxic, less harmful state (Kelly et al. 2006). The contaminants are mostly hydrophobic in nature and microbes take up these metal contaminants by forming a complex with biosurfactant secreted by them, having a strong ability to form ionic bonds with metals due to low interfacial tension (Thavasi 2011).

5.3 Transformation in redox state

The bioremediation reactions also involve the aerobic and anaerobic reactions in the soil controlled by the microbes. Aerobic reactions include oxygen atoms into reactions by different enzymes like monooxygenases, dioxygenases, hydroxylases, oxidative dehalogenases, etc. Some enzymes generate reactive oxygen such as ligninases or peroxidases. In the case of anaerobic reactions, heavy metals are used as a terminal electron acceptor to supply energy for the microbes. This technique changes the redox state of the metals and sometimes reduces their mobility in the contaminated soil and it is called immobilization.

The technique sometimes allows *ex situ* application of the chemicals in that contaminated site to initiate the reactions. It is called solidification (Evanko and Dzombak 1997). The heavy metals can be leached, precipitated, chelated or methylated but never can be destroyed. This transformation process is very important to change their redox state or to change the organic form to inorganic form so that the metals can be made less toxic, water-soluble and precipitated (Garbisu et al. 2001).

The microorganisms help the heavy metal to be oxidized in the oxidative environment, and nitrates and sulfates act as terminal electron accepters. In anaerobic conditions, the organic pollutants get oxidized by the microbes and at that time heavy metals act as terminal electron accepters (Lovley et al. 1988). The higher availability of the metals will accelerate the oxidation of the organic pollutants (Lovley et al. 1996, Spormann and Widdel 2000). The processes where metals act as terminal electron accepters are called dissimilatory metal reduction (Lovely et al. 2002). *Geobaccter* species reduces uranium from U^{+6} to U^{+4} leaving it in an insoluble state (Lovley et al. 1991).

5.4 Molecular mechanisms

The genes control all biological activities. The genes involved in the bioremediation process can be identified and inserted into other microbes by the genetic engineering process to make the bioremediation process more efficient. It redesigns the microbes and increases its potential to work more accurately and specifically. Scientists are interested to find out genes that can produce different coenzymes and siderophores for binding specific metals even in the adverse condition and dilute concentration (Penny et al. 2010). *Deinococcus geothemalis* is known to reduce the Hg at high temperatures showing the expression of gene mer operon from *E. coli* (Brim et al. 2003). *Cupriavidus metallidurans* (strain MSR33), a Hg resistant microorganism, is made capable of synthesizing organomercuriallyase protein (MerB) and mercuric reductase (MerA) for Hg biodegradation after inserting pTP6 plasmids within the genes (merB and merG) (Rojas et al. 2011). The gene JM109 in *E. coli* is genetically modified to express GST-PMT and Hg^{+2} transport system simultaneously, which increases the capability of Hg accumulation in very low concentration (Chen et al. 1997).

6. Phytoremediation

Phytoremediation is a bioremediation process that is cost-effective, eco-friendly and solar energy driven *in situ* process that uses various kinds of plants associated with microorganisms to clean, stabilize and transfer the pollutants in the contaminated site. Plants can be successfully used for cleaning radionuclides and pollutants of both organic and inorganic origin (Ali et al. 2013). Phytoremediation includes different techniques like phytoextraction, phytovolatilization, phytofiltration, phytodegradation and phytostabilization (Alkorta et al. 2004). The first process among them is phytoextraction, which is uptake of the different contaminants, translocating them into shoots and storing them in the tissues of shoots (Sekara et al. 2005). The next important process is phytofiltration, which is uptake of the contaminants by different plant parts like roots (rhizofiltration), seedlings (blastofiltration) or by excising plant shoots (caulofiltration) to reduce their movement into the groundwater (Mesjasz-Przybylowicz et al. 2004). Other processes are phytostabilization and phytoimmobilization. These processes reduce the bioavailability of the heavy metals so that it can prevent the contamination in the ground water and migration to the food chain (Erakhrumen et al. 2007). The immobilization happens in the same way as sorption, precipitation, the formation of organic complex and transformation of the redox state but all these processes happen in the rhizosphere controlled by the plants and related microorganisms (Barceló et al. 2003). The organic pollutants are degraded by the root exudates through the process of phytostimulation. The contaminants inside the plants are metabolized by the enzymes (oxygenaes, dehalogenases, etc.), which is a completely different process from the immobilization process of microorganisms (Vishnoi and Srivastava 2008). The volatile metals like Hg and Se are not completely immobilized by the plants, they rather get converted into different forms (from solid to gas) and are released into the atmosphere (Karami and Shamsuddin 2010). This process is known as phytovolatilization. This process can remove the contaminants temporarily from the soils as this process only allows the transformation of the heavy metals from one medium (soil and water) to another medium (air), so they can change the medium anytime. All these processes are shown briefly in the Figure 2.

Among the plants known for phytoremediation, hyperaccumulator is very popular due to its usefulness. The criteria for hyperaccumulation vary with the types of pollutant heavy metals like 100 mg kg^{-1} for Cd and 1000 mg kg^{-1} for Cu, Co, Cr, and Pb. The shoot to soil ratio of the metal should be higher than 1 in the case of hyperaccumulators (Baker et al. 1994). The hyperaccumulators should have a high tolerance towards the heavy metal concentration in their biomass (Prasad and Freitas 2003). There are some plants known for hyperaccumulation of heavy metals like *Arabidopsis halleri* and *Solanumnigram* L. for Cd accumulation (Wei et al. 2005), *Populous deltoids* for Hg accumulation (Che et al. 2003), *Brassica juncea* and *Astragalusbisulcatus* for Se (Bitther et al. 2012), *Populuscanescens* for Zn, etc. However, hyperaccumulator plants are very few, very slow

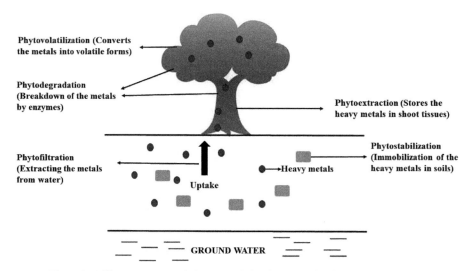

Figure 2. Different processes of phytoremediation for controlling heavy metal pollution.

growing, and low in biomass yield which restricts their application where quick remediation is needed (Xiao et al. 2010). These restrictions can be handled by applying growth-promoting rhizobacteria or arbuscular mycorrhiza (Wei et al. 2003). The root exudates containing carbohydrates, flavonoids, amino acids, etc. can accelerate the microbial activity, which produces an enzyme named ACC deaminase that reduces the of ethylene level in the soil, promoting an environment for healthier root development of the plants (Kuiper et al. 2004, Glick et al. 1998). *Kluyveraascorbate* SUD 165 is a Ni-resistant bacteria which lowers the level of ethylene and promotes the growth of *Brasicacampestris* (Burd et al. 1998).

7. Biotechnology in bioremediation

Recombinant DNA technology has been used to alter the genetic materials of microorganisms or plants to create genetically modified microorganisms or plants, more efficient and specific than the previous versions (Sayler and Ripp 2000). They are important as their ability to sustain in the adverse condition is more than the normal strain, the development of "microbial biosensor" is possible which can be used to detect the contamination accurately in a short periods of time, and many microorganisms associated with plants can increase the rate of bioremediation by increasing the rate of phytochelation and degradation of the metals (Divya et al. 2011). Genetically modified *E. coli* and *Moreaxella* sp. can accumulate 25 times more Cd and Hg than their wild type expressing a gene phytochelatin 20 on the cell surface (Bae et al. 2001, 2003). Following genetic modification, *P. fluorescens* expresses Phytochelatin synthase (PCS) and *E. coli* expresses Hg^{2+} transporter which increases removal of Ni and Hg, respectively (Zhao et al. 2005, Lopez et al. 2002, Sriprang et al. 2003). The problem is that the genetically engineered microorganisms face competition with the native microorganisms for survival (Wu et al. 2006).

The main problem with phytoremediation is the accumulation of the heavy metals and their metabolites within the tissues of the plants which is harmful to the plants and after the death of the plants again there is a chance to reenter into the atmosphere. If these plants are genetically modified with the genes of those bacteria which are capable of degrading heavy metals, then the metals could be degraded inside the plant tissues. The plants will be made capable of producing different metal chelators such as metallothineins and phytochetains, which will help the plant to uptake and accumulate more heavy metals from the soil (Ruis and Daniell 2009). High biomass yielding plants like poplar, willow, and jatropha can be used as heavy metal accumulators as well as the plants that can be used for energy production. But if these plants are burned after heavy metal accumulation, it will release the metals to the atmosphere which will transfer the problem from soil to air. Poplar trees are genetically modified to synthesize mercuric reductase and γ-gltamylsysteine, which increases the Hg, Cd and Cu accumulation and degradation inside the plant tissues, respectively, to ensure the production of healthy biomass for further application (Bittsanszkya et al. 2005, Gullner et al. 2001, Abhilash et al. 2012).

Mostly, the contaminated sites have multiple pollutants, which are very difficult for the plants to control efficiently. The rhizosphere needs some energy to fight with this situation. Biostimulation is a technique to increase the microbial activity in the rhizosphere by the addition of growth stimulants. Bioaugmentation is the addition of selected and cultured microorganisms to the rhizosphere to remove the contamination.

8. Nano-technology in bioremediation

Nanoparticles are more effective than the microorganisms as they can cover a huge area and can also reduce the processing time. The enhancement of the microbial activity by applying nanoparticles for the removal of the contaminants is called "nanobioremediation". Different biological cells are used for the preparation of the nano-particles because of their small size, cells can be easily genetically modified and they can be easily cultured under controlled conditions. Different polymers and magnetosomes are used as these macromolecules are easily converted to nanostructures. Different proteins can be used like virus-like protein (VLP) and tailored metal particles (Sarikaya et al. 2003). US Department of Energy (DOE) has taken an initiative to clean the radioactive waste by using a radioactive-resistant organism *Deinococcus radiodurans* (Brim et al. 2000, Smith et al. 1998).

9. Conclusion

Development and industrialization are synonymous. They are unstoppable and have an adverse effect on the environment. The conventional methods have failed to control the situation, and are not user-friendly. Bioremediation is the best hope left in our hands, though it has some restrictions. Several organisms are out there that cannot break the toxic contaminants successfully. Many of them cannot tolerate the adverse environmental condition. In order to improve these situations, scientists should be encouraged to find out other successful techniques like biotechnology. GEMs are more efficient than their wild type. Their outer protein membranes are modified by biotechnology so that they can adsorb more toxic metals. They are more element-specific and they are also able to break the contaminants into metabolites. Bioremediation will be the future of pollution control, but it must make sure that it is completely non-toxic to the environment.

References

Abhilash, P.C., Powell, J.R., Singh, H.B. and Singh, B.K. 2012. Plant–microbe interactions: Novel applications for exploitation in multipurpose remediation technologies. Trends Biotechnol. 30: 416–420.
Agency for Toxic Substances and Disease Registry (ATSDR). 1990. Toxicological Profile for Silver. U.S. Department of Health and Human Services, Public Health Service: Atlanta, GA, USA.
Ahluwalia, S.S. and Goyal, D. 2007. Microbial and plant derived biomass for removal of heavy metals from wastewater. Bioresour. Technol. 98: 2243–2257.
Akar, T., Tunali, S. and Cabuk, A. 2007. Study on the characterization of lead (II) biosorption by fungus *Aspergillus parasiticus*. Appl. Biochem. Biotech. 136: 389–406.
Ali, H., Khan, E. and Sajad, M.A. 2013. Phytoremediation of heavy metals—Concepts and applications. Chemosphere 91: 869–881.
Alkorta, I., Hernández-Allica, J., Becerril, J.M., Amezaga, I., Albizu, I. and Garbisu, C. 2004. Recent findings on the phytoremediation of soils contaminated with environmentally toxic heavy metals and metalloids such as zinc, cadmium, lead, and arsenic. Rev. Environ. Sci. Biotechnol. 3: 71–90.
Bae, W., Mehra, R.K., Mulchandani, A. and Chen, W. 2001. Genetic engineering of *Escherichia coli* for enhanced uptake and bioaccumulation of mercury. Appl. Environ. Microbiol. 67: 5335–5338.
Bae, W., Wu, C.H., Kostal, J., Mulchandani, A. and Chen, W. 2003. Enhanced mercury biosorption by bacterial cells with surface-displayed MerR. App. Environ. Microbiol. 69: 3176–3180.

Baker, A.J.M., McGrath, S.P., Sidoli, C.M.D. and Reeves, R.D. 1994. The possibility of *in situ* heavy metal decontamination of polluted soils using crops of metal accumulating plants. Resour. Conserv. Recycl. 11: 42–49.

Barceló, J. and Poschenrieder, C. 2003. Phytoremediation: principles and perspectives. Contrib. Sci. 2: 333–344.

Bitther, O.P., Pilon-Smits, E.A.H., Meagher, R.B. and Doty, S. 2012. Biotechnological approaches for phytoremediation. pp. 309–328. *In*: Arie Altman, A. and Hasegawa, P.M. (eds.). Plant Biotechnology and Agriculture. Academic Press: Oxford, UK.

Bittsanszkya, A., Kömives, Gullner, G., Gyulai, G., Kiss, J., Heszky, L., Radimszky, L. and Rennenberg, H. 2005. Ability of transgenic poplars with elevated glutathione content to tolerate zinc(2+) stress. Environ. Int. 31: 251–254.

Brim, H., McFarlan, S.C., Fredrickson, J.K., Minton, K.W., Zhai, M., Wackett, L.P. and Daly, M.J. 2000. Engineering *Deinococcus radiodurans* for metal remediation in radioactive mixed waste environments. Nat. Biotechnol. 18: 85–90.

Brim, H., Venkateshwaran, A., Kostandarithes, H.M., Fredrickson, J.K. and Daly, M.J. 2003. Engineering *Deinococcus geothermalis* for bioremediation of high temperature radioactive waste environments. App. Environ. Microbiol. 69: 4575–4582.

Burd, G.I., Dixon, D.G. and Glick, B.R. 1998. A plant growth-promoting bacterium that decreases nickel toxicity in seedlings. Appl. Environ. Microbiol. 64: 3663–3668.

Carter, P., Cole, H. and Burton, J. 2006. Bioremediation: Successes and shortfalls. pp. 1–14. *In*: Proceedings of Key International Conference and Exhibition for Spill Prevention, Preparedness, Response and Restoration (Interspill). London, UK, 23 March.

Che, D., Meagher, R.B., Heaton, A.C., Lima, A., Rugh, C.L. and Merkle, S.A. 2003. Expression of mercuric ion reductase in Eastern cottonwood (*Populusdeltoides*) confers mercuric ion reduction and resistance. Plant Biotechnol. J. 1: 311–319.

Chen, S. and Wilson, D.B. 1997. Genetic engineering of bacteria and their potential for Hg^{2+} bioremediation. Biodegradation (8): 97–103.

Comte, S., Guibaud, G. and Baudu, M. 2008. Biosorption properties of extracellular polymeric substances (EPS) towards Cd, Cu and Pb for different pH values. J. Hazard. Mater. 151: 185–193.

Costa, C.N., Meurer, E.J., Bissani, C.A. and Selbach, P.A. 2006. Contaminantes e poluentes do solo e do ambiente. *In*: Fundamentos de química do solo. Porto Alegre: Evangraf.

D'Amore, J.J., Al-Abed, S.R., Scheckel, K.G. and Ryan, J.A. 2005 Methods for speciation of metals in soils: A review. J. Environ. Qual. 34: 1707–1745.

Das, N., Vimala, R. and Karthika, P. 2008. Biosorption of heavy metals—An overview. Indian J. Biotechnol. 7: 159–169.

Divya, B. and Deepak Kumar, M. 2011. Plant-microbe interaction with enhanced bioremediation. Res. J. BioTechnol. 6: 72–79.

Erakhrumen, A.A. 2007. Phytoremediation: An environmentally sound technology for pollution prevention, control and remediation in developing countries. Educ. Res. Rev. 2: 151–156.99.

Evanko, C.R. and Dzombak, D.A. 1997. Remediation of metals-contaminated soil and groundwater. Environ. Sci. 412: 1–45.

Fang, L., Wei, X., Cai, P., Huang, Q., Chen, H., Liang, W. and Rong, X. 2011. Role of extracellular polymeric substances in Cu(II) adsorption on *Bacillus subtilis* and *Pseudomonas putida*. Bioresour. Technol. 102: 1137–1141.

Fang, L., Huang, Q., Wei, X., Liang, W., Rong, X., Chen, W. and Cai, P. 2010. Microcalorimetric and potentiometric titration studies on the adsorption of copper by extracellular polymeric substances (EPS), minerals and their composites. Bioresour. Technol. 101: 5774–5779.

Gan, S., Lau, E.V. and Ng, H.K. 2009. Remediation of soils contaminated with polycyclic aromatic hydrocarbons (PAHs). J. Hazard. Mater. 172: 532–549.

Garbisu, C. and Alkorta, I. 2001. Phytoextraction: A cost-effective plant-based technology for the removal of metals from the environment. Bioresour. Technol. 77: 229–236.

Glick, B.R., Penrose, D.M. and Li, J. 1998. A model for the lowering of plant ethylene concentrations by plant growth-promoting bacteria. J. Theor. Biol. 190: 63–68.

Guine, V., Spadini, L., Sarret, G., Muris, M., Dedolme, C., Gaudet, J.-P. and Martins, J.M.F. 2006. Zinc sorption to three gram-negative bacteria: Combined titration, modeling and EXAFS study. Environ. Sci. Technol. 40: 1806–1813.

Gullner, G. 2001. Enhanced tolerance of transgenic poplar plants overexpressing gamma-glutamylcysteine synthetase towards chloroacetanilide herbicides. J. Exp. Bot. 52: 971–979.

Haritash, A.K. and Kaushik, C.P. 2009. Biodegradation aspects of polycyclic aromatic hydrocarbons (PAHs): A review. J. Hazard. Mater. 169: 1–15.

Hussein, H., Farag, S. and Moawad, H. 2004. Isolation and characterization of *Pseudomonas* resistant to heavy metals contaminants. Arab. J. Biotechnol. 7: 13–22.

Kapoor, A. and Viraraghvan, T. 1995. Fungal biosorption—An alternative treatment option for heavy metal bearing wastewater: A review. Bioresour. Technol. 53: 195–206.

Karami, A. and Shamsuddin, Z.H. 2010. Phytoremediation of heavy metals with several efficiency enhancer methods. Afr. J. Biotechnol. 9: 3689–3698.

Kelly, D.J.A., Budd, K. and Lefebvre, D.D. 2006. The biotransformation of mercury in pH-stat cultures of microfungi. Can. J. Bot. 84: 254–260.

Kinya, K. and Kimberly, L.D. 1996. Current use of bioremediation for TCE cleanup: Results of a survey. Remediat. J. 6: 1–14.

Kuiper, I., Lagendijk, E.L., Bloemberg, G.V. and Lugtenberg, B.J.J. 2004. Rhizoremediation: A beneficial plant-microbe interaction. Mol. Plant-Microbe Interact. 17: 6–15.

Lombi, E. and Gerzabek, M.H. 1998. Determination of mobile heavy metal fraction in soil: Results of a pot experiment with sewage sludge. Commun. Soil Sci. Plant Anal. 29: 2545–2556.

Lopez, A., Lazaro, N., Morales, S. and Margues, A.M. 2002. Nickel biosorption by free and immobilized cells of *Pseudomonas fluorescens* 4F39: A comparative study. Water Air Soil Pollut. 135: 157–172.

Lovley, D.R. and Phillips, E.J.P. 1988. Novel mode of microbial energy metabolism: Organic carbon oxidation to dissimilatory reduction of iron or manganese. Appl. Environ. Microbiol. 54: 1472–1480.

Lovley, D.R., Philips, E.J., Gorby, Y.A. and Landa, E.R. 1991. Microbial reduction of uranium. Nature 350: 413–416.

Lovley, D.R., Coates, J.D., Blunt-Harris, E.L., Philips, E.J.P. and Woodward, J.C. 1996. Humic substances as electron acceptors for microbial respiration. Nature 382: 445–448.

Lovely, D.R. 2002. Dissimilatory metal reduction: From early life to bioremediation. ASM News 68: 231–237.

Mesjasz-Przybylowicz, O.J., Nakonieczny, M., Migula, P., Augustyniak, M., Tarnawska, M.M., Reimold, W.U., Koeberl, C., Przyby, O.W. and Owacka, E.G. 2004. Uptake of cadmium, lead, nickel and zinc from soil and water solutions by the nickel hyperaccumulator *Berkheyacoddii*. Acta Biol. Cracov. Bot. 46: 75–85.

Onwubuya, K., Cundy, A., Puschenreiter, M., Kumpiene, J. and Bone, B. 2009. Developing decision support tools for the selection of "gentle" remediation approaches. Sci. Total Environ. 407: 6132–6142.

Penny, C., Vuilleumier, S. and Bringel, F. 2010. Microbial degradation of tetrachloromethane: Mechanisms and perspectives for bioremediation. FEMS Microbiol. Ecol. 74: 257–275.

Perpetuo, E.A., Souza, C.B. and Nascimento, C.A.O. 2011. Engineering bacteria for bioremediation. pp. 605–632. *In*: Carpi, A. (ed.). Progress in Molecular and Environmental Bioengineering—From Analysis and Modeling to Technology Applications. Rijeka: InTech.

Pinedo-Rivilla, C., Aleu, J. and Collado, I.G. 2009. Pollutants biodegradation by fungi. Curr. Org. Chem. 13: 1194–1214.

Prasad, M.N.V. and Freitas, H.M.D.O. 2003. Metal hyperaccumulation in plants—Biodiversity prospecting for phytoremediation technology. Electro. J. Biotechnol. 6: 285–321.

Rajendran, P., Muthukrishnan, J. and Gunasekaran, P. 2003. Microbes in heavy metal remediation. Indian J. Exp. Biol. 41: 935–944.

Rojas, L.A., Yanez, C., Gonzalez, M., Lobos, S., Smalla, K. and Seeger, M. 2011. Characterization of the metabolically modified heavy metal-resistant *Cupriavidusmetallidurans* strain MSR33 generated for mercury bioremediation. PLoS One 6: e17555.

Ruis, O.N. and Daniell, H. 2009. Genetic engineering to enhance mercury phytoremediation. Curr. Opin. Biotechnol. 20: 213–219.

Salem, H.M., Eweida, E.A. and Farag, A. 2000. Heavy metals in drinking water and their environmental impact on human health. pp. 542–556. *In*: ICEHM 2000; Cairo University: Giza, Egypt.

Sarikaya, M., Tamerler, C., Jen, A.K., Schulten, K. and Baneyx, F. 2003. Molecular biomimetics: Nanotechnology through biology. Nat. Mater. 2: 577–585.

Sayler, G.S. and Ripp, S. 2000. Field applications of genetically engineered microorganisms for bioremediation process. Curr. Opin. Biotechnol. 11: 286–289.

Sekara, A., Poniedzialeek, M., Ciura, J. and Jedrszczyk, E. 2005. Cadmium and lead accumulation and distribution in the organs of nine crops: implications for phytoremediation. Pol. J. Environ. Stud. 14: 509–516.

Sikkema, J., de Bont, J.A. and Poolman, B. 1995. Mechanisms of membrane toxicity of hydrocarbons. Microbiol. Rev. 59: 201–222.

Smith, M.D., Lennon, E., McNeil, L.B. and Minton, K.W. 1998. Duplication insertion of drug resistance determinants in the radioresistant bacterium *Deinococcus radiodurans*. J. Bacteriol. 170: 2126–2135.

Spormann, A.M. and Widdel, F. 2000. Metabolism of alkylbenzenes, alkanes, and other hydrocarbons in anaerobic bacteria. Biodegradation 11: 85–105.

Sriprang, R., Hayashi, M., Ono, H., Takagi, M., Hirata, K. and Murooka, Y. 2003. Enhanced accumulation of Cd^{2+} by a *Mesorhizobium* sp. transformed with a gene from *Arabidopsis thaliana* coding for phytochelatin synthase. Appl. Environ. Microbiol. 69: 79–796.

Sumner, M.E. 2000. Beneficial use of effluents, wastes, and biosolids. Commun. Soil Sci. Plant Anal. 31: 1701–1715.

Sutherland, R.A., Tack, F.M.G., Tolosa, C.A. and Verloo, M.G. 2000. Operationally defined metal fractions in road deposited sediment, Honolulu, Hawaii. J. Environ. Qual. 29(5): 1431–1439.

Thavasi, R. 2011. Microbial biosurfactants: From an environment application point of view. J. Bioremed. Biodegrad. 2: Article 104e.

Tunali, S., Akar, T., Oezcan, A.S., Kiran, I. and Oezcan, A. 2006. Equilibrium and kinetics of biosorption of lead(II) from aqueous solutions by *Cephalosporiumaphidicola*. Sep. Purif. Technol. 47: 105–112.

Vishnoi, S.R. and Srivastava, P.N. 2008. Phytoremediation-green for environmental clean. *In*: Proceedings of the 12th World Lake Conference, pp. 1016–1021.

Wei, S., Zhou, Q., Zhang, K. and Liang, J. 2003. Roles of rhizosphere in remediation of contaminated soils and its mechanisms. Ying Yong Sheng Tai XueBao 14: 143–147.

Wei, S.H., Zhou, Q.X. and Wang, X. 2005. Cadmium-hyperaccumulator *Solanum nigrum* L. and its accumulating characteristics. Environ. Sci. 26: 167–171.

Wu, C.H., Wood, T.K., Mulchandani, A. and Chen, W. 2006. Engineering plant-microbe symbiosis for rhizoremediation of heavy metals. Appl. Environ. Microbiol. 72: 1129–1134.

Wuana, R.A. and Okieimen, F.E. 2011. Heavy metals in contaminated soils: A review of sources, chemistry, risks and best available strategies for remediation. ISRN Ecol. 2011: Article 20.

Xiao, X., Luo, S.L., Zeng, G.M., Wei, W.Z., Wan, Y., Chen, L., Guo, H.J., Cao, Z., Yang, L.X., Chen, J.L. and Xi, Q. 2010. Biosorption of cadmium by endophytic fungus (EF) *Microsphaeropsis* sp. LSE10 isolated from cadmium hyperaccumulator *Solanum nigrum* L. Bioresour. Technol. 101: 1668–1674.

Zhao, X.W., Zhou, M.H., Li, Q.B., Lu, Y.H., He, N., Sun, D.H. and Deng, X. 2005. Simultaneous mercury bioaccumulation and cell propagation by genetically engineered *Escherichia coli*. Process Biochem. 40: 1611–1616.

7
Mechanism of Heavy Metal Hyperaccumulation in Plants

Supriya Tiwari

1. Introduction

There has been a growing concern worldwide over the increased human interventions which have significantly affected the various ecosystem services. Heavy metal contamination is one of the emerging problems that seek immediate attention. Chemically "heavy metals" refer to those entities of the periodic table whose atomic mass is greater than 20 and specific gravity more than 5. However, in biological terms, heavy metals include a series of metals and metalloids that are toxic to cells even at low concentrations.

Some of the heavy metals such as Fe, Zn, Cu, Mg, Ni and Mo form important structural component of several enzymes and proteins and are essential for normal physiological functioning and plant development. These heavy metals are termed as essential heavy metals. However, certain heavy metals do not have any biological or physiological role and are termed as non-essential. Cd, Sb, Cr, Pb, Ar, Co, Ag, Se and Hg belong to this category. Although, heavy metal, essential or non-essential, play an important role in plant developmental processes, their toxic effects cannot be ruled out. It is not only the non-essential heavy metals that are toxic at low concentrations, the essential heavy metals can also produce toxic effects when their concentration crosses the supra optimal levels (Rascio and Navari-Izzo 2011). Heavy metals may bring about alterations in numerous physiological processes by inactivating enzymes, blocking functional groups of metabolically important molecules or by disrupting membrane integrity (Hossain et al. 2012). Enhanced production of reactive oxygen species (ROS) is a common consequence of heavy metal stress which exposes the cells to oxidative stress leading to lipid peroxidation, biological macromolecules deterioration, membrane dismantling, ion leakage and DNA strand cleavage (Hossain et al. 2012, Rascio and Navari-Izzo 2011).

Certain plants have the potential to withstand high concentration of heavy metals and can survive at high heavy metal doses, which are otherwise detrimental or lethal to most of the plant species (Leitenmaier and Kupper 2013). Such plants have developed alternate adaptive mechanisms for accumulating/tolerating high concentration of heavy metals. This ability of the plants to clean up their surrounding environment is termed as phytoremediation. This technique is environment friendly, and a potentially cheap unobstructive approach for decontaminating polluted soil, water

---

Department of Botany, Institute of Science, Banaras Hindu University, Varanasi-221005, UP, India.
Email: supriyabhu@gmail.com

and air by trace metals. Based upon their potential to withstand the metal contamination in the medium they grow, plants can be categorized as:

1. Indicator plants: The internal heavy metal contamination in the cell is a linear function of the bioavailable heavy metal concentration in the soil. These plants are highly sensitive and can be used as the indicator of metals in the soil.
2. Excluder plants: These plants are able to tolerate heavy metals into the soil up to a threshold concentration by preventing their accumulation in the cell. This exclusion is achieved either by blocking the uptake in the roots (Lux et al. 2011) or by active efflux pump (van Hoof et al. 2001). These plants retain and detoxify most of the heavy metals in root tissues with a minimum translocation to the above ground parts specifically leaves whose cells are more sensitive to the phytotoxic effects of heavy metals (Hall 2002). Most of the metal tolerant plants belong to this group.
3. Hyperaccumulators: These plants have a capacity to take up exceedingly large amounts of heavy metals from the soil and accumulate it in above ground parts without any symptoms of phytotoxicity (Reeves 2006).

2. Accumulation of heavy metals

Unlike excluders, hyperaccumulators translocate the heavy metals taken from the soil to the above ground parts, especially leaves at concentrations 100–1000 folds higher than that found in non hyperaccumulator species. Several researches have suggested threshold concentrations for some heavy metals that can be accumulated by the hyperaccumulators (Boyd 2004). These threshold concentrations are 10,000 $\mu g\ g^{-1}$ for Mn and Zn, 1,000 $\mu g\ g^{-1}$ for Ni, Cu and Se, and 100 $\mu g\ g^{-1}$ for Cd, Cr, Pb, Co, Al and As (Boyd 2004). Metal hyperaccumulation is not a common feature in terrestrial higher plants. About 450 angiosperm species belonging to 45 families, and a few pteridophytes have been identified as heavy metal hyperaccumulators, accounting less than 0.2% of all known species (Rascio and Navari-Izzo 2011) (Table 1). Among the various heavy metals, Ni is reportedly the most preferred metal for hyperaccumulation as more than 75% of the hyperaccumulator plants accumulate this metal in their above ground tissues (Milner and Kochian 2008, Verbruggen et al. 2009). Cd, however, is considered to be the least preferred metal for hyperaccumulation (Rascio and Navari-Izzo 2011). *Arabidopsis thalleri*, *Sedum alfredii*, *Thlaspi praecox*, *Thlaspi caerulescens* and *Solanum nigrum* are important Cd hyperaccumulators (Sun et al. 2006). Cd accumulators have currently been discovered in 11 families with the highest occurrence among Brassicaceae (43%), followed by Asteraceae (14%), while the remaining are distributed nearly equally in various families (Qui et al. 2012). Hyperaccumulators belonging to the family Brassicaceae are potential hyperaccumulators of trace metal (Zn, Ni, Mn, Cu, Co and Cd), metalloids (As) and non-metal (Se) (Verbruggen et al. 2009). *Elsholtzia splendens*, *Thlaspi caerulescens*, *Brassica juncea*, Alyssum sp. are important hyperaccumulators of the family Brassicaceae (Milner and Kochian 2008, Assuncao et al. 2010, Rascio and Navari-Izzo 2011). Several cereal crops such as maize (*Zea mays*), sorghum (*Sorghum bicolor*) and alfalafa (*Medicago sativa*) also act as metal accumulators. Species accumulating selenium are distributed in generas of different families like Fabaceae, Asteraceae, Brassicaceae, Scrophulariaceae and Chenopodiaceae (Reeves and Baker 2000). *Isatis cappadocica* and *Herperis persica* belonging to Brassicaceae, along with ferns belonging to *Pteris*, are recognized as potential accumulators of As (Karimi et al. 2009).

The hyperaccumulator species are distributed in a wide range of distantly related families which indicates that hyperaccumulation trait has evolved independently (Rascio and Navari-Izzo 2011). Most of the hyperaccumulators behave as "strict metallophytes" and can survive only in metalliferous soils, whereas the "facultative metallophytes" can occur in non-metalliferous habitats as well, but thrive more dominantly in metal enriched areas (Assuncao et al. 2003). It has been reported that certain species include both metallicolous and non-metallicolous populations. In

Table 1. Table showing important hyperaccumulator plants for a few heavy metals.

S. No	Metal accumulated	Number of families accumulating the metal	Plant	Family	Reference
1.	Ni	52	Alyssum obovatum	Brassicaceae	Prasad (2005)
			Alyssum oxycarpum	Brassicaceae	Prasad (2005)
			Alyssum peltarioides	Brassicaceae	Prasad (2005)
			Alyssum pinifolium	Brassicaceae	Prasad (2005)
			Alyssum pterocarpum	Brassicaceae	Prasad (2005)
			Alyssum robertianum	Brassicaceae	Prasad (2005)
			Alyssum penjwinensis	Brassicaceae	Prasad (2005)
			Alyssum samariferum	Brassicaceae	Prasad (2005)
			Alyssum bertolonii	Brassicaceae	Li et al. (2003)
			Alyssum caricum	Brassicaceae	Li et al. (2003)
			Alyssum lesbiacum	Brassicaceae	Ingle et al. (2005)
			Alyssum murale	Brassicaceae	Abou-Shanab et al. (2006)
			Alyssum serpyllifolium	Brassicaceae	Ma et al. (2011)
			Brassica juncea	Brassicaceae	Boyd and Martens (1998)
			Pseudosempervirum Sempervium	Brassicaceae	McCutcheon and Schnoor (2003)
			Pseudosempervirum aucheri	Brassicaceae	McCutcheon and Schnoor (2003)
			Cuscuta californica	Cuscutaceae	Boyd and Martens (1998)
			Helianthus annuus	Asteraceae	McCutcheon and Schnoor (2003)
			Hybanthus floribundus	Violaceae	Reeves and Baker (2000)
			Thlaspi caerulescens	Brassicaceae	McCutcheon and Schnoor (2003)
			Commelina benghalensis	Commelinaceae	Chandra et al. (2017)
			Phragmites cummunis	Poaceae	Chandra et al. (2017)
2.	Cu	20	Ocimum centraliafricanum	Lamiaceae	Baker and Walker (1989)
			Brassica juncea	Brassicaceae	McCutcheon and Schnoor (2003)
			Vallisneria americana	Hydrocharitaceae	McCutcheon and Schnoor (2003)
			Eichhornia crassipes	Pontederiaceae	McCutcheon and Schnoor (2003)
			Humaniastrum robertii	Lamiaceae	Baker and Brooks (1989)
			Helianthus annuus	Asteraceae	Baker and Brooks (1989)
			Larrea tridentate	Zygophyllaceae	Baker and Brooks (1989)
			Lemna minor	Lemnaceae	McCutcheon and Schnoor (2003)

Table 1 Contd. ...

...Table 1 Contd.

S. No	Metal accumulated	Number of families accumulating the metal	Plant	Family	Reference
			Pistia stratiotes	Araceae	Odjegba and Fasidi (2004)
			Parthenium hysterophorus	Asteraceae	Malik et al. (2010)
			Ipomea alpine	Convolvulaceae	Baker and Walker (1989)
3.	As	1	*Agrostris capillaries*	Poaceae	McCutcheon and Schnoor (2003)
			Agrostris castellana	Poaceae	McCutcheon and Schnoor (2003)
			Agrostris tenerrima	Poaceae	Prasad (2005)
			Cyanoboletus pulverulentus	Bolete mushroom	Baker and Brooks (1989)
			Pteris vittata	Pteridophyte	Tu et al. (2002)
			Corrigiola telephifolia	Caryophyllaceae	Garcia-Salgado et al. (2012)
4.	Cd	6	*Athyrium yokoscense*	Athyriaceae	McCutcheon and Schnoor (2003)
			Avena strigosa	Poaceae	Uraguchi et al. (2006)
			Brassica juncea	Brassicaceae	Bennetta et al. (2003)
			Crotolaria juncea	Fabaceae	Uraguchi et al. (2006)
			Hydrilla verticillata	Hydrocharitaceae	McCutcheon and Schnoor (2003)
			Salix viminalis	Salicaceae	Schmidt (2003)
			Spirodelia polyrhiza	Araceae	McCutcheon and Schnoor (2003)
			Tagetes erecta	Asteraceae	Uraguchi et al. (2006)
			Vallisnaria spiralis	Hydrocharitaceae	McCutcheon and Schnoor (2003)
			Azolla pinnata	Azollaceae	Rai (2008)
			Cannabis sativa	Cannabaceae	Girdhar et al. (2014)
			Commelina benghalensis	Commelinaceae	Chandra et al. (2017)
			Ranunculus sceleratus	Ranunculaceae	Chandra et al. (2017)
			Rorippa globosa	Brassicaceae	Girdhar et al. (2014)
			Solanum nigrum	Solonaceae	Girdhar et al. (2014)
5.	Zn	9	*Agrostis castellana*	Poaceae	McCutcheon and Schnoor (2003)
			Athyrium yokoscense	Athyriaceae	McCutcheon and Schnoor (2003)
			Brassica juncea	Brassicaceae	Bennetta et al. (2003)
			Brassica napa	Brassicaceae	McCutcheon and Schnoor (2003)

Table 1 Contd. ...

...Table 1 Contd.

S. No	Metal accumulated	Number of families accumulating the metal	Plant	Family	Reference
			Helianthus annuus	Asteraceae	Schidmt (2003)
			Eichhornia crassipes	Pontederiaceae	McCutcheon and Schnoor (2003)
			Salvinia molesta	Salviniaceae	McCutcheon and Schnoor (2003)
			Spirodela polyrhiza	Araceae	McCutcheon and Schnoor (2003)
			Alternanthera philoxeroides	Amaranthaceae	Chandra et al. (2017)
			Chenopodium album	Amaranthaceae	Malik et al. (2010)
			Croton bonplandianum	Euphorbiaceae	Chandra et al. (2017)
			Sacchrum munja	Poaceae	Chandra et al. (2017)
			Ricinus communis	Euphorbiaceae	Giordani et al. (2005)
6.	Pb	6	Drosera rotundifolia	Droseraceae	Fontem Lum et al. (2015)
			Eleusine indica	Poaceae	Fontem Lum et al. (2015)
			Euphorbia cheiradenia	Euphorbiaceae	Chehregani and Malayeri (2007)
			Rumex dentatus	Polygonaceae	Chandra et al. (2017)
			Ambrosia artemisiifolia	Asteraceae	Fiegl et al. (2011)
			Armeria maritime	Plumbaginaceae	Fiegl et al. (2011)
			Azolla filiculoides	Salviniaceae	McCutcheon and Schnoor (2003)
			Brassica oleraceae	Brassicaceae	Fiegl et al. (2011)
			Festuca ovina	Poaceae	Fiegl et al. (2011)
			Imopoea trifida	Convolvulaceae	Schmidt (2003)
			Triticum aestivum	Poaceae	Fiegl et al. (2011)
7.	Cr	-	Phragmites cummunis	Poaceae	Chandra et al. (2017)
			Pistia stratiotes	Araceae	Odjegba and Fasidi (2004)
			Salsola kali	Saliaceae	Gardea-Torresdey et al. (2005)
			Dicoma niccolifera	Asteraceae	McCutcheon and Schnoor (2003)
			Medicago sativa	Fabaceae	McCutcheon and Schnoor (2003)
8.	Mn	16	Sacchrum munja	Poaceae	Chandra et al. (2017)
			Rumex dentatus	Polygonaceae	Chandra et al. (2017)
			Austromyrtus bidwillii	Myrtaceae	Bidwell et al. (2002)
			Commelina benghalensis	Commelinaceae	Chandra et al. (2017)
			Chengiopanax sciadophylloides	Araliaceae	Baker and Walker (1990)
			Macadamia neurophylla	Proteaceae	Baker and Walker (1990)

certain Zn hyperaccumulators like *Arabidopsis halleri* and *Thlaspi caerulescens*, hyperaccumulation is a constitutive trait of the species and occurs strictly in all populations (Bert et al. 2000, Escarre et al. 2000). However, in other Zn accumulators like *Sedum alfredii*, this trait is not constitutive at species level, but only confined to metallicolous populations (Verbruggen et al. 2009). In another study, it was found that *Noccaea caerulescens* accumulates high amount of Zn and > 2% Cd as well. However, its accumulation of Cu is even less than the non-accumulator species (Mijovilovich et al. 2009). Similarly, the accumulation of As and Mn in this plant is even less than the excluder species (Martinez-Alcala et al. 2013). These studies clearly indicate that hyperaccumulation ability is largely dependent upon the population and the nature of the metal.

Metal accumulation is an eco-physiological adaptation to metalliferous soils. It has been a topic of interest for the researchers to investigate how and why plants evolved with metal hyperaccumulation capabilities. Some researchers think that hyperaccumulation process performs several beneficial functions for plants (Boyd 2004). Hanson et al. (2004) reported that Se hyperaccumulated by *Brassica juncea* enhanced the resistance of the plants towards biotic stress like fungal and pathogen attack. It has been observed that the perennial hyperaccumulators elevated the levels of metals in their canopy covered soil via the production of metal enriched litter (Boyd 2004). This phenomenon, termed as elemental allelopathy, prevents the development of other less tolerant species in their domain, thus ensuring their dominance for obtaining natural resources (Boyd 2004).

Several hypotheses have been put forward to explain the evolution of hyperaccumulators, which include the following:

1. Plants may hyperaccumulate trace metals because storing large quantities of metals may be a way of metal tolerance and disposal.
2. Hyperaccumulators may act as a source of osmotic resistance to drought (Kachenko et al. 2011).
3. Hyperaccumulation may be a mechanism of defense against pathogens and herbivores (Horger et al. 2013).
4. Elemental allelopathy may reduce competition from nearby species (El Mehdawi et al. 2011).
5. Hyperaccumulation may be an accidental occurrence (Rascio and Navari-Izzo 2011).

In addition to the entire proposed hypothesis, hyperaccumulation also relies upon hypertolerance, an essential characteristic that allows the hyperaccumulators to avoid heavy metal poisoning (Leitenmaier and Kupper 2013). Although hyperaccumulation and hypertolerance are independent traits (Leitenmaier and Kupper 2013), this "independence" is limited because hyperaccumulator would poison itself if accumulated metal is not be tolerated via specific mechanisms (Ferot et al. 2010, Willems et al. 2010).

3. Mechanism of hyperaccumulation

The degree of hyperaccumulation of heavy metals can vary significantly in different species or even in populations or different ecotypes (Cappa et al. 2014). However, all hyperaccumulators show a few characteristic traits which distinguish them from the non-accumulators. These common traits are:

1. Enhanced capacity of heavy metal absorption from the soil.
2. Faster and more effective root shoot translocation of metals.
3. Higher potential of detoxification and sequestration of heavy metal ions.

Studies have shown that genes responsible for accomplishing the key steps of hyperaccumulation are common to both hyperaccumulators and non hyperaccumulators, but are differently expressed and regulated in the two types of plants (Verbruggen et al. 2009). The ability of hyperaccumulators, however, depends upon the uptake capacity and intracellular transportation of plants. Figure 1 gives an account of the processes involved in hyperaccumulation.

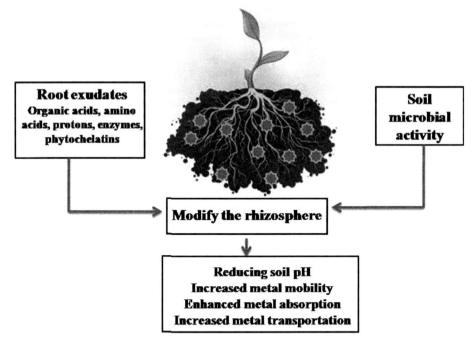

Figure 1. Different processes involved in the bioactivation of heavy metals in the rhizosphere.

3.1 Heavy metal uptake

One of the most significant features of hyperaccumulating plants is their extraordinary capacity to absorb metals from the soil (Ma et al. 2001, Yang et al. 2006). The bioavailability and uptake of heavy metals in the soil are significantly affected by the metal content, pH, organic substance and other elements present in the rhizosphere. Most of the heavy metals have low mobility in soils and are not easily absorbed by plant roots (Knight et al. 1994).

3.1.1 Bioactivation of rhizosphere

Many of the hyperaccumulating plants show a tendency to increase the heavy metal uptake from the soil via the bioactivation of rhizosphere (Figure 1). Hyperaccumulating plants bring about changes at the soil root interphase by releasing organic and inorganic compounds (Fonia et al. 2017). The root exudates may include organic acids, phytochelatins, amino acids, protons, enzymes, etc. and may not only improve the metal bioavailability in the rhizosphere but also affect the number and activity of microorganisms by reducing soil pH or by producing chelators and siderophores (Abou-Shanab et al. 2003, Wenzel et al. 2003). Secretion of protons by the roots of hyperaccumulators may reduce the pH, thereby increasing the metal dissolution (Bernal et al. 1994). Studies have demonstrated that Cu accumulating plant species, *Elsholtzia splendens,* showed a lower pH in the rhizosphere than in the bulk soil (Peng et al. 2005). Secretion of organic acids from the roots of hyperaccumulator plants can enhance the absorption from the soil by mobilizing the heavy metals present in the rhizosphere (Krishnamurti et al. 1997). Cieslinski et al. (1998) observed many low molecular organic acids such as acetic acid and succinate in the rhizosphere of Cd accumulating genotype (Arcola). Root exudates of Zn/Cd hyperaccumulating plant species *Sedum alfredi* Hance could extract more Zn and Pb from the contaminated soil (Li et al. 2012). The secretion of amino acid histidine in the root extracts of *Allysum* enhanced the transport and hyperaccumulation of Ni (Kramer et al. 1996).

Rhizosphere is populated by large concentration of microorganisms which play a significant role in increasing the bioavailability of various heavy metal ions for uptake (Figure 1). *Xanthomonas maltophyta* catalyzes the transformation of several toxic metal ions including Cr^{6+}, Pb^{2+}, Hg^{2+},

110 *Bioremediation Science: From Theory to Practice*

Au^{3+}, Te^{4+}, Ag^+ and oxyanions such as SeO_4^- (Weber et al. 2004, Zhao et al. 2002). Another bacterium, *Shewanella alga,* is known to increase the mobility of As, thus enhancing its uptake by hyperaccumulators (Lombi et al. 2001). Certain organisms are known to enhance Zn accumulation in the shoots of *Thlaspi caerulescens* by facilitating an increase in solubility of non-labile Zn in the soil, thus enhancing its bioavailability (Liu 2008). Soil microorganisms are also known to create certain organic exudates which significantly increase the bioavailability of certain metals like Mn^{2+} and Cd^{2+} (Hall 2002).

3.1.2 *Metal transporters*

Uptake of heavy metals by root cells is mediated by specific metal transport proteins (Verbruggen et al. 2009, Rascio and Navari-Izzo 2011). Metal transporters generally have broad substrate specificities and are encoded by genes of different families like ZIP (Zinc and Iron Regulated Transporter Proteins), HMA (Heavy Metal transporting ATPase), MATE (Multidrug And Toxin Efflux), YSL (Yellow Stripe 1 Like), MTP (Metal Tolerance Protein), etc. (Figure 2). Studies have shown that increased Zn uptake by *Thlaspi caerulescens*, and *Arabidopsis halleri* roots, as compared to their related non hyperaccumulator species, can be attributed to the constitutive overexpression of some genes belonging to the ZIP family, and coding for Zn transporters located in the plasma membrane (Kotrba et al. 2009). The expression of ZIP genes in non hyperaccumulating species is regulated, i.e., they express only under Zn deficient conditions (Assuncao et al. 2001). In hyperaccumulating species, ZIP genes are expressed at high Zn concentrations as well, persisting at high Zn availability (Assuncao et al. 2001, Weber et al. 2004). Rascio and Navar-Izzo (2011) reported a decline in Cd uptake in presence of Zn in Cd/Zn hyperaccumulator *Arabidopsis halleri* and most ecotypes of *Thlaspi caerulescens*. This observation clearly suggests that Cd uptake is largely mediated by Zn transporters with a strong preference for Zn over Cd (Zhao et al. 2002). However, in another Zn accumulator ecotype of *T. caerulescens*, Cd uptake remains unaffected by Zn concentration indicating the presence of an efficient and independent Cd transport system (Lombi et al. 2001).

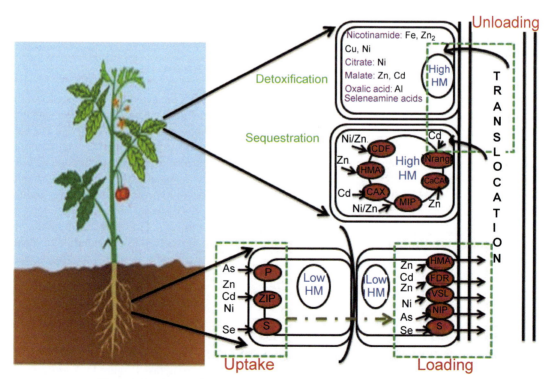

Figure 2. Mechanism of heavy metal uptake, transport and accumulation in plants.

Similarly, the preference of Zn over Ni in some Zn/Ni hyperaccumulators supplied with the same concentrations of both the heavy metals strongly suggests that Zn transporters are involved in Ni uptake by the roots (Assuncao et al. 2000, Kotrba et al. 2009, Assuncao et al. 2010, Rascio and Navari-Izzo 2011).

Transport of As in form of arsenate is mediated by phosphate/arsenate transporters associated with the plasma membranes in the root cells of As hyperaccumulator *Pteris vittata* (Meharg and Whittaker 2002). Again, the density of these transporters is more in hyperaccumulating species as compared to non hyperaccumulators probably due to constitutive gene overexpression (Calle et al. 2005). He et al. (2016) reported an aquaporin gene PvTIP4 to mediate As uptake in *Pteris vittata*. However, in addition to the role of phosphate/arsenate transporters, arsenic uptake also depends upon the plants' ability to increase the As bioavailability in rhizosphere by reducing pH via root exudation of organic carbon (Gonzaga et al. 2009).

Similarly, transport of Selenium (Se), in the form of selenate, is facilitated by sulphate transport owing to their chemical analogy and high affinity of the sulphate transporters towards Se (Shibagaki et al. 2002, Hirai et al. 2003). It has been observed that in hyperaccumulator species such as *Astragalus bisulcatus* (Fabaceae) and *Stanleya pinnata* (Brassicaceae), the Se/S ratio in shoots is much higher than their non hyperaccumulator counterparts (Galeas et al. 2007).

3.2 Root-shoot translocation of heavy metals

Contrary to the non hyperaccumulators which store heavy metals in their below ground parts, hyperaccumulators store the absorbed heavy metals in their above ground parts through bulk flow of metals via xylem, from roots to shoots (Grenan 2009). Storage of heavy metals in above ground parts is the characteristic feature of the hyperaccumulators (Kupper and Kroneck 2005). Lovy et al. (2013) have shown that in the Cd/Zn hyperaccumulator *Noccaea caerulescens*, 86% of the absorbed Cd is translocated to the shoots.

The first step towards the allocation of the heavy metals to shoots is the translocation of the absorbed metals from root symplast to xylem apoplast (Lietenmaier and Kupper 2013). This necessitates the availability of heavy metals for xylem loading, which is facilitated through their low sequestration and easy efflux out of the vacuoles of the root cells (Rascio and Navari-Izzo 2011). In the Cd/Zn hyperaccumulator *Noccaea caerulescens*, compared to their related non hyperaccumulator *Noccaea arvense*, the hyperaccumulator showed reduced sequestration into the root vacuoles, which was associated with higher root to shoot translocation efficiency of *Noccaea caerulescens* (Zhao et al. 2006). Similar observations were also recorded for hyperaccumulator and non hyperaccumulator relatives of *Sedum alfredii* growing on highly Cd/Zn contaminated soils (Lu et al. 2013). A lower sequestration into root vacuoles accounted for the enhanced As translocation in hyperaccumulator compared with non hyperaccumulator species of *Pteris* (Poynton et al. 2004).

The feature of enhanced xylem loading in the hyperaccumulators is attributed to the over expression of certain genes which are also present in non hyperaccumulators but only in their normal or down regulated forms (Kotrba et al. 2009, Rascio and Navari-Izzo 2011). The P_{-18} type ATPases, HMAs, MATE, YSL are few of the important genes coding for xylem loading and heavy metal transport (Axelsen and Palmgren 1998). It was reported that HMA4 is strongly over expressed in the roots of Cd/Zn hyperaccumulator plants *Arabidopsis halleri* and *Noccaea caerulescens* (Bernard et al. 2004, Weber et al. 2004). HMA4 expression is up regulated when these plants are exposed to high levels of Cd and Zn and is down regulated in non hyperaccumulator species (Leitenmaier et al. 2011). The up regulation of HMA4 is responsible for efflux of Cd and Zn from root symplast to xylem vessels (Willems et al. 2010). In addition, HMA4 enhances the expression of genes belonging to ZIP family, thus enhancing the metal uptake by the root cells (Willems et al. 2007, 2010). Over expression of HMA4 in shoot cells indicates its involvement in xylem unloading as well (Cracium et al. 2012). It has been observed that overexpression of HMA4 in non hyperaccumulators results in their poisoning due to lack of any specific mechanism to detoxify Cd and Zn reaching the shoots (Hanikenne et al. 2008). HMA4 plays an important role in transport of Cd and Zn in case of rice (*Oryza sativa*) as well (Takahashi et al. 2012). FDR3 genes belonging to MATE family is constitutively overexpressed in the roots of *Thlaspi caerulescens* and *Arabidopsis halleri* (Talke et al. 2006). TcYSL3, TcYSL5 and YSL7 genes of YSL family mediate loading and unloading of Ni complexed with nicotinamine in *Thlaspi caerulescens* (Rascio and Navari-Izzo 2011).

3.3 Detoxification/sequestration of heavy metals

After translocating the absorbed heavy metals to the above ground parts, hyperaccumulators have to store them in such a way that it does not interfere with the normal metabolic activities of the plants (Figure 2). Greater efficiency in detoxification and sequestration are the important traits of hyperaccumulators, which allow them to accumulate high concentration of heavy metals in their cells without any phytotoxic effect (Rascio and Navari-Izzo 2011). Another interesting feature is that the accumulation of heavy metals usually occurs in leaves which are the sites of photosynthesis (Rascio and Navar-Izzo 2011). Hyperaccumulation avoids the phytotoxic effects of heavy metals by following any of the two strategies.

3.3.1 Sequestration in shoot vacuoles

The main purpose of sequestration of heavy metals is to remove them from their metabolically active cytoplasm and moving them to inactive compartments, mainly cell wall and vacuoles (Leitenmaier and Kupper 2013). Vacuoles are the most preferred sites for heavy metal sequestration because they contain enzymes like phosphatases, lipases, proteinases, which are not targeted by heavy metals (Carter et al. 2004). Further, the large vacuoles of epidermal cells are preferred as those cells do not harbor chloroplasts, thus minimizing the probability of inhibiting photosynthesis (Kupper et al. 2000, Frey et al. 2000). Preferential storage of hyperaccumulated metals in epidermis has been shown for a majority of hyperaccumulator species and for elements as chemically diverse as Al, As, Cd, Ni, Se and Zn (Freeman ct al. 2006, Cosio et al. 2005). If the storage capacity of epidermis is exceeded, storage occurs in mesophyll cells (Kupper et al. 2009). Cell wall too is metabolically active and is the first site at which heavy metal encounters the cell, but its role in hyperaccumulation is still controversial (Fonia et al. 2017). However, Carr et al. (2003) have shown that in *Camellia sinensis* (tea), Al is accumulated mostly in cell walls, with a very low concentration inside the cells.

Comparative transcriptome analysis between hyperaccumulator and non hyperaccumulator species has demonstrated that the process of sequestration is also dependent upon the constitutive expression of genes that encodes proteins operating in heavy metal transfer across tonoplast and in excluding them from cytoplasm (Rascio and Navari-Izzo 2011). Protein families involved in vacuolar sequestration are Nramps (Natural resistance associated macrophage proteins), CDFs (Cation diffusion facilitator), HMA (Heavy metal ATPase), CaCA (Ca^{2+}/cation antiportor), ABC (ATP binding cassatte), etc. (Verbruggen et al. 2009).

CDFs are also called as metal transporter proteins (MTPs) and contain members involved in the transport of bivalent cations such as Zn^{2+}, Fe^{2+}, Cd^{2+}, Co^{2+} and Mn^{2+} from cytoplasm to vacuoles (Verbruggen et al. 2009). MTP1, a gene encoding for tonoplast specific protein, is highly over expressed in the leaves of Zn/Ni hyperaccumulators and play an important role in enhancing Zn accumulation (Drager et al. 2004, Gustin et al. 2009). It has been observed that Zn transport into vacuoles may initiate Zn deficiency response, which in turn enhances the heavy metal uptake and translocation via increased expression of Zn transporters in hyperaccumulator plants (Gustin et al. 2009). Persant et al. (2001) have shown that MTP members also facilitate the vacuole storage of Ni in *Thlaspi goesingense*. Kim et al. (2004), however, have shown that TgMTP1 is localized in the plasma membrane and mediates the efflux of both Ni and Zn from cytoplasm. Over expression of ShMTP is responsible for vacuolar accumulation of Mn in tropical legume *Stylosanthes hamata* (Delhaize et al. 2003).

The over expression of HMA3 coding for vacuolar P_{1B}-ATPase was shown to play significant role in Cd accumulation in *Noccaea caerulescens* (Ueno et al. 2011). MHX protein of CaCA family is shown to play an important role in Zn vacuolar storage in *Arabidopsis halleri* (Elbaz et al. 2006). The over expression of Nramp's was associated with Cd accumulation in *Noccaea caerulescens* (Takahashi et al. 2011).

Sequestration of heavy metals in mesophyll cells is an exceptional phenomenon and has been observed in a few hyperaccumulator species (Leitenmaier and Kupper 2013). In Zn/Cd hyperaccumulator *Arabidopsis hallerii*, although Zn is sequestered in trichomes and epidermal cells, a major portion of Cd is stored in mesophyll cells (Kupper et al. 2000). Accumulation of Cd in the mesophyll cells is considered to be responsible for the toxic responses shown by *Arabidopsis hallerii* at much lower concentration as compared to other Cd hyperaccumulators like *Noccaea caerulescens*. *Sedum alfredii*, a Cd/Zn accumulator, is yet another example where Cd sequestration occurs in mesophyll cells beside pith and cortex of the stem (Tian et al. 2011). However, the thick and succulent leaves of *Sedum alfredii* have exceptionally large vacuoles in the mesophyll that allow a large amount of Cd to be stored, without any toxic effect as compared to *Arabidopsis hallerii* (Leitenmaier and Kupper 2013). *Gossia bidwilli*, *Virotia neurophylla*, *Macadamia integrifolia* and *Macadamia tetraphylla* are a few Mn hyperaccumulators where Mn sequestration occurs in multiple palisade cell layers (Fernando et al. 2006a, b).

3.3.2 Detoxification of heavy metals

Apart from sequestration, detoxification of heavy metals is another strategy adopted by hyperaccumulators to avoid toxicity. Heavy metal toxicity is avoided by binding them to certain metal binding proteins or ligands. The association of heavy metals with ligands prevents the persistence of heavy metal as free ions in the cytoplasm (Rascio and Navari-Izzo 2011). The best known ligands for this purpose are thiols including glutathione, phytochellatins and metallothioneins and non thiols such as histidine and nicotinamide. Non thiols are more significant in hyperaccumulators rather than high molecular mass ligands like phytochelatins because of the excessive amount of sulphur and high metabolic cost that this kind of chelation requires (Schat et al. 2002, Rabb et al. 2004).

Histidine is an important amino acid residue forming metal binding sites in metalloprotein and was first shown to bind Ni in Ni hyperaccumulator *Alyssum lesbiacum* (Kramer et al. 1996), but later it was found to be involved in hyperaccumulation of Zn as well (Kupper et al. 2004). Nicotinamide has a strong affinity for binding Fe, Zn, Cu and Ni, and is found in high concentration in hyperaccumulators like *Alyssum halleri* and *Noccaea caerulescens* (Rascio and Navari-Izzo 2011). Small ligands such as organic acids also play a significant role in detoxification of heavy metals. Citrate is the main ligand which binds Ni in the leaves of *Thlaspi goesingense* (Kramer et al. 2000), while citrate and acetate bind Cd in the leaves of *Solanum nigrum* (Sun et al. 2006). A significantly high concentration of citrate ligands has been reported in *Phaseolus vulgaris*, *Crotalaria cobalticola*, *Raufolia serpentine* and *Silene cucubalis* upon their exposure to high levels of Ni and Co (Boyd 2007). Most of the Zn in *Arabidopsis halleri* and Cd in *Thlaspi caerulescens* are complexed with malate (Sarret et al. 2002, Salt et al. 1999) and Al with oxalic acids in the roots of buckwheat (*Fagopyrum esculentum*).

Metal hyperaccumulators generally have a high concentration of glutathione, cysteine and O-acetylserine as compared to non hyperaccumulators (Rascio and Navari-Izzo 2011). Glutathione can play a direct role in metal chelation and can also be a substrate for biosynthesis of phytochelatins (Rauser 1999). Cysteine also serves as building blocks of phytochelatins and play an important role in detoxification of As and Cd (Leitenmaier and Kupper 2013). The major detoxification strategy in Se hyperaccumulators is to get rid of selenoaminoacids, mainly selenocysteine, derived from selenate assimilation in leaf chloroplasts. Selenoaminoacids are misincorporated in proteins instead of sulphur amino acid, resulting in Se toxicity. The detoxification occurs through methylation of Se-Cys to a harmless non protein amino acidmethyloselenocysteine in the reaction catalysed by selenocysteine methylotransferase, which is constitutively expressed and activated only in leaves of hyperaccumulator species (Sors et al. 2009).

4. Significance of hyperaccumulation

The characteristic feature of heavy metal hyperaccumulation in plants has been effectively utilized as a successful tool in phytoremediation. Phytoremediation has emerged as an eco-friendly soil remediation technique that utilizes plant species. The naturally occurring heavy metal hyperaccumulator plants growing in metal enriched soils can accumulate 100–1000 folds higher level of metals, making them potential entities for phytoremediation (Navari-Izzo and Quatracci 2001, Kidd et al. 2009, Schwitzuebel et al. 2009). *Thlaspi* growing on Ni contaminated soils can accumulate about 3% of its dry matter and *Thlaspi caerulescens* can act as good hyperaccumulator of Cd and Zn, remediating as much as 60 kgZnh^{-1} and 80 kgCdh^{-1} from the soil (Robinson et al. 1998). *Pteris vittata* can accumulate as much as 22 gAskg^{-1} in its frond dry weight, with a bio concentration factor of 87, removing 26% of soil's initial As (McGrath and Zhao 2003). Common buckwheat (*Fagopyrum esculentum*) is quite effective in phytoremediation of Pb, which is largely immobile in soil and extraction rate is limited by solubility and diffusion to the root surface (Leitenmaier and Kupper 2013). Tamura et al. (2005) have shown that buckwheat can accumulate as much as 4.2 mgg^{-1} dry weight of Pb in the shoots. Similarly, *Phytolacca acinosa* can act as a potential hyperaccumulator of Mn, when grown in Mn rich soils (Xue 2004).

However, hyperaccumulators have limited potential of phytoremediation because most of them are metal selective and can be used only in their natural habitats. The small biomass, shallow root systems and slow growth rates limit the speed of metal removal from the soils (Rascio and Navari-Izzo 2011). Hyperaccumulation can take decades to clean up the contaminated sites. Studies have shown that to decrease Zn concentration from 440 to 300 mgZnkg^{-1} in the soil, nine cropping of *Thlaspi caerulescens* would be required and 28 years of cultivation of this plant would be needed to remove 2100 mgZnkg^{-1} from a soil (McGarth et al. 1993, Kidd and Monterroso 2005).

5. Use of transgenic plants as hyperaccumulators

A biotechnological approach has been utilized to improve the potential of hyperaccumulator growth rate through selective breeding or by transfer of hyperaccumulator genes to high biomass species. Recently, various genetically modified plants have been raised by transferring genes from one plant to another. The transfer of phytochelatin synthetase (PCS) gene from *Cyanodon dactylon* to tobacco enhanced the amount of phytochelatin 3.88 fold and subsequently enhanced Cd accumulation by 3 folds (Li and Chen 2006). Similarly, in garlic (*Allium sativum*) (Zhang et al. 2005), overexpression of PCS favored the stress mitigation caused by heavy metal stress. Studies have shown that transgenic plants enhanced the metal transformation (from toxic to non-toxic form) 10 times more efficiently as compared to wild type (Kotrba et al. 2009). Bioengineered plants, tolerant to the presence of toxic levels of metals like Cd, Zn, Cr, Cu, Pb, As and Se, have been used for phytoremediation (Bennet et al. 2003, Kawashima et al. 2004). Transgenic *Brassica juncea* showed enhanced uptake of Se and enhanced Se tolerance than the wild species (Van Huysen et al. 2004). In an effort to overcome the short biomass of hyperaccumulators, somatic hybrids have been generated between *Thlaspi caerulescens* and *Brassica napus* which can be used as potential hyperaccumulators of Zn (Brewer et al. 1999) and Pb (Gleba et al. 1999).

Although the biotechnological aspect to develop new transgenic hyperaccumulators seems promising, it is not a much explored area. More experiments need to be performed under field conditions to bring out the actual potential of bioengineered hyperaccumulators.

6. Conclusions and future prospects

Human interferences with the biosphere have worsened the problem of heavy metal pollution. Heavy metal toxicity in soils has resulted in impeding growth and development of plants. Hyperaccumulator plants can act as potential tools for heavy metal remediation of the soil. An understanding of the mechanism of heavy metal uptake and their sequestration/detoxification will help us to investigate the possibilities of using them to remove metals from contaminated or natural metalliferous soils. The information obtained on the genes involved in transfer and sequestration of heavy metals in hyperaccumulators will open the opportunities to transfer these specific genes, through biotechnological processes, into plants with high biomass promising species. However, in spite of significant works done on this aspect in the recent years, the complexity and mechanism of hyperaccumulation still needs much exploration. Another important aspect is to search for and characterize other hyperaccumulator species, cultivate them and, using different agronomic management practices, enhance their plant growth and metal uptake by selective breeding and gene manipulation.

References

Abou-Shanab, R.A., Angle, J.S., Delorme, T.A., Chaney, R.L., van Berkum, P., Moawad, H., Ghanem, K. and Ghozlan, H.A. 2003. Rhizobacterial effects on nickel extraction from soil and uptake by *Alyssum murale*. N. Phytol. 158(1): 219–224. [doi:10.1046/j.1469-8137.2003.00721.x].

Abou-Shanab, R.A.I., Angle, J.S. and Chaney, R.L. 2006. Bacterial inoculants affecting nickel uptake by *Alyssum murale* from low, moderate and high Ni soils. Soil Biol. Biochem. 38: 2882–2889.

Assunção, A.G.L., Martins, P.D.C., De Folter, S., Vooijs, R., Schat, H. and Aarts, M.G.M. 2001. Elevated expression of metal transporter genes in three accessions of the metal hyperaccumulator *Thlaspi caerulescens*. Plant Cell Environ. 24: 217–226.

Assuncão, A.G.L., Schat, H. and Aarts, M.G.M. 2003. *Thlaspi caerulescens*, an attractive model species to study heavy metal hyperaccumulation in plants. N. Phytol. 159: 351–360.

Assunção, A.G.L, Herrero, E., Lin, Y.F., Huettel, B., Talukdar, S., Smaczniak, C., Immink, R.G., Van Eldik, M., Fiers, M., Schat, H. and Aarts, M.G. 2010. *Arabidopsis thaliana* transcription factors bZIP19 and bZIP23 regulate the adaptation to zinc deficiency. Proc. Natl. Acad. Sci. U.S.A. 107: 10296–10301.

Axelsen, K.B. and Palmgren, M.G. 1998. Inventory of the super family of P-Type ion pumps in Arabidopsis. Plant Physiol. 126: 696–706.

Baker, A.J.M. and Brooks, R.R. 1989. Terrestrial higher plants which hyperaccumulate metallic elements: A review of their distribution, ecology and phytochemistry. Biorecovery 1: 81–126.

Baker, A.J.M. and Walker, P.L. 1989. Ecophysiology of metal uptake by tolerant plants. pp. 155–177. *In*: Shaw, A.J. (ed.). Heavy Metal Tolerance in Plants: Evolutionary Aspects. Boca Raton, FL: CRC Press.

Bennet, L.E., Burkhead, J.L., Hae, K.L., Terry, N., Pilon, M. and Pilon-Smits, E.A.H. 2003. Analysis of transgenic Indian mustard plants for phytoremediation of metal-contaminated mine tailings. J. Environ. Qual. 32: 432–440.

Bennetta, L.E., Burkheada, J.L., Halea, K.L., Terry, N., Pilona, M. and Pilon-Smits, E.A.H. 2003. Analysis of transgenic indian mustard plants for phytoremediation of metal-contaminated mine tailings. J. Environ. Qual. 32(2): 432. doi:10.2134/jeq2003.0432.

Bernal, M.P., McGrath, S.P., Miller, A.J. and Baker, A.J.M. 1994. Comparison of the chemical changes in the rhizosphere of the nickel hyperaccumulator Alyssum murale with the non-accumulator Raphanus sativus. Plant Soil 164: 251–259.

Bernard, C., Roosens, N., Czernic, P., Lebrun, M. and Verbruggen, N. 2004. A novel CPx-ATPase from the cadmium hyperaccumulator Thlaspi caerulescens. FEBS Lett. 569: 140–148. doi: 10.1016/j.febslet.2004.05.036.

Bert, V., Macnair, M.R., De Laguerie, P., Saumitou-Laprade, P. and Petit, D. 2000. Zinc tolerance and accumulation in metallicolous and nonmetallicolous populations of *Arabidopsis halleri* (Brassicaceae). New Phytol. 146: 225–233.

Bidwell, S.D., Woodrow, I.E., Batianoff, G.N. and Sommer-Knudsen, J. 2002. Hyperaccumulation of manganese in the rainforest tree Austromyrtus bidwillii (Myrtaceae) from Queensland, Australia. Funct. Plant Biol. 29: 899–905.

Boyd, R.S. and Martens, S.N. 1998. The significance of metal hyperaccumulation for biotic interactions. Chemoecology 8: 1–7.

Boyd, R.S. 2004. Ecology of metal hyperaccumulation. New Phytol. 162: 563–567.

Boyd, R.S. 2007. The defence hypothesis of elemental hyperaccumulation: status, challenges and new directions. Plant Soil 293: 153–176.

Brewer, E.P., Saunders, J.A., Angle, J.S., Chaney, R.L. and McIntosh, M.S. 1996. Somatic hybridization between the zinc accumulator *Thlaspi caerulescens* and *Brassica napus*. Theor. Appl. Genet. 99: 761–771.

Caille, N., Zhao, F.J. and McGrath, S.P. 2005. Comparison of root absorption, translocation and tolerance of arsenic in the hyperaccumulator *Pteris vittata* and the non hyperaccumulator *Pteris tremula*. New Phytol. 165: 755–761.

Cappa, J.J. and Pilon-Smith, E.A.H. 2014. Evolutionary aspects of elemental hyperaccumulation. Planta. 239(2): 267–275.

Carr, H.P., Lombi, E., Küpper, H., McGrath, S.P. and Wong, M.H. 2003. Accumulation and distribution of aluminium and other elements in tea (*Camellia sinensis*) leaves. Agronomie 23: 705–710. doi: 10.1051/agro:2003045.

Carter, C., Pan, S., Zouhar, J., Avila, E.L., Girke, T. and Raikhel, N.V. 2004. The vegetative vacuole proteome of *Arabidopsis thaliana* reveals predicted and unexpected proteins. Plant Cell 16: 3285–3303. doi: 10.1105/tpc.104.027078.

Chandra, R. and Kumar, V. 2017. Phytoextraction of heavy metals by potential native plants and their microscopic observation of root growing on stabilized distillery sludge as a prospective tool for *in situ* phytoremediation of industrial waste. Environ. Sci. Pollut. Res. 24(3): 2605–2619.

Chandra, R., Yadav, S. and Yadav, S. 2017. Phytoextraction potential of heavy metals by native wetland plants growing on chlorolignin sludge of pulp and paper industry. Ecol. Eng. 98: 134–145.

Chehregani, A. and Malayeri, B.E. 2007. Removal of heavy metals by native accumulator plants. Int. J. Agric. Biol. 9: 462–465.

Cieslinski, G.K.C., Van Rees, J., Szmigielska, A.M., Krishnamurti, G.S.R. and Huang, P.M. 1998. Low-molecular weight organic acids in rhizosphere soils of durum wheat and their effect on cadmium bioaccumulation. Plant Soil 203: 109–117.

Cosio, C., De Santis, L., Frey, B., Diallo, S. and Keller, C. 2005. Distribution of cadmium in leaves of *Thlaspi caerulescens*, J. Exp. Bot. 56: 565–575.

Craciun, A.R., Meyer, C.-L., Chen, J., Roosens, N., De Groodt, R., Hilson, P. and Verbruggen, N. 2012. Variation in HMA4 gene copy number and expression among Noccaea caerulescens populations presenting different levels of Cd tolerance and accumulation. J. Exp. Bot. 63: 4179–4189. doi: 10.1093/jxb/ers104.

Dräger, B.D., Desbrosses-Fonrouge, A.G., Krach, C., Chardonnens, A.N., Meyer, R.C. and Saumitou-Laprade, P. 2004. Two genes encoding *Arabidopsis halleri* MTP1 metal transport proteins co-segregate with zinc tolerance and account for high MTP1 transcript levels. Plant J. 39: 425–439.

Elbaz, B., Shoshani-Knaani, N., DavidAssael, O., Mizrachy-Dagri, T., Mizrahi, K., Saul, H., Brook, E., Berezin, I. and Shaul, O. 2006. High expression in leaves of the zinc hyperaccumulator Arabidopsis halleri of AhMHX, a homolog of an *Arabidopsis thaliana* vacuolar metal/proton exchanger. Plant Cell Environ. 29: 1179–1190. doi: 10.1111/j.1365-3040.2006.01500.x.

ElMehdawi, A.F., Quinn, C.F. and Polin-Smith, E.A.H. 2011. Effects of selenium hyperaccumulation on plant–plant interactions: evidence for elemental allelopathy. New Phytol. 191(1): 120–131.

Emmanuel, Delhaize, Tatsuhiko Kataoka, Diane M. Hebb, Rosemary G. White and Peter R. Ryan. 2003. Genes encoding proteins of the cation diffusion facilitator family that confer manganese tolerance. The Plant Cell 15: 1131–1142.

Escarré, J., Lefebvre, C., Gruber, W., Leblanc, M., Lepart, J., Riviere, Y. and Delay, B. 2000. Zinc and cadmium hyperaccumulation by *Thlaspi caerulescens* from metalliferous and nonmetalliferous sites in Mediterranean area: implication for phytoremediation. New Phytol. 145: 429–437.

Fernando, D.R., Bakkaus, E.J., Perrier, N., Baker, A.J.M., Woodrow, I.E., Batianoff, G.N. and Collins, R.N. 2006a. Manganese accumulation in the leaf mesophyll of four tree species: a PIXE/EDAX localization study. New Phytol. 171: 751–758. doi: 10.1111/j.1469-8137.2006. 01783.x.

Fernando, D.R., Batianoff, G.N., Baker, A.J.M. and Woodrow, I.E. 2006b. *In vivo* localisation of manganese in the hyperaccumulator Gossia bidwillii (Benth.) N. Snow & Guymer (Myrtaceae) by cryoSEM/EDAX. Plant Cell Environ. 29: 1012–1020.

Fiegl, J.l., Bryan P. McDonnell, Jill A. Kostel, Mary E. Finster and Dr. Kimberly Gray. 2011. A resource guide: The phytoremediation of Lead to urban residential soils. Site adapted from a report from Northwestern University.

Fontem, Lum, A., Ngwa, E.S.A., Chikoye, D. and Suh, C.E. 2015. Phytoremediation potential of weeds in heavy metal contaminated soils of the Bassa Industrial Zone of Douala, Cameroon. Int. J. Phytoremediation 16(3): 302–319.

Freeman, J.L., Zhang, L.H., Marcus, M.A., Fakra, S., McGrath, S.P. and Pilon-Smits, E.A.H. 2006. Spatial imaging, speciation and quantification of Se in the hyperaccumulator plants *Astragalus bisulcatus* and *Stanleya pinnata*, Plant Physiol. 142: 124–134.

Frérot, H., Faucon, M.P., Willems, G., Godé, C., Courseaux, A., Darracq, A., Verbruggen, N. and Saumitou-Laprade, P. 2010. Genetic architecture of zinc hyperaccumulation in *Arabidopsis halleri*: the essential role of QTLx environment interactions. New Phytol. 187: 355–367.

Frey, B., Keller, C., Zierold, K. and Schulin, R. 2000. Distribution of Zn in functionally different leaf epidermal cells in the hyperaccumulator *Thlaspi caerulescens*. Plant Cell Environ. 23: 675–687.

Galeas, M.L., Zhang, L.H., Freeman, J.L., Wegner, M. and Pilon-Smits, E.A.H. 2007. Seasonal fluctuations of selenium and sulphur accumulation in selenium hyperaccumulators and related nonaccumulators. New Phytol. 173: 517–525.

Garcia-Salgado, S., Garcia-Casillas, D., Quijano-Nieto, M.A. and Bonilla-Simon, M.M. 2012. Arsenic and heavy metal uptake and accumulation in native plant species from soil polluted by mining activities. Water Air Soil Pollut. 223: 559–572.

Gardea-Torresdey, J.L., De la Rosa, G., Peralta-Videa, J.R., Montes, M., Cruz-Jimenez, G. and CanoAguilera, I. 2005. Differential uptake and transport of trivalent and hexavalent chromium by tumbleweed (Salsola kali). Arch. Environ. Contam. Toxicol. 48: 225–232.

Giordani, C., Cecchi, S. and Zanchi, C. 2005. Phytoremediation of soil polluted by nickel using agricultural crops. Environ. Manage. 36: 675–681.

Girdhar, M., Sharma, N.R., Rehman, H., Kumar, A. and Mohan, A. 2014. Comparative assessment for hyperaccumulatory and phytoremediation capability of three wild weeds. Biotechnology 4: 579–589.

Gleba, D., Borisjuk, N.V., Borisjuk, L.G., Kneer, R., Poulev, A., Skarzhinskaya, M., Dushenkov, S., Logendra, S., Gleba, Y.Y. and Raskin, I. 1999. Use of plant roots for phytoremediation and molecular farming. Proc. Natl. Acad. Sci. U.S.A. 96: 5973–5977.

Gonzaga, M.I., Ma, L.Q., Santos, J.A. and Matias, M.I. 2009. Rhizosphere characteristics of two arsenic hyperaccumulating *Pteris* ferns. Sci. Total Environ. 407: 4711–4716.

Grennan, A.K. 2009. Identification of genes involved in metal transport in plants. Plant Physiol. 149: 1623–1624.

Gustin, J.L., Loureiro, M.E., Kim, D., Na, G., Tikhonova, M. and Salt, D.E. 2009. MTP1-dependent Zn sequestration into shoot vacuoles suggests dual roles in Zn tolerance and accumulation in Zn hyperaccumulating plants. Plant J. 57: 1116–1127.

Hall, J.L. 2002. Cellular mechanisms for heavy metal detoxification and tolerance. J. Exp. Bot. 53: 1–11.

Hanikenne, M., Talke, I.N., Haydon, M.J., Lanz, C., Nolte, A., Motte, P., Kroymann, J., Weigel, D. and Krämer, U. 2008. Evolution of metal hyperaccumulation required cisregulatory changes and triplication of HMA4. Nature 453: 391–395.

Hanson, B., Lindblom, S.D., Loeffler, M.L. and Pilon-Smits, E.A.H. 2004. Selenium protects plants from phloem-feeding aphids due to both deterrence and toxicity. New Phytol. 162: 655–662.

He, Z., Yan, H., Chen, Y., Shen, H., Xu, W., Zhang, H., Shi, L., Zhu, Y.G. and Ma, M. 2016. An aquaporin PvTIP4;1 from Pteris vittata may mediate arsenite uptake. New Phytol. 209: 746–761.

Hirai, M.Y., Fujiwara, T., Awazuhara, M., Kimura, T., Noji, M. and Saito, K. 2003. Global expression profiling of sulphur-starved Arabidopsis by DNA microarray reveals the role of O-acetyl-l-serine as a general regulator of gene expression in response to sulphur nutrition. Plant J. 33: 651–663.

Hörger, A.C., Fones, H.N. and Preston, G.M. 2013. The current status of the elemental defense hypothesis in relation to pathogens. Front. Plant Sci. 4(395): 1–11.

Hossain, M.A., Hasanuzzaman, M. and Fujita, M. 2011. Coordinate induction of antioxidant defense and glyoxalase system by exogenous proline and glycinebetaineis correlated with salt tolerance in mung bean. Front. Agric. China 5: 1–14. doi: 10.1007/s11703-010-1070-2.

Ingle, R.A., Smith, J.A. and Sweetlove, L.J. 2005. Responses to nickel in the proteome of the hyperaccumulator plant Alyssum lesbiacum. Biometals 18(6): 627–641.

Kachenko, A.G., Bhatia, N. and Singh, B. 2011. Influence of drought stress on Ni hyperaccumulating shrub *Hybanthus floribundus* (Lindl) F. Muell subsp. *floribundus*. Int. J. Plant Sci. 17(3): 315–322.

Karimi, N., Ghaderian, S.M., Raab, A., Feldmann, J. and Meharg, A.A. 2009. An arsenic accumulating, hypertolerant brassica, *Isatis cappadocica*. New Phytol. 184: 41–47.

Kawashima, C.G., Noji, M., Nakamura, M., Ogra, Y., Suzuki, K.T. and Saito, K. 2004. Heavy metal tolerance of transgenic plants over-expressing cysteine synthase. Biotechnol. Lett. 26: 153–157.

Kidd, P. and Monterroso, C. 2005. Metal extraction by *Alyssum serpyllifolium* spp. Lusitanicum on mine-spoil soils from Spain. Sci. Total Environ. 336: 1–11.

Kidd, P., Barceló, J., Bernal, M.P., Navari-Izzo, F., Poschenrieder, C., Shilev, S., Clemente, R. and Monterroso, C. 2009. Trace element behaviour at the root–soil interface: implications in phytoremediation. Environ. Exp. Bot. 67: 243–259.

Kim, D., Gustin, J.L., Lahner, B., Persans, M.W., Baek, D., Yun, D.J. and Salt, D.E. 2004. The plant CDF family member TgMTP1 from the Ni/Zn hyperaccumulator Thlaspi goesingense acts to enhance efflux of Zn at the plasma membrane when expressed in *Saccharomyces cerevisiae*. Plant J. 39: 237–251.

Knight, B., Zhao, F.J., McGrath, S.P. and Shen, Z.G. 1994. Zinc and cadmium uptake by hyperaccumulator Thlaspi caerulescens in contaminated soils and its effects on the concentration and chemical speciation of metals in soil solution. Plant Soil 197: 71–78.

Kotrba, P., Najmanova, J., Macek, T., Ruml, T. and Mackova, M. 2009. Genetically modified plants in phytoremediation of heavy metal and metalloid soil and sediment pollution. Biotechnol. Adv. 27: 799–810.

Krämer, U., Cotter-Howells, J.D., Charnock, J.N., Baker, A.J.M. and Smith, J.A.C. 1996. Free histidine as a metal chelator in plants that accumulate nickel. Nature 379: 635–638.

Krämer, U., Pickering, I.J., Prince, R.C., Raskin, I. and Salt, D.E. 2000. Subcellular localization and speciation of nickel in hyperaccumulator and non-accumulator *Thlaspi* species. Plant Physiol. 122: 1343–1354.

Krishnamurti, G.S.R., Cieśliński, G., Huang, P.M. and Van Rees, K.C.J. 1997. Kinetics of cadmium release from soils as influenced by organic acids: Implication in cadmium availability. Environ. Qual. 26: 271–277.

Küpper, H., Lombi, E., Zhao, F.J. and McGrath, S.P. 2000. Cellular compartmentation of cadmium and zinc in relation to other elements in the hyperaccumulator *Arabidopsis halleri*. Planta 212: 75–84.

Küpper, H., Mijovilovich, A., MeyerKlaucke, W. and Kroneck, P.M.H. 2004. Tissue- and agedependent differences in the complexation of cadmium and zinc in the Cd/Zn hyperaccumulator Thlaspi caerulescens (Ganges ecotype) revealed by X-ray absorption spectroscopy. Plant Physiol. 134: 748–757. doi: http://dx.doi.org/10.1104/pp.103.032953.

Küpper, H. and Kroneck, P.M.H. 2005. Heavy metal uptake by plants and cyanobacteria. pp. 97–142. *In*: Sigel, A., Sigel, H. and Sigel, R.K.O. (eds.). Metal Ions in Biological Systems. Vol. 44 (New York: Marcel Dekker, Inc.).

Küpper, H., Mijovilovich, A., Götz, B., Küpper, F.C. and Meyer-Klaucke, W. 2009. Complexation and toxicity of copper in higher plants (I): Characterisation of copper accumulation, speciation and toxicity in Crassula helmsii as a new copper hyperaccumulator. Plant Physiol. 151: 702–714. doi: 10.1104/pp.109.139717.

Leitenmaier, B., Witt, A., Witzke, A., Stemke, A., Meyer-Klaucke, W., Kroneck, P.M. and Küpper, H. 2011. Biochemical and biophysical characterisation yields insights into the mechanism of a Cd/Zn transporting ATPase purified from the hyperaccumulator plant *Thlaspi caerulescens*. Biochim. Biophys. Acta 1808: 2591–2599. doi: 10.1016/j.bbamem.2011.05.010.

Leitenmaier, B. and Küpper, H. 2013. Compartmentation and complexation of metals in hyperaccumulator plants. Front. in Plant Sci. 4: 1–13.

Li, H.Y. and Chen, Z.S. 2006. The influence of EDTA application on the interactions of cadmium, zinc, and lead and their uptake of rainbow pink (Dianthus chinensis). J. Hazardous Mater. 137: 1710–1718.

Li, T., Xu, Z., Han, X., Yang, X. and Sparks, D.L. 2012. Characterization of dissolved organic matter in the rhizosphere of hyperaccumulator Sedum alfredii and ts effect on the mobility of zinc. Chemosphere 88: 570–576. doi: 10.1016/j.chemosphere.2012.03.031.

Li, Y.M., Chaney, R., Brewer, E., Roseberg, R., Angle, J.S., Baker, A., Reeves, R. and Nelkin, J. 2003. Development of a technology for commercial phytoextraction of nickel: Economic and technical consideration. Plant and Soil 246: 107–115.

Lombi, E., Zhao, F.J., McGrath, S.P., Young, S.D. and Sacchi, G.A. 2001. Physiological evidence for a high-affinity cadmium transporter highly expressed in a *Thlaspi caerulescens* ecotype. New Phytol. 149: 53–60.

Lovy, L., Latt, D. and Sterckeman, T. 2013. Cadmium uptake and partitioning in the hyperaccumulator Noccaea caerulescens exposed to constant Cd concentrations throughout complete growth cycles. Plant Soil 362: 345–354. doi: 10.1007/s11104012-1291-7.

Lu, L., Tian, S., Zhang, J., Yang, X., Labavitch, J.M., Webb, S.M., Latimer, M. and Brown, P.H. 2013. Efficient xylem transport and phloem remobilization of Zn in the hyperaccumulator plant species Sedum alfredii. New Phytol. 198: 721–731. doi: 10.1111/nph.12168.

Lux, A., Martinka, M., Vaculík, M. and White, P.J. 2011. Root responses to cadmium in the rhizosphere: a review. J. Exp. Bot. 62: 21–37. doi: 0.1093/jxb/erq281.

Ma, L.Q., Komar, K.M., Tu, C., Zhang, W. and Cai, Y. 2001. A fern that hyperaccumulates arsenic. Nature 409: 579.

Ma, Y., Prasad, M.N.V., Rajkumar, M. and Freitas, H. 2011. Plant growth promoting rhizobacteria and endophytes accelerate phytoremediation of metalliferous soils. Biotechnol. Adv. 29: 248–258.

Malik, R.N., Husain, S.Z. and Nazir, I. 2010. Heavy metal contamination and accumulation in soil and wild plant species from industrial area of Islamabad. Pakistan J. Bot. 42(1): 291–301.

Martínez-Alcalá, I., Hernández, L.E., Esteban, E., Walker, D.J. and Bernal, M.P. 2013. Responses of Noccaea caerulescens and Lupinus lbusintrace elements-contaminated soils. Plant Physiol. Biochem. 66: 47–55. doi: 10.1016/j.plaphy.2013.01.017.

McCutcheon, S.C. and Schnoor, J.L. 2003. Phytoremediation: Transformation and Control of Contaminants. John Wiley & Sons, Inc., New Jersey.

McGrath, S.P. and Zhao, F.J. 2003. Phytoextraction of metals and metalloids from contaminated soils. Curr. Opin. Biotechnol. 14: 277–282.

Meharg, A.A. and Hartley-Whitaker, J. 2002. Arsenic uptake and metabolism in arsenic resistant and non-resistant plant species. New Phytol. 154: 29–42.

Mijovilovich, A., Leitenmaier, B., Mayer-Klaucke, W., Kroneck, P.M.H., Götz, B. and Küpper, H. 2009. Complexation and toxicity of copper in higher plants II. Different mechanisms for copper vs. cadmium detoxification in the copper-sensitive cadmium/zinc hyperaccumulator *Thlaspi caerulescens* (Ganges ecotype). Plant Physiol. 151: 715–731. doi: 10.1104/pp.109.144675.

Milner, M.J. and Kochian, L.V. 2008. Investigating heavy-metal hyperaccumulation using *Thlaspi caerulescens* as a model system. Ann. Bot. 102: 3–13.

Navari-Izzo, F. and Quartacci, M.F. 2001. Phytoremediation of metals: tolerance mechanisms against oxidative stress. Minerva Biotechnol. 13: 23–83.

Odjegba, V.J. and Fasidi, I.O. 2004. Accumulation of trace elements by Pistia stratiotes: Implications for phytoremediation. Ecotoxicology 13: 637–646.

Peng, H.Y., Yang, X. and Jiang, L. 2005. Copper phytoavailability and uptake by *Elsholtzia splendens* from contaminated soil as affected by soil amendments. J. Environ. Sci. Health 40: 839–856.

Persant, M.W., Nieman, K. and Salt, D.E. 2001. Functional activity and role of cation-efflux family members in Ni hyperaccumulation in *Thlaspi goesingense*. Plant Biol. 98: 9995–10000.

Poynton, C.Y., Huang, J.W.W., Blaylock, M.J., Kochian, L.V. and Ellass, M.P. 2004. Mechanisms of arsenic hyperaccumulation in *Pteris* species: root As influx and translocation. Planta 219: 1080–1088.

Prasad, M.N.V. 2005. Nickelophilous plants and their significance in phytotechnologies. Braz. J. Plant Physiol. 17(1): 113–128.

Raab, A., Feldman, J. and Meharg, A.A. 2004. The nature of arsenic–phytochelatin complexes in Holcus lanatus and Pteris cretica. Plant Physiol. 134: 1113–1122.

Rai, P.K. 2008. Phytoremediation of Hg and Cd from industrial effluent using an aquatic free floating macrophytes Azolla pinnata.Int. J. of Phytoremed. 10: 430–439.

Rascio, N. and Navari-Izzo, F. 2011. Heavy metal hyperaccumulating plants: How and why do they do it? And what makes them so interesting. Plant Sci. 180: 169–181.

Rauser, W.E. 1999. Structure and function of metal chelators produced by plants: the case for organic acids, amino acids, phytin and metallothioneins. Cell Biochem Biophys. 31: 19–48.

Reeves, R.D. and Baker, A.J.M. 2000. Metal-accumulating plants. pp. 193–229. *In*: Raskin, I. and Ensley, B.D. (eds.). Phytoremediation of Toxic Metals: Using Plants to Clean up the Environment. John Wiley & Sons.

Reeves, R.D. 2006. Hyperaccumulation of trace elements by plants. pp. 1–25. *In*: Morel, J.L., Echevarria, G. and Goncharova, N. (eds.). Phytoremediation of Metal-Contaminated Soils, NATO Science Series: IV: Earth and Environmental Sciences. Springer, NY.

Robinson, B.H., Leblanc, M. and Petit, D. 1998. The potential of Thlaspi caerulescens for phytoremediation of contaminated soils. Plant Soil 203: 47–56.

Salt, D.E., Prince, R.C., Baker, A.J.M., Raskin, I. and Pickering, I.J. 1999. Zinc ligands in the metal accumulator *Thlaspi caerulescens* as determined using X-ray absorption spectroscopy. Environ. Sci. Technol. 33: 713–717.

Sarret, G., Saumitou-Laprade, P., Bert, V., Proux, O., Hazemann, J.L., Traverse, A., Marcus, M.A. and Manceau, A. 2002. Forms of zinc accumulated in the hyperaccumulator Arabidopsis halleri. Plant Physiol. 130: 1815–1826.

Schat, H., Llugany, M., Vooijs, R., Hartley-Whitaker, J. and Bleeker, P.M. 2002. The role of phytochelatins in constitutive and adaptive heavy metal tolerances in hyperaccumulator and non-hyperaccumulator metallophytes. J. Exp. Bot. 53: 2381–2392.

Schmidt, U. 2003. Enhancing phytoextraction: the effect of chemical soil manipulation on mobility, plant accumulation, and leaching of heavy metals. J. Environ. Qual. 32(6): 1939–54.

Schwitzuébel, J.P., Kumpiene, J., Comino, E. and Vanek, T. 2009. From green to clean: a promising approach towards environmental remediation and human health for the 21st century. Agrochimica, LIII-N 4: 209–237.

Shibagaki, N., Rose, A., McDermott, J.P., Fujiwara, T., Hayashi, H., Yoneyama, T. and Davies, J.P. 2002. Selenate-resistant mutants of *Arabidopsis thaliana* identify Sultr1; 2, a sulfate transporter required for efficient transport of sulfate into roots. Plant J. 29: 475–486.

Sors, T.G., Martin, C.P. and Salt, D.E. 2009. Characterization of selenocysteine methyltransferases from Astragalus species with contrasting selenium accumulation capacity. Plant J. 59: 110–122.

Sun, R., Zhou, Q. and Jin, C. 2006. Cadmium accumulation in relation to organic acids in leaves of *Solanum nigrum* L. as a newly found cadmium hyperaccumulator. Plant Soil 285: 125–134.

Takahashi, R., Ishimaru, Y., Senoura, T., Shimo, H., Ishikawa, S., Arao, T., Nakanishi, H. and Nishizawa, N.K. 2011. The OsNRAMP1 iron transporter is involved in Cd accumulation in rice. J. Exp. Bot. 62: 4843–4850. doi: 10.1093/jxb/err136.

Takahashi, R., Bashir, K., Ishimaru, Y., Nishizawa N.K. and Nakanishi, H. 2012. The role of heavy metal ATPases, HMAs, in zinc and cadmium transport in rice. Plant Signal. Behav. 7: 1605–1607. doi: 10.4161/psb.22454.

Talke, I.N., Hanikenne, M. and Krämer, U. 2006. Zinc-dependent global transcriptional control, transcriptional deregulation, and higher gene copy number for genes in metal homeostasis of the hyperaccumulator Arabidopsis halleri. Plant Physiol. 142: 148–167.

Tamura, H., Honda, M., Sato, T. and Kamachi, H. 2005. Pb hyperaccumulation and tolerance in common buckwheat (*Fagopyrum esculentum* Moench). J. Plant Res. 118: 355–359.

Tian, S., Lu, L., Labavitch, J., Yang, X., He, Z., Hu, H., Sarangi, R., Newville, M., Commisso, J. and Brown, P. 2011. Cellular sequestration of cadmium in the hyperaccumulator plant species Sedum alfredii. Plant Physiol. 157: 1914–1925. doi: 10.1104/pp.111. 183947.

Tu, C., Ma, L.Q. and Bondada, B. 2002. Arsenic accumulation in the hyperaccumulator Chinese brake and its utilization potential for phytoremediation. J. Environ. Qual. 31(5): 1671.

Ueno, D., Milner, M.J., Yamaji, N., Yokosho, K., Koyama, E., Clemencia Zambrano, M., Kaskie, M., Ebbs, S., Kochian, L.V. and Ma, J.F. 2011. Elevated expression of TcHMA3 plays a key role in the extreme cadmium tolerance in a Cd-hyperaccumulating ecotype of Thlaspi caerulescens. Plant J. 66: 852–862. doi: 10.1111/j.1365313X.2011.04548.x.

Uraguchi, S., Izumi, Watanabe, Akiko, Yoshitomi, Masako, Kiyono and Katsuji Kuno. 2006. Characteristics of cadmium accumulation and tolerance in novel Cd-accumulating crops, *Avena strigosa* and *Crotalaria juncea*. J. Exp. Bot. 57(12): 2955–2965.

Van, Hoof, N.A.L.M., Koevoets, P.L.M., Hakvoort, H.W.J., Ten Bookum, W.M., Schat, H., Verkleij, J.A.C. and Ernst, W.H.O. 2001. Enhanced ATP-dependent copper efflux across the root cell plasma membrane in copper-tolerant Silene vulgaris. Physiol. Plant. 113: 225–232. doi: 10.1034/j.1399-3054.2001.1130210.x.

Van, Huysen, T., Terry, N. and Pilon-Smits, E.A.H. 2004. Exploring the selenium phytoremediation potential of transgenic Indian mustard over-expressing ATP sulfurylase or cystathionine-synthase. Int. J. Phytoremed. 6: 111–118.

Verbruggen, N.C., Hermans and Schat, H. 2009. Molecular mechanisms of metal hyperaccumulation in plants. New Phytol. 181: 759–776.

Weber, M., Harada, E., Vess, C., von Roepenack-Lahaye, E. and Clemens, S. 2004. Comparative microarray analysis of *Arabidopsis thaliana* and *Arabidopsis halleri* root identifies nicotinamine synthase, a ZIP transporter and other genes as potential metal hyperaccumulation factors. Plant J. 37: 269–281.

Wenzel, W.W., Bunkowski, M., Puschenreiter, M. and Horak, O. 2003. Rhizosphere characteristics of indigenous growing nickel hyperaccumulator and excluder plants on serpentine soil. Environ. Poll. 123: 131e138.

Willems, G., Dräger, D.B., Courbot, M., Godé, C., Verbruggen, N. and Saumitou-Laprade, P. 2007. The genetic basis of zinc tolerance in the metallophyte Arabidopsis halleri ssp. halleri (Brassicaceae): an analysis of quantitative trait loci. Genetics 176: 659–674.

Willems, G., Frerot, H., Gennen, J., Salis, P., Saumitou-Laprade, P. and Verbruggen, N. 2010. Quantitative trait loci analysis of mineral element concentrations in the *Arabidopsis halleri* × *Arabidopsis lyrata petraea* F2 progeny on cadmium-contaminated soil. New Phytol. 187: 368–379.

Xue, S.G. 2004. Manganese uptake and accumulation by the hyperaccumulator plant *Phytolacca acinosa* Roxb (Phytolaccaceae). Environ. Poll. 131: 393–399.

Yang, X.E., Li, T.Q., Long, X.X., Xiong, Y.X., He, Z.L. and Stoffella, P.J. 2006. Dynamics of zinc uptake and accumulation in the hyperaccumulating and nonhyperaccumulating ecotypes of *Sedum alfredii* Hance. Plant Soil 284: 109–119.

Zhang, H., Xu, W., Guo, J., He, Z. and Ma, M. 2005. Coordinated responses of phytochelatins and metallothioneins to heavy metals in garlic seedlings. Plant Sci. 69: 1059–1065.

Zhao, F.J, Hamon, R.E., Lombi, E., McLaughlin, M.J. and McGrath, S.P. 2002. Characteristics of cadmium uptake in two contrasting ecotypes of the hyperaccumulator *Thlaspi caerulescens*. J. Exp. Bot. 53: 535–543.

Zhao, F.J., Jiang, R.F., Dunham, S.J. and McGrath, S.P. 2006. Cadmium uptake, translocation, and tolerance in the hyperaccumulator *Arabidopsis halleri*. New Phytol. 172: 646–654. doi: 10.1111/j.14698137.2006.01867.x.

Biological Indicators for Monitoring Soil Quality under Different Land Use Systems

Bisweswar Gorain[1,*] *and Srijita Paul*[2]

1. Introduction

Soil quality or health differs from air or water quality as the former is specified by its dynamic chemical, physical, biological and even ecological properties, whereas the latter are related to the concentration of specific contaminants with well-defined threshold limits. Moreover, when referring to air or water quality, there is hardly any well-defined ideal state due to the limitless number of environmental scenarios (Sojka and Upchurch 1999), unlike when soil quality is referred to. Since soil quality is central to agriculture, environment and other disciplines, authors have been defining soil quality from different perspectives from time to time (Parr et al. 1992, Doran and Parkin 1994, Harris et al. 1996, Karlen et al. 1997). This becomes evident when one searches for the term 'soil quality' on internet search engine as more than 14,000 publications can be found since the 1940s, signifying the importance linked to this subject. This also reflects the tremendous effort of the soil scientists to objectify the subjective nature of soil quality, i.e., to quantify the quality of a soil (Sojka and Upchurch 1999). This is the reason the term continues to lack a universal definition and researchers define it according to specific situations or their interests. Perhaps, due to this, the "concept of quality" is much pertinent to use instead of focusing on soil quality as such and as Sojka and Upchurch (1999) affirm, anything that can be defined infinitely is actually indefinable. However, one must agree that maintenance of soil quality is critical for ensuring the sustainability of the environment and the biosphere despite the difficulty of defining soil quality in a universal framework (Smith et al. 1993, Arshad and Martin 2002). Rigorous studies on this particular aspect becomes more important as fertile agricultural soils are recently becoming more vulnerable to natural or environmental degradation than never before, often by erosion, leaching, or even without anthropogenic intervention (Popp et al. 2000).

However, from agricultural sustainability and soil health point of view, soil quality is defined as the capacity of a soil to function within its natural ecosystem boundaries to support healthy plants and animals, maintain or enhance air or water quality and support human health and habitation (Karlen et al. 1997) and is often described based on indices. A soil quality index may be defined as

---

[1] ICAR-CSSRI, RRS Bharuch, Gujarat, India.
[2] SAMETI, Ramakrishna Mission Ashrama, Narendrapur, Kolkata, W.B., India.
* Corresponding author: gorainbisweswar@gmail.com

the minimum set of interrelated parameters that provides quantitative data on the capacity of a soil to carry out one or more functions. A soil quality indicator is a sensitive and measurable parameter that provides soil the capacity to carry out a given function (Acton and Padbury 1993). Studies on soil quality indices are complex in nature as diverse physical, chemical, microbiological and biochemical properties need to be integrated statistically to reach to such quality indices (Papendick and Parr 1992, García et al. 1994, Halvorson et al. 1996).

2. Importance and characteristics of sensitive soil indicators in relation to soil quality and functions

2.1 Importance of soil indicators

Soil indicators are important parameters that explain the soil quality at a given time and which provide reference to measure trends and patterns of soil quality changes with cropping systems, management practices and land use systems. It also helps in assessing the soil quality parameters in relation to the precautionary levels so that pertinent actions may be taken before the deterioration of soil quality. The relationship between soil quality and soil functions are also described by the indicators (Bastida et al. 2008). Even though soil indicators are specific to land use systems, cropping systems and management practices, there are certain characteristics which are common for all situations. Thus, the characteristics of a good soil indicator, in general, may be ascribed as follows:

- It should be sensitive to land use under study.
- It should explain relationship with soil function.
- It should have ease and reliability of measurement.
- It should explain spatial and temporal patterns of soil quality variation.
- It should be sensitive to changes in soil management.
- It should be comparable during routine sampling and monitoring.
- It should involve simple skills for interpretation.

2.2 Sensitive indicators of soil quality in relation to soil functions

The important soil functions are as follows: biodiversity maintenance, food, feed, fibre and timber production, water and solute flow, filtering and buffering of different chemicals, nutrient cycling and providing structural support. There are sensitive indicators suited for specific soil functions (Table 1).

Table 1. Sensitive biological indicators (in bold letters) specific to soil functions.

Soil functions	Indicators
Production	OC, N, pH, EC, Al, Avl. water, **weed species**, soil strength
Biodiversity	OC, pH, bases, **basal respiration**, pH, EC, crusts, soil water tension
Water and solute flow	Tillage, aggregate stability, **earthworms, termites**, porosity, bulk density, soil structure
Filtering and buffering	**Basal respiration**, OC, texture, CEC, **microbial biomass**, chemical constituents' concentration
Nutrient cycling	OC, N, CEC, **basal respiration, microbial biomass, POM, PMN, enzyme activities (dehydrogenase, urease, phosphatase)**
Structural support	Soil texture, structure, BD, landscape slope, aggregate stability

[OC-organic carbon, EC-electrical conductivity, CEC-cation exchange capacity, N-nitrogen, POM-particulate organic matter, PMN-potentially mineralizable nitrogen].

3. Approaches for quantitative estimation of soil quality

The quantitative estimation of soil quality in terms of an index value is highly significant due to the fact that it provides a holistic idea about the soil health, production potential and environmental sustainability. The different approaches for quantitative estimation of soil quality (Doran and Parkin 1994) are as follows:

- **Comparative assessment:** In this process, the quality of a soil is assessed in comparison to a reference soil with similar genesis and mineralogy but different management practices.
- **Dynamic assessment:** It deals with the changes in soil physical, chemical and biological properties with time. The soil properties are analyzed before and after imposition of certain treatments and the temporal variations are measured.
- **Regression analysis:** In this statistical approach, some of the sensitive indicators of soil (soil organic carbon, aggregate stability, enzymatic activity, etc.) are correlated with certain management practices (fertilizer management, water availability) and the data are statistically processed to infer the inter-dependence of these variables.
- **Pedotransfer functions:** This method is used to assess the efficacy of soil indicators through meta-statistical approach. In this technique, a parameter of interest (indicator) is quantified from known parameters using linear and non-linear functions with empirical values.
- **Standardize scoring functions based on threshold limits and base line values:** In this process, the different indicators of soil are weighed based on their sensitivity to soil quality under specific management goals. Later, they are summed up depending on their threshold values and in reference to the baseline. The index thus obtained gives an idea about soil quality. Unlike the above mentioned processes, this takes into account all the sensitive indicators based on their importance for the determination of soil quality.
- **Principal component analysis (PCA):** In recent times, it is the most widely used technique to assess the soil quality. In this process, a minimum data set is prepared with different related indicators and they are interpreted with certain scores depending on their role in soil quality maintenance. Later, the scores are integrated to arrive at an index value (Figure 1) which represents the soil quality (Andrews et al. 2004).

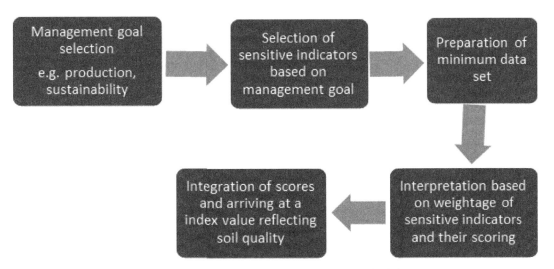

Figure 1. Flowchart depicting assessment of soil quality index.

4. Sensitive indicators (physical, chemical and biological) of soil quality specific to objectives of experiments over the years

Soil quality indicators are primarily classified into three broad categories: physical, chemical and biological. The physical indicators deal with the orientation of pores and soil particles, thereby governing root growth, germination of seeds and rate of plant growth. The indicators include soil depth, bulk density, porosity (macro and micro), soil strength, aggregate stability, texture, structure and compaction.

Chemical indicators include soil reaction, salinity and sodicity, organic matter content, nutrient availability, cation exchange capacity, nutrient cycling rate, and the concentration of contaminants such as heavy metals, organic compounds, radioactive substances, etc. These indicators are helpful in assessing the continuum of toxic substances in soil-water-plant-animal system. Physical and chemical indicators are more static in nature as compared to the biological indicators which are more dynamic.

Biological indicators include measurements of soil macro-organisms, e.g., earthworm, nematodes, termites, ants and microorganisms, e.g., microbial biomass, fungi, actinomycetes, or lichens for their role in soil development and conservation, nutrient cycling and specific soil fertility. Biological indicators also comprise of metabolic processes as well as metabolites such as metabolic quotient (qCO_2; ratio of respiration to microbial biomass), glomalin and ergosterol concentrations, enzymes, e.g., cellulases, arylsulfatase, phosphatases, etc. related to specific functions of substrate degradation or mineralization of organic N, S or P. These indicators are regarded as the biological fingerprints of past soil. Several authors have tried to point out indicators sensitive to land use changes and management practices. The indicators depend upon the climatic conditions as well as inherent soil properties, types of crop grown as well as the management practices (Bastida et al. 2008).

Table 2. Objective specific sensitive indicators of soil quality used by authors over the years.

Objectives	Indicators used	Authors
Evaluation of soil quality under climax vegetation	MBC, PMN, phosphatase, b-glucosidase and urease activities	Trasar-Cepeda et al. 1998
Soil quality evaluation under forest management	BD, water table depth, N mineralization, aeration, microbial activity	Burguer and Kelting 1999
Integrated pollution index: soil quality of heavy metal contaminated sites	Content of heavy metals (Cu, Ni, Pb and Zn)	Chen et al. 2006
Microbiological Degradation Index: soil quality assessment of semiarid degraded lands	Dehydrogenase, urease activities, soil respiration, water soluble C	Bastida et al. 2008
Relative soil stability index: effects of herbicide on soil functional stability	Arylsulphatase, b-glucosidase, urease	Becaert et al. 2006
Integrated Fertility Index: evaluation of soil fertility	Organic matter, total N, P and K, available P and K, alkali-hydrolysable N, electrical conductivity	Pang et al. 2006
Biological Quality Index: variation in relation to the ecosystem degradation	Soil respiration, carboxymethyl cellulase, β-D-glucosidase and dehydrogenase activity	Armas et al. 2007
Soil Quality Index: Evaluation of recuperation of hydrocarbon contaminated soils by nutrient applications, surfactants or soil agitation	Microbial biomass C, respiration, dehydrogenase activity and earthworms	Dawson et al. 2007
Evaluate environmental soil quality of forest soils under natural vegetation without human intervention	β-glucosidase, urease and phosphatase activities, microbial biomass C	Zornoza et al. 2007

5. Important biological indicators sensitive to soil quality assessment

5.1 Earthworm species and population

The population and diversity of earthworm is an important biological indicator of soil quality. They are instrumental in organic matter decomposition, particularly by processing the soil in their guts. They support aggregate stability, improve water-holding capacity, and moderate pore size and infiltration rate (Stockdill 1982). In the southern agricultural region of Saint-Lawrence Valley-Canada, significant relationship was observed between two endogen earthworm species and soil aggregate stability as soil quality indicators. The presence of *Aporrectodea caliginosa* was correlated with silt and clay contents, while *Allolobophora chlorotica* was related with organic amendments (Lapied et al. 2009). Incorporation of organic matter in soil as amendments regulates soil temperature and aeration. Earthworms have been regarded as anthropogenic land use indicators. In Netherlands, they are used as a biological indicator of soil quality, whereas in Germany as an indicator for soil biological site classification, a practical way to define the use of soil, based on the structure of the earthworm community, their abundance and biomass (Rombke et al. 2005). These organisms are widely recognized indicators of soil health due to their sensitivity to pesticides and heavy metals (Cikutovic et al. 1999), and organo-metallic compounds (pentachlorophenol) (Booth et al. 2000, Bunn et al. 1996). The science of earthworms in soil quality improvement is still in progress. Suitable research techniques are required for exploring the knowledge obtained in the micro-scale within upper soil layers to a field scale or landscape scale. Furthermore, suitable framework is required to understand the role of earthworms in the biogeochemical cycles.

5.2 Ants and termites

Ants are considered as important invertebrates as they are very good indicators of soil quality. The ants are preferred for the study due to their abundance, macroscopic size, and sensitivity to soil degradation. It has been observed that soils from ant nests (*Messor andrei*) contain more nutrients and harbor major groups of soil micro fauna and flora, e.g., bacteria, fungi, nematodes, miscellaneous eukaryotes and microarthropods as compared to adjacent non-ant soil from a semi-arid, serpentine grassland in California. The relevance of ants as biological indicators exist particularly in restoration processes after adverse soil impacts (e.g., mining, dumping of hazardous wastes) as their prevalence depends on the composition and diversity of plant communities projecting them as a better predictor of soil health than plant species diversity (Andersen et al. 2002, Boulton et al. 2003). Moreover, association of ants with termites have been regarded as indicators of land recovery due to the enhanced carbon and nutrient levels post processing of soil organic matter through their enzymatic systems (De Bruyn and Conacher 1990). Cammeraat et al. (2002) analyzed the effect of *Messor bouvieri* (seed harvesting ants) on soil fertility, infiltration of water, soil structure, and hydrophobicity of semi-arid soils in Spain and they reported that ant nests had a lower soil reaction, higher organic carbon concentrations and inorganic nutrients, greater structural stability, and significantly higher infiltration rate than the soils of the adjacent areas.

Termites are also used as soil quality indicators. Roose-Amsaleg et al. (2005) studied the nests of soil-feeding termite (*Cubitermes*) of different ages (fresh to mature to old) in tropical rain forest of Lopé-Gabón Africa and found significantly higher concentration of potassium and phosphorous, clay and fine silt, organic matter, water retention capacities and cation exchange capacity in mature nests as compared to control soils. In savanna ecosystems in Colombia, Mora et al. (2005) studied the dynamics of 10 soil enzymes (xylanase, amylase, cellulase, α-glucosidase, β-glucosidase, β-xylosidase, *N*-acetyl-glucosaminidase, alkaline and acidphosphatases, and laccase) to characterize the functional diversity of soil-feeding termite (*Ruptitermes* sp.) and soil pellets formed by two species of leaf-cutting ants (*Acromyrmex landolti* and *Atta laevigata*) in comparison to a control soil; it was observed that some of the soil pellets have unique enzymatic profile than the control soil and the diversity of these structures and species is related to different pathways for the decomposition of organic matter (Jimenez et al. 2008).

5.3 Metabolic substances (ergosterol and glomalin)

Several compounds are found in soil, e.g., sterols, antibiotics, protein, enzymes, etc. as a result of microbe mediated catabolic and anabolic processes in soil. The two most important of them with respect to soil health monitoring include ergosterol and glomalin.

5.3.1 Ergosterol

It is predominantly considered as the principal sterol secreted endogenously from fungi, actinomycetes, and some of the microalgae. The concentration of ergosterol is a critical indicator of fungal proliferation on organic compounds and its mineralization efficiency (Battilani et al. 1996). It was demonstrated that ergosterol content was not affected by heavy metals concentrations (Cu 80 ppm, Zn 50 ppm or Cd 10 ppm) and fungicides (Thiram 3 ppm or pentachlorophenol 1.5 ppm) even at concentrations that reduce the metabolic activity by 18% to 53% (pollutant stressed cultures), while Zineb, a fungicide at a concentration of (25 ppm), reduced the ergosterol content significantly. Similarly, a significant correlation was observed between fungal hyphae and ergosterol concentration in pastures and arable soils. It was also demonstrated through electron microscopy that a beneficial role of fungi in thixotropy (a physical process involving the orientation of the claymicelles) also exists in soil (Barajas et al. 2002, Molope et al. 1987). Puglisi et al. (2003) worked out the concentrations of cholesterol, sitosterol, and ergosterol in some agricultural soils (hazelnut irrigated with contaminated water and intensive horticulture) and reported that crop rotation does not affect the concentration of these sterols, but ergosterol hastened the metabolic activity in soils with industrial contamination. In the Pacific region, Joergensen and Castillo (2001) found positive correlations between qCO_2 and ergosterol to biomass C. They opined that the low soil microbial availability was attributed to the low soil organic matter and phosphorus leading to poor soil fertility and health in Nicaragua. The reduced fungal population caused P deficiency to crops probably due to the inability of plant roots to access immobile nutrients for plant growth.

5.3.2 Glomalin

It is an important fungal component which is hydrophobic and proteinaceous in nature (Wright and Upadhyay 1996). Glomalin as glomalin-related soil protein (GRSP) is reported to improve soil stability by avoiding water mediated defloculation (Wright et al. 2008, Wright and Upadhyay 1998). A good correlation between glomalin concentration and the amount of water stable aggregates (WSA) was observed. Glomalin-related soil protein (GRSP) being gelatinous in nature seals most soil pores, thereby hindering penetration of water in soil aggregates (Wright and Upadhyaya 1998, Rillig 2004, Harner et al. 2004, Rillig 2004). GRSP is often used as a biochemical marker in soil due to its stability even with negative management effects (Rosier et al. 2006) on mycorrhizal fungi, viz. tillage, and inclusion of fallow into crop rotation. Bedini et al. (2009) used isolates of *Glomus mosseae* and *Glomus intraradices* for inoculating *Medicago sativa* plants in a microcosm experiment and reported enhanced soil aggregate stability as mean weight diameter (MWD) of macro aggregates of 1–2 mm diameter, in inoculated soils compared to non-inoculated ones. They also reported a strong positive correlation between GRSP concentration and soil aggregate stability with mycorrhizal root volume and a weak correlation with total root volume.

5.4 Microbial biomass carbon (MBC)

Microorganisms are widely recognized as sensitive soil quality indicators. They play a crucial role in nutrient cycling and energy flow (Li and Chen 2004) and are sensitive to intercropping, organic matter addition, and management practices (64) affecting soil structure stabilization (Suman and Gaur 2006). Microbial biomass carbon is a meager but active fraction of soil organic matter and is of immense interest in soil fertility due to its sensitivity to management practices than the bulk organic matter (Janzen 1987). It acts as a pool of plant nutrients and is vital in governing the nutrient availability to plants. Several authors (Azmal et al. 1996, Sridevi et al. 2003) have reported a significant enhancement in microbial biomass following incorporation of crop residues. Post residue incorporation, microbial biomass-C (MBC) enhanced and reached maximum value of MBC (as per the studies of Azmal et al. 1996, Sridevi et al. 2003). Ocio and Brooks (1990) reported that addition of straw, as compared to control, improved the microbial biomass by 87.5% in a sandy loam soil and by about 50% in a clay soil. Malik et al. (1998) observed a large increase in microbial biomass during the early stages of rice crop on wheat straw and green manure application in a rice–wheat cropping system. This also resulted in synchronization of N release from soil and its uptake by plants. Patra et al. (1992) opined that biomass C content varied with type of crop residues as they found higher MBC in wheat straw amended soil compared to cowpea; however, for microbial biomass N (MBN), it was vice-versa. Azmal et al. (1997) showed an immediate increase in microbial biomass C and N on rice straw incorporation under aerobic incubations in a clay loam soil, which reached maxima after 7 days of each application (2 g C kg^{-1} soil as rice straw at an interval of 1.5 months), and decreased thereafter. The maximum biomass formation reached its optima after the second application, signifying the limited capacity of soil to incorporate biomass. In rice-lentil (Lens esculenta) crop rotation under dry land conditions, microbial C was maximum (408–420 µg g^{-1}) with the application of wheat straw (10 t ha^{-1}) and fertilizer, followed by only straw (360–392 µg g^{-1}) and only fertilizer treatments (238–246 µg g^{-1}) (Singh and Singh 1991). Straw incorporation coupled with fertilizer treatment showed 77% more microbial biomass C accumulation over control. The rapid initial increase of microbial activity may probably be attributed to the faster catabolism of simple C compounds contained in crop residues.

5.5 Microbial biomass nitrogen (MBN)

Microbial Biomass N is the assimilated N in the body of the microbes. It is in a state of dynamic equilibrium and represents a significant proportion of the total soil N, which is relatively constant. The biomass N increased from 46 µg N in control to 80 µg N g^{-1} soil by 5th day and remained at this level at 20th day on application of labeled wheat straw to a clay loam soil (Ocio et al. 1991). Bremer and Van Kessel (1992) worked on the microbial C and N dynamics after incorporating ^{14}C- and ^{15}N-labeled lentil and wheat straw under field conditions in a sandy loam soil. In their experiment, 65 to 81% of added ^{15}N was traced in residue which suggested that microbial biomass minimizes losses of N during the lag phases of crop demand and may supply the same during the log phase. Kushwaha et al. (2000) showed increase in MBC from 214–264 µg g^{-1} during crop growth from in straw removal treatment and from 368–503 µg g^{-1} in straw incorporation treatment after two annual rice-barley crop cycles. The MBC and MBN were enhanced by 48% and 60%, respectively, in residue retained over residue devoid plots. Microbial biomass may behave as slow-release N fertilizer. Bird et al. (2001) observed significantly higher soil microbial biomass on straw incorporated plots compared to straw devoid plots. As soil microbial biomass is a major source of plant available N, straw incorporation increased the soil N status as well as humic substances.

5.6 Potentially mineralizable N

Potentially mineralizable nitrogen (PMN) is an important biological indicator which shows the capacity of a soil microbial community to mineralize nitrogen present in soil organic residues into one of the plant available forms of N, i.e., NH^{4+}. Mijangos et al. (2006) studied the sensitivity of soil biological parameters, viz. soil enzymes, potentially mineralizable nitrogen, soil respiration, abundance of earthworms, and microbial community metabolic profiles on the fertilization and tillage practices in a field trial with forage corn. They observed higher biological efficiency in organically fertilized, no-tillage plots, compared to conventional tillage and mineral fertilization. The values of potentially mineralizable N, basal respiration, substrate induced respiration, and earthworm abundance thus recorded were 23.1, 35.3, 55.7, and 226.9 per cent more in organically fertilized, no-tillage plots than in inorganically fertilized, conventional tillage plots, respectively. They concluded that potentially mineralizable nitrogen as well as soil enzymes play a key role as early indicators of soil quality changes compared to conventional physicochemical parameters as they were more sensitive to monitor the relatively subtle changes in soil properties caused by these management practices.

5.7 Enzyme activities

Enzymes are the biochemical agents that catalyze the soil metabolic processes. In the past, majority of researches focused on the microflora in soil and hardly any focus was put on the extracellular enzymes of microbial origin formed after decomposition of organic matter. The principal functions of enzymes in soil include decomposition of organic matter, catalyzing metabolic reactions, assisting various life processes of microorganisms in soils, stabilization of soil structure, humus formation, nutrient cycling, facilitating an early indication of the soil history and effects of agricultural management (Ceron and Melgarejo 2005, Kandeler et al. 2006). For these, they are used as soil biological indicators from the 80s. Due to their intimate association with several indicators of biogeochemical cycles, decomposition of organic matter and soil remediation processes, they can be used to assess the physical, chemical and biological properties and therefore the quality of a soil (Gelsomino et al. 2006). Enzymes are widely recognized as good indicators because of their (a) close relation to organic matter decomposition, soil physical characteristics, microbial activity and biomass in the soil, (b) provide early information about changes in soil quality, and (c) rapid assessibility (Dick 1996, Nielsen and Winding 2001, Eldor 2007). It is due to the fact that enzymes in soil are produced both intra and extracellularly from microbes (bacteria, fungi, plants, and a range of macro invertebrates) and are associated to different substrates like alive or dead cells, clays or/ and humic molecules and assay laboratory conditions, it is essential to optimize the protocols for enzymatic activity determination. Since extraction of enzymes from soils is a difficult task and they lose their integrity easily (Gianfreda and Ruggiero 2006, Verchot and Borelli 2005, Dick 1996), utmost care should be taken regarding temperature, incubation period, pH buffer, ionic strength of the solution, and substrate concentration before their assessment (Gutierrez et al. 2008, Eldor 2007).

5.7.1 β-Glucosidase

It is widely observed in soil system and its presence has been detected in soil, fungi and plants. It has been used as a key soil quality indicator due to its crucial role on cellulose degradation, releasing glucose as energy source for the maintenance of metabolically active microbial biomass in soil (Dick 1996, Sotres 2005). It also helps in releasing energy upon degradation of labile carbon in soil thereby stabilizing recalcitrant C pools (Knight and Dick 2004). However, due to its reduced efficiency in the presence of heavy metals, it is a less advocated indicator of soil quality in heavy metal contaminated soil (Makoi and Ndakidemi 2008). In free state in soil system, it usually has a small lifespan, as it is easily prone to degradation, denaturation and irreversible inhibition. A portion of these free enzymes get adsorbed on soil clay minerals, gets incorporated with the humic particles and therefore lose stability and stay in soil in less catalytically active state (Marx et al. 2005, Burns 1982).

5.7.2 Phosphatase

Phosphorus is a primary nutrient crucial for plant growth (especially roots) and activation of several enzymes in plant system. The availability of this element is largely governed by soil pH. In acid soils, it is fixed as Fe-P and Al-P and in calcareous soils as Ca-P, thereby making it bio-unavailable (Dick et al. 2000). These fixed P are solubilized by soil microbes, e.g., phosphate solubilizing bacteria, viz. *Bacillus polymyxa, Pseudomonas striata*, fungi, viz. *Aspergillus awamori, Penicillium digitatum*. They release low molecular weight organic acids, which release the fixed P into soil solution through the production of extracellular enzymes as phosphatases (Sundara and Hari 2002). Phosphatases hydrolyze esters and anhydrides of phosphoric acid. Its activity depends on extracellular concentration of phosphatase enzymes, which can be in free state in soil solution, adsorbed in the humic fraction or clay minerals (Sundara and Hari 2002, Turner and Haygarth 2005). Since phosphomonoesterases are active both under acidic and alkaline soil reactions and solubilize low molecular weight P-compounds, viz. nucleotides, sugar phosphates and polyphosphates, they are the most studied soil enzymes (Makoi and Ndakidemi 2008) and thus are widely used as soil quality indicators. A strong correlation was observed between phosphatase activity and soil properties such as pH, total N, organic P and clay content by Turner and Haygarth (2005) while working in temperate grassland, suggesting it to be a good soil biological indicator.

5.7.3 Dehydrogenase

Dehydrogenase is an important soil enzyme under oxidoreductase group which oxidizes a substrate through the reduction of an electron acceptor. In soil, determination of dehydogenase enzyme activity provides a large amount of information about soil fertility as well as soil health. These enzymes are an integral part of microorganisms and are instrumental in organic matter oxidation; nevertheless, its activity is weakly correlated with microbial respiration or microbial biomass (Dick 1996). In spite of this, it is considered as a soil quality indicator for its involvement in electron transport systems of oxygen metabolism in an intracellular environment (living cells) to show its activity (Kandeler and Dick 2007). Unlike hydrolases (β-Glucosidase, urease, phosphatase), it is inactive in extracellular environment and therefore cannot provide information about soil degeneration processes; its activity depends on management practices and/or climatic conditions (Kandeler and Dick 2007). Dehydrogenase activity is related to living and active cells, but presence of heavy metals hinders its activity through the catalysis of assay by extracellular phenol oxidase and alternative electron acceptors such as nitrate and humic substances (Tate 2002, Speir and Ross 2002). Speir and Ross (2002) and Kandeler and Dick (2007) observed that presence of Cu adversely affected dehydrogenase activity.

5.7.4 Urease

These enzymes hydrolyze urea into CO_2 and NH_3 and consequently reduce soil pH through production of carbonic acid and hasten N losses through volatilization of gaseous NH_3. This enzyme is of major concern as urea is the most widely used nitrogenous fertilizer in developing countries like India and the N availability to plants is regulated by this enzyme. However, since majority of N fertilizer is lost through volatilization and leaching, role of urease enzymes to the total losses of N is meager (Makoi and Ndakidemi 2008). Studies on other related enzymes, e.g., ammonia monooxygenase (AMO), are in their infancy and sufficient data about their interaction in soil is still to be obtained (Gutierrez et al. 2009). For this reason, it is not included as a soil quality indicator but this membrane-bound enzyme could give insights on nitrification rates and the efficiency of nitrification inhibitors, instead of quantifying nitrate *per se*. Gutiérrez et al. (2009) observed the spatial variability of AMO in a paddy field in Chile and reported its reduced activity in fallow and dry soils. However, a positive correlation was found with soil available N suggesting its usefulness in studying nitrification, denitrification and volatilization processes in soil. Urease has widely been used as a soil quality indicator due to its sensitivity to management practices, especially organic fertilization, soil tillage, cropping history, organic matter content, soil depth, management practices, heavy metals as well as environmental factors, such as temperature and pH (Yang et al. 2006, Saviozzi et al. 2001). This enzyme, of extracellular origin, represents up to 63% of total soil enzymatic activity. Several authors have reported its relation to soil microbial community as well as soil properties (physical, chemical and biological) (Corstanje et al. 2007). The stability of this enzyme is dependent on organo-metallic complexes and humic substances, which makes them resistant to denaturing agents such as heat and other proteolytic scenarios. The better understanding of this enzyme will provide efficient ways to urea fertilizer management, especially in warm humid regions, submerged soils and under irrigated agriculture (Makoi and Ndakidemi 2008).

5.8 Species and population of arthropods

The abundance and diversity of invertebrate communities like arthropods are sensitive to chemical and physical soil characteristics. Soil biological quality is assessed through indices (Shannon, Menhinick, Simpsonand Pielou indices). Santorufo et al. (2012) found Acarina, Collembola, Enchytraeids and Nematoda to be more resistant taxa to the urban environment. They reported collembolans to be more sensitive to changing soil properties. Maleque et al. (2009) found arthropod populations as sensitive bioindicators for the sustainable management of conifer plantations. The structural elements of forests (e.g., deadwood, coarse woody debris, understory vegetation, herbaceous plants, flowering plants, nectar plants, leaf litter) affect soil health which consequently influences the arthropod communities. They opined that monitoring arthropods as bioindicators may be a cost effective technique for assessing sustainable forest management plans.

6. Sensitive biological indicators in different land use systems based on management goals

6.1 Sensitive biological indicators in rice-wheat cropping system in a Typic Haplustept (productivity and environmental protection)

In a long-term experiment (7 years), conducted at the Indian Agricultural Research Institute farm (semi-arid, sub-tropical), New Delhi, important biological, chemical and physical indicators of soil quality were assessed and were later unified to a soil quality index (SQI). Two tillage methods, three water management techniques and nine nutrient management practices were evaluated for the study. For the identification of critical soil indicators with two primary goals—productivity (PCASQI-P) and environmental protection (PCSQI-EP), principal component analysis (PCA) was carried out and a soil quality index (SQI) was calculated using the Soil Management Assessment Framework (SMAF). The results indicated that management goal significantly influenced indicator selection and fluctuations in the index values of these indicators can provide early warning against deterioration of soil quality (Bhaduri and Purakayastha 2014). The sensitive indicators for rice and wheat with productivity and environmental protection are illustrated in Table 3.

Table 3. Sensitive biological indicators in rice-wheat cropping system based on management goals.

Crops	Management goals	Sensitive indicators
Rice	Productivity	MBC, PMN
	Environmental protection	MBC, PMN, DHA
Wheat	Productivity	PMN, DHA
	Environmental protection	MBC, PMN, DHA

[MBC-Microbial biomass carbon, PMN-Potentially mineralizable nitrogen, DHA-dehydrogenase activity].

6.2 Soil microbial biomass C and N and their mineralization rates as sensitive indicators in woody and grassland communities (sustainability and environmental protection)

In the subtropical Rio Grande Plains of southern Texas and northern Mexico, Mc Culley et al. (2004) studied soil biological indicators, viz. soil respiration, potential C and N mineralization rates, microbial biomass in soil and nitrification rates in grassland and adjoining woody plant communities. The larger C and N pool sizes in the woody communities recorded higher annual soil respiration (SR) (745 compared to grasslands 611 g C m^{-2}yr^{-1}), greater soil microbial biomass C (444 compared to 311 mg Ckg^{-1} soil), potential rates of N mineralization (0.9 compared to 0.6 mg N kg^{-1} ·d^{-1}) and nitrification (0.9 compared to 0.4 mg N kg^{-1} ·d^{-1}). The mean residence time of surface SOC in wooded communities (11 years) was higher than that of grassland communities (6 years). It was reported that both labile and recalcitrant pools of SOC and total N increased when there was a shift from grassland to woody vegetation due to the higher flux as well as accumulation of C and N in the latter. The study concluded that soil respiration, potential C and N mineralization rates, and microbial biomass in soil were sensitive indicators of soil quality in woody plant communities and grasslands.

6.3 Glomalin as a soil biological indicator in undisturbed forest site vs. restored sites (soil sustainability)

Vasconcellos et al. (2013) studied arbuscular mycorrhizal fungi (AMF) and glomalin related soil protein (GRSP) to understand their potential as biological indicators of soil quality in an undisturbed forest site and three other sites at different stages of recovery after reforestation for 20, 10 and 5 years. They observed that AMF species distribution, total-GRSP (T-GRSP) and easily extractable-GRSP (EE-GRSP) have influence on soil physical, chemical and microbiological attributes. A positive correlation between EE-GRSP and T-GRSP as well as total carbon, nitrogen, dehydrogenase and urease activity, microbial biomass carbon and microbial biomass nitrogen was recorded, especially in summer (Table 4). Macroporosity had a positive effect and soil bulk density had a negative correlation with EE-GRSP, signifying usage of either EE-GRSP or T-GRSP as biological indicator depending on the soil characteristics and management. The study demonstrated the effect of recovery period, seasonality and other soil attributes on AMF and GRSP distribution and indicated the use of glomalin as a potential soil biological indicator in the Atlantic forest of Brazil.

Table 4. Correlation coefficient between Glomalin content and soil biological indicators.

Variables	EE-Glomalin		T-Glomalin	
	Summer	Winter	Summer	Winter
MBC	0.48	0.43	0.42	0.22
MBN	0.57	0.62	0.41	0.40
Acid phosphatase	0.42	0.38	0.36	0.064 ns
Urease	0.55	0.29	0.43	0.11 ns
Dehydrogenase	0.66	0.44	0.63	0.16 ns

6.4 Soil biological indicators under forest, extensive rubber plantations (jungle rubber), rubber and oil palm monocultures (sustainability and environmental protection)

Guillaume et al. (2016) studied soil biological indicators (basal respiration, microbial biomass, acid phosphatase) in Ah horizons from rainforests, jungle rubber, rubber (*Hevea brasiliensis*) and oil palm (*Elaeis guineensis*) plantations in Sumatra. The negative impact of land-use changes on the measured biological indicators increased in the following order: forest < jungle rubber < rubber < oil palm. Microbial C use efficiency was not dependent on land use systems. The basal respiration and SOC was non-linearly related, i.e., SOC losses reduced microbial activity. Therefore, a meager reduction in C content under oil palm as compared to rubber plantations caused a great reduction in microbial activity. They concluded that the biological indicators thus studied can quantitatively assess resilience of agroecosystems with various use intensities and therefore can monitor soil quality.

6.5 Earthworm biomass as a sensitive biological indicator to tillage practices (yield and sustainability)

A modelling approach was used with production as management goal to predict the effect of agricultural management practices (pesticide applications and tillage) on soil functioning through earthworm populations. It was found that zero and reduced tillage practices can enhance crop yields while sustaining natural ecosystem functions. The results thus obtained suggested that conventional tillage practices have prolong effects on soil biota than pesticide control, provided the pesticide has a shorter half-life. They reported that an increase in soil organic matter could increase the recovery rate of earthworm populations (Johnston et al. 2015). The earthworm biomass (gm^{-2}) as well as crop yield decreased in the following sequence: zero tillage (27.1) > reduced (40) > conventional tillage (49.7) (Table 5).

Table 5. Earthworm biomass as a sensitive biological indicator of soil quality in different tillage practices.

Tillage practice	Earthworm biomass (gm^{-2})	Productivity (tha^{-1})
Conventional	27.1	2.18
Reduced	40.0	2.63
Zero	49.7	2.95

6.6 Soil biological indicators under long term fertilizer and manure use in a sub-tropical inceptisol (productivity and sustainability)

Masto et al. (2007) analyzed microbial biomass carbon (MBC), microbial quotient (MBC/SOC), soil respiration, dehydrogenase and phosphatase activities in a 31 years field experiment involving manure and inorganic fertilizers treatments in a maize (*Zea mays* L.)–wheat (*Triticum aestivum* L.)–cowpea (*Vigna unguiculata* L.) rotation at the Indian Agricultural Research Institute, New Delhi, India and observed that the application of farmyard manure (FYM) plus NPK fertilizer significantly improved microbial biomass (124–291 $mgkg^{-1}$) and microbial quotient from 2.88 to 3.87. Besides, soil respiration, dehydrogenase and phosphatase enzyme activities were also increased with FYM application. The response of MBC to FYM+100% NPK compared to 100% NPK was significantly higher than that for soil respiration (6.24 vs. 6.93 $mlO_2\ g^{-1}\ h^{-1}$), indicating MBC to be more sensitive indicator to FYM addition than soil respiration. Dehydrogenase activity increased as NPK rates were increased from 50% to 100%, but at 150% NPK, it showed a decrease. Phosphatase activity was more sensitive to season or crop type as compared to fertilizer treatment, although both MBC and phosphatase activity were increased with balanced fertilization. Calleja-Cervantes et al. (2015) worked in a vineyard with 13 years of continued application of composted organic wastes and observed changes in soil quality characteristics. They concluded that continuous long-term application of organic amendments affects soil quality positively through the augmentation of microbial activity.

6.7 Soil biological indicators sensitive to metal pollution (soil pollution assessment)

Epelde et al. (2008) studied the growth of *Thlapsi caerulescens* in metal contaminated soils and recorded shoot metal concentration as high as 337 mg Cd, 5670 mg Zn and 76.6 mg Pb per kg of dry plant weight. *T. caerulescens*, under heavy metal contaminated condition, showed significant effect on soil biological parameters. In metal polluted sites, *T. caerulescens* recorded 154, 115, 140, 37 and 164 per cent increase in the b-glucosidase, arylsulphatase, acid phosphatase, alkaline phosphatase and urease activity, respectively. They concluded that soil enzymes are sensitive biological indicators to evaluate the process of metal phytoextraction.

6.8 Mycorrhizal fungi fatty acid methyl ester (FAME) biomarkers, soil microbial biomass and enzymes as sensitive indicators to cover crop and tillage in a silty loam soil (environmental protection)

Mbuthia et al. (2015) studied the effect of long term (31 years) tillage (till and no-till), cover crops (Hairy vetch; *Vicia villosa* and winter wheat; *Triticum aestivum*, and a no cover control) and N-rates (0, 34, 67 and 101 kg N ha^{-1}) on soil microbial community structure, activity and resultant soil quality index using the soil management assessment framework (SMAF) scoring index under continuous cotton production on a silt loam soil in West Tennessee. Significantly, greater abundance of Gram positive bacteria, actinomycetes, mycorrhizae fungi fatty acid methyl ester (FAME) biomarkers were observed in no-till as compared to tilled soil. The saprophytic fungal FAME biomarkers were significantly less abundant under no-till treatments resulting in a lower fungi to bacteria ratio. Different enzymes associated with C, N and P cycling (b-glucosidase, b-glucosaminidase, and phosphodiesterase) showed significantly greater activities under no-till as compared to tilled soil. Mycorrhizae fungi biomarkers showed significant reduction under increased N-rate and were much less under vetch cover crop compared to wheat and no-cover, suggesting them as sensitive indicators to crop types. Consequently, the total organic carbon (TOC) and b-glucosidase SMAF quality scores were significantly greater under no-till compared to till and under the vetch compared to wheat and no cover treatments, resulting in a significantly greater soil quality index (SQI) in the latter. Their results indicate that under a long-term no-till system and use of cover crops under a low biomass monoculture crop production system (cotton), mycorrhizal biomarkers, and soil enzymes like b-glucosidase, b-glucosaminidase, and phosphodiesterase are sensitive indicators to assess soil quality.

7. Conclusions

Biological indicators of soil quality are more sensitive to soil orders, land use changes, cropping systems and management practices due to their dynamic nature. However, to obtain a soil quality index, the physical and chemical properties are also equally important. The sensitivity of different biological indicators were found to be specific to the management goals, e.g., production, sustainable soil management, environmental protection, etc. Microbial biomass carbon (MBC), potential mineralizable nitrogen (PMN) and dehydrogenase activity (DHA) were found to be the sensitive biological indicators in a rice–wheat cropping system in a Typic Haplustept. The microbial biomass and enzyme activity were found to be the sensitive biological indicators under different tillage practices in a Typic Calcixerept. Good correlation was observed between Glomalin and other soil biological indicators, indicating it to be a good soil biological indicator in difference stages of restored soils in the Atlantic forest of Brazil. Soil ergosterol concentration acted as an indicator of fungal biomass in a boreal forest soil. Biological indicators of soil health, such as soil enzymes and community level physiological profiles, were found to be valid tools to evaluate the success of a continuous metal phytoextraction process. The microbial biomass carbon (MBC) and basal respiration were higher in forest ecosystem followed by rubber and oil palm plantations, confirming them as suitable biological indicators sensitive to land use systems. Biological indicators such as MBC, soil respiration, DHA and PA activities increased with long term manure and fertilizer application.

References

Acton, D.F. and Padbury, G.A. 1993. A conceptual framework for soil quality assessment and monitoring. A Program to Assess and Monitor Soil Quality in Canada. Soil Quality Evaluation Summary. Res. Branch Agric. Ottawa, Canada. Soil Sci. 171: 210–222.

Andersen, A., Hoffmann, B., Muller, W. and Griffiths, A. 2002. Using ants as bioindicators in land management: simplifying assessment of ant community responses. J. App. Eco. 39: 8–17.

Andrews, S.S., Karlen, D.L. and Cambardella, C.A. 2004. The soil management assessment framework: a quantitative soil quality evaluation method. Soil Sci. Soc. Am. J. 68: 1945–1962.

Arshad, M.A. and Martin, S. 2002. Identifying critical limits for soil quality indicators in agro-ecosystems. Agric. Ecosyst. Environ. 88: 153–160.

Azmal, A.K.M., Marumoto, T., Shindo, H. and Nishiyama, M. 1996. Mineralization and changes in microbial biomass in water-saturated soil amended with some tropical plant residues. Soil Sci. Plant Nutr. 42: 483–492.

Azmal, A.K.M., Marumoto, T., Shindo, H. and Nishiyama, M. 1997. Changes in microbial biomass after continuous application of azolla and rice straw in soil. Soil Sci. Plant Nutr. 43: 811–818.

Barajas, M., Hassan, M., Tinoco, R. and Vazquez, R. 2002. Effect of pollutants on the ergosterol content as indicator of fungal biomass. J. Microbiol. Methods 50: 227–236.

Bastida, F., Zsolnay, A., Hernandez, T. and Garcia, C. 2008. Past, present and future of soil quality indices: A biological perspective. Geoderma. 147: 159–171.

Battilani, P., Chiusa, C., Trevisan, C. and Ghebbioni, C. 1996. Fungal growth and ergosterol content in tomato fruits infected by fungi. Italian J. Food Sci. 4: 283–90.

Bedini, S., Pellegrino, E., Avio, L., Pellegrini, S., Bazzoffi, P., Argese, E. and Giovannetti, M. 2009. Changes in soil aggregation and glomalin-related soil protein content as affected by the arbuscular mycorrhizal fungal species *Glomus mosseae* and *Glomusintraradices*. Soil Bio. Biochem. 41(7): 1491–1496.

Bhaduri, D. and Purakayastha, T.J. 2014. Long-term tillage, water and nutrient management in rice–wheat cropping system: Assessment and response of soil quality. Soil Tillage Res. 144: 83–95.

Bird, J.A., Howarth, W.R., Eagle, A.J. and Van Kessel, C. 2001. Immobilization of fertilizer nitrogen in rice: Effects of straw management practice. Soil Sci. Soc. Am. J. 65: 1143–1152.

Booth, L.H., Hodge, S. and O'Halloran, K. 2000. The use of enzyme biomarkers in Aporrectodeacaliginosa (Oligochaeta; Lumbricidae) to detect organophosphate contamination: a comparison of laboratory tests, mesocosms and field studies. Environ. Toxic. Chem. 19: 417–422.

Boulton, A., Jaffee, B. and Scow, K. 2003. Effects of a common harvester ant (Messorandrei) on richness and abundance of soil biota. App. Soil Eco. 23(3): 257–265.

Bremer, E. and Van Kessel, C. 1992. Plant available nitrogen from lentil and wheat residues during a subsequent growing season. Soil Sci. Soc. Am. J. 56: 1155–1160.

Bunn, K., Thompson, H. and Tarrant, K. 1996. Effects of agrochemicals on the immune systems of earthworms. Bulletin Environ. Cont. and Toxic. 57: 632–639.

Burns, R. 1982. Enzyme activity in soil: Location and a possible role in microbial activity. Soil Bio. Biochem. 14: 423–427.

Calleja-Cervantes, M.E., Gonz_alez, Ignacio Irigoye and Pedro M. Sergio Menendez. 2015. Thirteen years of continued application of composted organic wastes in a vineyard modify soil quality characteristics. Soil Bio. Biochem. 90: 241–254.

Cammeraat, L., Willott, S., Compton, S. and Incoll, L. 2002. The effects of ants' nests on the physical, chemical and hydrological properties of a rangeland soil in semi-arid Spain. Geoderma. 105(1-2): 1–20.

Ceron, L. and Melgarejo, L. 2005. Enzimas del suelo: Indicadores de salud y calidad. Acta Biologica Colombiana. 10(1): 5–18.

Cikutovic, M.F., Goven, L., Venables, A. and Giggleman, B. 1999. Wound healing in earthworms Lumbricusterrestris: a cellular-based biomarker for assessing sublethal chemical toxicity. Bulletin Environ. Cont. and Toxic. 62: 508–514.

Corstanje, R., Schulin, R. and Lark, R. 2007. Scale-dependent relationships between soil organic matter and urease activity. European J. Soil Sci. 58(5): 1087–1095.

De Bruyn, L. and Conacher, A. 1990. The role of termites and ants in soil modification—A review. Australian J. Soil Res. 28(1): 55–93.

De Bruyn, L. 1999. Ants as bioindicators of soil function in rural environments. Agric. Ecosystems Env. 74(1-3): 425–441.

Dick, R. 1996. Soil enzyme activities as integrative indicators of soil health. pp. 121–56. *In*: Doran, J. and Jones, A. (eds.). Methods for Assessing Soil Quality. Madison, Wisconsin: Soil Science Society of America, Inc.

Dick, W., Cheng, L. and Wang, P. 2000. Soil acid and alkaline phosphatase activity as pH adjustment indicators. Soil Bio. Biochem. 32: 1915–1919.

Doran, J.W. and Parkin, T.B. 1994. Defining and assessing soil quality. pp. 3–21. *In*: Doran, J.W., Coleman, D.C., Bezdicek, D.F. and Stewart, B.A. (eds.). Defining Soil Quality for a Sustainable Environment. SSSA Special Pub., vol. 34. Soil Sci. Soc. Am., Madison, Wisconsin, USA.

Ebersberger, D., Niklaus, P. and Kandeler, E. 2003. Elevated carbon dioxide stimulates N-mineralization and enzymes activities in calcareous grassland. Soil Bio. Biochem. 35: 965–972.

Eldor, P. 2007. Soil Microbiology, Ecology and Biochemistry. Tercera, Eldor, P. (ed.). Chennai, India: Academic Press.

Epelde, L., Becerril, J.M., Javier Hernández-Allica, Oihana Barrutia and Carlos Garbisu. 2008. Functional diversity as indicator of the recovery of soil health derived from Thlaspi caerulescens growth and metal phytoextraction. App. Soil Eco. 39: 299–310.

García, C., Hernández, T. and Costa, F. 1994. Microbial activity in soils under Mediterranean environmental conditions. Soil Bio. Biochem. 26: 1185–1191.

Gelsomino, A., Badalucco, L., Ambrosoli, R., Crecchio, C., Puglisi, E. and Meli, S. 2006. Changes in chemical and biological soil properties as induced by anthropogenic disturbance: a case study of agricultural soil under recurrent flooding by wastewaters. Soil Bio. Biochem. 38: 2069–2080.

Gianfreda, L. and Ruggiero, P. 2006. Enzyme activities in soil. pp. 257–311. *In*: Nannipieri, P. and Smalla, K. (eds.). Nucleic Acids and Proteins in Soil. Springer, Heidelberg.

Guillaume, T., Maranguit, D., Murtilaksono, K. and Kuzyakov, Y. 2016. Sensitivity and resistance of soil fertility indicators to land-use changes: New concept and examples from conversion of Indonesian rain forest to plantations. Ecol. Indic. 67: 49–57.

Gutierrez, V., Pinzon, A., Casas, J. and Martinez, M. 2008. Determinacion de la actividadcelulolitica del sueloproveniente de cultivos de *Stevia rebaudiana* Bertoni. Agronomia Colombiana. 26(3): 497–504.

Gutierrez, V., Martinez, M. and Ortega, R. 2009. Spatial variability of the activity of β-glucosidase, acid phosphatase and ammonia monooxygenase (AMO) enzymes in a rice soil in Chile. Footprints in the Landscape: Sustainability through Plant and Soil Science. Pittsburgh: Section 97–104.

Halvorson, J.J., Smith, J.L. and Papendick, R.I. 1996. Integration of multiple soil parameters to evaluate soil quality: a field experiment example. Biol. Fertil. Soils. 21: 207–214.

Harner, M.J., Ramsey, P.W. and Rillig, M.C. 2004. Protein accumulation and distribution in floodplain soils and river foam. Eco. Letters 7: 829–836.

Harris, R.F., Karlen, D.L. and Mulla, D.J. 1996. A conceptual framework for assessment and management of soil quality and health. pp. 61–82. *In*: Doran, J.W. and Jones, A.J. (eds.). Methods for Assessing Soil Quality. SSSA Spec Publ., vol. 49. SSSA, Madison, Wisconsin.

Janzen, H.H. 1987. Soil organic matter characteristics after long-term cropping in various spring wheat rotations. Can. J. Soil Sci. 67: 845–856.

Jimenez, J, Decaens, T. and Lavelle, P. 2008. C and N concentrations in biogenic structures of a soil-feeding termite and a fungus growing ant in the Colombian savannas. Applied Soil Eco. 40(1): 120–128.

Joergensen, R. and Castillo, X. 2001. Interrelationships between microbial and soil properties in young volcanic ash soils of Nicaragua. Soil Bio. Biochem. 33: 1581–1589.

Johnston, A.L.A., Sibly, R.M., Hodson, M.E., Tania Alvarez and Pernille Thorbek. 2015. Effects of agricultural management practices on earthworm populations and crop yield: validation and application of a mechanistic modelling approach. J. App. Eco. 52: 1334–1342.

Kandeler, E., Mosier, A., Morgan, J., Milchunas, D., King, J., Rudolph, S. and Tscherko, D. 2006. Response of soil microbial biomass and enzyme activities to the trescient elevation of carbon dioxide in a semi-arid grassland. Soil Bio. Biochem. 38: 2448–2460.

Kandeler, E. and Dick, R. 2007. Soil enzymes: spatial distribution and function in agroecosystems. pp. 263–79. *In*: Benckiser, G. and Schnell, S. (eds). Biodiversity in Agricultural Production Systems. Boca Raton, Fl: Taylor & Francis Group.

Karlen, D.L., Mausbach, M.J., Doran, J.W., Cline, R.G., Harris, R.F. and Schuman, G.E. 1997. Soil quality: a concept, definition, and framework for evaluation. Soil Sci. Soc. Am. J. 61: 4–10.

Knight, T. and Dick, R. 2004. Differentiating microbial and stabilized β-glucosidase activity relative to soil quality. Soil Bio. Biochem. 36: 2089–2096.

Kushwaha, C.P., Tripathi, S.K. and Singh, K.P. 2000. Variations in soil microbial biomass and N availability due to residue and tillage management in a dryland rice agroecosystem. Soil Till. Res. 56: 153–166.

Lapied, E., Nahmani, J. and Rousseau, G. 2009. Influence of texture and amendments on soil properties and earthworm communities. Appl. Soil Eco. 43: 241–249.

Li, X. and Chen, Z. 2004. Soil microbial biomass C and N along a climatic transect in the Mongolia steppe. Bio. Fertility Soils 39: 344–351.

Makoi, J. and Ndakidemi, P. 2008. Selected soil enzymes: Examples of their potential roles in the ecosystem. African J. Biotech. 7: 181–191.

Maleque, M.A., Maeto, K. and Ishii, H.T. 2009. Arthropods as bioindicators of sustainable forest management, with a focus on plantation forests. Appl. Entomol. Zool. 44(1): 1–11.

Malik, V., Kaur, B. and Gupta, S.R. 1998. Soil microbial biomass and nitrogen mineralization in straw incorporated soils. pp. 557–565. *In*: Ecological Agriculture and Sustainable Development, Proceedings of the International Conference on Ecological Agriculture: Towards Sustainable Development. Vol. 1. Chandigarh, India.

Marx, M.-C., Kandeler, E., Wood, M., Wermbter, N. and Jarvis, S.C. 2005. Exploring the enzymatic landscape: distribution and kinetics of hydrolytic enzymes in soil particle-size fractions. Soil Bio. Biochem. 37: 35–48.

Masto, R.E., Chhonkar, P.K., Singh, D. and Patra, A.K. 2007. Soil quality response to long-term nutrient and crop management on a semi-arid Inceptisol. Agric. Ecosyst. Environ. 18: 130–142.

Mbuthia, L.W., Veronica Acosta-Martínez, DeBruyn, J., Schaeffer, S., Tyler, D., Odoi, E., Mpheshea, M., Walker, F. and Eash, N. 2015. Long term tillage, cover crop and fertilization effects on microbial community structure, activity: Implications for soil quality. Soil Biol. Biochem. 89: 24–34.

Mc Culley, R.L., Archer, S.R., Boutton, T.W., Honsf, M. and Zuberer, D.A. 2004. Soil respiration and nutrient cycling in wooded communities developing in grassland. Eco. 85(10): 2804–2817.

Mijangos, I., Roberto Pérez, Albizu, I. and Garbisu, C. 2006. Effects of fertilization and tillage on soil biological parameters. Enzyme Microbial Tech. 40: 100–106.

Molope, M., Grieve, I. and Page, E. 1987. Contribution by fungi and bacteria to aggregate stability: microbial aspects. J. Soil Sci. 38: 71–77.

Mora, P., Miambi, E., Jimenez, J., Decaens, T. and Rouland, C. 2005. Functional complement of biogenic structures produced by earthworms, termites and ants in the neotropical savannas. Soil Bio. Biochem. 37(6): 1043–1048.

Ocio, J.A. and Brooks, P.C. 1990. An evaluation of methods for measuring the microbial biomass in soils following recent additions of wheat straw and the characterization of the biomass that develops. Soil Bio. Biochem. 22: 685–694.

Ocio, J.A., Brooks, P.C. and Jenkinson, D.S. 1991. Field incorporation of rice straw and its effect on soil microbial biomass and soil inorganic N. Soil Bio. Biochem. 23: 171–176.

Papendick, R.I. and Parr, J.F. 1992. Soil quality—the key to a sustainable agriculture. Am. J. Altern. Agric. 7: 2–3.

Parr, J.F., Papendick, R.I., Hornik, S.B. and Meyer, R.E. 1992. Soil quality: attributes and relationship to alternative and sustainable agriculture. Am. J. Altern. Agric. 7: 5–10.

Patra, D.D., Bhandari, J.C. and Misra, A. 1992. Effect of plant residues on the size of microbial biomass and nitrogen mineralization in soil: Incorporation of cowpea and wheatstraw. Soil Sci. Plant Nutr. 38: 1–6.

Popp, J.H., Hyatt, D.E. and Hoag, D. 2000. Modeling environmental condition with indices: a case study of sustainability and soil resources. Ecol. Model. 130: 131–143.

Preger, A., Rillig, M., Johns, A., Du Preez, C., Lobe, I. and Amelung, W. 2007. Losses of glomalin-related soil protein under prolonged arable cropping: A chronosequence study in sandy soils of the South African Highveld. Soil Bio. Biochem. 39(2): 445–453.

Puglisi, E., Nicelli, M., Capri, E., Trevisan, E. and Del Re, A. 2003. Cholesterol, s-sitosterol, ergosterol, and coprostanol in agricultural soils. J. Environmental Quality 32: 466–471.

Rillig, M.C. 2004. Arbuscular mycorrhizae, glomalin and soil aggregation. Canadian J. Soil Sci. 84: 355–363.

Rombke, J., Jansch, S. and Didden, W. 2005. The use of earthworms in ecological soil classification and assessment concepts. Ecotoxic. and Environ. Safety 62(2): 249–65.

Roose-Amsaleg, C., Mora, P. and Harry, M. 2005. Physical, chemical and phosphatase activities characteristics in soil-feeding termite nests and tropical rainforest soils. Soil Bio. Biochem. 37(10): 1910–1917.

Rosier, C., Hoye, A. and Rillig, M. 2006. Glomalin-related soil protein: Assessment of current detection and quantification tools. Soil Bio. Biochem. 38(8): 2205–11.

Santorufo, L., Van Gestel, C.A.M., Annamaria Rocco and Giulia Maisto. 2012. Soil invertebrates as bioindicators of urban soil quality. Env. Poll. 161: 57–63.

Saviozzi, A., Levi-Minzi, R., Cardelli, R. and Riffaldi, R. 2001. A comparison of soil quality in adjacent cultivated, forest and native grassland soils. Plant Soil. 233: 251–259.

Shannon, D., Sen, A. and Johnson, D. 2002. A comparative study of the microbiology of soil structure under organic and conventional regimes. Soil Use Manage. 18: 274–283.

Singh, M. and Singh, T.B. 1991. Influence of organic and inorganic amendments, modified urea, and application methods on ammonia volatilization in saturated calcareous soil. Int. Rice Res. Newsl. 16(2): 17–18.

Smith, J.L., Halvorson, J.J. and Papendick, R.I. 1993. Using multivariable-indicator kriging for evaluating soil quality. Soil Sci. Soc. Am. J. 57: 743–749.

Sojka, R.E. and Upchurch, R.R. 1999. Reservation regarding the soil quality concept. Soil Sci. Soc. Am. J. 63: 1039–1054.

Sotres, F., Cepeda, C., Leiros, M. and Seoane, S. 2005. Different approaches to evaluating soil quality using biochemical properties. Soil Bio. Biochem. 37: 877–887.

Speir, T.W. and Ross, D.J. 2002. Hydrolytic enzyme activities to assess soil degradation and recovery. pp. 407–31. *In*: Burns, R. and Dick, R. (eds.). Enzymes in the Environment: Activity, Ecology and Applications. New York: Marcel Dekker.

Sridevi, S., Katyal, J.C., Srinivas, K. and Sharma, K.L. 2003. Carbon mineralization and microbial dynamics in soil amended with plant residues and residue fractions. J. Indian Soc. Soil Sci. 51: 133–139.

Stockdill, S. 1982. Effects of introduced earthworms on the productivity of New Zealand pastures. Pedobiologia. 24(1): 29–35.

Suman, A., Menhi, L., Singh, A. and Gaur, A. 2006. Microbial biomass turnover in India subtropical soil under different sugarcane intercropping systems. Agron. J. 98: 698–704.

Sundara, B. and Hari, V. 2002. Influence of phosphorus solubilizing bacteria on the changes in soil available phosphorus and sugar cane and sugar yields. Crops Res. 77: 43–49.

Tate, R.L. 2002. Microbiology and enzymology of carbon and nitrogen cycling. pp. 227–248. *In*: Burns, R.G. and Dick, R.P. (eds.). Enzymes in the Environment: Activity, Ecology and Applications. New York: Marcel Dekker.

Turner, B. and Haygarth, P. 2005. Phophatase activity in temperate pasture soils: potential regulation of labile organic phosphorous turnover by phosphodiesterase activity. Sci. Total Environ. 344: 37–46.

Vasconcellos, R.L.F., Bonfim, J.A., Dilmar Baretta and Elke, J.B.N. 2013. Cardoso arbuscular mycorrhizal fungi and glomalin-related soil protein as potential indicators of soil quality in a recuperation gradient of the Atlantic forest in Brazil. Land Degrad. Develop.

Verchot, L. and Borelli, T. 2005. Application of para-nitrophenol (pNP) enzyme assays in degraded tropical soils. Soil Bio. Biochem. 37: 625–633.

Wright, S.F. and Upadhyaya, A. 1996. Extraction of an abundant and unusual protein from soil and comparison with hyphal protein of arbuscular mycorrhizal fungi. Soil Sci. 161: 575–586.

Wright, S.F. and Upadhyaya, A. 1998. A survey of soils for aggregate stability and glomalin, a glycoprotein produced by hyphae of arbuscular mycorrhizal fungi. Plant Soil. 198: 97–107.

Wright, S.F., Green, V.S. and Cavigelli, M.A. 2008. Glomalin in aggregate size classes from three different farming systems. Soil Till. Res. 94: 546–549.

Yang, Y.Z., Liu, S., Zheng, D. and Feng, S. 2006. Effects of cadmium, zinc and lead on soil enzyme activities. J. Environ. Sci. 18(6): 1135–1141.

9

Aromatic Plants as a Tool for Phytoremediation of Salt Affected Soils

B.B. Basak,[1,]* *Smitha G.R.,*[2] *Anil R. Chinchmalatpure,*[3] *P.K. Patel*[1] *and Prem Kumar B.*[4]

1. Introduction

Soil, a non-renewable resource, is central to all primary production system. 'Salt-affected' is a general term used for soils which contain soluble salts or exchangeable sodium and/or both, in such amounts that can retard growth and development of plants. Such soils cause reduction in crop yield and are required to be managed and remediated for sustainable agriculture. Mostly, salt-affected soils exist in arid and semi-arid regions but are also found in some humid to sub-humid climatic areas, where conditions are favourable for their development. Over the years, in countries like India, due to high population rate, the landmass has suffered from different types of extreme pressure on crop lands and also from degradations. In arid and semi arid region, indiscriminate use of canal water for irrigation often causes accumulation of harmful salts in soil which limits crop productivity. Salt affected soils are distributed in 120 countries covering 1074 M ha and it reduced crop productivity to 7–8% at the global scale (Yadav 2003). In India, about 6.74 M ha (Table 1) land has been affected by salinity and alkalinity problems in different parts of country comprising Uttar Pradesh, Gujarat, West Bengal, Rajasthan, Punjab and Haryana (Figure 1). These types of land are either lying barren or producing less than their normal potential. Most of the traditional agricultural crops on majority of such soils either failed to grow or would give very poor return without substantial investment on reclamation of these lands. Sometimes, the problem of secondary salinization is more serious, since it usually represents losses of productive agricultural lands. The vast occurrence of soil salinity and alkalinity is a serious factor adversely affecting targeted crop production and economic utilization of land resources. This necessitates either development of land by suitable reclamation and management practices or its proper utilization by growing salt tolerant crops.

Phytoremediation is one of the important options for reclamation of salt affected soil. "Phytoremediation is the application of plant-controlled interactions with groundwater and organic

---

[1] ICAR-Directorate of Medicinal and Aromatic Plants Research, Anand 387310, India.
[2] Division of Floriculture and Medicinal Crops, ICAR-Indian Institute of Horticultural Research, Hessaraghatta Lake post, Bengaluru-560089, India.
[3] ICAR-Central Soil Salinity Research Institute, Regional Research Station, Bharuch 392012, India.
[4] Dept. of Soil Science and Agricultural Chemistry, Anand Agricultural University, Anand 388110, India.
* Corresponding author: biraj.ssac@gmail.com

Table 1. Extent and distribution of salt affected soil in India and in the world.

Salt affected soils	Saline soils (million ha)	Alkali soils (million ha)	Coastal saline soils (million ha)	Total (million ha)
India	1.711	3.788	1.246	6.744
World	676.8	239.28	157.92	1074

Source: www.cssri.org; FAO (2008).

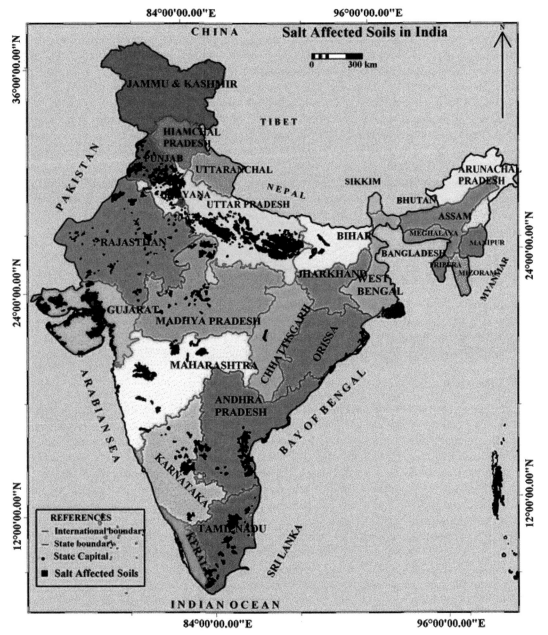

Figure 1. Distribution of salt affected soils in India (based on map prepared by NRSA, Hyderabad, CSSRI Karnal and LUP, Nagpur).

and inorganic molecules at contaminated sites to achieve site-specific remedial goals". Plants are known to reduce the pollution load of the environment by absorbing certain toxic metals or chemicals through their roots and translocating or accumulating them in less toxic forms in various parts. In phytoremediation method, natural conditions of the environment can be sustained. It is the least destructive method among the different types of remediation methods. Most plants, when grown in contaminated areas, uptake and translocate toxic elements to the harvestable parts. Polluted environments are remodeled using phytoremediation as a sustainable strategy to lower the pollution load.

India is endowed with natural abundance of diverse flora including enormously large number of aromatic plants which have the potential to grow in wide range of soil including salt affected soil and unculturable waste land. As a result of research and development carried out by various laboratories under Council of Scientific and Industrial Research (CSIR), Indian Council of Agricultural Research (ICAR) and the state and central universities, India has made significant progress in the production and processing of aromatic plants. Evidence from research being carried out in recent past reveals that some of these aromatic plants are well suited to saline and alkaline soils than the traditional agricultural crops because of their tolerance to salts and high benefit cost ratio. Some aromatic grasses, Vetiver (*Vetiveria zizanioides* (Linn) Nash), Palmarosa (*Cymbopogon martinii* var. motia), Lemongrass (*Cymbopogon flexuosus*) and Java citronella (*Cymbopogon winterianus*), appear suited for growth in salt affected soils, able to withstand salinity in soil and irrigation water to a great extent than traditional agriculture crops. Besides having a usually high benefit to cost ratio, aromatic grasses could be useful sources of high value, as essential oils that are in great demand in the cosmetic, pharmaceutical, and flavouring industries of India and abroad. In the following discussion, information regarding the use of aromatic grasses for phytoremediation in saline sodic condition has been presented.

2. Salt affected soils

Salt affected soils are designated as problematic soils. Salt affected soils are unproductive unless excess salts are reduced or removed. These soils are most extensively found in arid climates, but these soils are also found in coastal areas where soils are inundated by ocean or sea water. Such salt affected soils are categorized into various groups as hereunder.

2.1 Saline soil

Saline soils are defined as soils having electrical conductivity of the saturation extract greater than 4 dSm^{-1} (0.4 Sm^{-1} or 4 mmhos cm^{-1}) and an exchangeable sodium percentage (ESP) less than 15. The pH is usually less than 8.5. Formerly, these soils were called 'white alkali soils' because of surface crust of white salts. Saline soil is also called as '*Solonchak*' (Russian term). The process by which saline soils are formed is called "salinization". Saline soils occur mostly in arid or semi arid regions. Under humid conditions, soluble salts originally present in soil materials and those formed by weathering of rocks and minerals generally are percolated downward into the ground water and are transported ultimately by stream or oceans. Saline soils are, therefore, practically non-existent in humid regions, except when the soil has been subjected to sea water in river deltas and low lying lands near the sea. In arid regions, saline soils occur not only because there is less rainfall available to leach and transport the salt but also because of high evaporation rates, which further tend to concentrate the salts in soils and in surface waters. Restricted drainage is another factor that usually contributes to the salinization of soils and may involve the presence of high ground water table or low permeability of the soil which causes poor drainage by impeding the downward movement of water.

2.2 Sodic (Alkali) soil

Sodic soil is defined as a soil having electrical conductivity of the saturation extract less than 4 dSm^{-1} (0.4 Sm^{-1} or 4 mmhos cm^{-1}) and an exchangeable sodium percentage (ESP) greater than 15. The pH is usually between 8.5–10.0 (Table 2). Formerly, these soils were called 'black alkali soils' and the soils so formed is called '*solod, soloth or solonetz*' (Russian term). It is evident that soil colloids adsorb and retain cations on their surfaces. Cation adsorption occurs as a consequence of the electrical charges at the surface of the soil colloids. While adsorbed cations are combined chemically with the soil colloids, they may be replaced by other cations that occur in the soil solution. The reaction whereby a cation in solution replaces an adsorbed cation is called cation exchange and is expressed as milliequivalent per 100 g soil. Calcium and magnesium are the dominant cations found in the soil solutions and on the exchange complex of normal soil in arid regions. When excess soluble salts accumulate in these soils, sodium frequently becomes the dominant cation in the soil solution. In arid regions, as the soil solution becomes concentrated through evaporation or water absorption by plants, the solubility limits of calcium phosphate, calcium carbonate and magnesium carbonate are often exceeded, in which case they are precipitated with a corresponding increase in sodium concentration. Under such conditions, a part of the original exchangeable calcium and magnesium is replaced by sodium resulting alkali or sodic soils.

2.3 Saline-sodic soil

Saline-sodic soil is defined as a soil having electrical conductivity of the saturation extract greater than 4 dSm^{-1} (0.4 Sm^{-1} or 4 mmhos cm^{-1}) and an exchangeable sodium percentage (ESP) greater than 15 (Table 2). The pH is variable and usually above 8.5 depending on the relative amount of exchangeable sodium and soluble salts. When soils are dominated by exchangeable sodium, the pH will be more than 8.5 and when soils are dominated by soluble salts, the pH will be less than 8.5. These soils form as a result of combined processes of salinization and alkalization. If the excess soluble salts of these soils are leached downward, the properties of these soils may change markedly and become similar to those of sodic soil. As the concentration of these salts in the soil solution is lowered, some of exchangeable sodium hydrolyzes and forms sodium hydroxide (NaOH). This may change to sodium carbonate upon reaction with carbon dioxide absorbed from the atmosphere. On extensive leaching, the soil may become strongly alkaline, the particles disperse and the soils become unfavourable for the entry and movement of water and for tillage operation. At the same time, sodium toxicity to plants is increased. These soils sometimes contain gypsum and when it is subjected to intense leaching, calcium dissolves and the replacement of exchangeable sodium by calcium takes place concurrently with the removal of excess salts.

Table 2. Basic chemical properties of different salt affected soils.

Soil	EC (dSm^{-1})	pH	ESP	SAR
Saline soil	> 4.0	< 8.5	< 15.0	< 13.0
Sodic soil	< 4.0	> 8.5	> 15.0	> 13.0
Saline-sodic soil	> 4.0	< 8.5	> 15.0	> 13.0

3. Soil degradation processes

Soil degradation has become a serious problem in both rainfed and irrigated areas of India.

3.1 Salinization of soil

The sources of salts in soil are from the soil itself or also from ground water, irrigation water, canal water and tide water. The primary source of salts in soil is from rock weathering. During the process of rock weathering and soil formation, soluble salts are formed. Solute movement in the water is the determining factor in the soil salinization process. In the humid and sub humid regions with adequate rainfall, most of the soluble salts are leached down either at some depth below the surface or into the ground water. If the rainfall is not adequate and the evapotranspiration exceeds rainfall, leaching is not adequate to remove the soluble salts. Soil salinization is quite common in the arid and semi arid regions having an annual rainfall of less than 55 cm. Further, incrustation of salts on the land surface also occurs.

Fluctuating depth of ground water or the water table leads to salinity in soil. Soluble salts move upward along with rise in the water table. The salts are left behind when the water table recedes and accumulate at varying depth below the soil surface. Upward movement of soluble salts that are already accumulated at some depth of soil also takes place. The process of salinization is accelerated by rapid evaporation from the surface. The higher the depth of the water table, the higher is the rate of evaporation. The critical depth of the water table is defined as the depth of the water table above which soil solution can move upward by capillary action of the surface to cause soil salinization. This critical depth depends on the texture, water retention and transmission characteristics of soil in the profile. Soil salinization in the coastal area is due to the accumulation of salts from inundated sea water. Irrigation water containing high concentration of soluble salts, particularly sodium, leads to soil salinity, if proper drainage is not provided to leach the salts beyond the root zone. The soils which have salt accumulation at certain depths below the surface may also lead to salinity if the depth of irrigation water is inadequate to leach down the salts. Seepage from canal or irrigation channel may cause salinity. Soluble salts move along with seepage water and are accumulated at the water front.

3.2 Alkalization of soil

Formation of carbonates of Na and alkalization in the soil take place as a result of carbonation of alumino-silicate minerals in the presence of water. Sodium carbonate is highly soluble and its hydrolysis results in high alkalinity (pH up to 12).

In the presence of CO_2, the pH is lowered because of the formation of bicarbonates of Na, according to the reaction

$$Na_2CO_3 + H_2O + CO_2 \rightarrow 2NaHCO_3$$

The release of CO_2 with decomposition of organic matter in soil accentuates the process of $NaHCO_3$ formation in soil. In arid region, these reactions go on indefinitely, resulting in excessive accumulation of Na_2CO_3 and $NaHCO_3$ in the soil. With excessive evaporation and extreme arid conditions, carbonates and bicarbonates of sodium may accumulate in the soil as double salt crystals ($Na_2CO_3 \cdot NaHCO_3 \cdot 2H_2O$) or pure $NaHCO_3$.

With increase in the concentration of soil solution due to evapotranspiration, the solubility limits of calcium sulphate, calcium carbonate and magnesium carbonate are exceeded and hence they get precipitated. Therefore, the relative proportion of Na^+ ions with respect to Ca^{2+} and Mg^{2+} ions becomes high in soil solution, resulting in the increase of Na^+ ions on the exchange complex as per equilibrium and ESP increases to produce alkali soils. Most alkali soils, particularly in arid and semi-arid regions, contain $CaCO_3$ in the profile in some form, and constant hydrolysis of $CaCO_3$ sustains the release of OH^- ions in soil solution.

$$CaCO_3 + H_2O = Ca^{2+} + HCO_3^- + OH^-$$

The OH^- ions so released results in the maintenance of higher pH in calcareous alkali soils than that in non calcareous alkali soils. The increase in ESP can markedly affect the physicochemical properties of the soil. High sodium saturation in the soil results in the dispersion of clay particles, i.e., they tend to repel each other and remain independent of others. Highly hydrated Na^+ ions increase the zeta potential of the exchange sites, resulting in the repulsion of clay particles from each other. These soils therefore have very poor structure. The dispersed clay particles move downwards through soil pores and produce a dense or compact layer of very low permeability at some depth. Due to the decrease in hydraulic conductivity of alkali soils with increase in ESP, irrigation water stagnates at the surface of the field. Clogging of pores in the surface soils, followed by drying, results in the development of crust, which hinders seedling emergence and deteriorates air-water relations in the root zone. Alkali soils become very hard when dry and sticky when wet.

4. Phytoremediation of salt affected soils

Generally, chemical amendments are used to ameliorate sodic and saline-sodic soil by supplying readily available source of Ca^{2+} to replace excess Na^+ on the cation exchange complex. In this respect, amendments such as gypsum ($CaSO_4 \cdot 2H_2O$) supply soluble sources Ca^{2+} to the soil solution, which then replace excess Na^+ on the exchange complex. Application of chemical amendments, particularly gypsum and dolomite, for management of sodic soil is a century-old practice. However, there are some constraints with chemical ameliration of sodic soils in several developing countries because of (1) low quality of amendments containing a large fraction of impurities; (2) restricted availability of amendments; and (3) increased costs due to competing demand. On the other hand, scientific research and farmer's practices have demonstrated that sodic and saline-sodic soils can ameliorate through organic and plant based materials. The organic materials and the action of plant roots improve biological activity in the soil. The plant-assisted approach of amelioration of sodic and saline-sodic soils is also known as phytoremediation (Mishra et al. 2002, Qadir et al. 2002). The symonemous terminology of phytoremediation includes vegetative bioremediation, phytoamelioration, and biological reclamation.

In case of typical phytoremediation strategies of metal contaminated soils, the contaminated soil is exhausted by cultivation of specific plant species capable of hyper accumulating targeted ionic species in their biomass, thereby removing them from the soil (McGrath et al. 2002, Salt et al. 1998). In contrast, phytoremediation of sodic and saline-sodic soils is achieved by the ability of plants' roots to increase the dissolution rate of calcite, thereby resulting in enhanced level of Ca^{2+} in soil solution to effectively replace Na^+. The salinity levels in soil solution during phytoremediation improve soil structure through aggregate stability that facilitates the water movement through soil profile and enhances the amelioration process (Oster et al. 1999).

4.1 Phytoremediation mechanisms

4.1.1 Partial pressure of CO_2 in the root zone

The mechanisms of phytoremediation of salt affected soil involve the enhancement of dissolution rate of calcite (i.e., Ca^{2+} ions in soil solution) in the calcareous soils. This phytoremediation mechanism by regulating the partial pressure of CO_2 in the root zone (R_{PCO2}) involves dissolution and precipitation kinetics of calcite which is represented as

$$CaCO_3 + CO_2 \rightleftharpoons Ca^{2+} + 2HCO_3^-$$

This kinetics involves 3 processes:

a) Conversion of CO_2 in aqueous matrix (soil solution) into H_2CO_3 which then reacts with $CaCO_3$ is given by

$$CaCO_3 + H_2CO_3 \rightleftharpoons Ca^{2+} + 2\ HCO_3^-$$

b) Dissolution of H_2CO_3 into H^+ and HCO_3^- ions and $CaCO_3$ reacts with this H^+ ion produced.

$$CaCO_3 + H^+ \rightleftharpoons Ca^{2+} + HCO_3^-$$

c) Dissolution of $CaCO_3$ to Ca^{2+} and CO_3^{2-} ions through mineral hydrolysis. However, as calcite is less soluble, Ca^{2+} ion production through mineral dissolution is less compared to the processes discussed before.

$$CaCO_3 + H_2O \rightleftharpoons Ca^{2+} + CO_3^{2-} + H_2O$$

In the above dissolution reactions, Ca^{2+}, HCO_3^- and CO_3^{2-} are released into the soil solution. P_{CO2} may increase to a maximum level of 1 k Pa in aerobic soils, which is equivalent to 1% of the soil air by volume (Nelson and Oades 1998), while under anaerobic conditions of flooded soils it is much higher (Narteh and Sahrawat 1999, Ponnamperuma 1972), where saturated conditions inhibit the escape of CO_2 to the atmosphere and increases P_{CO2} in the soil. In non-calcareous soils, an increase in CO_2 results in more of H^+ and thus pH reduces. However, no such decrease of pH to a great extent in calcareous soils was observed (Nelson and Oades 1998), since pH changes are buffered by the enhanced dissolution of calcite (Van den Berg and Loch 2000).

Also, other processes are involved in regulation of P_{CO2} other than root respiration, viz.

i) CO_2 production from oxidation of plant root exudates (polysaccharides, proteins) by microbes.
ii) Microbes produce organic acids which dissolve calcite.

All these individually or collectively increase CO_2 production and ultimately Ca^{2+} availability increases, which replaces exchangeable Na^+ at a rate higher than P_{CO2} of atmosphere.

4.1.2 Proton release by plant roots

The release of H^+ from plant roots decreases the pH of the rhizosphere. As nitrogen source, when plants are supplied with ammonia (NH_4^+), it acidifies, while with nitrate (NO_3^-) it alkalizes the rhizosphere (Marschner and Romheld 1983, Schubert and Yan 1997). Though legumes rely on symbiotic N_2-fixation and acidify the soil (Schubert et al. 1990), the acidification mechanism has been studied only under acid soils rather than its role in remediation of salt affected soils. The chemical reaction involved in this is similar to that of dissolution mechanism of calcite due to increased P_{CO2}.

An electrochemical gradient develops between soil and root due to proton release and cation uptake increases net H^+ release which facilitates active H^+ pumping (Schubert and Yan 1997). Cytosolic pH increases due to H^+ release and organic anion synthesis is induced. The organic anion is thus a measure of net H^+ release at the root–soil and its complement in crop or tree litter is called ash alkalinity (Jungk 1968). Plants growing under high base status produce H^+ at the root–soil interface since they have high ash alkalinity. Therefore, such species that adapts to sodic soil conditions releases H^+ and can enhance the rate of calcite dissolution. Thus, proton release by plant can be effectively employed for salty soil amelioration by proper crop management, which facilitates high amounts of CO_2 and H^+ in root zone.

4.1.3 Physical effects of roots

Importance of plant roots lies in the fact that it maintains soil structure, macrospore formation and improves soil porosity by biopores or structural crusts (Czarnes et al. 2000, Oades 1993, Pillai and McGarry 1999, Yunusa and Newton 2003). Production of polysaccharides and fungal hyphae at root soil interface improves aggregate stability (Boyle et al. 1989, Tisdall 1991). Deep rooted plants withstand salinity and sodicity up to certain levels by facilitating the leaching of Na^+ to the deeper soils layers. This is evident from the improvement in soil structure by deep rooted legumes and perennial grasses (Tisdall 1991). Phytoremediation effects of different crop rotation with and without gypsum was studied by Ilyas et al. (1993) on low permeable hard saline soil (pHs = 8.8, ECe = 5.6 dS m^{-1}, SAR = 49). Results showed that alfalfa roots penetrated as deep as 1.2 m in the gypsum-treated plots as compared to 0.8 m in untreated plots and caused a twofold increase in saturated hydraulic conductivity (Ks). Similar increase in Ks was seen with sesbania-wheat-sesbania rotation up to 0.4 m depth.

Although deep tillage has been effective in ameliorating sub-soils with low porosity, the benefits are short lived (Cresswell and Kirkegaard 1995). Biological drilling by plant roots was found as an alternative to deep tillage for the amelioration of dense sub-soils (Elkins 1985). Two stages in biological drilling are: (1) Creation of macro pores in the subsoil by the roots resulting in improved water and gas flow and (2) Benefits for the next crop (Cresswell and Kirkegaard 1995, Elkins 1985).

4.1.4 Salt removal by plant biomass

Phytoremediation of salt affected soil can also be done by removal of above ground biomass as they accumulate salts and Na^+ in their shoots. Halophytes are highly salt-resistant crops that also accumulate high amounts of salts and Na^+ in their shoots. Salt concentration in leaf ash of atriplex was as high as 390 g salt kg^{-1} under salt affected soils (Malcolm et al. 1988), while under rangeland condition it was 130–270 g salts kg^{-1} (Hyder 1981).

Phytoremediation by salt accumulated shoot removal is only up to certain extent that it cannot ameliorate salt affected soils completely. This is evident from the research findings of Barrett-Lennard (2002) that it would require about 20 consecutive years under non-irrigated conditions to remove half of the initial content of salts by halophytic crops having annual productivity of 10 t ha^{-1}. But halophytes such as artiplex hardly produce more than 2 mg ha^{-1} annually (Barrett-Lennard et al. 1990). Therefore, under non irrigated conditions, effect of halophytes is minimal to ameliorate salt affected soils, while under irrigated conditions there is excess Na^+ removal by enhanced calcite dissolution making salt removal by shoots less effective comparatively. This is because salt removal by shoots is less than salts actually accumulated through irrigation water. For example, kallar grass (*Leptochloa fusca* L.) has forage salt levels of 40–80 g kg^{-1} under soil salinity level of 20 dSm^{-1}. When they are irrigated with 10^7 litre ha^{-1} water having salinity of 1.5 dSm^{-1}, salt added would be 9.6 t ha^{-1} but only 1–2 t ha^{-1} salt is removed by forage. So, a better way to remove salinity is by leaching the salts to greater depths than removal of shoot biomass.

4.2 Improvement of soil properties through phytoremediation

The ability of different plant species used in phytoremediation of sodic and saline-sodic soils has been found to be highly variable. In general, plant species with greater biomass production potential together with the ability to withstand soil salinity and sodicity have been found suitable for soil amelioration (Qadir et al. 2002). The study revealed that Na^+ uptake by above ground biomass of several plant species constitutes 2–20% of the total salt uptake (Qadir et al. 2007). It was also found that Na^+ removal by plant biomass such as alfalfa would contribute to only 1–2% of the Na^+ removed during phytoremediation (Qadir et al. 2003). However, phytoremediation has two major advantages: (1) no financial liabilities to purchase chemical amendments, and (2) financial benefits from cultivation of crops during amelioration. In various parts of the world, several studies have been conducted aimed at amelioration of sodic and saline-sodic soils through phytoremediation approaches (Ahmad et al. 1990, Ghaly 2002). Apart from the amelioration of soil sodicity and salinity, this approach has been compared for its effect on soil drainage, nutrients dynamics and environmental benefit in terms of carbon sequestration (Garg 1998, Kaur et al. 2002).

4.2.1 Amelioration of soil sodicity

Various field experiments revealed that phytoremediation and chemical amelioration approaches are equally effective in terms of ability to decrease the soil sodicity and salinity. Field experiment indicated that amelioration efficiency of two grass species, Para grass (*Brachiaria mutica*) and Karnal grass (*Dichanthium annulatum*), was comparable with soil application gypsum @ 12.5 Mg ha^{-1} (Kumar and Abroal 1984). It was also found that amelioration efficiency of Kallar grass was greater in the pots leached after 6 days of harvesting and it was comparable with the gypsum-treated soil (Hamid et al. 1990). However, some field trials indicated that phytoremediation approaches were not successful because a salt-resistant crop was not the first crop of the rotation. Several crop rotations have been evaluated to ameliorate sodic soils. The study revealed that the entire crop rotation could ameliorate only upper 0.15 m of soil after 1 year as did amelioration by the gypsum treatment (Qadir et al. 2003). It is pertinent to note that growing rice in submerged soil has been recognized as promising phytoremediation approach for amelioration of moderate sodic and saline-sodic soils. Under submerged condition, rice rhizosphere accumulated CO_2 in the soil atmosphere to react and neutralize alkalinity (Qadir et al. 2002). Field experiment (n = 17) conducted in different parts of the world concluded that chemical amelioration is able to decrease 60% of the initial sodicity level, whereas 48% was observed in case of phytoremediation approaches (Singh and Singh 1989, Ahmad et al. 1990, Muhammed et al. 1990, Rao and Burns 1991, Helalia et al. 1992, Qadir et al. 1996, 2002, Ghaly 2002, Ahmad et al. 2006). In general, phytoremediation approach worked well on moderately sodic and saline-sodic soils provided: (1) excess irrigation was done to facilitate adequate leaching and (2) excess irrigation was applied when crop growth and P_{CO2} were at their peak. On such conditions, the potential of phytoremediation was comparable with gypsum application (Qadir et al. 2003).

4.2.2 Amelioration throughout the root zone

The anticipated zone of amelioration is an important parameter to determine relative efficiency of the chemical amelioration and phytoremediation approaches. In most of the comparative studies, it was found that amelioration occurred primarily in the root zone where chemical amendment (gypsum) was incorporated (Qadir et al. 1996, Ilyas et al. 1997). In case of phytoremediation, amelioration of sodic and saline-sodic soils occurs throughout the root zone. However, different plant species have variable capacity and depth of soil amelioration which is influenced by the root morphology (Ahmad et al. 1990, Akhter et al. 2003). So, deep-rooted crop with tap root system have advantages in terms of greater depth of soil amelioration. For example, alfalfa root can penetrate as deep as 1.2 m in the soil.

4.2.3 Soil nutrient dynamics

Apart from reducing soil salinity and sodicity levels of sodic and saline-sodic soils, phytoremediation provides additional benefits over other amelioration approaches. A 20-year study with tree plantation (*Prosopis juliflora, Acacia nilotica, Eucalpytus tereticornis* and *Albizia lebbeck*) on alkali soil resulted in considerable decrease in soil pH and increase in soil organic carbon (SOC) content, and available phosphorus (P) and potassium (K) in surface soil (Singh and Gill 1990). There was an increase in phosphorus (P), zinc (Zn) and copper (Cu) availability in the phytoremediation soils probably due to the production of root exudates and likely dissolution of some nutrient-coated calcite. Soil microbial biomass (MBC), dehydrogenase (DHA) and respiration are related to microbial populations and provide an index of the overall soil biological status. The levels of DHA in post amelioration soil were found higher in phytoremediated soils than gypsum treated soil. Permanent vegetation such as grasses resulted in significant increase in urease and DHA in alkali soils (Rao and Ghai 1985). The greater microbial activity in upper 0.6 m soil under the tree species (*A. nilotica, D. sissoo, P. juliflora* and *T. arjuna*) due to the increased soil organic C from leaf litter biomass decomposition (Garg 1998). The inorganic C of arid and semiarid soils is converted to organic form by plants through photosynthesis, and in soils through the reaction of CO_3^{2-} with decomposing organic matter (added via phytoremediation). Thus, the transfer of inorganic C from inorganic to organic form provides a better environment for C sequestration and soil quality (Sahrawat et al. 2005). The rate of C sequestration through this pathway ranged between 0.25–1.0 Mg C ha^{-1} year^{-1} (Wilding 1999). The amelioration of saline and saline-sodic soil through phytoremediation approaches could lead to both organic and inorganic C sequestration simultaneously.

5. Aromatic plant species for phytoremediation

Aromatic plants are those which produce essential oils having aromatic compounds and are widely used in cosmetics, toiletries, agarbattis, tooth paste, food products like confectionary, chocolate, ice cream, medicine, pharmaceuticals, etc. India is having unique advantage of growing majority of aromatic plant species due to diversity in its climate, soil, rainfall and geographical conditions. The growth rate of demand is 9% in domestic and 25% for export market. At present, India produces about 16,000–18,000 tons of essential oil out of 80,000 tons of world's production which is about 20–25% of total world's production. Most of the oil that have a good demand in the present market are: mints, lemongrass, rose oil, citronella oil, basil, geranium, palmarosa, eucalyptus, vetiver, jasmine, sandal wood, lavender, ginger oil, cinnamon, etc. (Figure 2).

Sources of essential oil: Essential oil or aromatic substances are present in one or more plant parts, such as flowers (rose, tuberose, violet, cananga, jasmine, narcissus, orange flower, ylang ylang), leaves (cinnaomon, cedar, patchouli, bitter orange, wintergreen, sweet worm wood, eucalyptus), wood (cedar, birch, rose, camphor, laurel, sandal), bark (birch, cascarilla, cassia, cedar, cinnamon, cypress), fruit peels (bergamot, citron, grapefruit, lemon, limes, mandarin, orange, tangerine), grass (citronella, ginger grass, lemongrass, palmarosa), seeds (ambrette, angelica, cardamom, carrot, croton, cumin, dill, mustard, parsley, bitter almond), dried leaves (cherry laurel, eucalyptus, nirouli, patchouli), dried fruits (coriander, anise, juniper, nutmeg), dried buds and berries (clove, cubeb, juniper, pimento), gums (elemi, galbanum, mastic, myrrh, styrax), roots and rhizomes (angelic, costus, ginger, valerian, vetiver), and herbs (basil, chamomile, dill, fennel, geranium, lavender marjoram, parsley, peppermint, rosemary, sage, spearmint, thyme, verbena, wormseed, wormwood).

India is a veritable emporium of essential oil bearing plants. It is one of the few countries in the world where aromatic plants of all types can be cultivated in one or other areas of the country because of the vast areas and the climatic conditions. Cultivation of these crops will help in diversifying our agriculture for new cash crops. There is a need to provide genuine quality of raw material for the user

Figure 2. Some important aromatic plants in demand: (a) Lemongrass (*Cymbopogon flexuosus*), (b) Palmarosa (*Cymbopogon martini* var. motia), (c) Citronella (*Cymbopogon winterianus*), (d) Sweet wormwood (*Artemisia annua*), (e) Vetiver (*Vetiveria zizanioides*), (f) Tulasi (*Ocimum tenuiflorum*), (g) Japanese mint (*Mentha arvensis*), (h) Sweet Basil (*Ocimum basilicum*), (i) Pathcouli (*Pogostemon cablin*), (j) Galangal (*Alpinia galangal*).

industries. Information regarding the ameliorative potential of aromatic crop species for improving salt affected soil is limited. Some of the aromatic grasses have the potential for hyper accumulation of salt, which helps in reclaiming the saline soil, e.g., palmarosa is a hyper salt accumulator. It has been reported that aromatic grasses such as palmarosa, and lemongrass (*C. flexuosus*) can be grown successfully on moderately alkaline soils having pH up to 9.0 and 9.5, respectively, while vetiver (*Vetiveria zizanioides*) withstands both high pH and salinity. These grasses not only produce essential oils used for industrial purpose but also ameliorate the degraded soil (Table 3). When palmarosa, lemongrass and vetiver were grown for two years on sodic soils having pH 10.6, 9.8 and 10.5, respectively, the reduction in pH was noticed as 9.4, 8.95 and 9.50 in each soil, respectively. The continuous growth of palmarosa may reduce the sodicity and improve the physico-chemical properties of sodic soils. Like other aromatic grasses, vetiver can reduce the sodicity and improve the physico-chemical properties of sodic soils (Patra et al. 1998, Xia and Shu 2003). Chand et al. (2014) conducted an experiment to assess the suitable and profitable medicinal and aromatic plants in different sodic soils. They obtained highest essential oil yield 154.25, 145.75, 145.00 and 144.10 kg ha^{-1} during different harvests under various sodic soils. Profitability of experiment indicated that palmarosa gave highest net returns of Rs. 71675, 62125, 61750 and 60160 ha^{-1} with more benefit: cost ratio of 1.97, 1.55, 1.54 and 1.47 followed by other plants under various sodic soils. On the basis of results, they showed that palmarosa was the most suitable and profitable crop under various ranges of sodic soils. Therefore, they concluded that if managed judiciously, sodic soils can be successfully utilized for growing of palmarosa, lemongrass and khus without using amendments.

Sinha et al. (2016) suggested that farmers/growers can raise palmarosa seedlings (varieties: PRC-1, Trishna and Tripta) in the nursery under normal soil condition and transplant these seedlings in soil having KCl salt concentration upto 50 mM except variety Trishna which can be grown only up to 100 mM KCl concentration. It has also been reported that sweet wormwood (*Artemisia annua*), which is commonly grown for both essential oil and medicinal properties, could withstand Exchangeable Sodium Percentage (ESP) as high as 55 (pH 9.6) (Kalaichelvi and Swaminathan 2009). Dagar et al. (2013) explored the possibilities of raising lemongrass on degraded calcareous soil using saline water up to EC 8.6 dSm^{-1} without build up of soil salinity if normal rainfall occurs once in 3–4 years.

An experiment was conducted in glazed pots with artificially prepared sodic soils having different ESP levels (16, 55, 65, 75 and 85 ESP) at CIMAP, Lucknow to evaluate the ameliorative potential of high value crop palmarosa for reclamation of sodic soil. The results revealed that growth attributes and herb yield decreased significantly with the increased ESP level. However, essential oil yield increased significantly at ESP-55 over ESP-16. The sodium (Na) concentration in plant tissue increased with increased soil ESP level, but decreased concentrations of calcium (Ca), magnesium (Mg) and potassium (K). This study indicated that cultivation of palmarosa in sodic soil would decrease the ESP level without the use of chemical amendments for reclamation of sodic soil (Kumar et al. 2004). The growth of selected cultivars of palmarosa, lemongrass and jamarosa on normal soil and sodic soil was evaluated to determine the tolerance and productivity of these three species in sodic environment (Patra et al. 1998). The effect of sodicity was less pronounced on essential oil yield and quality. Differences among cultivars of palmarosa and lemongrass in sensitivity to soil sodicity were noted. The growth and yield of jamarosa was the best among the three aromatic grasses evaluated. The effect of different amendments on sodic soil including normal soil on palmarosa was studied under field experiment (Singh et al. 2002). Palmarosa oil yield was increased when amendments were incorporated in the salt affected soil. It was concluded from the study that palmarosa is the best suitable aromatic crop in salt affected soil for good herb and oil yield as well to improve the salt affected soil. The herb yield of geranium was significantly increased with increasing soil ESP up to 16.0, and oil yield increased with increasing ESP up to 7.0. Further,

Table 3. Ameliorative potential of aromatic plants in degraded and marginal lands.

Aromatic plants	Soil category	Potential benefits	References
Palmarosa (*Cymbopogon martini* var. *motia*)	Sodic soil	Palmarosa crop has ameliorative potential and in the long term it may reclaim the sodic soil by reducing the pH and ESP and produce good amount of oil yield and income.	(Kumar et al. 2004)
	Salt affected soil	Palmarosa improves the salt affected soil by decreasing the pH and SAR. Improves the physicochemical properties and fertility of salt affected soil.	(Singh et al. 2002)
	Saline soils	Yield of palmarosa was not affected adversely up to the EC of 12 mmhos cm^{-1}, but it slightly decreased at EC of 16 mmhos cm^{-1} as compared to normal soil. Herbage yield and oil content were not affected by various salinity levels.	(Singh et al. 2004)
	Saline water irrigation	Palmarosa can tolerate irrigation water salinity up to 4.0 dSm^{-1} without adversely affecting herbage yield.	(Takankhar et al. 2018)
	Sewage sludge	*Cymbopogon martini* acts as hyper-accumulator and thus could be used for phytoremediation of sewage sludge.	(Singh et al. 2019)
Vetiver grass (*Vetiveria zizanioides*)	Saline water irrigation	Vetiver was well survived (93–95%) and remained unaffected by saline irrigation.	(Tomar and Minhas 2004)
	Sodic soil	Vetiver withstands both high pH and stagnation of water, and can successfully be grown without significant yield reduction in highly alkaline soils. Soil pH, EC and ESP can be reduced by growing vetiver grass.	(Dagar et al. 2004)
	Sodic soil	Vetiver could withstand soil alkalinity up to pH 9.5. Herb and oil yield of vetiver was not significantly affected up to pH 9.5.	(Anwar et al. 1996)
	Sodic water irrigation	Vetiver grass can withstand sodic irrigation water and increase the biomass yield and oil yield.	(Prasad et al. 1999)
Lemon grass (*Cymbopogon flexuous*)	Calcareous soil and saline water irrigation	Results indicated the possibilities of raising lemon grass on degraded calcareous soil using saline water up to EC 8.6 dSm^{-1}.	(Dagar et al. 2013)
	Salinity stress	Successfully grown with salinity upto 15 dSm^{-1}.	(Kumar and Chauhan 2017)
	Sodic soil	Lemon grass could successfully be grown on moderately alkaline soils of pH up to 9.2. Soil pH and ESP decreased under influence of lemongrass.	(Dagar et al. 2004)
Citronella (*Cymbopogon winterianus*)	Saline soil	Citronella can tolerate salinity up to 5 dSm^{-1} and reduction in yield moderately up to 10 dSm^{-1}.	(Chauhan and Kumar 2014)
Scented Geranium (*Pelargonium graveolens*)	Sodic soil	Geranium can tolerate the soil sodicity stress of ESP up to 20.0.	(Prasad et al. 2006)
Basil (*Ocimum Basilicum*)	Sodicity stress	Basil is highly tolerant to soil sodicity stress of ESP up to 36.	(Prasad et al. 2007)

increase in soil ESP decreased the yield. When ESP level increased in the soil, sodium concentration was increased in plant roots and shoot tissue of the geranium. This study suggested that geranium is slightly or moderately tolerant of soil sodicity stress condition (Prasad et al. 2006).

The suitability to saline irrigation on the relative performance of aromatic grasses, viz. citronella (*Citronella java*), lemon grass (*Cymbopogon citrates* Stapf), palmarosa [*Cymbopogon martini* (Roxb.) Wats.], and vetiver [*Vetiveria zizanioides* (L.) Nash] was evaluated by Tomar and Minhas (2004). On an average, vetiver produced the maximum bio-mass (90.9 tonnes ha^{-1} dry weight basis), followed by palmarosa (29.1 tonne ha^{-1}) and lemongrass (16.1 tonnes ha^{-1}). However, citronella could not survive. Reduction in yield with saline irrigation ranged from 24 to 29%, whereas it ranged from 3 to 21% with alternation in irrigation with canal and saline waters, vetiver being least affected. The effects of residual sodium carbonate (RSC) in irrigation water on soil sodication and yield and cation composition of palmarosa (*Cymbopogon martinii* Roxb. Wats) and lemongrass (*Cymbopogon flexuosus* Steud Wats) indicated that the increasing RSC in irrigation water significantly increased the pH, electrolyte conductivity (ECe) and SARe of the soil and, hence, considerably decreased the herb and oil yield of both the palmarosa and lemongrass. The concentration of Na increased significantly and K and Ca decreased with increase in RSC of irrigation water in vegetative tissues of both species. The lemongrass accumulates significantly greater amount of Na in shoot tissues as compared to palmarosa and it failed to survive at high RSC after 21 months of transplanting (Prasad et al. 2001).

6. Perspectives

Aromatic grasses were found to be promising for phytoremediation of sodic and saline-sodic soils. A relative performance of amelioration with that of chemical amendments highlights the effective role of aromatic grass in amelioration of sodic and saline-sodic soils. Phytoremediation has been found to be beneficial in several aspects: (1) no financial liabilities for chemical amendments, (2) financial benefits from aromatic crops during amelioration, (3) more uniform and greater zone of amelioration (4) improvement in soil properties and (5) environmental benefits in terms of C sequestration. However, phytoremediation approach is effective only when used on moderately sodic and saline-sodic soils. Moreover, the amelioration process through phytoremediation is very slow and it requires calcite to be present in the soil. So, the feasibility of phytoremediation is limited when soil is highly saline or sodic. Considering the challenge associated with amelioration of sodic and saline-sodic soils and environmental consequences, it is high time to consider the soil as an useful natural resource rather than environmental burden. In this context, phytoremediation through cultivation of aromatic crops is a cost effective intervention for amelioration of these soils. The economic importance of aromatic crops as a candidate for phytoremediation of sodic and saline-sodic soils of resource poor farmers is also realized. So, phytoremediation can be more effective if selected plants have economic value or local utilization at the farm level. However, in future, we need to consider the economic value of improved soils.

References

Ahmad, N., Qureshi, R.H. and Qadir, M. 1990. Amelioration of a calcareous saline-sodic soil by gypsum and forage plants. Land Degrad. Dev. 2(4): 277–284.

Ahmad, S.A.G.H.E.E.R., Ghafoor, A.B.D.U.L., Qadir, M.A.N.Z.O.O.R. and Aziz, M.A. 2006. Amelioration of a calcareous saline-sodic soil by gypsum application and different crop rotations. Int. J. of Agriculture and Biol. 8: 142–146.

Akhter, J., Mahmood, K., Malik, K.A., Ahmed, S. and Murray, R. 2003. Amelioration of a saline sodic soil through cultivation of a salt-tolerant grass Leptochloa fusca. Environ. Conservation 30(2): 168–174.

Anwar, M., Patra, D.D. and Singh, D.V. 1996. Influence of soil sodicity on growth, oil yield and nutrient accumulation in vetiver (*Vetiveria zizanioides*). Ann. Arid Zone 35(1): 49–52.

Barrett-Lennard, E.G., Warren, B.E. and Malcolm, C.V. 1990. Agriculture on saline soils direction for the future. pp. 37–45. *In*: Myers, B.A. and West, D.W. (eds.). Revegetation of Saline Land. Institute for Irrigation and Salinity Research, Tatura, Victoria, Australia.

Barrett-Lennard, E.G. 2002. Restoration of saline land through revegetation. Agricultural Water Manage. 53: 213–226.

Boyle, M., Frankenberger, W.T. Jr. and Stolzy, L.H. 1989. The influence of organic matter on aggregation and water infiltration. J. of Production Agriculture 2: 290–299.

Chauhan, Nishant and Kumar, Dheeraj. 2014. Effect of salinity stress on growth performance of *Citronella java*. Int. J. of Geology, Agriculture and Environ. Sci. 2(6): 11–14.

Cresswell, H.P. and Kirkegaard, J.A. 1995. Subsoil amelioration by plant-roots-the process and the evidence. Soil Res. 33(2): 221–239.

Czarnes, S., Hallett, P.D., Bengough, A.G. and Young, I.M. 2000. Root- and microbial-derived mucilages affect soil structure and water transport. Europ. J. of Soil Sci. 51: 435–443.

Dagar, J.C., Tomar, O.S., Kumar, Y. and Yadav, R.K. 2004. Growing three aromatic grasses in different alkali soils in semi arid regions of northern India. Land Degrad. Dev. 15: 143–151.

Dagar, J.C., Tomar, O.S., Minhas, P.S. and Kumar, M. 2013. Lemongrass (*Cymbopogon flexuosus*) productivity as affected by salinity of irrigation water, planting method and fertilizer doses on degraded calcareous soil in a semi-arid region of northwest India. Indian J. of Agricultural Sci. 83(7): 734–738.

Elkins, C.B. 1985. Plant roots as tillage tools. Tillage Machinery Systems as Related to Cropping Systems. Proceedings of the International Conference on Soil Dynamics 3: 519–523, June 17–19, 1985. Auburn, AL.

FAO, IIASA, ISRIC, ISS-CSA and JRC. 2008. Harmonized World Soil Database (version 1.0), FAO, Rome, Italy and IIASA, Laxenburg, Austria.

Garg, V.K. 1998. Interaction of tree crops with a sodic soil environment: potential for rehabilitation of degraded environments. Land Degrad. Dev. 9(1): 81–93.

Ghaly, F.M. 2002. Role of natural vegetation in improving salt affected soil in northern Egypt. Soil and Tillage Res. 64(3-4): 173–178.

Gritsenko, G.V. and Gritsenko, A.V. 1999. Quality of irrigation water and outlook for phytomelioration of soils. Eurasian Soil Sci. 32: 236–242.

Hamid, A., Chaudhry, M.R. and Ahmad, B. 1990, November. Biotic reclamation of a saline-sodic soil. pp. 73–86. *In*: Proceedings of Symposium on Irrigation Systems Management/Research.

Helalia, A.M., El-Amir, S., Abou-Zeid, S.T. and Zaghloul, K.F. 1992. Bio-reclamation of saline-sodic soil by Amshot grass in Northern Egypt. Soil and Tillage Res. 22(1-2): 109–115.

Hyder, S.Z. 1981. Preliminary observations on the performance of some exotic species of *Atriplex* in Saudi Arabia. J. of Range Management 34: 208–210.

Ilyas, M., Miller, R.W. and Qureshi, R.H. 1993. Hydraulic conductivity of saline-sodic soil after gypsum application and cropping. Soil Sci. Society of America J. 57: 1580–1585.

Ilyas, M., Qureshi, R.H. and Qadir, M.A. 1997. Chemical changes in a saline-sodic soil after gypsum application and cropping. Soil Tech. 10(3): 247–260.

Jungk, A. 1968. Die Alkalinität der Pflanzenasche als Maß fü̈r den Kationenü̈berschuß in der Pflanze. Zeitschrift fü̈r Pflanzenernä̈hrung und Bodenkunde. J. of Plant Nutrition and Soil Sci. 120: 99–105.

Kalaichelvi, K. and Swaminathan, A.A. 2009. Alternate land use through cultivation of medicinal and aromatic plants—A review. Agric. Rev. 30(3): 176–183.

Kaur, B., Gupta, S.R. and Singh, G. 2002. Bioamelioration of a sodic soil by silvopastoral systems in northwestern India. Agrofor. Syst. 54(1): 13–20.

Kumar, A. and Abrol, I.P. 1984. Studies on the reclaiming effect of Karnal-grass and para-grass grown in a highly sodic soil. Indian J. Agric. Sci.

Kumar, Dheeraj and Chauhan, Nishant. 2017. Effect of salinity stress on growth performance of lemongrass. International Journal of Engineering Research & Technol. 5(12): 1–4.

Kumar, Dinesh., Singh, Kambod., Chauhan, H.S., Prasad, Arun., Beg, S.U. and Singh, D.V. 2004. Ameliorative potential of palmarosa for reclamation of sodic soils. Commun. Soil Sci. Plant Anal. 35(9&10): 1197–1206.

Malcolm, C.V., Clarke, A.J., D'Antuono, M.F. and Swaan, T.C. 1988. Effects of plant spacing and soil conditions on the growth of five *Atriplex* species. Agriculture, Ecosyst. and Environ. 21: 265–279.

Marschner, H. and Romheld, V. 1983. *In vivo* measurement of root-induced pH changes at the soil-root interface: Effect of plants species and nitrogen source. J. of Plant Nutrition and Soil Sci. 111: 241–251.

McGrath, S.P., Zhao, J. and Lombi, E. 2002. Phytoremediation of Metals, Metalloids, and Radionuclides. Agronomy-V.75, IACR-Rothamsted, UK.

Mishra, A., Sharma, S.D. and Khan, G.H. 2002. Rehabilitation of degraded sodic lands during a decade of Dalbergia sissoo plantation in Sultanpur district of Uttar Pradesh, India. Land Degrad. Dev. 13(5): 375–386.

Muhammed, S., Ghafoor, A., Hussain, T. and Rauf, A. 1990. Comparison of biological, physical and chemical methods of reclaiming salt-affected soils with brackish groundwater. pp. 35–42. *In*: Proceedings of the Second National Congress of Soil Science. December 20–22, 1988. Soil Science Society of Pakistan, Faisalabad, Pakistan.

Narteh, L.T. and Sahrawat, K.L. 1999. Influence of flooding on electrochemical and chemical properties of West African soils. Geoderma 87: 179–207.

Nelson, P.N. and Oades, J.M. 1998. Organic matter, sodicity, and soil structure. pp. 51–75. *In*: Sumner, M.E. and Naidu, R. (eds.). Sodic Soil: Distribution, Management and Environmental Consequences. Oxford University Press, New York.

Oades, J.M. 1993. The role of biology in the formation, stabilisation and degradation of soil structure. Geoderma 56: 377–400.

Oster, J.D., Shainberg, I. and Abrol, I.P. 1999. Reclamation of salt affected soils. pp. 659–691. *In*: Skaggs, R.W. and van Schilfgaarde, J. (eds.). Agricultural Drainage. ASA-CSSA-SSSA, Madison, WI.

Patra, D.D., Anwar, M., Tajuddin and Singh, D.V. 1998. Growth of aromatic grasses in sodic and normal soils under subtropical conditions. J. of Herbs, Spices & Medicinal Plants 5(2): 11–20.

Pillai, U.P. and McGarry, D. 1999. Structure repair of a compacted vertisol with wet/dry cycles and crops. Soil Science Society of America J. 63: 201–210.

Ponnamperuma, F.N. 1972. The chemistry of submerged soils. Adv. Agron. 24: 29–96.

Prasad, A., Kumar, D. and Singh, D.V. 1999. Effect of sodic water irrigation on yield and cation composition of vetiver (*Vetiveria zizanioides* (L.) Nash.). J. of Spices and Aromatic Crops 8(1): 89–92.

Prasad, Arun., Chattopadhyay, Amitabha., Chand, Sukhmal., Naqvi, A.A. and Yadav, A. 2006. Effect of soil sodicity on growth, yield, essential oil composition, and cation accumulation in rose-scented geranium. Communications in Soil Sci. and Plant Analysis 37: 1805–1817.

Prasad, A., Dinesh, Kumar and Singh, D.V. 2001. Effect of residual sodium carbonate in irrigation water on the soil sodication and yield of palmarosa (*Cymbopogon martini*) and lemongrass (*Cymbopogon flexuosus*). Agricultural Water Manage. 50(3): 161–172.

Prasad, A., Lal, R.K., Chattopadhyay, A., Yadav, V.K. and Yadav, A. 2007. Response of basil species to soil sodicity stress. Communications in Soil Sci. and Plant Analysis 38: 2705–2715.

Qadir, M., Qureshi, R.H., Ahmad, N. and Ilyas, M. 1996. Salt-tolerant forage cultivation on a saline-sodic field for biomass production and soil reclamation. Land Degrad. Dev. 7(1): 11–18.

Qadir, M., Qureshi, R.H. and Ahmad, N. 2002. Amelioration of calcareous saline sodic soils through phytoremediation and chemical strategies. Soil Use and Manage. 18(4): 381–385.

Qadir, M., Steffens, D., Yan, F. and Schubert, S. 2003. Sodium removal from a calcareous saline sodic soil through leaching and plant uptake during phytoremediation. Land Degrad. Dev. 14(3): 301–307.

Qadir, M., Oster, J.D., Schubert, S., Noble, A.D. and Sahrawat, K.L. 2007. Phytoremediation of sodic and saline-sodic soils. Advan. in Agronomy 96: 197–247.

Rao, D.L.N. and Ghai, S.K. 1985. Urease and dehydrogenase activity of alkali and reclaimed soils. Soil Res. 23(4): 661–665.

Rao, D.L.N. and Burns, R.G. 1991. The influence of blue-green algae on the biological amelioration of alkali soils. Biol. and Fertility of Soils 11(4): 306–312.

Sahrawat, K.L., Bhattacharyya, T., Wani, S.P., Chandran, P., Ray, S.K., Pal, D.K. and Padmaja, K.V. 2005. Long-term lowland rice and arable cropping effects on carbon and nitrogen status of some semi-arid tropical soils. Current Sci. 2159–2163.

Salt, D.E., Smith, R.D. and Raskin, I. 1998. Phytoremediation. Annual Review of Plant Biol. 49: 643–668.

Singh, G., Pankaj, U., Ajayakumar, P.V. and Verma, R.K. 2019. Phytoremediation of sewage sludge by *Cymbopogon martinii* (Roxb.) Wats. var. motia Burk. grown under soil amended with varying levels of sewage sludge. Int. J. Phytoremed. 1–11.

Schubert, E., Schubert, S. and Mengel, K. 1990. Effect of low pH of the root medium on proton release, growth, and nutrient uptake of field beans (*Vicia faba*). Plant Soil 124: 239–244.

Schubert, S. and Yan, F. 1997. Nitrate and ammonium nutrition of plants: Effect on acid/base balance and adaptation of root cell plasmalemma H^+ ATPase. J. of Plant Nutrition and Soil Sci. 160: 275–281.

Singh, G.B. and Gill, H.S. 1990. Raising trees in alkali soils. Wasteland News 6: 15–18.

Singh, M.V. and Singh, K.N. 1989. Reclamation techniques for improvement of sodic soils and crop yield. Indian J. of Agricultural Sci. 59(8): 495–500.

Singh, R.P., Singh, R.S., Singh, R.K. and Singh, Ajeet. 2002. Effect of palmarosa on improvement of salt affected soils. Symposium no 33. WCSS, Thailand.

Singh, R.P., Singh, R.S. and Singh, R.K. 2004. Cultivation of aromatic plants in saline soils. Bharatiya Vaigyanik Evam Audyogik Anusandhan Patrika 12(1): 114–115.

Sinha, G., Mali, H., Ram, G., Srivastava, D.K. and Kumar, B. 2016. Impact of salt stress on different varieties of palmarosa during seed germination. J. of Essential Oil-bearing Plants 19(4): 1025–1030.

Subash Chand, Sanjay Kumar and Gautam, P.B.S. 2014. Suitability and profitability of medicinal and aromatic plants on different sodic soils. Crop Res. 48(1-3): 92–94.

Takankhar, V.G., Karanjikar, P.N., Tayde, V.V. and Patil, R.G. 2018. Performance of palmarosa cultivars under saline condition. Int. J. of Current Microbiol. and Appl. Sci. 6: 2452–2458.

Tisdall, J.M. 1991. Fungal hyphae and structural stability of soil. Australian J. of Soil Res. 29: 729–743.

Tomar, O.S. and Minhas, P.S. 2004. Relative performance of aromatic grasses under saline irrigation. Indian J. of Agronomy 49(3): 207–208.

Van den Berg, G.A. and Loch, J.P.G. 2000. Decalcification of soils subject to periodic waterlogging. Europ. J. of Soil Sci. 51: 27–33.

Wilding, L.P. 1999. Comments on manuscript by Lal, R., Hassan, H.M. and Dumanski, J. pp. 146–149. *In*: Rosenberg, N.J., Izauralde, R.C. and Malone, E.L. (eds.). Carbon Sequestration in Soils: Science, Monitoring and Beyond. Battelle Press, Columbus.

Xia, H. and Shu, W. 2003. Application of the Vetiver system in the reclamation of degraded land. Proc. ICV-3, Guangzhou, China, 419–27.

Yadav, J.S.P. 2003. Managing soil health for sustained high productivity. J. of the Indian Society of Soil Sci. 51: 448–485.

Yunusa, I.A.M. and Newton, P.J. 2003. Plants for amelioration of subsoil constraints and hydrological control: The primer-plant concept. Plant Soil 257: 261–281.

10

Microbial Mediated Biodegradation of Plastic Waste

An Overview

Rajendra Prasad Meena,[1,*] *Sourav Ghosh,*[2] *Surendra Singh Jatav,*[3] *Manoj Kumar Chitara,*[4] *Dinesh Jinger,*[5] *Kamini Gautam,*[6] *Hanuman Ram,*[1] *Hanuman Singh Jatav,*[2] *Kiran Rana,*[7] *Surajyoti Pradhan*[8] *and Manoj Parihar*[1]

1. Introduction

Plastics are synthetic as well as semi-synthetic polymers, which develop mainly from hydrocarbons fuels (crude oil, coal and natural gas) and some other components (Saminathan et al. 2014, Ahmed et al. 2018). The word "plastic" comes from the Greek language "plastikos", which means the material that can modify into any shape. Generally, plastics are characterized by high molecular weights and long chains of hydrocarbons (Iram et al. 2019) largely derived from petroleum, coal, natural gas and petrochemical derived materials. The commonly used plastic polymers (Table 1) in our daily life are polybutylene succinate (PBS), polyhydroxy butyrate (PHB), polyure thane (PUR), polyethylene (PE), polycaprolactone (PCL), polystyrene (PS), polyvinyl chloride (PVC), polylactic acid or polylactide (PLA), polyethylene terephthalate (PET), polyhydroxy alkanoate (PHA), polypropylene (PP), etc. (Muhamad et al. 2015, Yoshida et al. 2016, Ahmed et al. 2018). Plastic is an integral part of our daily routine life that can't be omitted. The wide use of plastics starting from domestic, agriculture and industrial purposes are commercially available (Sivan 2011). The demand of plastic is rising day by day considering it as an integral component of daily life. The large scale production of plastic was initiated during 1950, which had a 20 fold enhancement after 1964 (Iram 2019). In 2014, global estimation of plastic production was 311 million tons (Urbanek et al. 2015, 2018), which further increased and reached to 359 million tons in 2018. Out of this

---

[1] ICAR-Vivekananda Parvatiya Krishi Anusandhan Sansthan, Almora, Uttarakhand 263601, India.
[2] ICAR-Directorate of Onion and Garlic Research, Rajgurunagar-410505, Maharashtra, India.
[3] Department of Soil Science and Agricultural Chemistry, Institute of Agricultural Science, Banaras Hindu University, Varanasi 221005, UP, India.
[4] Department of Plant Pathology, College of Agriculture, GBPUAT, Panatnagar-263145, Uttarakhand, India.
[5] ICAR-Indian Institute of Soil and Water Conservation Dehradun-248195, Uttarakhand, India.
[6] ICAR-Indian Grassland and Fodder Research Institute, Jhansi-284003, Uttar Pradesh, India.
[7] Department of Agronomy, Institute of Agricultural Science, Banaras Hindu University, Varanasi-221005, UP, India.
[8] Orissa University of Agriculture and Technology, KrishiVigyan Kendra Sonepur-767017, Odisha.
* Corresponding author: rajagroicar@gmail.com

Table 1. Classification of plastic based on biodegradability.

Plastic Type	Structure	Use/property	Degrading microbes	References
Biodegradable				
Polylactic Acid (PLA)		Packaging paper, textiles and geotextiles, crop covers, compost bags, binder fibers, fiberfill and medicinal discipline for bone fracture internal fixation devices	Amycolatopsis, Thermoactinomyces sp., Saccharotrix Thermomyceslanuginosus, Aspergillus fumigatus, Mortierella sp., Doratomycesmicrosporus	Pranamuda et al. (1997), Teeraphatpornchai et al. (2003) Tokiwa et al. (2009) Karamanlioglu et al. (2014)
Polyhydroxyal-kanoates (PHAs)		Plasticizer, packaging materials, build up chiral compounds, razors, packaging bags, paper coatings, utensils fertilizers, insecticides, synthesizing containers for shampoos, cosmetics, and hygiene products, single-medical devices	Pseudomonas fluorescens, Pseudomonas putida, Pseudomonas aeruginosa, Pseudomonas sp.	Colak and Güner (2004), Bhatt et al. (2008)
Polyhydroxyb-utyrate (PHB)		Medicinal applications	Entrobacter sp., Bacillus sp., Gracilibacillus sp. Pseudomonas lemoignei	Volova et al. (2010) Kumaravel et al. (2010)
Poly(hydroxyb-utyrate-co-valerate) (PHBV)		Transparent barriers Orthopedic device Control released drugs Specialty packaging	Actinomadura sp. Microcossus sp., Bacillus sp.	Shah et al. (2010) Shah et al. (2007)
Polycaprolacto-ne (PCL)		Biomedical field. Several polymeric devices like pellets, microcapsules, nanoparticles, microspheres, films, and implants	*Penicillium Aspergillus* sp. *Clostridium*	Tokiwa et al. (2009) Abou-Zeid et al (2001)
Polybutylene succinate (PBS)		Electronics, food packaging materials, bowls, cups, plates. Plastics industry, shopping bags, and agriculture films	Amycolatopsis sp., Streptomyces sp, Paenibacillus sp., Paenibacillus amylolyticusn Purpureocillium sp., Cladosporium sp., Aspergillus fumigatus, Aspergillus niger, Fusariumsolani	Teeraphatpornchai et al. (2003), Penkhrue et al. (2015), Ishii et al. (2008), Abe et al. (2010), Li et al. (2011), Penkhrue et al. (2015)

Table 1 Contd. ...

...Table 1 Contd.

Plastic type	Structure	Use/property	Degrading microbes	References
Polybutylene adipate terephthalate (PBAT)		Packaging materials, rubbish bags. Good flexibility in rigid bioplastics	Thermomonosporafusca, Thermobifidafusca	Witt et al. (2001) Kleeberg et al. (2005)
Non-Biodegradable				
Polystyrene		Yoghurt containers, egg boxes, glassy surface, fast food trays, hard vending cups, disposable cutlery, brittle seed trays, high clarity, coat hangers, low cost brittle toys	Exiguobacterium sp.	Yang et al. (2015)
Polyethylene		Bags, water bottles, food packaging, film, toys, pipes, motor oil bottles	Zalerionmaritimum Phormidium sp., Rivularia Pseudophormidium sp., Phormidium sp.	Paco et al. (2017) Zettler et al. (2013) Oberbeckmann et al. (2014)
Polypropylene		Lunch boxes, margarine containers, yogurt pots, syrup bottles, prescription bottles, plastic bottle caps, potato crisp bags, biscuit wrappers	Phormidium sp., Rivularia	Zettler et al. (2013)
Polyvinyl Chloride (PVC)		Credit cards, carpet backing, wire and cable sheathing, synthetic leather products, guttering pipes and fittings	Not found	
Polyurethane		Footwear, automotive, furniture, and bedding. Building insulation. Refrigerators' coatings and adhesives	Cladosporium, Alternaria genus	(Álvarez-Barragán et al. 2016, Matsumiya et al. 2010)

whole production, only a small fraction is recycled. In 2008, worldwide global plastic consumption was 260 million metric tons and currently, 359 million metric tons plastic is being produced, which is expected to double in next 20 year and possibly quadruple by 2050 (World Economic Forum 2016). Six billion tons of plastics has been produced from 1950 to 2018 worldwide (Goel and Tripathi 2019). It is estimated that the production of plastic products account for 8% of global oil production (Goel and Tripathi 2019). The recycling rate varies country wise, mainly depending on the nature of plastic and management policy of the country. The recycle rate of countries like Greece, Malta and Cyprus is below 20% with poor scientific disposal, which causes environmental pollution with high rates of plastics ending in landfill. Nine countries in Europe had banned landfill practice from 1996 to 2006 and gained incineration rates up to 95% up to 2014 with proper recycling (Plastics Europe 2016). Plastic waste is a growing concern which needs to be addressed properly to reduce land pollution for better environmental safety. Since last 40 years, scientific community is trying to discover the viable and effective alternatives for better management of plastic pollution. Among the various strategies, microbial degradation could be an efficient approach which includes direct uptake of plastic fragments by microbes for nutritional purpose or indirectly via enzymatic degradation. Many studies have found that *Pseudomonas aeruginosa*, *P. fluorescens* and *Penicillium simplicissimum* as are effective bacterial and fungal isolate to degrade the plastic waste. Waste decomposition by different kind of microbes could be sound strategy, which can manage the plastic problem to a certain level (Ahmed et al. 2018). Considering the huge importance of these bio-organism, present chapter provides a scientific outlook on current scenario, classification, mechanism and factors involved in biodegradation of plastic wastes.

2. Current status of plastic pollution

One of the most serious environmental pollution that we face today is plastic pollution which affects our environment and health badly. Plastic is being utilized on a large scale in transport, cables, plastic packaging, telecommunications, clothing, etc. The extensive use of this polymer has led to the accumulation of plastic in our environment. Now this plastic which has been discarded is accumulated and damaging our terrestrial and aquatic biosphere (Richard et al. 2009). Plastics are produced from petroleum in the same way as refined gasoline, with the entire process releasing harmful gases like carbon monoxide, ozone, benzene, hydrogen sulfide and methane into our environment (Goel and Tripathi 2019). Similarly, burning of plastics and plastic products also releases carbon dioxide leading to global warming. Plastic on degradation releases toxic chemicals like polystyrene and bisphenol A (BPA) which gets mixed with soil and water (Knight 2012) causing pollution and impacting human health. Moreover, additives present in plastics are chemicals and have also been reported to disturb endocrine glands' functions (Alabi et al. 2019). Ingestion, contact and inhalation of the plastic additives cause serious health issues in human and animals. Besides this, consumption of micro-plastics by marine animals is also detrimental. Marine wildlife is influenced by plastic pollution through entanglement, ingestion and bioaccumulation and changes the integrity and functioning of wildlife habitats.

The plastic debris are classified as mega debris (> 100 mm), macro-debris (> 20 mm diameter), meso-debris (5–20 mm) and micro-debris (< 5 mm) and these debris are estimated to have longevity of thousands years (Barnes et al. 2009). Micro-plastics are the major pollutants deteriorating the ecosystem; they are either produced as design or as result of degradation of macro-plastic. Rising human population and its rising demand for plastics and plastic products has led to its greater accumulation in the environment. Single-use plastic accounts for almost 40% of the overall plastic usage (Worm et al. 2017). According to an estimate, 13 million tons of plastic bags reach ocean, killing 100,000 marine lives (Alabi et al. 2019) and by 2050, oceans will have more plastics than fish in terms of weight. Leaching of plastic additives like coloring materials, additives, and heavy metals through degradation of plastics leads to soil and water contamination and pollution. Therefore, this is high time to reduce plastic usage by avoiding single-use plastics or by raising awareness among the

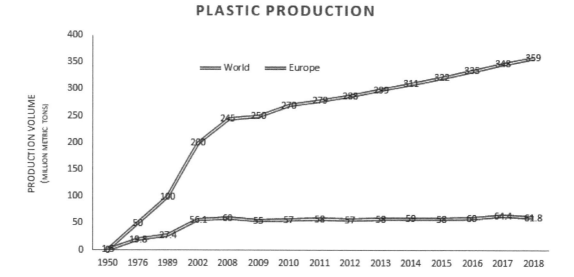

Figure 1. Global plastic production (source: Statista 2019).

people on hazardous effects of plastic pollution. Moreover, reuse and recycling of plastic material is also needed along with searching alternative biodegradable options for the scientific management of plastic.

3. Classification of plastics based on biodegradability

Based on biodegradability, the fossil-based and bio-based plastics can be classified into two groups such as non-biodegradable and biodegradable plastic (Table 1). Non-biodegradable plastics are high molecular weight derivative of hydrocarbon and petroleum compound with high stability and do not readily enter into the degradation cycles of the biosphere (Vijaya and Reddy 2008, Ghosh et al. 2013), for example, polypropylene (PP), polystyrene (PS), polyvinyl chloride (PVC), polyethylene terephthalate (PET), polyethylene (PE) and polyurethane (PUR) (Ahmed et al. 2018). Biodegradable plastics depend upon the degree of biodegradability and microbial assimilation (Wackett and Hershberger 2001). This process involves the degradation of complex polymer compound into smaller compounds in the presence of enzymes which are secreted by microorganisms (Artham and Doble 2008), for example, starch (Chattopadhyay et al. 2011), Polyhydroxyalkanoates (Shimao 2001), Polylactic acid (Ikada and Tsuji 2000), Polyethylene succinate (Hoang et al. 2007) and Polycaprolactone (Wu 2005). Biodegradable plastic is further classified into two categories, which is bio-based biodegradable plastic and fossil-based biodegradable plastic.

3.1 Bio-based biodegradable plastic

Bio-based biodegradable plastics are derived from renewable resources and are considered as eco-friendly plastics which can be completely degraded biologically after their application (Kale et al. 2007), such as cellulose, starch and starch-based polymers. Starch is the most commonly used bio-based polymer for the production of biodegradable plastics. The starch contains some important properties such as high richness, ready availability, inexpensiveness and biodegradability under certain environmental condition and is utilized frequently to synthesize bio-based biodegradable plastics (Chattopadhyay et al. 2011, Kyrikou and Briassoulis 2007, Nanda et al. 2010). The main constituent of starch is amylopectin and amylase polymers, which makes it a viable substitute. Various microorganisms have been reported to degrade bio-based polymers under both anaerobic and aerobic conditions (Shah et al. 2008), for example, *Variovoraxparadoxus*, *Comamonas* sp., *Aspergillus fumigatus*, *Acidovoraxfaecilis* and *P. lemoignei*. There are two types of bio-based biodegradable plastics, viz. Polyhydroxyalkanoates (PHA) and Polylactic acid (PLA) (Elbanna et al. 2004).

3.1.1 Polyhydroxyalkanoates

Polyhydroxyalkanoates are bio-based biodegradable plastics obtained by bacterial fermentation of sugar and lipids (Shimao 2001). Due to their biodegradability, it can be used in packaging, medical, pharmaceutical industries (Philip et al. 2007), fast food service materials, disposable medical tools, etc. (Flieger et al. 2003). The biodegradation of PHA by microorganisms, i.e., *Bacillus*, *Burkholderia*, *Nocardiopsis*, *Cupriavidus*, *Mycobacterium* and *Micromycetes,* etc. involves both aerobic and anaerobic degradation mechanisms (Boyandin et al. 2013).

3.1.2 Polylactic acid

Polylactic acid is produced from corn starch, tapioca roots, or sugarcane. It has been used extensively in medicine because of the ability of the polymer to be incorporated into human and animal bodies (Ikada and Tsuji 2000). It is the most important among the bio-based biodegradable plastics because of its availability, biodegradability, and good mechanical attributes.

3.2 *Fossil-based biodegradable plastics*

Fossil-based biodegradable plastics are mainly used in the packaging industry, and these are non-biodegradable plastics which cause a serious problem to the environment (Hoshino et al. 2003, Vert et al. 2002). Its degradation takes more time, which involves various microbes and enzymes under different environmental conditions (Chen and Patel 2011, Shah et al. 2008, Chen 2010, Mir et al. 2017), for example, enzyme glycosidase (*A. flavus*) is involved in polycaprolactone (PCL) degradation (Tokiwa et al. 2009).

3.2.1 Polyethylene succinate

It is thermoplastic polyesters which are developed by the process of copolymerization of ethylene oxide and succinic anhydride or via ethylene glycol and succinic acid poly-condensation (Hoang et al. 2007). It is used in plastic films production for agriculture sector, paper coating agent, and for the manufacturing of shopping bags. The degradation of this polymer is reported by *Pseudomonas* sp. AKS2 (Tribedi and Sil 2014), *Bacillus* and *Paenibacillus* genera (Tezuka et al. 2004, Tokiwa et al. 2009).

3.2.2 Polycaprolactone

It is a fossil-based biodegradable polymer that can easily be degraded by aerobic and anaerobic microorganisms. It is highly flexible and biodegradable, which is made up with partially crystalline polyester mingled with other copolymers (Wu 2005). The polycaprolactone (PCL) is commonly used in packaging material, biomedical, catheters and blood bags (Wu 2005). It is degraded by the microbial lipases and esterases (Karakus 2016), viz. *Rhizopusdelemar C. botulinum* (Tokiwa et al. 2009). The *Aspergillus* sp. ST-01 has also been found efficient in PCL degradation into butyric, succinic, caproic, and valeric acids (Sanchez et al. 2000).

4. Mechanism of plastic biodegradation

Oxidative degradation is the principal mechanism for the degradation of plastics. These mechanisms reduce the molecular weight of the material. In this mechanism of plastic biodegradation, microorganisms stick with polymers and eventually colonize on the surface. This enzyme-based hydrolysis of plastics occur in two steps: firstly, the enzymes attach to the polymer substrate followed by hydrolytic division (Figure 2). The extracellular and intracellular enzymes that are produced by the microbes convert the polymers into monomer, dimer, and oligomer. Degradation

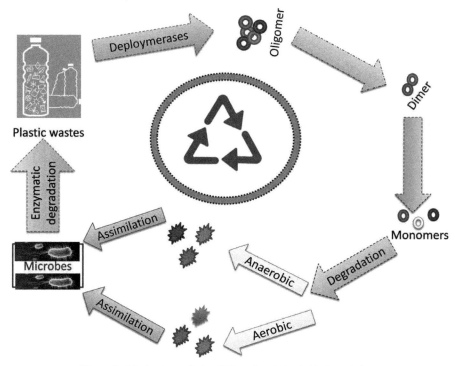

Figure 2. Mechanisms of microbial mediated plastic biodegradation.

products of polymers like oligomers, dimers, and monomers are much low in molecular weight and are eventually converted to CO_2 and H_2O by mineralization (Tokiwa et al. 2009). The by-products produced during conversion enter into the microbial cell and can be utilized as the energy source (Shimao 2001). Under aerobic conditions, bacteria use oxygen as electron acceptor and form other organic compounds along with CO_2 and H_2O as last products (Priyanka and Archana 2012). Under anaerobic conditions, nitrate (NO_3^-), sulfate (SO_4^{2-}), metals such as iron (Fe_3^+) and manganese (Mn_4^+), or even CO_2 play the role of electrons acceptor from the degraded contaminant (National Research Council 1993). The byproducts of anaerobic respiration are nitrogen gas (N_2), hydrogen sulfide (H_2S), metals (in reduced form), and methane (CH_4), which largely depends on electron accepting species.

5. Microbes involved in plastic biodegradation

Microorganisms are bestowed with enzymes which play a vital role in destruction of environmental contaminants. They are so tiny that they can reach and contact contaminants easily and efficiently. Optimum quantities of nutrients are required for their work efficiency and metabolism. Bacteria and fungi often produce extracellular enzymes which easily degrade the various sorts of bio and fossil-based plastics (Shah et al. 2014). Bacteria and fungi degrade these plastic polymers into CO_2 and H_2O through various metabolic and enzymatic mechanisms. The microbial species and strains affect the catalytic activity of enzymes and finally degradation process. Various types of enzymes are known to degrade many polymer types. Microbial enzymes accelerate the biodegradation rate of plastics very effectively without causing any harm to the environment. Enzymes are present in living cell of every organism and hence in all microbes. Different types of microbes produce different kinds of enzymes in different quantities. For instance, *Brevibacillus* spp., and *Bacillus* spp., produce proteases enzyme responsible for degradation of polyethylene (PE) (Sivan 2011). Fungi are known to degrade the lignin frequently as they produce laccases enzymes which catalyze the aromatic and non-aromatic compounds (Mayer and Staples 2002). Lignin and manganese-dependent peroxidases (LiP and MnP, respectively) and laccases are major lignin olytic enzymes (Hofrichter et al. 2001). Lignin-degrading fungi and manganese peroxidase, partially purified from the strain of *Phanerochaetechryso sporium,* also help in the degradation of high-molecular weight PE under nitrogen and carbon limited conditions (Shimao 2001). The list of plastic degradation by various microbes and their enzymes has been given in Table 2.

Table 2. Microbes and enzymes responsible for degradation of specific polymer groups.

Polymer	Degrading enzymes	Microbes involved
Polyethylene terephthalate (PET)	PET hydrolase and tannase (cutinase, lipases, carboxylesterase)	Actinobacteriae.g *Thermobifida, Thermomonospora, Ideonellasakaiensis,* Saccharomonospora
	Cutinase and Lipase	*Pseudomonas pseudoalcaligenes, Pseudomonas pelagia*
	Fungal cutinase	*Fusarium, Humicola*
	PET esterase	*Desulfovibriofructosivorans*
Polyurethanes (PUR)	Lipase	*Pseudomonas chlororaphis, Pseudomonas putida, Candida rugosa*
	Esterase	*Comamonas acidovorans, Aspergillus flavus*
	Cutinase	*Thermobifida*
Polyethylene (PE)	Unidentified	*Pseudomonas, Ralstonia and Stenotrophomonas Staphylococcus, Streptomyces, Bacillus*
	Laccase	*Aspergillus, Cladosporium, Penicillium*
Polyamide (PA) – Nylon	Hydrolases	Pseudomonas, Achromatobacter
	Dehydrogenase	*Bacillus cereus, Bacillus sphaericus, Vibrio furnisii and Brevundimonasvesicularis*
	Peroxidase	White rot fungi
Polystyrene (PS)	Styrene monooxygenase, Styrene oxide isomerase, Phenylacetaldehyde dehydrogenase	Brown-rot fungi – Gloeophyllumtrabeum White rot fungi – *Pleurotusostreatus, Phanerochaetechrysosporium, Trametes versicolor* *Pseudomonas, Xanthobacter, Rhodococcus, Corynebacterium*

Source: Danso et al. (2019).

6. Plastic degrading enzymes and their role

The key polymers (plastics) that are manufactured and are of significance to our economy are: Polyurethane (PUR), Polyethylene (PE), Polyamide (PA), Polyethylene terephthalate (PET), Polystyrene (PS), Polyvinylchloride (PVC) and Polypropylene (PP). Plastic degradation is a change in the properties—tensile strength, color, shape, etc.—of a plastic polymer or polymer-based product under the influence of one or more environmental factors such as heat, light, and chemicals such as acids, alkalis, salts or biological enzymes. Three major mechanisms for degradation of plastics are photodegradation, thermooxidative degradation and biodegradation. The first two types of degradation mechanisms are abiotic and the last one is biotic in nature. Biodegradation of polymers occurs through various mechanisms like solubilization, dissolution, hydrolysis and enzyme-catalyzed degradation and can occur under aerobic or anaerobic conditions. Microbes mediated degradation of plastics is facilitated by secreting enzymes which act on long chain polymer surface converting them into monomers (enzymatic biodegradation).

The microbial degradation of synthetic polymers is an extremely sluggish process, stemming from the high molecular weight of the polymer fiber, the resilient C-C bonds and the exceedingly hydrophobic surface, which is very difficultly attacked by the enzymes. Particularly, the different polymers with varying molecular weights having amorphous and crystalline nature resulting in their

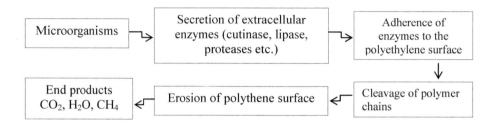

difference in degradability as well as susceptibility to enzymes. Enzymatic hydrolysis of polyesters can be grouped under:

A. enzymatic surface modification (hydrolyze the surface polymer chain) – hydrolases, such as lipases, carboxylesterases, cutinases, and proteases and

B. enzymatic hydrolysis (substantial degradation of the building blocks of PET) – cutinases and some other hydrolases.

First step of enzymatic biodegradation is hydrolysis and addition of a functional group which enhances the hydrophobicity of polymer. Subsequently, the polymers are broken down to monomers and enter the microbes through their semi-permeable membranes. Even after hydrolysis, if the polymers' size exceeds the threshold, these are first depolymerized which allows them to penetrate through the cell membrane and are thereafter broken down by intracellular enzymes.

The hydrolytic reaction is first initiated by class of enzymes known as 'hydrolase' comprising cutinase, phosphatase, laccase, esterase, lipase, glycosidase and many more (Bano et al. 2017). These enzymes have a characteristic α/β-hydrolase fold and the catalytic triad is comprised of a serine, a histidine and an aspartate residue (Wei et al. 2014). They can also comprise numerous disulfide bonds produced by cysteine residues, which help in thermal stability and specific binding to polymers such as PET.

6.1 Cutinases

Cutinases are involved in (catalyze) hydrolysis of cutin which are aliphatic polyesters and are found in plant cuticle structure. Under the super family of α/β hydrolases, this category of polyester hydrolases acts upon several polyester plastics (Wei and Zimmermann 2017a, b). Depending on their homology, origin, and structure, plastic-degrading cutinases can be divided into fungal and bacterial. Cutinases from fungal origin such as *Thermomycesinsolens, Fusarium* and *Humicola* are useful and show excellent activity in the hydrolysis and surface alterations of polyethylene terephthalate (PET) films and fibers (Zimmermann and Billig 2010), due to their remarkable activity and thermal stability at 70°C near the glass transition temperature of PET (Ronkvist et al. 2009). Bacterial cutinases capable of hydrolyzing PET have been segregated from various *Thermobifida species* (Then et al. 2015), *Thermomonosporacurvata* (Wei et al. 2014), *Saccharomonosporaviridis* (Kawai et al. 2014), *Ideonellasakaiensis* (Yoshida et al. 2016), as well as the metagenome isolated from plant compost (Sulaiman et al. 2012). The bacterium, *Ideonellasakaiensis* 201-F6, exhibits rare capability to thrive on PET as a major carbon and energy source and secretes PETase (PET-digesting enzyme) leading to its biodegradation (Yoshida et al. 2016).

6.2 Lipases

Lipases belonging to α/β hydrolases are capable of hydrolyzing aliphatic polyesters or aliphatic-aromatic co-polyesters (Herzog et al. 2006). Lipases from *Thermomyceslanuginosus* degrade PET and polytrimethylene terephthalate (Ronkvist et al. 2009). Lipases exhibit slow hydrolytic activity against PET in comparison to cutinases, possibly due to their lid type assembly, shelling the buried hydrophobic catalytic centre, which restricts the entry of aromatic polymeric substrates to the active site of the enzyme (Zimmermann and Billig 2010). Lipases from *T. lanuginosus* (Eberl et al. 2009) and *Candida antarctica* (Carniel et al. 2017) also degrade low-molecular-weight degradation products of PET. The PueB lipase from *Pseudomonas chlororaphis* as well as from *Fusarium solani* and *Candida ethanolica* are reported to be involved in metabolism of Polyurethanes (PUR).

6.3 Carboxylesterases

Bacteria such as *Thermobifidafusca*, *Bacillus licheniformis* and *Bacillus subtilis* contain enzyme carboxylesterases instigating hydrolysis of PET oligomers and their analogues (Barth et al. 2016). The carboxylesterase TfCa from *T. fusca* removes water-soluble components from high-crystalline PET fibers (Zimmermann and Billig 2010). Strains of *Comamonasacidovorans* TB-35 produced a PUR-active esterase enzyme, with hydrophobic PUR surface binding properties essential for PUR degradation. Fungi belonging to the *Cladosporiumclado sporioides* complex, 249 including the species *C. pseudocladosporioides*, *C. tenuissimum*, *C. asperulatum*, *C. montecillanum*, *Aspergillus fumigatus* and *Penicillium,* secreted esterases responsible for the degradation of poly butylene adipate-co-terephthalate (PBAT) (Wallace et al. 2017). Polyesterase acting on aromatic polyesters (primarily PET) was first reported by Muller et al. (2005) in *Thermobifidafusca.*

6.4 Proteases

Proteases obtained from *Pseudomonas chlororaphis* and *Pseudomonas fluorescens* degrade polyester polyurethanes (PU) (Matsumiya et al. 2010). In addition to the microbial enzymes, other proteases active against PU include papain, a cysteine protease from papaya, which can hydrolyze amide and urethane bonds and elastase obtained from porcine pancreas (Labow et al. 1996).

6.5 Lignin-modifying enzymes

Laccases, manganese peroxidases and lignin peroxidases involved in the degradation of lignin (Ruiz-Dueñas and Martinez 2009) also have role in the biodegradation of polyethylene (Krueger et al. 2015). A laccase isolated from *Trametes versicolor* degraded a high-molecular weight PE (PE-HMW) membrane in the presence of 1-hydroxybenzotriazole, which mediated the oxidation of non-phenolic substrates by the enzyme (Fujisawa et al. 2001). A Penicillium derived laccase, potentially involved in PE breakdown, was reported by Sowmya et al. (2015). A microbial peroxidase extracted from white rot fungus was shown to act on high molecular weight nylon fibers (Danso et al. 2019).

Limited information is still available on the known variety of enzymes and microbes acting on synthetic polymers. Consequently, effort has been made towards identification of organisms capable of acting on the most dominant polymers. The major constraint lies in the initial breaking of the high molecular weight crystalline structure of the robust polymers. Enzymatic degradation of polymers and its commercial implementation are of prime importance and require extensive research and funding that would allow the effective management of plastic thereby reducing environmental pollution.

7. Factors affecting biodegradability of plastic

Plastic biodegradation is the function of microbes to convert polymers (organic substrates) into small molecular weight fragments that can be further degraded to water and carbon dioxide (Ahmed et al. 2018). Among the various factors affecting plastic biodegradation, physical and chemical properties of polymers and environmental conditions are of prime importance (Figure 3).

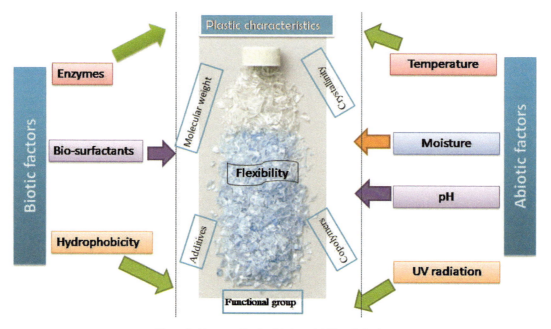

Figure 3. Factors affecting biodegradability of plastics.

7.1 Polymer characteristics

7.1.1 Molecular weight

The molecular weight plays a critical role in biodegradation of plastic as microbial colonization depends on surface features that allow microbes to establish a locus from which they expand their growth. Molecular weight is inversely related to biodegradation as polymer biodegradability decreases with increase in molecular weight (Tokiwa et al. 2009).

7.1.2 Biosurfactants

Compounds which are amphiphilic in nature and produced on living surface are called biosurfactants. The biodegradation process of plastic is enhanced by the addition of biosurfactants due to the presence of specific functional groups (Auras et al. 2004).

7.1.3 Shape and size

The polymers' properties like shape and size play critical role in the biodegradation process. The polymers having large surface area degrade faster than small surface area. Certain types of standards are fixed for various types of polymers having different shape and size (Tokiwa et al. 2009).

7.1.4 Additives

Polymer additives are usually low molecular weight organic compounds that provide help to microbes to form colonization due to their ease of biodegradation. Non-polymeric contaminants such as dyes (waste of catalysts used for the polymerization and additives conversion products) or filler materials affect the biodegradation ability of microbes (Ahmed et al. 2018). It is well known that thermal stability would be reduced when lingo-cellulosic filler increases in the sample. The major factors responsible for the thermal stability of the composite system are the dispersal and interfacial adhesion between the lingo-cellulosic filler and the thermoplastic polymer system (Yang et al. 2005). Likewise, metals serve as excellent pro-oxidants in polyolefin manufacturing of polymers sensitive to thermo-oxidative degradation.

7.1.5 Polymer crystallinity

Polymer crystallinity can play a sound role as it has been observed that colonization of microbes to the surface of polymer occurs and utilizes polymer substances in amorphous sections of the polymer surface.

7.2 Environmental factors

7.2.1 Moisture

Moisture is the most important factor responsible for the growth and multiplication of microorganisms, so it plays a major role in the biodegradation of polymers (Ahmed et al. 2018). Sufficient quantity of moisture is needed for activation of microbes and the hydrolytic capacity of microbes is also enhanced with increase in moisture content (Iram et al. 2019).

7.2.2 pH

The rate of hydrolysis reaction of polymer can alter through changing the acidic or basic condition. Change in the pH also affects the growth and multiplication of microorganism and ultimately the biodegrading of polymers.

7.2.3 Temperature

Biodegradability of polymers is significantly influenced by the polymer's softening temperature. Polymers having higher melting point are less prone to biodegradation. Efficient enzymatic biodegradation occurs with polymers having low melting point (Tokiwa 2009). For example, purified lipase of *R. delemar* capably hydrolyzed polyesters like PCL which were showing low melting points (Tokiwa and Calabia 2004). Different enzymes showing unique active sites have the ability to biodegrade various types of polymer substrates, for example, straight chain polyesters, obtained from diacid monomers with 6 to 12 C-atoms, have been degraded faster by enzymes produced by fungal species *A. flavus* and *A. nigeras* compared to straight chain polyesters produced by other monomer (Kale et al. 2007). Polymers made from the petro chemical sources, because of their hydrophobicity and 3D structure, cannot be easily degraded in the environment (Yamada-Onodera et al. 2001).

8. Conclusion

Now, it's clear that plastic use in our daily life is inevitable and their demand is an ever increasing trend. So there is urgent need to formulate the bio-based biodegradable polymer to maintain the environmental health. In addition to bio-based biodegradable polymer, new and potential microbial isolates to degrade these bio or fossil based polymer need to be characterized with possible mechanism to exploit them to the maximum. The technology of biodegradation of plastics employing microbial enzymes such as cutinases, lipases, hydrolases, etc. should be fine-tuned and commercialized at the earliest. Moreover, the higher use of biodegradable plastics in various industries and other sectors must be positively correlated with proper waste management and littering control mechanism in order to achieve higher environmental safety and sustainability. Alternative to synthetic plastic, there is an urgent need to generate bioplastics having properties such as lower durability and faster decomposability in order to create a cleaner and safer environment.

References

Abe, M., Kobayashi, K., Honma, N. and Nakasaki, K. 2010. Microbial degradation of polybutylene succinate by *Fusariumsolani* in soil environments. Polym. Degrad. Stabil. 95: 138–143.

Abou-Zeid, D.M., Muller, R.J. and Deckwer, W.D. 2001. Degradation of natural and synthetic polyesters under anaerobic conditions. Polym. Degrad. Stabil. 86: 113–126.

Ahmed, T., Shahid, M. and Azeem, F. 2018. Biodegradation of plastics: current scenario and future prospects for environmental safety. Environ. Sci. Pollut. Res. 25: 7287–7298.

Alabi, A.O., Ologbonjaye, K., Awosolu, O. and Alalade, E.O. 2019. Public and environmental health effects of plastic wastes disposal: a review. J. Toxicol. Risk. Assess. 5(2): 1–13.

Alvarez-Barragan, J., Dominguez-Malfavon, L., Vargas-Suarez, M., Gonzalez-Hernandez, R., Aguilar-Osorio, G. and Loza-Tavera, H. 2016. Biodegradative activity of selected environmental fungi on a polyester polyurethane varnish and polyether polyurethane foams. Appl. Environ. Microbiol. 82: 5225.

Artham, T. and Doble, M. 2008. Biodegradation of aliphatic and aromatic polycarbonates. Macromol. Biosci. 8(1): 14–24.

Auras, R., Harte, B. and Selke, S. 2004. An overview of polylactides as packaging materials. Macromol. Biosci. 4: 835–864.

Bano, K., Kuddus, M., Zaheer, M.R, Zia, Q., Khan, M.F., Gupta, A. and Aliev, G. 2017. Microbial enzymatic degradation of biodegradable plastics. Curr. Pharm. Biotechno. 18(5): 429–440.

Barth, M., Honak, A., Oeser, T., Wei, R., Belisário-Ferrari, M.R., Then, J., Schmidt, J. and Zimmermann, W. 2016. A dual enzyme system composed of a polyester hydrolase and a carboxylesterase enhances the biocatalytic degradation of polyethylene terephthalate films. Biotechnol. J. 11(8): 1082–1087.

Bhatt, R., Shah, D., Patel, K.C. and Trivedi, U. 2008. PHA–rubber blends: synthesis, characterization and biodegradation. Bioresour. Technol. 99: 4615–4620.

Boyandin, A.N., Prudnikova, S.V., Karpov, V.A., Ivonin, V.N., Do N.L., Nguyen, T.H., Le, T.M.H., Filichev, N.L., Levin, A.L., Filipenko, M.L., Volova, T.G. and Gitelson, I.I. 2013. Microbial degradation of polyhydroxyalkanoates in tropical soils. Int. Biodeter. Biodegr. 83: 77–84.

Carniel, A., Valoni, É., Junior, J.N., da Conceição Gomes, A. and de Castro, A.M. 2017. Lipase from Candida antarctica (CALB) and cutinase from Humicolainsolens act synergistically for PET hydrolysis to terephthalic acid. Process. Biochem. 59: 84–90.

Chattopadhyay, S.K., Singh, S., Pramanik, N., Niyogi, U.K., Khandal, R.K., Uppaluri, R., Ghoshal, A.K. 2011. Biodegradability studies on natural fibers reinforced polypropylene composites. J. of Appl. Poly. Sci. 121(4): 2226–2232.

Chen, G.Q. 2010. Introduction of bacterial plastics PHA, PLA, PBS, PE, PTT, and PPP. pp. 1–16. In: Chen, G.Q., (ed.). Plastics from Bacteria. Natural Functions and Applications. Springer: Berlin/Heidelberg, Germany.

Chen, G.Q. and Patel, M.K. 2011. Plastics derived from biological sources: present and future: a technical and environmental review. Chem. Rev. 112: 2082–2099.

Colak, A. and Guner, S. 2004. Polyhydroxyalkanoate degrading hydrolase-like activities by *Pseudomonas* sp. isolated from soil. Int. Biodeter. Biodegr. 53: 103–109.

Danso, D., Chow, J. and Streit, W.R. 2019. Plastics: Microbial degradation, environmental and biotechnological perspectives. Appl. Environ. Microbiol, pp. AEM-01095.

Eberl, A., Heumann, S., Brückner, T., Araujo, R., Cavaco-Paulo, A., Kaufmann, F., Kroutil, W. and Guebitz, G.M. 2009. Enzymatic surface hydrolysis of poly (ethylene terephthalate) and bis (benzoyloxyethyl) terephthalate by lipase and cutinase in the presence of surface-active molecules. J. Biotechnol. 143(3): 207–212.

Elbanna, K., Lutke-Eversloh, T., Jendrossek, D., Luftmann, H. and Steinbuchel, A. 2004. Studies on the biodegradability of polythioester copolymers and homopolymers by polyhydroxyalkanoate (PHA)-degrading bacteria and PHA depolymerases. Arch. Microbiol. 182(2-3): 212–225.

Flieger, M., Kantorova, M., Prell, A., Rezanka, T. and Votruba, J. 2003. Biodegradable plastics from renewable sources. Folia. Microbiol. 48(1): 27–44.

Fujisawa, M., Hirai, H. and Nishida, T. 2001. Degradation of polyethylene and nylon-66 by the laccase-mediator system. J. Polym. Environ. 9(3): 103–108.

Ghosh, S.K., Pal, S. and Ray, S. 2013. Study of microbes having potentiality for biodegradation of plastics. Environ. Sci. Pollut. Res. 20(7): 4339–4355.

Goel, M. and Tripathi, N. 2019. Strategies for controlling plastics pollution in India: policy paper. Climate change research institute in association with India International Center, New Delhi India. pp. 1–27.

Herzog, K., Müller, R.J. and Deckwer, W.D. 2006. Mechanism and kinetics of the enzymatic hydrolysis of polyester nanoparticles by lipases. Polym. Degrad. Stabil. 91(10): 2486–2498.

Hoang, K.C., Tseng, M. and Shu, W.J. 2007. Degradation of polyethylene succinate (PES) by a new thermophilic Microbispora strain. Biodegradation 18(3): 333–342.

Hofrichter, M., Lundell, T. and Hatakka, A. 2001. Conversion of milled pine wood by manganese peroxidase from *Phlebiaradiata*. Appl. Environ. Microbiol. 67: 4588–4593.

Hoshino, A., Tsuji, M., Ito, M., Momochi, M., Mizutani, A., Takakuwa, K., Higo, S., Sawada, H. and Uematsu, S. 2003. Study of the aerobic biodegradability of plastic materials under controlled compost. pp. 47–54. In: Chielline, E. and Solaro, R. (eds.). Biodegradable Polymers and Plastics, Springer.

Ikada, Y. and Tsuji, H. 2000. Biodegradable polyesters for medical and ecological applications. Macromol. Rapid. Commun. 21(3): 117–132.

Iram, D., Riaz, R.A. and Iqbal, R.K. 2019. Usage of potential micro-organisms for degradation of plastics. J. Environ. Biol. 4(1): 007–0015.

Ishii, N., Inoue, Y., Tagaya, T., Mitomo, H., Nagai, D. and Kasuya, K. 2008. Isolation and characterization of polybutylene succinate-degrading fungi. Polym. Degrad. Stabilil. 93: 883–888.

Kale, G., Kijchavengkul, T., Auras, R., Rubino, M., Selke, S.E. and Singh, S.P. 2007. Compostability of bioplastic packaging materials: an overview. Macromol. Biosci. 7(3): 255–277.

Karakus, K. 2016. Polycaprolactone (PCL) based polymer composites filled wheat straw flour. Kastamonu Üniversitesi Orman Fakültesi Dergisi 16(1): 264–268.

Karamanlioglu, M., Houlden, A. and Robson, G.D. 2014. Isolation and characterisation of fungal communities associated with degradation and growth on the surface of polylactic acid (PLA) in soil and compost. Int. Biodeter. Biodegr. 95: 301–310.

Kawai, F., Oda, M., Tamashiro, T., Waku, T., Tanaka, N., Yamamoto, M., Mizushima, H., Miyakawa, T. and Tanokura, M. 2014. A novel Ca_2^+ activated, thermostabilized polyesterase capable of hydrolysing polyethylene terephthalate from Saccharomonospora viridis AHK190. Appl. Microbiol. Biotechnol. 98: 10053–10064.

Kleeberg, I., Welzel, K., VandenHeuvel, J., Muller, R.J. and Deckwer, W.D. 2005. Characterization of a new extracellular hydrolase from Thermobifidafusca degrading aliphatic-aromatic copolyesters. Biomacromolecules 6: 262–270.

Knight, G. 2012. Plastic pollution. Raintree Publisher, London.

Krueger, M.C., Harms, H. and Schlosser, D. 2015. Prospects for microbiological solutions to environmental pollution with plastics. Appl. Microbiol. Biotechnol. 99(21): 8857–8874.

Kumaravel, S., Hema, R. and Lakshmi, R. 2010. Production of polyhydroxybutyrate (bioplastic) and its biodegradation by *Pseudomonas lemoignei* and *Aspergillus niger*. E. J. Chem. 7: S536–S542.

Kyrikou, I. and Briassoulis, D. 2007. Biodegradation of agricultural plastic films: a critical review. J. Polymers. Environ. 15(2): 125–150.

Labow, R.S., Erfle, D.J. and Santerre, J.P. 1996. Elastase-induced hydrolysis of synthetic solid substrates: poly (ester-urea-urethane) and poly (ether-urea-urethane). Biomaterials 17(24): 2381–2388.

Li, F., Hu, X., Guo, Z., Wang, Z., Wang, Y., Liu, D., Xia, H. and Chen, S. 2011. Purification and characterization of a novel polybutylene succinate-degrading enzyme from *Aspergillus* sp. XH0501-a. World J. Microbiol. Biotech. 27: 2591–2596.

Matsumiya, Y., Murata, N., Tanabe, E., Kubota, K. and Kubo, M. 2010. Isolation and characterization of an ether-type polyurethane-degrading micro-organism and analysis of degradation mechanism by *Alternaria* sp. J. Appl. Microbiol. 108: 1946–1953.

Mayer, A.M. and Staples, R.C. 2002. Laccase: new functions for an old enzyme. Phytochemistry 60(6): 551–565.

Mir, S., Asghar, B., Khan, A.K., Rashid, R., Shaikh, A.J., Khan, R.A. and Murtaza, G. 2017. The effects of nanoclay on thermal, mechanical and rheological properties of LLDPE/chitosan blend. J. Polym. Eng. 37(2): 143–149

Muhamad, W. and Naimatul, A.W. 2015. Microorganism as plastic biodegradation agent towards sustainable environment. Advan. Environ. Biol. 9: 8–13.

Müller, R.J., Schrader, H., Profe, J., Dresler, K. and Deckwer, W.D. 2005. Enzymatic degradation of poly (ethylene terephthalate): rapid hydrolyse using a hydrolase from T. fusca. Macromol. Rapid. Commun. 26(17): 1400–1405.

Nanda, S., Sahu, S. and Abraham, J. 2010. Studies on the biodegradation of natural and synthetic polyethylene by *Pseudomonas* spp. J. Appl. Sci. Environ. Manag. 14(2): 57–60.

Oberbeckmann, S., Loeder, M.G., Gerdts, G. and Osborn, A.M. 2014. Spatial and seasonal variation in diversity and structure of microbial biofilms on marine plastics in Northern European waters. FEMS Microbiol. Ecol. 90(2): 478–492.

Paco, A., Duarte, K., Da costa, J.P., Santos, P.S., Pereira, R., Pereira, M.E., Freitas, A.C., Duarte, A.C. and Rocha-Santos, T.A. 2017. Biodegradation of polyethylene micro plastics by the marine fungus *Zalerionmaritimum*. Sci. Total Environ. 586: 10–15.

Penkhrue, W., Khanongnuch, C., Masaki, K., Pathom-aree, W., Punyodom, W. and Lumyong, S. 2015. Isolation and screening of biopolymer-degrading microorganisms from northern Thailand. World J. Microb. Biot. 31: 1431–1442.

Philip, S., Keshavarz, T. and Roy, I. 2007. Polyhydroxyalkanoates: biodegradable polymers with a range of applications. J. Chem. Technol. Biot. 82(3): 233–247.

Plastics Europe. 2016. Plastics—the facts 2016. Düsseldorf. http://www.plasticseurope.org. Accessed 13 Oct 2017.

Pranamuda, H., Tokiwa, Y. and Tanaka, H. 1997. Polylactide degradation by an *Amycolatopsis* sp. Appl. Environ. Microbiol. 63: 1637–1640.

Priyanka, N. and Archana, T. 2012. Biodegradability of polythene and plastic by the help of microorganism: a way for brighter future. J. Anal. Toxicol. 1: 12–15

Richard, C., Thompson, C.J., Moore, F.S. and Vom, S. 2009. Plastics, the environment and human health: current consensus and future trends and Shanna H. Swan. Philosophical Transactions of the Royal Society B: Biological Sci. 364: 2153–2166.

Ronkvist, Å.M., Xie, W., Lu, W. and Gross, R.A. 2009. Cutinase-catalyzed hydrolysis of poly (ethylene terephthalate). Macromolecules 42(14): 5128–5138.

Ruiz-Dueñas, F.J. and Martínez, Á.T. 2009. Microbial degradation of lignin: how a bulky recalcitrant polymer is efficiently recycled in nature and how we can take advantage of this. Microb. biotechnol. 2(2): 164–177.

Saminathan, P., Sripriya, A., Nalini, K., Sivakumar, T. and Thangapandian, V. 2014. Biodegradation of plastics by *Pseudomonas putida* isolated from garden soil samples. J. Adv. Botany Zool. 1(3): 34–38.

Sanchez, J.G., Tsuchii, A. and Tokiwa, Y. 2000. Degradation of polycaprolactone at 50°C by a thermotolerant *Aspergillus* sp. Biotechnol. Lett. 22(10): 849–853.

Shah, A.A., Hasan, F., Hameed, A. and Ahmed, S. 2007. Isolation and characterisation of poly 3-hydroxybutyrate-co-3-hydroxyvalerate degrading actinomycetes and purification of PHBV depolymerase from newly isolated *Strepto verticillium kashmirense* AF1. Ann. Microbiol. 57: 583–588.

Shah, A.A., Hasan, F., Hameed, A. and Ahmed, S. 2008. Biological degradation of plastics: a comprehensive review. Biotechnol. Advan. 26(3): 246–265.

Shah, A.A., Hasan, F. and Hameed, A. 2010. Degradation of poly 3-hydroxybutyrate-co-3-hydroxyvalerate by a newly isolated *Actinomadura* sp. AF-555, from soil. Int. Biodeter. Biodegr. 64: 281–285.

Shah, A.A., Kato, S., Shintani, N., Kamini, N.R. and Nakajima-Kambe, T. 2014. Microbial degradation of aliphatic and aliphatic-aromatic co-polyesters. Appl. Microbiol. Biotechnol. 98(8): 3437–3447.

Shimao, M. 2001. Biodegradation of plastics. Curr. Opinion in Biotechnol. 12(3): 242–247.

Sivan, A. 2011. New perspectives in plastic biodegradation. Curr. Opin. Biotech. 22(3): 422–426.

Sowmya, H.V., Krishnappa, M. and Thippeswamy, B. 2015. Degradation of polyethylene by Penicilliumsimplicissimum isolated from local dumpsite of Shivamogga district. Environ. Dev. Sustain. 17(4): 731–745.

Sulaiman, S., Yamato, S., Kanaya, E., Kim, J.J., Koga, Y., Takano, K. and Kanaya, S. 2012. Isolation of a novel cutinase homolog with polyethylene terephthalate-degrading activity from leaf-branch compost by using a metagenomic approach. Appl. Environ. Microbiol. 78(5): 1556–1562.

Teeraphatpornchai, T., Nakajima-Kambe, T., Shigeno-Akutsu, Y., Nakayama, M., Nomura, N., Nakahara, T. and Uchiyama, H. 2003. Isolation and characterization of a bacterium that degrades various polyester-based biodegradable plastics. Biotechnol. Lett. 25: 23–28.

Tezuka, Y., Ishii, N., Kasuya, K.I. and Mitomo, H. 2004. Degradation of poly (ethylene succinate) by mesophilic bacteria. Polym. Degrad. Stabil 84(1): 115–121.

Then, J., Wei, R., Oeser, T., Barth, M., Belisário-Ferrari, M.R., Schmidt, J. and Zimmermann, W. 2015. Ca^{2+} and Mg^{2+} binding site engineering increases the degradation of polyethylene terephthalate films by polyester hydrolases from Thermobifidafusca. Biotechnol. J. 10(4): 592–598.

Tokiwa, Y. and Calabia, B.P. 2004. Review degradation of microbial polyesters. Biotechnol. Lett. 26(15): 1181–1189.

Tokiwa, Y., Calabia, B.P., Ugwu, C.U. and Aiba, S. 2009. Biodegradability of plastics. Int. J. Mol. Sci. 10(9): 3722–3742.

Tribedi, P. and Sil, A. 2014. Cell surface hydrophobicity: a key component in the degradation of polyethylene succinate by *Pseudomonas* sp. AKS2. J. Appl. Microbiol. 116(2): 295–303.

Urbanek, K., Rymowicz, W. and Miro, A.M. 2015. Isolation and characterization of Arctic microorganisms decomposing bioplastics. Bioplastics, 1–11.

Urbanek, K., Rymowicz, W. and Miro, A.M. 2018. Degradation of plastics and plastic degrading bacteria in cold marine habitats plastic. Appl. Microbial. Biotechnol. 102: 7669–7678.

Vert, M., Santos, I.D., Ponsart, S., Alauzet, N., Morgat, J.L., Coudane, J. and Garreau, H. 2002. Degradable polymers in a living environment: where do you end up? Polym. Int. 51(10): 840–844.

Vijaya, C. and Reddy, R.M. 2008. Impact of soil composting using municipal solid waste on biodegradation of plastics. Indian J. Biotechnol. 7: 235–239.

Volova, T.G., Boyandin, A.N., Vasiliev, A.D., Karpov, V.A., Prudnikova, S.V., Mishukova, O.V., Boyarskikh, U.A., Filipenko, M.L., Rudnev, V.P., Ba Xuan, B., Vit Dung, V. and Gitelson, I.I. 2010. Biodegradation of polyhydroxyalkanoates (PHAs) in tropical coastal waters and identification of PHA-degrading bacteria. Polym. Degrad. Stabil 95: 2350–2359.

Wackett, L. and Hershberger, C. 2001. Biocatalysis and Biodegradation: Microbial Transformation of Organic Compounds. Washington, D.C.: ASM Press.

Wallace, P.W., Haernvall, K., Ribitsch, D., Zitzenbacher, S., Schittmayer, M., Steinkellner, G., Gruber, K., Guebitz, G.M. and Birner-Gruenberger, R. 2017. PpEst is a novel PBAT degrading polyesterase identified by proteomic screening of Pseudomonas pseudoalcaligenes. Appl. Microbiol. Biotechnol. 101(6): 2291–2303.

Wei, R., Oeser, T. and Zimmermann, W. 2014. Synthetic polyester-hydrolyzing enzymes from thermophilic actinomycetes. Adv. Appl. 89: 267–305.

Wei, R. and Zimmermann, W. 2017a. Microbial enzymes for the recycling of recalcitrant petroleum-based plastics: how far are we? Microbial Biotechnol. 10(6): 1308–1322.

Wei, R. and Zimmermann, W. 2017b. Biocatalysis as a green route for recycling the recalcitrant plastic polyethylene terephthalate. Microbial Biotechnol. 10(6): 1302–1307.

Witt, U., Einig, T., Yamamoto, M., Kleeberg, I., Deckwer, W.D. and Muller, R.J. 2001. Biodegradation of aliphatic–aromatic copolyesters: evaluation of the final biodegradability and eco-toxicological impact of degradation intermediates. Chemosphere 44: 289–299.

World Economic Forum. 2016. The New Plastic Economy, Rethinking the Future of Plastics. 1–36.

Worm, B., Lotze, H.K., Jubinville, I., Wilcox, C. and Jambeck, J. 2017. Plastic as a persistant marine pollutant. Annu. Rev. Environ. Resour. 42: 1–26.

Wu, C.S. 2005. A comparison of the structure, thermal properties, and biodegradability of polycaprolactone/chitosan and acrylic acid grafted polycaprolactone/chitosan. Polymer. 46(1): 147–155.

Yamada-Onodera, K., Mukumoto, H., Katsuyaya, Y., Saiganji, A. and Tani, Y. 2001. Degradation of polyethylene by a fungus, Penicillium Simplicissimum YK. Polym. Degrad. Stabil. 72(2): 323–327.

Yang, H.S., Wolcott, M., Kim, H.S. and Kim, H.J. 2005. Thermal properties of lignocellulosic filler-thermoplastic polymer bio-composites. J. Therm. Anal. Calorim. 82(1): 157–160.

Yang, Y., Yang, J., Wu, W.M., Zhao, J., Song, Y., Gao, L., Yang, R. and Jiang, L. 2015. Biodegradation and mineralization of polystyrene by plastic-eating mealworms: part 2. Role of gut microorganisms. Environ. Sci. Technol. 49: 12087–12093.

Yoshida, S., Hiraga, K., Takehana, T., Taniguchi, I., Yamaji, H., Maeda, Y., Toyohara, K., Miyamoto, K., Kimura, Y. and Oda, K. 2016. A bacterium that degrades and assimilates poly (ethylene terephthalate). Science 351(6278): 1196–1199.

Zettler, E.R., Mincer, T.J. and Amaral-Zettler, L.A. 2013. Life in the Bplastisphere: microbial communities on plastic marine debris. Environ. Sci. Technol. 47(13): 7137–7146.

Zimmermann, W. and Billig, S. 2010. Enzymes for the biofunctionalization of poly (ethylene terephthalate). pp. 97–120. *In*: Biofunctionalization of Polymers and their Applicat. Springer, Berlin, Heidelberg.

11

Agrochemical Contamination of Soil
Recent Technology Innovations for Bioremediation

Suryasikha Samal and *C.S.K. Mishra**

1. Introduction

Pesticides and chemical fertilizers are important tools in modern agriculture to improve the quantity and quality of food production. These agrochemicals minimize the economic losses caused by weeds, insects, and diseases. However, extensive and unscientific application of agrochemicals in the crop fields and its consequent ecotoxicity has called for special attention of scientific community all over the world to look for effective and sustainable mechanisms for their minimal use. The environmental impact of hyper doses of pesticides and chemical fertilizers has been invariably detrimental. Adverse impact of these agricultural inputs could cause serious damage to soil biota which play vital role in maintaining soil health. The toxic intensity of such hazardous chemicals varies with time, dose, organism characteristics, environmental presence and nature of chemicals. Their presence in environment determines the risk of persistence (Ahmad and Ahmad 2014, Uqab et al. 2016).

The consequence of ecological maladies caused by recurrent use of the diverse types of pesticides has necessitated the development and implementation of new technologies to reduce or eliminate their use or residues. Earlier, conventional technologies such as landfills, recycling, pyrolysis, etc. were used to neutralize pesticide toxicity, but these methods subsequently proved to be less efficient and cost prohibitive. Besides, these methods led to formation of toxic intermediates which are detrimental to the environment. Persistence of pesticides and chemical fertilizers in the environment might occur due to their physico-chemical properties or absence of organisms able to degrade them. Various physical factors such as light, temperature or humidity could lead to loss of toxic substances up to some extent by either volatilization or degradation process. Degeneration process could be enriched by presence of certain organisms in soil. Hence, these biota could be used to improve elimination of the undesirable pollutants from the environment through bioremediation. The ability of organisms to bioremediate pesticides is mainly based on their biodegradation activity of complex synthetic compounds (Paul et al. 2005, Jain et al. 2005). Bioremediation of agrochemicals is an eco-friendly as well as cost-effective option available to eradicate environmental contamination.

Department of Zoology, Odisha University of Agriculture and Technology, Bhubaneswar-751003, India.
Email: suryasikha.777@gmail.com
* Corresponding author: cskmishra@yahoo.com

2. Impact of pesticides and fertilizers on soil

Based on different sources of contamination, soil pollutants are mainly categorised into various categories such as sewage sludge and industrial wastes, agrochemicals and metals, polycyclic aromatic hydrocarbons and radioactive substances. In addition, several anthropogenic activities have been observed to generate soil pollutants, the medium mostly gets affected due to direct application of agrochemicals.

Zhang et al. (2011) documented the outrageous consumption of agrochemicals in major countries. As per the reports available, pesticide consumption in the world has presently reached 2 million tonnes. European countries such as Italy, Spain, UK and the USA were reported as major consumers of pesticides. The utilization rate of Europe was 45% followed by USA (24%) and rest of the world (25%). Pesticide consumption in Asia has also reached an alarming level. Among the Asian countries, the pesticide application rate in China is the highest followed by Korea, Japan and India. In India, after the first green revolution post independence, the rate of agrochemical application has increased significantly in agricultural fields for high yield crop varieties. Presently, India is considered as the largest producer of pesticides in Asia and ranks 12th in world (Abhilash and Singh 2009).

The demand of pesticides increased in India during 1960s in the form of the chlorinated hydrocarbons and insecticides which was gradually replaced with organophosphates due to its faster natural degradation. Unregulated usage of these organophosphates in agricultural fields has threatened the soil ecosystem. Impacts of soil toxicity and loss of fertility due to pesticides have been reported in different areas of India (Kumar et al. 2016, Khajuria 2016). Insecticides such as organochlorines, organophosphates, carbamates, and pyrethroids are conventionally applied in different parts of the country. Besides the use of organic herbicides, the demand for synthetic herbicides such as 2,4-D, glyphosate, butachlor, nitrogen based chemical fertilizers such as urea, ammonium nitrate, ammonium sulphate, phosphates in the form of nitro-phosphate and potassium fertilizers has increased manyfold. Some by products of industrial wastes such as phosphogypsum and paper mill sludge, which are used to fortify the soil, could also prove detrimental to the soil biota at elevated concentrations (Nayak et al. 2011, Nicolopoulou-Stamati et al. 2016, Mishra and Samal 2018, Samal et al. 2019a, b).

Agrochemicals can remain in soils and sediments for a long time and enter the food chain directly or percolate down to the water table. These chemicals could directly accumulate within the adipose tissue of animals or through biomagnification in higher trophic level organisms, such as mammals and cause health hazards over time because of the enhanced levels of toxic compounds within the body (Ortiz-Hernández et al. 2014). Accumulation of pesticides in soil may affect the soil fauna like nematodes, microarthropods, earthworms, and diverse types of microbes.

Agrochemicals at high concentrations are likely to harm natural predators of pests and pollinators. Pesticide contamination in soil poses a serious threat to both below ground and above ground organisms and reduces the population of millions of tiny organisms including protozoa, fungi, bacteria, and soil arthropods such as Collembola, Hymenoptera and Arachnida (Aktar et al. 2009, Geiger et al. 2010). Some authors observed histopathological alterations in tissue of the earthworm *E. fetida, E. kinneari* exposed to various concentrations of Dimethoate, fluorides and herbicides (Sharma and Satyanarayan 2011, Lakhani et al. 2012). Alterations in histo-morphological structures, metabolic and antioxidant enzymes like lactate dehydrogenase, acetylcholinesterase and catalase of *Eudrilus eugeniae* exposed to elevated concentrations of pesticides monocrotopohos and glyphosate have been reported (Samal et al. 2019a). Severe skin lesions and setal conformational alterations have been observed due to pesticide exposures (Figure 1). Serious damage to musculature of the body wall in different species of earthworms has been found due to agrochemical hypertoxicity (Figure 2).

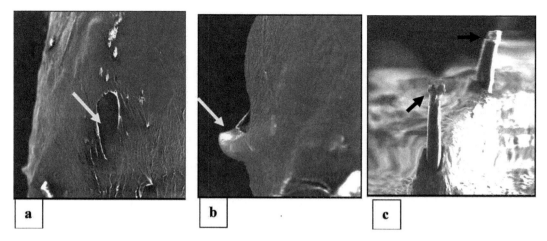

Figure 1. Scanning electron microscopy photograph of *E. eugeniae* exposed to elevated concentration of monocrotophos and urea. (a) skin lesion exposed to monocrotophos, (b) setal damage with monocrotophos, (c) setal damage with glyphosate (Samal et al. 2019a).

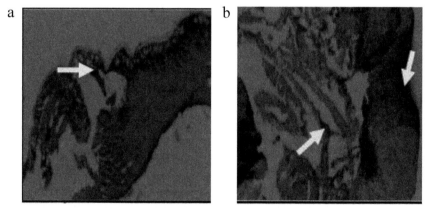

Figure 2. Muscular anomalies in the body wall of the earthworm *E. eugeniae* exposed to high concentrations of agrochemicals. (a) monocrotophos, (b) glyphosate (Samal et al. 2019a). The arrow mark indicates degeneration of muscle fibres at high concentrations of the pesticides.

3. Remediation strategies of hazardous agrochemicals

3.1 Physical remediation methods

A number of physical and chemical methods are available to remove contaminants from soil. Non biological methods such as low temperature thermal desorption has often been used to remediate pesticide-contaminated sites. This technology involves the removal of organic compounds including pesticides using temperature and volatilization. It requires highly specialized facilities and demands comparatively high cost. Air sparging method is more often used to remove hydrocarbons by injecting large volumes of pressurized air into contaminated soil, removing volatile organic compounds that might otherwise be removed by carbon filtering systems. Another popular method is incineration which is applied to the contaminated soil using heat and oxygen as the oxidizing compounds. Each technology has its advantage as well as limitation for the treatment of specific contaminant (Frazar 2000, Yao et al. 2012).

3.2 Biological methods

Compensation of high cost technologies can be done by replacing the physical and chemical remediation techniques with "bioremediation" techniques. Utilization of biota to remediate the contamination from the medium is a superior concept. The soil organisms not only indicate the toxicity but also eliminate those contaminants from trophic spheres.

Discrimination between natural biodegradation and bioremediation is a difficult task. The process of bioremediation accelerates the rate of the natural microbial degradation of pollutants by providing these microorganisms with nutrients, carbon sources or electron donors. The process demands the addition of indigenous microorganisms having characteristics to degrade the desired contaminant at a faster rate. Production of H_2O and CO_2 without producing the toxic intermediates is the unique feature of bioremediation (Frazar 2000).

4. Bioremediation and its types

Bioremediation technologies can be broadly classified as *in situ* or *ex situ*. *In situ* bioremediation involves treating the contaminated material at the site, while *ex situ* deals with the removal of the contaminated material to be treated outside the site. *Ex situ* bioremediation technologies include bioreactors, biofilters, and land farming, whereas *in situ* methods include bioventing, biosparging, biostimulation and liquid delivery systems. *In situ* technology is more popular due to its less equipment requirement, lower cost and eco-friendly nature. However, this treatment has limited application in fields. Bioremediation processes may be either aerobic or anaerobic based on the contaminated site and types of contamination (Master et al. 2002, Chowdhury et al. 2012).

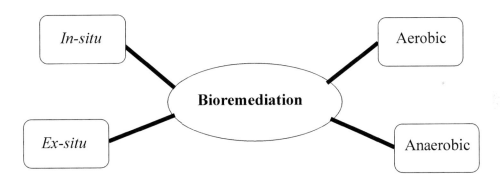

4.1 Principles of bioremediation

Micro organisms are mostly used as bioremediation tool because of their ability to degrade different substrates than the natural carbon sources. Capability of adaptation through mutation is an important characteristic of micro-organisms, which in turn helps to develop the ability of degrading toxic or complex compounds, probably because of the evolution of more adequate transport systems.

Biodegradation of xenobiotic compound involves the breakdown of toxic compounds to less complex compounds and ultimately to water and CO_2 by the action of microbes. The complete breakdown of pesticides into inorganic compounds is known as biomineralization. The degradation process may be complete or partial, which conventionally leads to formation of less toxic organic compounds, referred to as partial biodegradation. Metabolic activities of bacteria, fungi, actinobacteria, and plants play significant role in the degradation process. Although most of the pesticides are biodegradable, certain pesticides cannot degrade easily and are called recalcitrant. The degradation process of xenobiotic compounds depends upon the physical, chemical, and microbiological characteristics of the soil and the chemical properties of the pollutants. Pesticide degradability decreases as molecular weight and degree of branching increase. Pesticide metabolism completes in three steps. It includes phase I, the transformation of the original toxic compound through oxidation, reduction, or hydrolysis reactions. The second phase which includes the conjugation of pesticide or its metabolites to sugars and aminoacids, results in more water solubility and lesser toxicity in compounds. Phase III involves the conversion of metabolites into less toxic secondary conjugates and stimulation of intra- or extracellular enzymes such as hydrolases, peroxidases, and oxygenases by soil fauna (Chino-Flores et al. 2011, Abdel-Razek et al. 2013).

4.2 Application of soil biota in bioremediation of agrochemicals

Soil of agricultural field is a territory of several microorganisms, plants and animals. Each and every species of ecosystem contributes equally to maintain the functionality of this habitat. These biota also play an important role in eliminating toxic substances such as pesticides and chemical fertilizers from the fields. The biological remediation mainly includes microbial, animal and phytoremediation.

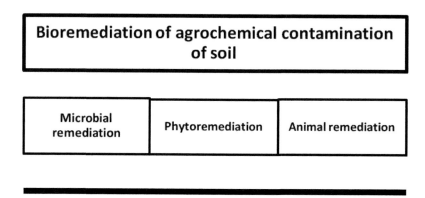

4.2.1 Microbial bioremediation

Microbial degradation of pesticide occurs through consumption of pesticide as a food source by bacteria or fungi. Each gram of soil contains millions of microbes. The use of microbial metabolic potential for eliminating soil pollutants has greater advantage over other commonly used physico-chemical strategies. The remediation process by living organisms is carried out by two different ways:

a) Biostimulation and
b) Bioaugmentation.

Biostimulation process stimulates the metabolic activity of indigenous microbial populations by the addition of specialized nutrients such as N or P, etc. and suitable physiological conditions which ultimately degrade the contaminants (Evans and Hedger 2001, Trindade et al. 2005). Biostimulation treatment could be necessary for the enzyme inducers and the co-metabolic substrates in the pesticide degradation pathways (Plangklang and Reungsang 2010). Bioaugmentation process is the introduction of exogenous microbes with specific catabolic abilities into the contaminated site to eliminate the toxic contaminants. It has been used to degrade a wide range of chemical contaminants such as ammonia, hydrogen sulphide, and insecticides. Certain factors such as temperatures, pH levels, adequate soil moisture, aeration (oxygen) and amount of adsorption influence microbial degradation process. Adsorbed pesticides are slowly degraded as these are less available to some microorganisms.

Application of both biostimulation and bioaugmentation processes result in detoxification of toxic elements through biotransformation. Micro-organisms convert the toxic compounds into non-toxic compounds by the chemical reaction either aerobically or anaerobically. Anaerobic biotransformation is essentially used for degradation of organic compounds such as chlorinated hydrocarbons, polychlorinated phenols, and nitroaromatics, whereas aerobic biotransformation helps in transformation of thiocyanates, cyanates, aromatic hydrocarbons, gasoline monoaromatics, and methyl *tert*-butyl ether (Head and Oleszkiewich 2004, Perelo 2010).

Major species of bacteria that degrade the pesticides belong to genera *Flavobacterium, Arthobacter, Aztobacter, Burkholderia,* and *Pseudomonas*. The complete biodegradation of the pesticide involves the oxidation of the parent compound resulting in to carbon dioxide and water, which provides energy to microbes. To facilitate the degradation process of innate soil microbial population, certain additional pesticide degrading micro flora is recommended. Bioremediation of pesticides by microbes depends on the enzyme system but is also influenced by physical conditions such as temperature, pH and nutrients (Yao et al. 2015, Uqab et al. 2016). Kumar and Philip (2006) observed the biodegradation potential of three bacterial species such as *Staphylococcus* sp., *Bacillus circulans*-I, and *Bacillus circulans*-II to remediate pesticide endosulfan by mixed culture and pure culture.

Several fungal species like *Flammulina velupites, Stereum hirsutum, Coriolus versicolor, Dichomitus squalens, Hypholoma fasciculare, Auricularia auricula, Pleurotus ostreatus, Avatha discolor* and *Agrocybe semiorbicularis* help to degrade pesticides and release them into soil where it is susceptible to further degradation. These species have tremendous potential to remediate pesticides like triazine, phenylurea and chlorinated organophosphorous compounds. Reports are available on bioremediation potential of white rot fungi *Phanerochaete chrysosporium* and *Pleurotus pulmonarius* to degrade highly recalcitrant pesticides like the chlorinated triazine herbicide 2-chloro-4-ethylamine-6-isopropylamino-1,3,4-triazine (atrazine), HCH, dieldrin, diuron, aldrin, DDT, etc. *Phanerochaete chrysosporium* has been observed to degrade a number of toxic xenobiotics such as aromatic hydrocarbons (Benzo alpha pyrene, Phenanthrene, Pyrene), chlorinated organics (Alkyl halide insecticides, Chloroanilines, DDT, Pentachlorophenols, Trichlorophenol, Polychlorinated biphenyls, Trichlorophenoxyacetic acid), nitrogen aromatics (2,4-Dinitrotoluene, 2,4,6-Trinitrotoluene-TNT) and several miscellaneous compounds such as sulfonated azodyes (Yao et al. 2015).

The microorganisms used in bioremediation are either natural or genetically modified, known as "super strains". These engineered micro-organisms are developed to introduce the efficient catabolic genes to stimulate the degradation capacity in indigenous microorganisms. Microbes with their catabolic gene and enzymes have been isolated and identified by several workers to demonstrate the degradation of different pesticides like carbofuran, carbaryl, baygon or aldicarb (Singh 2008).

Enzymes actively take part in the degradation of pesticide compounds in environment, via biodegradation by soil microorganisms. The first bacterium was isolated from a soil sample of Philippines and identified as *Flavobacterium* sp. ATCC 27551 to degrade OP (organophosphate) compounds. Certain fungal enzymes, especially oxidoreductases, laccase and peroxidases, have potential to remove the polyaromatic hydrocarbons (PAHs) contaminants in terrestrial ecosystems (Watanabe et al. 2000, Uqab et al. 2016).

4.2.2 Phytoremediation

Phytoremediation is the application of plants in the field of detoxification of contaminants for environmental clean-up. Living green plants are able to fix or adsorb contaminants, and cleaning the contaminants or reducing the risk of toxicants. The phytoremediation process includes three major processes:

a) Phytotransformation and Phytostabilization,

b) Phytovolatilization and

c) Phytoextraction.

Phytotransformation is the process where plants transform organic contaminants into less toxic and more stable forms. This process includes phytodegradation, which is the metabolism of the organic contaminant by the plant enzymes. Enzyme secretion by plant roots breaks down the organic compounds and reduce their toxicity. This process helps to eliminate several organic contaminants like herbicides, trichloroethylene, and methyl tert-butyl ether. In the case of organic pollutants, such as pesticides, and other xenobiotic substances, certain plants, such as Cannas, render these substances non-toxic by their metabolism and chemical transformation of toxicants. Fixation of heavy metals from pollutants by plants through the adsorption, precipitation and reduction of root is done by phytostabilization. It reduces the bioavailability and prevents the contaminants from migrating into the food chain. Phytovolatilization includes transfer of heavy metals into volatile state through leaves. Phytoextraction is adsorbing the contaminants using tolerant and accumulating plants, and then transferring to the above ground parts. Plants that can uptake extremely high amounts of contaminants from the soil are called hyperaccumulators (Hussain et al. 2009, Ziarati and Alaedini 2014, Yao et al. 2015).

The detoxification process of plants can be compared with human liver, and is hence called "green liver". Plants secrete different enzymes having functional groups such as hydroxyl groups (-OH) to increase the polarity of xenobiotics, known as Phase I metabolism. In plants, certain enzymes such as peroxidases, phenoloxidases, esterases and nitroreductases carry out the detoxification process. In Phase II metabolism, plant biomolecules such as glucose and amino acids are conjugated with the polarized xenobiotic to further increase the polarity. The increased polarity makes the compounds less toxic and allows for easy transport along aqueous channels. In the later or final stage, a sequestration of the xenobiotic occurs within the plant as a complex structure to be safely stored, and does not affect the functioning of the plant. Often the plant and microbial populations provide nutrients for one another. Plants also help as catalysts for increasing microbial growth and activity, which subsequently increases the biodegradation potential. This process is often referred to as phytostimulation or plant-assisted bioremediation (Shen and Chen 2000, Wang et al. 2009, Rani and Dhania 2014).

The high uptake plants for phytoremediation could be screened on the basis of following characters as per rules of U.S. department (Wang and Wen 2001),

a) Plants must have high accumulating efficiency under the low contaminants concentration
b) Plants must accumulate many different kinds of contaminants
c) Having high growth potential along with pest and disease resistance ability

Genetically modified plants have been developed to enhance the remediation potential of plants to detoxify the pollutants. Several modified plants are used in pesticide remediation. Enzymes such as glyphosate oxidase, cytochrome P450 enzymes, a Rieske non-heme monooxygenase (named DMO) that has been expressed in *A. thaliana*, tomato, tobacco and soybean plants convert dicamba to 3,6-dichlorosalicyclic acid (Behrens et al. 2007). Aryloxyalkanoate dioxygenase enzymes (TfdA) also have been successfully expressed in corn for contaminants' remediation (patented for the degradation of 2,4-D and pyridyloxyacetate herbicides) (Scott et al. 2011, Rani and Dhania 2014). Plants like maize (*Zea mays*), giant foxtail (*Setaria faberi*), brinjal (*Solanum melongena*), spinach (*Spinacea oleracea*), radish (*Raphanus sativus*) and rice (*Oryza sativa*) can accumulate and transform some pesticides like DDT and benzene hexachloride. Basil (*Ocimum basilicum*) can also remediate endosulfan from soil, whereas barley (*Hordeum vulgare*) is able to translocate herbicide metolachlor into vacuoles (Velázquez-Fernández et al. 2012).

Phytoremediation is also an innovative technology that is gaining popularity due to its cost-effective, less disruptive and aesthetically-pleasing method of remediating contaminated sites. As herbicides are formulated to destroy plants, the use of phytoremediation to remediate them can be a difficult and complicated task.

4.2.3 Animal remediation

Animal remediation or zooremediation is applied in contaminated fields according to the characterization of some soil invertebrates based on their contaminant's adsorbing, degrading and removing ability. The field of animal remediation is usually limited to invertebrates owing to ethical concerns.

Among different invertebrates, earthworms are the most widely known animals used for both bioindication and bioremediation. These fauna have been widely used for land recovery, reclamation and rehabilitation to rectify sub-optimal soils such as poor mineral and open cast mining sites. Exotic species such as *Eisenia fetida* and *Eudrillus eugeniae* are highly suitable for the bioremediation of different pollutants like heavy metals, polycyclic aromatic hydrocarbons and pesticides. Besides direct application of earthworms in polluted site, digested material (vermicast) of these animals could be used to facilitate the remediation process (Anderson et al. 2003, Okeke et al. 2014). Several authors have observed the positive influence of soil animals like earthworms, enchytraeids and mites on utilization of organic compounds in their own metabolism, and enhancing metabolic activity of soil microbes (Haimi and Huhta 1987, Prakash et al. 2017). The part of remediation process using animals has not been thoroughly investigated.

5. Limitations of bioremediation

Remediation strategies using soil biota show different advantages compared to other treatments in elimination of organic pollutants from environment. Strategies like phytoremediation are self maintained to eliminate pollutants from contaminated site throughout the year. Unlike the above, remediation of pollutants by microbes and animals should be implemented at specific times. Organisms included in the process of bioremediation need the suitable environment to degrade the environmental pollutants. Mostly, *in situ* bioremediation using microbes needs the addition of oxygen and chemical fertilizers to the medium containing pollutants, which in turn creates toxic state for other organisms. Other limitations like the results of cost/benefit ratios, i.e., cost versus overall environmental impact do not lead in a very positive direction. The bioremediation process does not degrade all kinds of pollutants and may not give foolproof result in all agrochemicals. The molecules and enzymes involved in this process are partially compound specific and the mechanisms vary from one agrochemical to the other. Hence, proper application of bioremediation technology should be implemented in a very specific and careful manner.

6. Conclusion

The coordination of different organisms to remove toxicants from ecosystem can make the concept of bioremediation practicable. Micro-organisms like bacteria and fungi degrade the complex molecules by using them in their own metabolism and growth process. Plants and animals are not only directly involved in this contaminant degradation process but also facilitate microbial degradation. The biodegradation of toxic substances depends on the nature of organisms and the enzymes involved. In spite of several limitations, using living organisms to remove toxicants from soil is an eco-friendly but challenging concept. The recent advancement in science and technology will certainly open up new vistas to produce genetically modified organisms with significantly higher remediation efficiency to ensure decontamination of soil.

References

Abdel-Razek, M.A.R.S., Folch-Mallol, J.L., Perezgasga-Ciscomani, L., Sánchez-Salinas, E., Castrejón-Godínez, M.L. and Ortiz-Hernández, M.L. 2013 Optimization of methyl parathion biodegradation and detoxification by cells in suspension or immobilized on tezontle expressing the opd gene. J. Environ. Sci. Heal B. Pestic. Food Contam. Agric. Wastes 48: 449–461.

Abhilash, P.C. and Singh, N. 2009. Pesticide use and application: an Indian scenario. J. Hazard. Mater. 165(1-3): 1–12.

Ahmad, M. and Ahmad, I. 2014. Recent advances in the field of bioremediation. Biodegradation and Bioremediation Studium Press LLC. 1–42.

Aktar, W., Sengupta, D. and Chowdhury, A. 2009. Impact of pesticides use in agriculture: their benefits and hazards. Interdisciplinary Toxicol. 2(1): 1–12.

Anderson, R.T., Vrionis, H.A., Ortiz-Bernad, I., Resch, C.T., Long, P.E., Dayvault, R., Karp, K., Marutzky, S., Metzler, D.R., Peacock, A., White, D.C., Lowe, M. and Lovely, D.R. 2003. Stimulating the *in-situ* activity of geobacter species to remove uranium from the groundwater of a uranium contaminated aquifer. J. of Appl. Environ. Microbiol. 69: 5884–5891.

Behrens, M.R., Mutlu, N., Chakraborty, S., Dumitru, R., Jiang, W.Z., LaVallee, B.J. and Weeks, D.P. 2007. Dicamba resistance: enlarging and preserving biotechnology-based weed management strategies. Science 316(5828): 1185–1188.

Chino-Flores, C., Dantán-González, E., Vázquez-Ramos, A., Tinoco-Valencia, R., Díaz-Méndez, R., Sánchez-Salinas, E., and Ortiz-Hernández, M.L. 2011. Isolation of the *opdE* gene that encodes for a new hydrolase of *Enterobacter* sp. capable of degrading organophosphorus pesticides. Biodegradation 23: 387–397.

Chowdhury, S., Bala, N.N. and Dhauria, P. 2012. Bioremediation—a natural way for cleaner environment. Int. J. Pharmaceut. Chem. Biol. Sci. 2: 600–611.

Evans, C.S. and Hedger, J.N. 2001. Degradation of plant cell wall polymers. pp. 1–26. In British Mycological Society Symposium Series (Vol. 23).

Frazar, C. 2000. The bioremediation and phytoremediation of pesticide-contaminated sites. US Environmental Protection Agency. 1–50.

Geiger, F., Bengtsson, J., Berendse, F., Weisser, W.W., Emmerson, M., Morales, M.B. and Eggers, S. 2010. Persistent negative effects of pesticides on biodiversity and biological control potential on European farmland. Basic Appl. Ecol. 11(2): 97–105.

Haimi, J. and Huhta, V. 1987. Comparison of composts produced from identical wastes by "vermistabilization" and conventional composting. Pedobiologia. 30: 137–144.

Head, M.A. and Oleszkiewicz, J.A. 2004. Bioaugmentation for nitrification at cold temperatures. Water Res. 38: 523–530.

Hussain, S., Siddique, T., Arshad, M. and Saleem, M. 2009. Bioremediation and phytoremediation of pesticides: recent advances. Critical Rev. Environ. Sci. Technol. 39(10): 843–907.

Jain, R.K., Kapur, M., Labana, S., Lal, B., Sarma, P.M., Bhattacharya, D. and Thakur I.S. 2005. Microbial diversity: application of microorganisms for the biodegradation of xenobiotics. Current Sci. 101–112.

Khajuria, A. 2016. Impact of nitrate consumption: case study of Punjab, India. J. Water Resour. Prot. 8(02): 211.

Kumar, M. and Philip, L. 2006. Enrichment and isolation of a mixed bacterial culture for complete mineralization of endosulfan. J. Environ. Sci. Health B 41: 81–96.

Kumar, S., Kaushik, G. and Villarreal-Chiu, J.F. 2016. Scenario of organophosphate pollution and toxicity in India: A review. Environ. Sci. and Pollut. Res. 23(10): 9480–9491.

Lakhani, L., Khatri, A. and Choudhary, P. 2012. Effect of dimethoate on testicular histomorphology of the earthworm *Eudichogaster kinneari* (Stephenson). Int. Res. J. of Biological Sci. 1: 77–80.

Master, E.R., Lai, V.W.M., Kuipers, B., Cullen, W.R. and Mohn, W.W. 2002. Sequential anaerobic-aerobic treatment of soil contaminated with weathered Aroclor 1260. Environ. Sci. Technol. 36: 100–103.

Mishra, C.S.K. and Samal, S. 2018. Phosphogypsum: Agricultural Utilization and Implications. pp. 68–69. *In*: Stefan Worsley (ed.). Fertilizer Focus. Argus Media Ltd., London, WC1X 8NL, UK (In Press).

Nayak, S., Mishra, C.S.K., Guru, B.C. and Rath, M. 2011. Effect of phosphogypsum amendment on soil physico-chemical properties, microbial load and enzyme activities. J. Environ. Biol. 32: 613–617.

Nicolopoulou-Stamati, P., Maipas, S., Kotampasi, C., Stamatis, P. and Hens, L. 2016. Chemical pesticides and human health: the urgent need for a new concept in agriculture. Front. Public Health 4: 148. doi: 10.3389/fpubh.2016.00148.

Okeke, J.J., Isreal, O.I. and Akubukor, F.C. 2014. Role of invertebrates as bioremediators—a review. Am. J. Life. Sci. Res. 3(1): 13–20.

Ortiz-Hernández, M.L., Rodríguez, A., Sánchez-Salinas, E. and Castrejón-Godínez M.L. 2014. Bioremediation of soils contaminated with pesticides: experiences in Mexico. pp. 69–99. *In*: Analia Alvarez and Marta Alejandra Polti (eds.). Bioremediation in Latin America. Springer, Cham.

Paul, D., Pandey, G., Pandey, J. and Jain, R.K. 2005. Accessing microbial diversity for bioremediation and environmental restoration. TRENDS in Biotechnol. 23(3): 135–142.

Perelo, L.W. 2010. Review: *In situ* and bioremediation of organic pollutants in aquatic sediments. J. Hazard Mater. 177: 81–89.

Plangklang, P. and Reungsang, A. 2010 Bioaugmentation of carbofuran by Burkholderia cepacia PCL3 in a bioslurry phase sequencing batch reactor. Process Biochem. 45: 230–238.

Prakash, S., Selvaraju, M., Ravikumar, K. and Punnagaiarasi, A. 2017. The role of decomposer animals in bioremediation of soils. pp. 57–64. *In*: Prashanthi, M., Sundaram, R., Jeyaseelan, A. and Kaliannan T. (eds.). Bioremediation and Sustainable Technologies for Cleaner Environment. Springer, Cham.

Rani, K. and Dhania, G. 2014. Bioremediation and biodegradation of pesticide from contaminated soil and water—a noval approach. Int. J. Curr. Microbiol. Appl. Sci. 3(10): 23–33.

Samal, S., Mishra, C.S.K. and Sahoo, S. 2019a. Setal-epidermal, muscular and enzymatic anomalies induced by certain agrochemicals in the earthworm Eudrilus eugeniae (Kinberg). Environ. Sci. and Pollut. Res. 26(8): 8039–8049.

Samal, S., Samal, R.R., Mishra, C.S.K. and Sahoo. S. 2019b. Setal anomalies in the tropical earthworms *Drawida willsi* and *Lampito mauritii* exposed to elevated concentrations of certain agrochemicals: An electron micrographic and molecular docking approach. Environ. Technol. and Innovat. 15: 100391.

Scott, C., Begley, C., Taylor, M., Pandey, G., Momiroski, V., French, N. and Bajet, C. 2011. Free-enzyme bioremediation of pesticides: A case study for the enzymatic remediation of organophosphorous insecticide residues. Chapter 11. pp. 155–174. *In*: Kean S. Goh, Brian L. Bret, Thomas L. Potter and Jay Gan (eds.). Pesticide Migration Strategies for Surface Water Quality. 1075, ACS Publication.

Sharma, V.J. and Satyanarayan, S. 2011. Effects of selected heavy metals on the histopathology of different tissues of earthworm *Eudrilus eugeniae*. Environ. Monitoring and Assess. 180: 257–267.

Shen, Z.G. and Chen, H.M. 2000. Bioremediation of heavy metal polluted soils. Rural Eco-Environment 16(2): 39–44.

Singh, D.K. 2008. Biodegradation and bioremediation of pesticide in soil: concept, method and recent developments. Indian J. of Microbiol. 48(1): 35–40.

Trindade, P.V.O., Sobral, L.G., Rizzo, A.C.L., Leite, S.G.F. and Soriano, A.U. 2005. Bioremediation of a weathered and recently oil-contaminated soils from Brazil: A comparison study. Chemosphere 58: 515–522.

Uqab, B., Mudasir, S. and Nazir, R. 2016. Review on bioremediation of pesticides. J. Bioremediat. Biodegrad. 7(3): 1–5.

Velázquez-Fernández, J.B., Martínez-Rizo, A.B., Ramírez-Sandoval, M. and Domínguez-Ojeda, D. 2012. Biodegradation and bioremediation of organic pesticides. In Pesticides-recent trends in pesticide residue assay. IntechOpen. DOI: 10.5772/48631.

Wang, H.F., Zhao, B.W., Xu J. et al. 2009. Technology and research progress on remediation of soils contaminated by heavy metals. Environ. Sci. and Manage. 34(11): 15–20.

Wang, J.L. and Wen, X.H. 2001. Environmental Biotechnology. Beijing: Tsinghua University Press.

Watanabe, K., Watanabe, K., Kodama, Y., Syutsubo, K. and Harayama, S. 2000. Molecular characterization of bacterial populations in petroleum-contaminated groundwater discharged from underground crude oil storage cavities. Appl. Environ. Microbiol. 66(11): 4803–4809.

Yao, Z.T., Ji, X.S., Sarker, P.K., Tang, J.H., Ge, L.Q., Xia, M.S. and Xi, Y.Q. 2015. A comprehensive review on the applications of coal fly ash. Earth Sci. Reviews 141: 105–121.

Yao, Z., Li, J., Xie, H. and Yu, C. 2012. Review on remediation technologies of soil contaminated by heavy metals. Procedia Environ. Sci. 16: 722–729.

Zhang, W., Jian, F. and Ou, J. 2011. Global pesticide consumption and pollution: with China as a focus. Proceed. of the Int. Academy of Ecol. and Environ. Sci. 1(2): 125.

Ziarati, P. and Alaedini, S. 2014. The phytoremediation technique for cleaning up contaminated soil By Amaranthus sp. Environ. Anal. Toxicol. 4(2): 1–4.

12

Bioremediation of Pesticides with Microbes:

Methods, Techniques and Practices

Rakesh Kumar Ghosh,[1] Deb Prasad Ray,[1] Ajoy Saha,[2] Neethu Narayanan,[3] Rashmita Behera[4] and Debarati Bhaduri[4]*

1. Introduction: Pesticides in the environment

The pesticide has empowered human civilization not only to secure agricultural produces including food, fodder, and fibre during crop cultivation, but also over the storage period. The application of pesticides has played a significant role to secure the supply of food for the growing population of this planet by controlling the pests. Ina broad definition, the term 'pest' includes all biotic sources harming agricultural products either during production or storing phase and, generally it covers various insects, pathogens, and weed along with some other damage-causing organisms like nematodes, rodents, etc. It has been estimated that the major loss of agricultural products is nearly 14, 13, and 13%, which is contributed by insects, pathogens, and weeds, respectively. Non-application of pesticides may increase loss in agricultural production up to 78, 54, and 42% for fruits, vegetables, and cereals, respectively (UN 2015). Further, global warming under the scenario of climate change is assumed to add momentum in the frequency and quantum of pests and subsequently a higher loss of agricultural produce. It has been reported that a one-degree increase in temperature may result in 10–25% more loss in major crops like rice, wheat, and maize (Deutsch et al. 2018). Hence, pesticides have gained attention to protect agricultural produce from various pests, and pesticide sciences have gone through various changes to develop pesticide molecules of better performance against pests. Pesticides can be grouped into five generations, which started with first-generation products (before 1940), mostly broad-spectrum inorganic compounds and few botanical extracts, followed by 2nd generation (during 1940–1960) products covering organochlorines (DDT, HCH, Mirex, Toxaphene, cyclodienes, etc.), organophosphates and carbamates, 3rd generation (after the 1970s) candidates including molting hormones, chitin synthesis inhibitors, juvenile hormones, etc., 4th generation products of antifeedants and pheromones and 5th generation members like

---

[1] ICAR-Natural Institute of Natural Fibre Engineering and Technology, Kolkata, India.
[2] Research Centre of ICAR-Central Inland Fisheries Research Institute, Bengaluru, India.
[3] ICAR-Indian Agricultural Research Institute, New Delhi, India.
[4] ICAR-National Rice Research Institute, Cuttack, India.
* Corresponding author: debarati.ssiari@gmail.com

novel natural products, brain hormone antagonists, etc. Without pesticide applications, the loss in production of major crops like wheat, maize, rice, soybeans, potato, and cotton, etc. may go to the extent of 20 to 40% (Oerke 2006). Now, the application of pesticides has become a mandatory practice in the modern intensive agricultural system. Presently, the total pesticide production across the world has increased from 0.2 million tonnes to 5 million tonnes with an annual growth rate of 11%. As per one report, Asia leads in the average pesticide usage (3.62 kg/ha), followed by America (3.39 kg/ha), Europe (1.67 kg/ha), Oceania (1.17 kg/ha) and Africa (0.31 kg/ha) (FAOSTAT 2017). Out of various pesticides, the share of synthetic pesticides from the 2nd and 3rd generation is still more than 4th or 5th generation pesticides. Indiscriminate and injudicious applications of pesticides have polluted the entire ecosystem.

Researchers have found that only 7–10% of applied pesticides reach target/pest, and more than 90% of pesticides reach mostly soil (Figure 1). Soil acts as a sink for received pesticides and pesticides undergo various transformation processes like sorption, volatilisation, degradation (photo, chemical, and microbial) and leaching to groundwater (Ghosh and Singh 2013). Some portions of soil-sorbed pesticide reach open water bodies by surface runoff, soil erosion, or sub-surface drainage, and may come back again to the same/new soil site while irrigating the crops with contaminated water (Carvalho et al. 2003). Hence, open water bodies act as the second sink of pesticide residues, after soil. Some part of pesticides which are prone to volatilisation/fumigation may travel long distances and contaminate new areas by condensation/precipitation process. Researchers have tracked toxaphene residues in Canada's Great Lake which traveled from South USA (Li and Jin 2013) or residues of chlorpyriphos in Arctic ice which were sprayed in Central America (Garbarino et al. 2002). Indiscriminate use of pesticides has resulted in contamination of various components of the environment and several reports across the world have drawn attention. Pesticides may adversely affect the population dynamics, sex ratio, size at birth, swimming pattern and physiology of phytoplankton (Ikram and Shoaib 2018) and zooplanktons (Hanazato 2001), morphological and behavioural abnormalities in invertebrates dwelling in soil (Frampton et al. 2006) and water

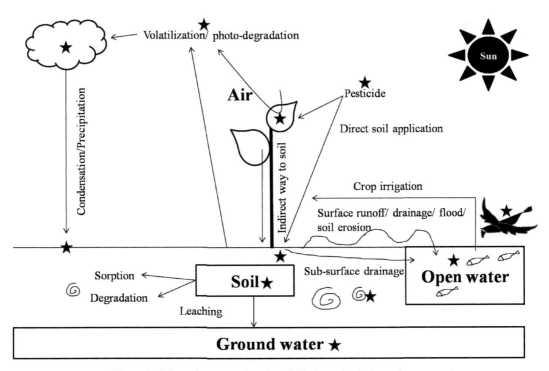

Figure 1. Schematic presentation of pesticide dynamics in the environment.

(Schäfer et al. 2012), various morphological defects in vertebrates including fishes (Allison et al. 1964, Robinson et al. 2015), birds (Goulson 2014), reptiles (Khan and Law 2005) including humans. In humans, pesticides may pose risks of respiratory diseases, cancer, tumor, neurotoxicity, child development, fertility, reproduction (Osman 2011, Mostafalou and Abdollahi 2013), etc. This is the result of severe contamination of soil, air and water, and various trophic levels of the food chain with pesticide residue (Burnett and Welford 2007, Chourasiya et al. 2015, Witczak and Abdel-Gawad 2014, Nag and Raikwar 2011). Lately, Bhaduri et al. (2018) have also comprehensively reviewed the common and potential bioindicators under soils polluted by pesticides.

Two pre-emergence herbicides, pendimethalin and oxyfluorfen, were found to stimulate soil microbial biomass carbon, fluorescein diacetate hydrolysing activity, alkaline phosphatase and ammonification rates in peanut grown soil, while dehydrogenase activity, acid phosphatase, nitrification rate and available phosphorous was adversely affected (Saha et al. 2015). Further, increased soil alkaline phosphatase and decreased acid phosphatase activities was observed after applying two post-emergence herbicides (imazethapyr and quizalofop-p-ethyl) while stimulated soil ammonification and nitrification rates indicates that these herbicides had distinct effects on nitrogen and phosphorus dynamics in soil both soil process (Saha et al. 2016a). In another study, tebuconazole application at field rate (FR) and 2FR resulted in a short-lived and transitory toxic effect while the disturbance was persistent at higher rate (10FR). It showed stimulating effect on soil ammonification and nitrification rates, and microbial biomass C, but was more toxic to soil ergosterol which is an indicator of the presence of viable fungi (Saha et al. 2016b).

2. Bioconcentration, bioaccumulation and biomagnification of pesticide

The presence of pesticide residues in soil/water and their ill/adverse effects on various organisms have drawn global attention. These man-made plant protection chemicals, namely, organochlorines like dichlorodiphenyltrichloroethane (DDT) along with its metabolites like (dichlorodiphenyldichloroethane [DDD] and dichlorodiphenyldichloroethylene [DDE]), hexachlorocyclohexane (HCH), benzene hexachloride (BHC), cyclodienes like—aldrin and dieldrin, toxaphene and polychlorinatedbiphenyls (PCBs) have been reported to be bioaccumulative and bio-toxic (Mostafalou and Abdollahi 2013, Robinson et al. 2015). Nine organochlorine pesticides, namely, DDT with DDE and DDD, HCB, aldrin, dieldrin, endrin, heptachlor, chlordane, toxaphene, and mirex, have been identified as persistent organic pollutants (POPs) in the 12th Stockholm Convention due to their prolonged persistence, wide distribution across various environmental components (soil, air, and water), lipophilic nature and ability to accumulate in various trophic levels, resulting in higher accumulation in higher levels of the food chain and are also toxic to animals and humans. Indiscriminate and non-judicious use of these pesticides has resulted in the worldwide presence of their residue, even in Arctic areas (Garbarino et al. 2002). These are the result of bioconcentration, bioaccumulation, and biomagnifications of toxic residues of pesticides. Bioconcentration is known as the concentration of a contaminant taken up by the aquatic organisms, provided that water is the only source of contaminant. The bioconcentration factor (BCF) is calculated as the concentration of pesticide residue in biota and water. Bioconcentration of a pesticide residue is controlled by 3 factors, namely, (i) physico-chemical property of the pesticide molecule, (ii) condition of the aquatic organism, and (iii) environmental condition. Physico-chemical properties include, mainly, water solubility (WS), octanol/water partition coefficient (K_{ow}) and molecular size. The BCF can be expressed as follows:

$\log BCF = a \log K_{ow}$ (or WS) + b

Pesticides with low WS and higher K_{ow} and bigger molecular size (diameter < 15A0) have higher BCF (Dimitrov et al. 2002). Among the physiological conditions of aquatic organisms, parameters like size, biochemical constituents (mainly lipid), and metabolism will change with the growth stage of the organism, which will certainly result in variations in uptake, elimination, and bioconcentration profiles (Swackhamer and Skoglund 1993). The environmental condition including pH, salinity, temperature of water, and quantity of dissolved/adsorbed organic matter significantly affects the bioconcentration. Bioconcentration of chlorinated pesticides has been reported to increase with an increase in temperature for various species of fishes, algae, mussels, etc. (Nawaz and Kirk 1995, Fisher et al. 1999). Now, bioaccumulation means the intake of contaminants from both food/dietary sources and water. Factors responsible for bioaccumulation include dissolved/particulate organic carbon in water, nature of button sediments, and physiological condition of aquatic organisms. Further, biomagnifications combine both processes of bioconcentration and bioaccumulation resulting in an increase in contaminants' concentration at higher trophic levels. During the trophic transfer of pesticide residues, the trophic transfer coefficient (TTC) is an important factor. It can be calculated as the ratio of the concentration of pesticide residue in the consumers' tissue to the pesticide concentration in the food. When TTC ≥ 1, biomagnification is expected to occur. Pesticides like DDT (TTC > 8), DDE (TTC > 9.7), and toxaphene (> 4.7) have a severe issue of biomagnifications in various food chains (Kay 1984, Evans et al. 1991). Residues of 13 organochlorine pesticides were detected in 36 species belonging to three lakes of north-eastern Louisiana, USA, and pesticide residue concentrations were more in tertiary consumers like green-backed heron (Butorides striatus), and snakes, spotted gar (*Lepisosteus oculatus*), and largemouth bass (*Micropterus salmoides*) as compared to the secondary consumers like bluegill (Lepomis macrochirus), yellow-crowned night-heron (*Nycticorax violaceus*), blacktail shiner (Notopis venustus), etc. and lowest levels were detected in primary consumers like crayfish (*Orconectes lancifer*) and threadfin shad (*Dorosoma petenense*), etc. (Niethammer et al. 1984). Biomagnification of dichlorodiphenyltrichloroethane (DDT) along with its metabolite (4,4'-DDE) was detected in four fish species (Clarias gariepinus, Oreochromis niloticus, Tilapia zillii, and Carassius auratus) from Lake Ziway, Rift Valley, Ethiopia (Deribe et al. 2013). Recently, biomagnifications of DDTs, HCHs, and chlordanes in a food chain from Zhoushan Fishing Ground, China, have been reported with TTC value ranging from 4.17–9.77 (Zhou et al. 2018). Biomagnification of pesticides in non-target organisms and its detrimental effects on the physiological, morphological, and behavioural nature of amphibian, bird, etc. are well documented. Organochlorine pesticides inhibit gamma-aminobutyric acid (GABA) receptors in the brain and hamper the central nervous system of birds. Further, thinning of egg-shell by DDE, a metabolite of DDT, is another reason for the lower bird population. Though organophosphate and carbamate pesticides do not have bioaccumulation issues, they are acute poisons. Pesticides have been reported to alter the hormonal balances, reproduction nature, birth defects, metabolism, immune suppression, and behaviour of birds (Oriss et al. 2000). Biomagnification of pesticides has resulted in adverse effects in humans also like the occurrence of cancer, tumor, neurotoxicity, child development, fertility, reproduction, etc. (Mostafalou and Abdollahi 2013). The overall impact of pesticides' biomagnifications in terms of environmental and societal damages was nearly $12 billion per year alone in the USA. It includes wildlife loss (bird, fish, and others) of $2.2 billion, followed by water contamination worth $2 billion, loss due to development of pesticide resistance in pests costing around $1.5 billion, affecting public health valued $1.1 billion and crop losses worth $1.1 billion (Pimentel 2009). Hence, biomagnification of pesticides is a global problem and it needs to be addressed seriously for a safe and sustainable environment.

3. Environmental remediation of pesticides

The techniques used for remediation of other organic pollutants can also be extended for remediation of pesticide-contaminated soil (Castelo-Grande et al. 2010). Nevertheless, some particular method has been developed due to the specific condition that arises as a result of the contamination of soil by pesticides. As pesticides contaminated the soil through diffuse pollution, so the remediation measures have to be different as compared to point-source contamination. On the contrary, since pesticides are used in agricultural soil, maintenance of soil properties is important and aggressive technologies used for remediation of industrial polluted soils cannot be replicated in agricultural soils. Furthermore, choosing of appropriate remediation method which can protect the human and environmental health is also important as the level of pesticides contamination may exceed the standard safe limit. Remediation techniques involving physical, chemical, and biological methods including adsorption, oxidation, catalytic degradation, membrane filtration, and biological treatment have been evolving continuously for remediation of pesticide-contaminated environment.

Based on the level of pesticide contamination and their risk, money and time involved, remediation of contaminated soil may be carried out *in situ*, or the contaminated matrix may be transported to special reactors or vessels (*ex situ*) where treatment will be carried out.

In general, the following is the broad categorization of pesticide remediation technology which is also given in Figure 2. These are:

- Chemical and physical methods or a combination of both
- Biological methods: bioremediation approach
- Thermal destruction methods

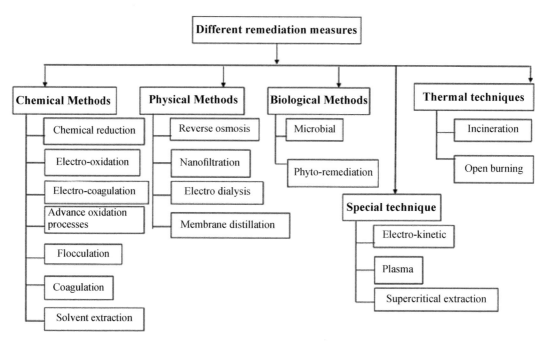

Figure 2. Flow chart of different remediation measures for pesticide contamination.

3.1 Chemical and physical remedial techniques

The objective of remediation involving chemical and/or physical methods is to transform the chemical environment of the pesticides so that it can not enter into the different elements of soil system including plants, groundwater, or soil organisms. Such preventive steps include decreasing mobility or utilizing any change of chemical constituents of the pesticides. Here we are discussing the different physico-chemical methods involved in the remediation of pesticide-contaminated soil and the biological method is given in detail in the subsequent sections.

3.1.1 Chemical reduction

In redox reactions, one reactant loses electrons (is oxidized), and the other gains electrons (is reduced). Degradation of pesticides by reduction is enhanced by aerobic environments. Nanoscale zero-valent iron (nZVI) is the most commonly used chemical reductant for halogenated, more particularly chlorinated pesticides in soil. Several reports are available where reductive dehalogenation of persistent pesticides on the surface of zero-valent iron (ZVI) becomes a promising technique for treating the contaminated soil (Sayles et al. 1997, El-Temsah et al. 2016). The addition of Fe or Al salts to zero-valent iron (nZVI) enhances the metolachlor degradation by creating pH and redox conditions that favor the formation of green rusts (Satapanajaru et al. 2003). However, caution should be taken to avoid the negative environmental effects of nZVI when it is used in an open environment.

3.1.2 Chemical oxidation

The main purpose of this method is to mineralize the pesticides in contaminated soil to CO_2, water, and inorganics, or converting them to non or less toxic metabolites. Ozone, hydrogen peroxide, hypochlorites, chlorine, and chlorine dioxide are the most commonly used oxidizing agents (Pavel and Gavrilescu 2008). However, sometime these conventional oxidizing agent may not be able to completely degrade the pesticides and consequently, they are combined with iron salts, semiconductors (such as TiO_2) and/or ultraviolet-visible light irradiation for better remediation and the process is known as "advanced oxidation process" (AOPs) (Giménez et al. 2015). The use of different approaches in the AOP field for pesticide degradation has been studied by several workers (Parker et al. 2017, Komtchou et al. 2017, Saylor et al. 2019). However, the success of this method in the real field will provide a reference for its applications.

3.1.3 Soil washing

The main principle of this technique is to separate the contaminants like pesticides from soils and sediments by using physical, chemical techniques or a combination of both the techniques. It involves the extraction of pesticides from contaminated soil with the help of a solution that efficiently transfers the pesticides from solid soil phase to liquid solvent phase. It can be considered as an off-site technology. Here the dug soil sample is processed in special extractor units which can be considered as a soil–liquid extraction operation unit, where the pesticides are transferred from the solid soil to the liquid washing fluid. For pesticide remediation by soil washing techniques, surfactant solutions are the most commonly used liquid phases. The efficiency of the combination of soil washing and electrolysis with diamond electrodes to remove atrazine using sodium dodecyl sulfate (dos Santos et al. 2015a, b) as a surfactant can be cited. Jinzhong et al. (2017) treated the organochlorine pesticides (OCPs) contaminated soils by washing soil with triton X–100 (TX–100) surfactant coupled with adsorption treatment of the solution with activated carbon. Cotillas et al. (2018) reported that the addition of electrolytic processes during the soil washing process efficiently enhances the removal of pesticides from soil.

3.1.4 Chemical extraction–solvent extraction

Supercritical fluid extraction (SFE) can be applied for the extraction of pesticides from contaminated soil. This method is modern and has a high solvency and recovery capacity, being composed of a principal solvent agent, for instance, the methanol, often used as an aid to carbon dioxide (CO_2) (Castelo-Grande et al. 2005). Pesticides get solubilized in CO_2 when it passes through contaminated soil and the collected solvent is disposed of safely. Solvent extraction can be used to recover a wide variety of substances. Due to the high solvency power of CO_2, the method can be applied for remediation of a wide variety of pesticides without disturbing soil characteristics, such as structures and nutrients. However, remediation potential depends on soil properties (pH, moisture, organic matter content) and type of extraction method used. The selective extraction of persistent organic pollutants including DDT reported (Bielská et al. 2013) the reduction in remediation when total organic matter and other soil properties in the analyzed soils were higher.

3.1.5 Soil flushing

This is an *in situ* chemical method of remediation of pesticide-contaminated soil. Here, pesticides are extracted from the contaminated soil by injection of fluid. When fluid is injected into the contaminated soil, the adsorbed pesticides get desorbed from the soil and retrieved in the fluid. The process is continuous as the contaminated fluid is purified and re-injected or circulated to extract the pesticides and, re-extracted, and re-purified. Chemistry of pesticides particularly water solubility and volatilization influences significantly the success of this remediation method (dos Santos et al. 2016). However, this method performs poorly or fails in low permeable soils and with pesticides adsorbed strongly on the solid soil surface.

4. Microbial degradation of pesticides

Microbial degradation is a powerful technology for the remediation of polluted sites. This methodology is highly cost-effective and offers remediation of the contaminated soil, sediment, sludge, groundwater, etc. (Parte et al. 2017). Microbes have an innate ability to degrade pesticides which is the basis for degradation of contaminants by microorganisms (bioremediation). Microbes having pesticide degradation potential include bacteria, especially actinomycetes and cyanobacteria, algae, and fungi. The microbes having the ability to degrade pesticides were isolated from contaminated sites. The major source of these microbes is the soil where these contaminants were applied for crop production or protection purposes. Table 1 summarizes the different microbes involved in the bioremediation of pesticides.

4.1 Pesticide degradation strategies by microbes

There are different strategies exhibited by microbes for the degradation of pesticides. They involve: (a) co-metabolism, (b) commensalism and mutualism, (c) catabolism and (d) gratuitous biodegradation. In *co-metabolism*, the microbes transform the pesticide, which is a non-growth substrate, along with its natural metabolic functions. A non-growth substrate cannot serve as a sole source of carbon and energy for a pure culture of a bacterium, and hence cannot support cell division. Some of the recalcitrant pesticides can be degraded by co-metabolism. Many microbial cell-bound and extracellular enzymes catalyze the transformation of pesticides. For example, *Pseudomonas putida* strain PP3 as such cannot metabolize MCA (monochloroacetate), but while metabolizing MCPA (monochloropropionate), the microbe catalyzes dehalogenation of MCA (Tewari et al. 2012). *Commensalism* is the interaction of two different microbial populations that live together in which one population is benefitted from the interaction while the other is not affected. In *mutualism*, the interaction between the two species is mutually beneficial. In *catabolism*, the organic molecule is utilized as a source of nutrition and energy. Sometimes, the microbes get adapted to the use of some pesticides as the sole source of carbon or nitrogen leading to enhanced microbial degradation of the applied pesticides. This occurs as a result of soil enrichment in microbial species. This trend is recently seen in pesticides like carebofuran, 2,4-D, and atrazine. Biostimulation techniques are usually combined with enhanced biodegradation. Biostimulation is the addition of nutrients, electron donors, or acceptors to stimulate the microbial population existing in natural soil. Bioaugmentation is also paired with enhanced biodegradation. In this technique, specific microorganisms are introduced for enhanced biodegradation of target molecules. In *gratuitous biodegradation*, enzymes secreted by microbes are able to degrade the pesticides other than its natural substrate.

4.2 Biochemical mechanisms in microbial degradation of pesticides

4.2.1 Oxidation

Oxidation is the first step in the biotransformation of pesticides. These reactions are generally mediated by oxidative enzymes such as cytochrome P450. Cytochrome P450 catalyses monooxygenase reactions resulting in hydroxylation. Along with P450, other enzymes are also involved in microbial

Table 1. Bioremediation of pesticides by microbes.

Microbes involved	Pesticides	Reference
Bacteria		
Sphingomonaspaucimobilis	hexachlorocyclohexane	Pal et al. 2005
Sphingobacterium sp.	DDT	Fang et al. 2010
P. aeruginosa	endosulfan	Jayshree and Vasudevan 2007
Stenotrophomonasmaltophilia *Rhodococcuserythropolis*	endosulfan	Kumar et al. 2008
Citrobacteramalonaticus	chlordecone	Chaussonnerie et al. 2016
P. aeruginosa *Stenotrophomonasmaltophilia* *B. atrophaeus* *Citrobacteramolonaticus* *Acinetobacterlowffii*	endosulfan	Ozdal et al. 2016
Raoultella sp.	dimethoate	Liang et al. 2010
Proteus vulgaris *Vibrio* sp. *Serratia* sp. *Acinetobacter* sp.	dichlorovos	Agarry et al. 2013
Spingomonas sp.	chlorpyrifos	Li et al. 2007
Streptomycetes sp.	chlorpyrifos	Briceno et al. 2012
Pseudomonas sp.	profenofos	Malghani et al. 2009
Bacillus sp.	cypermethrin	Sharma et al. 2016
Pseudomonas stutzeri	dichlorovos	Parte et al. 2017
Rhodococcus sp.	acetamiprid	Phugare and Jadhav 2015
Burkholderiacepacia	imidacloprid metribuzin	Madhuban et al. 2011
Sphingomonas sp. *Arthrobacter* sp.	Carbofuran	Kim et al. 2004
Pseudomonas sp.	oxamyl	Rousidou et al. 2016
Pseudomona sp.	carbaryl	Trivedi et al. 2016
Fungi		
Phlebiatremellosa *Phlebiabrevispora* *Phlebiaacanthocystis*	heptachlor	Xiao et al. 2011
Aspergillus niger	endosulfan	Bhalerao and Puranik 2007
Mortierella sp.	endosulfan	Shimizu 2002
Fusariumoxysporum *Lentinulaedodes* *Penicillium brevicompactum Lecanicillium saksenae*	difenoconazole terbuthylazine pendimethalin	Shi et al. 2012
Trichodermaviride *T. harzianum*	pirimicarb	Eapen et al. 2007
Algae		
Chlamydomonasreinhardtii	prometryne	Jin et al. 2012
Chlamydomonasreinhardtii	fluroxypyr	Zhang et al. 2011
Chlamydomonasreinhardtii	isoproturon	Bi et al. 2012
Chlorococcum sp. *Scenedesmus* sp.	α-endosulfan	Sethunathan et al. 2004

degradation processes. They include peroxidase, polyphenol-oxidase, laccase, tyrosinase, etc. which catalyse the polymerization of various anilines and phenols.

4.2.2 Hydrolysis

The compounds containing amide, carbamate, and ester functional groups are generally metabolized by hydrolytic enzymes. Ester hydrolysis is carried out by esterases and different types of esterases have been reported in *Pseudomonas fluorescens*.

4.2.3 Carbon-Phosphorus bond cleavage reactions

C-P bonds are seen in organophosphorus compounds and are not easily degraded by different degradation processes. Bacteria such as *Escherichia coli* are involved in the degradation of C-P bonds through the C-P lyase enzyme.

4.2.4 Conjugation

Pesticide conjugation is a co-metabolic process in which an exogenous or endogenous natural compound is joined to a pesticide. Uridine diphosphate-glucosyl (UDPG) transferase enzyme mediates pesticide-glucose conjugation and pesticide-glucose ester conjugation reactions. *Xylosylation, alkylation, acylation,* and *nitrosation* are some of the microbial pesticide conjugation reactions. Conjugated pesticides are generally bound to plant cell walls and are difficult to extract. It is reported that some microbes can mineralize the bound residues and make them bioavailable.

4.3 Microbial enzymes in pesticide degradation

Enzymes play a major role in the degradation of pesticides by microbes. These enzymes were produced during different metabolic pathways of plants and microbes present in the soil. Engineered microbes were also used to produce enzymes that act on the pesticides leading to its degradation. Strains of genetically modified bacteria contain enzymes that degrade pesticides belonging to organophosphorus, carbamates, and pyrethroid groups (Javaid et al. 2016). A large number of enzymes such as dehydrogenases, cytochrome P450, dioxigenases, ligninases, etc. are involved in pesticide degradation. Hydrolysis is one of the major processes of degradation. Phosphotrioesterase is an enzyme involved in the hydrolysis of OP compounds. An enzyme, carbofuran hydrolase, is reported to carry out hydrolysis of the methylcarbamate linkage of carbamate pesticides. Table 2 summarizes some of the enzymes involved in pesticide degradation.

5. Microbial biodegradation methods

Microbial degradation techniques are broadly classified into two categories based on whether the degradation is carried out *in situ* or *ex situ*.

5.1 In situ microbial degradation

In Latin, *in situ* means "in the origin place". In this technique, the polluted substances are treated at the site of pollution. Compared to *ex situ* techniques, it is less costly since no excavation and transport of contaminants is required. But these techniques are a highly time-consuming and seasonal variation of microbes in the soil also affects the degradation processes. If the native microbes lack biodegradation ability, then genetically engineered microbes are introduced at the site of contamination to enhance the degradation process. There are different types of *in situ* degradation processes.

a) Bioventing

In this technique, the flow of oxygen is supplied to the unsaturated zones in the soil through controlled airflow stimulation to enhance the biodegradation process. In this technique, the biodegradation is enhanced by adding nutrients and moisture to the soil. The success of

Table 2. Enzymes involved in pesticide degradation.

Microbes	Enzymes	Pesticides
Klebsiella sp. *Alcaligenes* sp. *Staphylococcus* sp. *Pseudomonas* sp.	Dehalogenases	Organochlorine compounds
Flavobacterium *Pseudomonas* sp. *Agrobacterium radiobacter* *Alteromonas* sp. *Plesiomonas* sp. *Achromobacter* *Pseudaminobacter Ochrobactrum* *Brucella* *B. diminuta* *Flavobacterium* sp. *Pleurotusostreatus* *Aspergillus* sp. *Penicillium* sp.	Organophosphorus hydrolase (OPH) Organophosphorus acid anhydrolase (OPAA) Laccase Aspergillus enzyme (A-OPH) Penicillium enzyme (P-OPH)	Organophosphorus compounds
Achromobacter sp. *Pseudomonas* *Mesorhizobium* *Ralstonia* *Rhodococcus* *Ochrobactrum* *Bacillus*	Carbofuran hydrolase	Carbamate compounds
Serratia *Pseudomonas* *Aspergillus niger*	Carboxyl esterase Phosphotriesterase Pyrethroid hydrolase	Pyrethroids

Adapted from Parte et al. 2017.

bioventing based bioremediation techniques relies upon the uniform distribution of air in the unsaturated zone.

b) Bioslurping

Bioslurping involves vacuum enhanced pumping, soil vapour extraction, and bioventing by the indirect provision of oxygen simulating the enhanced biodegradation of pesticides. This technique is used to remediate soils contaminated with volatile and semi-volatile organic compounds.

c) Biosparging

Biosparging is very similar to bioventing with a difference that in this technique air is injected into a saturated zone causing upward movement of the volatile organic compounds to the unsaturated zone promoting biodegradation. Biosparging increases the contact between soil and groundwater. This technique is commonly used in aquifers contaminated with hydrocarbons.

d) Phytoremediation

This technique involves plant interaction in the contaminated sites resulting in the degradation of pesticides. Plant promoting rhizobacteria play a major role in phytoremediation techniques since it enhances biomass production.

e) Permeable reactive barriers

This is used for decontaminating groundwater contaminated with different pollutants. In this technique, a permanent or semi-permanent reactive barrier made up of zero-valent iron is used in the way of polluted groundwater. The polluted water is trapped and undergoes several reactions to give clean water.

f) Intrinsic bioremediation or natural attenuation

It is the passive remediation of contaminated sites without any human intervention. In this, both aerobic and anaerobic microbial degradation causes the removal of contaminants from the site. This technique is less expensive compared to other *in-situ* techniques.

5.2 Ex situ microbial degradation

In this technique, the pollutants are excavated from their sites and transported to other sites for treatment. These involve aerobic techniques. It is a costly process compared to *in situ* degradation processes. Based on the nature of contaminants, *ex situ* microbial degradation systems are broadly classified into two classes: (1) slurry phase microbial degradation and (2) solid-phase microbial degradation. In slurry phase degradation, contaminated soil is mixed with water to form a slurry and is treated in bioreactors or contained ponds or lagoons. In solid-phase degradation techniques, polluted soils are excavated and placed in piles for treatment. The soils are sprayed with water to maintain the moisture. There are different techniques under which *ex situ* microbial degradation takes place.

a) Biopile

In this technique, bioremediation is done by the above-ground piling of excavated contaminated soil, followed by nutrient addition and aeration to increase the microbial degradation of contaminants. This technique involves different components such as aeration, irrigation, nutrients and leachate collection systems, and a treatment bed.

b) Windrows

In this technique, a periodic turning of the polluted soil is done to enhance bioremediation by microbes. The periodic turning along with the addition of water increases aeration, uniform distribution of contaminants, and the rate of microbial degradation.

c) Bioreactor

In bioreactors, the polluted soil is fed and due to different biological reactions, contaminants are converted to specific products. In this, the different parameters such as pH, temperature, substrate, etc. can be controlled. But this technique is costly and labour intensive.

d) Land farming

This is the simplest form of bioremediation technique. This technique requires less equipment and the process is cost-effective. This technique can be *in situ* or *ex situ* based on the site of treatment. If the contaminated soil is excavated and treated on-site, it is called *in situ* otherwise it is *ex situ*. If the contaminants are present at a depth of less than 1 m, then it can be treated without excavation. Excavation needs to be done if the contaminants are present at a depth of more than 1.7 m. This technique is very simple to design and implement and efficient in treating a large volume of contaminated soil.

One of the advantages of *ex situ* treatment is that extensive initial assessment of the polluted site is not required which reduces the labour and cost. These techniques are generally faster, easy to control, and can be used to treat several contaminants (Azubuike et al. 2016).

Microbial degradation can take place under the aerobic and anaerobic conditions depending upon the availability of oxygen.

5.3 Aerobic microbial degradation

In aerobic degradation, the microbes utilise the atmospheric oxygen for its metabolic activities to mineralize pesticides. Different electron-withdrawing substituents such as chloro, nitro groups restrict the use of oxygenase enzyme by aerobic microbes to initiate the electrophilic attack on aromatic molecules. The microbes catalyse the hydrolysis reactions through co-metabolism. In some pesticides, anaerobic degradation works better than the aerobic degradation processes.

5.4 Anaerobic microbial degradation

In anaerobic degradation, microbes utilize the organic compounds present in soil as substrate for its energy needs. In this type of degradation, nucleophilic attack initiates the degradation of aromatic compounds. Unlike in aerobic degradation, the substituents such as chloro, nitro groups favour the attack of anaerobic microbes on aromatic compounds. The electron donating groups hinder the anaerobic transformation of aromatic compounds. The common degradation reactions in the anaerobic degradation of pesticides include: (a) replacement of hydrogen atoms with hydroxyl groups, (b) oxidation of S to SO_2 which is a common reaction resulting in the formation of epoxides, (c) addition/removal of methyl groups, (d) dechlorination, (e) migration of chlorine groups, (f) conversion of the nitro group to an amino group (nitrate reduction), etc. Generally, the halogenated compounds are easily degraded under anaerobic conditions.

6. Microbial bioremediation techniques and practices (some case studies, commercial products)

Being economic and ecologically sustainable, bioremediation with microbes is always regarded as the most important option for pesticide decontamination. However, various environmental factors may limit its application, particularly at the field level. The bioremediation technique at the lab scale may not succeed or fail at the field scale. The reason may be the difficulties in simulating the controlled laboratory condition to the field level. Several biotic and abiotic factors which control the microbes may also control the outcome of bioremediation experiment at field level. Another important factor is the bioavailability of pesticides, particularly persistent pesticides like DDT towards the microbes. Therefore, it is important to determine ways of increasing pesticide bioavailability. Table 3 lists some success stories where microbial bioremediation of pesticides was validated at the field level or mesocosm level. Strong et al. (2000) succeeded in 97% degradation of atrazine at the filed level by using recombinant *E. coli* overexpressing the atrazine chlorohydrolase gene derived from *Pseudomonas* sp. ADP. The addition of phosphate as a biostimulant increases the biodegradation of atrazine in soil plots. Sagarkar et al. (2013) carried out a mesocosm study (100 Kg soil) for biodegradation of atrazine herbicides and almost 90% atrazine degradation was observed with an atrazine degrading consortium comprising of 3 novel bacterial strains. John et al. (2018) used *Klebsiella* sp. isolated from pesticide-contaminated agricultural soil for *in situ* bioremediation of insecticide chlorpyrifos and demonstrated that the microbes can degrade the toxic chlorpyrifos into non-toxic products which increased the growth of soil microorganisms and dehydrogenase activity. Several field level studies for bioremediation of DDT contaminated site resulted in 68–95% removal of DDT.

Table 3. Microbial bioremediation of pesticides at field/mesocosm level.

Pesticides	Bioremediation method	Site of application/ Scale of treatment	Impact on removal rate (%)	References
DDT	Xenorem® Anaerobic-aerobic composting	Savannah River	95	Sasek et al. 2003, West Swiss Riders Chapter 2003
Atrazine	Consortium of 3 novel bacterial strains	Mesocosm	90	Sagarkar et al. 2014
Chlorpyrifos	*Klebsiella* sp.	Agricultural soil	82	John et al. 2018
DDT	Daramend®	Superfund site, Montgomery, Alabama	68	Environmental Protection Agency 2002

So, for on-site bioremediation, a systematic understanding of the surrounding nature and level of contamination of pesticides is important. Attention should be given in case of mixed pesticides contamination, since the degradation of one pesticide may stimulate or hinder the activity of microbes associated with the degradation of other pesticides. Prior knowledge is useful in choosing the best bioremediation strategy for field applications.

7. Biotechnological intervention in microbial remediation

Biotechnology is considered a prospective resource of safe, inexpensive, and effective methods for the bioremediation of contaminated sites including pesticides. Even though the use of microorganisms for bioremediation has gained some success, the use of biotechnological tools may help in gaining momentum. The genetic engineering approach may help in generating microbes with the best mix of biochemical pathways for remediation of contaminated sites. The recent development in the evolution of pesticide degradation pathways, along with the organization of catabolic genes involved, enabled the researcher to develop genetically engineered microbes with enhanced decontamination potential. Advancement of recombinant DNA technology gives a more clear understanding of the degradation process. Catabolic genes involved in the degradation pathway have been identified for several pesticides. By manipulating these degradation genes, attempts are made to generate hybrid pathways by developing microbial strains with enhanced degradation capabilities.

7.1 Genetically modified microorganism (GMO) for pesticide bioremediation

The use of microbes and their enzymes for the degradation of pesticides is considered as an eco-friendly and sustainable approach as they are self-sustainable and low-cost. However, the biodegradation potential of native microorganisms for different pesticides with diverse chemistry is very much limited. To overcome these constraints, designing transgenic microbes through genetic engineering approaches is a highly important step for enhancing the biodegradation of pesticides. Due to their adaptation to wider environmental conditions, genetically modified organisms (GMOs) offer better potential for faster degradation of pesticides and thus environmental remediation. Thus, this branch of biotechnology helps in the remediation of pesticide pollution by converting them into a non-toxic or low toxic form. Microbes have been continuously and consciously introduced into the environment for a specific reason. However, current knowledge of biotechnology helps in developing a new strain with thrilling capabilities. Here, through modification/alteration of genetic material, i.e., DNA, one particular organism is modified to get the necessary character and is usually called as genetically modified. Several terminologies are associated with this technology and among them, the most used are "gene technology," or "recombinant DNA technology" (RDT), or "genetic engineering," and the modified organism is known as "genetically modified," "genetically engineered," or "transgenic."

Wasilkowski et al. (2012) affirmed that integration of conventional microbiology, biochemistry, ecology along with genetic engineering led to effective measures for *in situ* bioremediation. A wide variety of genetically modified microorganisms has shown great potential for effective and faster degradation of pesticides into non-toxic metabolites (Rayu et al. 2017). A large number of native microorganisms have been genetically modified for faster biodegradation of pesticides. For remediation of soil contaminated with multiple pesticides, Cao et al. (2013) constructed a genetically engineered microorganism by fusing the organophosphorus hydrolase with INPNC (ice nucleation protein) of *Pseudomonas syringae* onto the cell surface of *Sphingobium japonicum* UT26, an HCH-degrader.

Several genes have been isolated and identified with an ability of faster degradation of pesticides (Table 4), and can be suitably used for the construction of genetically engineered microbes. Encoding of atrazine chlorohydrolase by gene atzA resulted in faster degradation of atrazine (Neumann et al. 2004). Atrazine bioremediation in field-scale was accomplished by using a killed and stabilized whole-cell suspension of recombinant *Escherichia coli* engineered to atrazine chlorohyrolase, where

Table 4. List of genes from different species involved in pesticide degradation.

Enzyme	Gene	Species	Degradation substrate	Reference
Dehydrochlorinase	linA2 gene	*Sphingomonas paucimobilis* B90	lindane (gamma-hexachlorocyclohexane)	Chaurasia et al. 2013
Atrazine chlorohydrolase	atzA	*Pseudomonas* sp. strain ADP	Atrazine	Neumann et al. 2004
Chlorpyrifos/ carbofuran hydrolase	*mpd, gfp* and *mcd*	*Pseudomonas putida* KT2440	Chlorpyrifos/carbofuran	Gong et al. 2016
Phosphotriesterase	ophc2	*Pseudomonas pseudoalcaligenes*	Methyl parathion	Gotthard et al. 2013
Methyl parathion hydrolase	mpd	*Pseudomonas putida* DLL-1 *Sphingomonas* sp. CDS-1	Methyl parathion and carbofuran	Liu et al. 2006
Esterase SulE		*Hansschlegelia zhihuaiae* S113	Sulfonylurea herbicide	Hang et al. 2012

dechlorination of atrazine resulted in the formation of non-toxic and non-phytotoxic metabolites (Strong et al. 2000). For degradation of hexachlorocyclohexane (HCH) and methyl parathion simultaneously, Lu et al. (2008) constructed genetically engineered *Sphingomonas* sp. BHC-A-mpd by overexpressing methyl parathion hydrolase gene (mpd) to *Sphingomonas* sp. BHC-A, a highly efficient HCH-degrader. Gu et al. (2006) constructed an engineered strain *P. putida* KT2440-DOP for degradation of organophosphorus pesticides along with some aromatic hydrocarbons. Yang et al. (2012) developed a genetically modified organism by overexpressing organochlorines (OCs) and organophosphates (OPs)-degradation gene linA and mpd, respectively, in *E. coli* for degradation of OCs and OPs simultaneously. An innovative approach for increasing the biodegradation of herbicides 2,4-D by native microorganism was achieved by the natural conjugative transfer of catabolic plasmids from *E. coli* to the indigenous soil bacteria (Top et al. 1998).

Other than the overexpression gene related to pesticide degradation, the fusion of protoplast is emerging as a promising technology for constructing multifunctional genetic strains where the gene for specific functionalities is transferred from one species to another and thus getting the benefit of both parents (Dillon et al. 2008). In the field of bioremediation of pesticides, this is a novel and reformative technique. By protoplast fusion of *Rhodococcus* sp. BX2 and *Acinetobacter* sp. LYC-1, Feng et al. (2013) constructed a functional strain F1 for faster degradation of bensulfuron-methyl and butachlor simultaneously. For simultaneous degradation of neonicotinoid pesticides chlorothalonil and acetamiprid, Wang et al. (2016) also developed a functional strain, AC, through a protoplast fusion of *Pseudomonas* sp. CTN-4 and *Pigmentiphaga* sp. strain AAP-1. With the advancement of recombinant DNA technologies, "suicidal GEMs" (S-GEMs) technology has also been emerged for remediation of contaminated sites (Paul et al. 2005, Kumar et al. 2013) which is safe and more efficient. However, survival and biodegradation ability of GMO in the field depends on the several biotic and abiotic factors in the environment. Factors like high clay and pH and moisture content increase the survivability of GMO, whereas factors like prolonged dry periods, presence of competing microorganisms, lytic bacteriophage, and protozoan predation will negatively affect the introduced microbes. Although a considerable number of GMOs have been constructed throughout the world for bioremediation of pesticides, their successful field-scale application is rare. Furthermore, constrain in upscaling the laboratory experiments, low bioavailability of pesticides to the induced microbes, and legislative problems related to the legal use of GMOs have prohibited wide-scale application.

Due to legal challenges and high expenses involved in transgenic research, this approach affects many companies and until now it is confined to academic and research institutes. Controversy related to the use and release of GMOs in the open environment restricts its use. Concerns have been expressed by several countries including India over the safety and ecological damage linked with the release of GMO. In India, release of GMOs for bioremediation in uncontrolled conditions is strictly prohibited. In the Indian scenario, for releasing GMOs, several step permission is required which starts from Institutional Bio-Safety Committee (IBSC) to Research Committee on Genetic Modification (RCGM) for field and other related tests and trials and GEAC (Genetic Engineering Approval Committee) approval is must for commercial releases. Since bioremediation with GMO is rarely practiced, the system as a consequence is not yet defined completely. Therefore, environmental and ecological concerns and regulatory constraints hamper the use of GMOs for field-scale bioremediation of pesticides.

7.2 Metagenomics and metaproteomics approach for evaluation of pesticide bioremediation

The indigenous microbial community establishes a complex ecological niche for remediation of the pesticide-contaminated site. But they are not readily cultured. Therefore, the analysis of the structural and functional composition of microbes is essential to determine their role during bioremediation of various pesticides. Concurrently, the knowledge of primary and secondary metabolites/protein will help in understanding the interaction between pesticide and microbe, how it works at the molecular level and response of organisms exposed to a pesticide-contaminated environment. The repeated or high dose of the application of pesticides may have negative effects on the ecosystem due to the formation of toxic metabolites which can disturb the endogenous metabolism as the microbes are exposed to changing environments. Furthermore, due to the persistent nature of pesticides, particularly organochlorine pesticides, there is a chance of partial mineralization by microbes which lead to toxic metabolites accumulation. Consequently, compared to the traditional method, there is a necessity of a more accurate method for pesticide bioremediation. Metagenomics and proteometabolomics approach may create a system that can help to understand the site-specific microorganism during the active bioremediation process. Omics based approaches have great potential for microbial detection and community analysis. These omics-based techniques have a great impact on the bioremediation potential of microbes for pesticide degradation.

The main prospect for the researchers is to find out the major mechanisms involved in bioremediation processes for the degradation of pesticides. The integration of functional genomics, proteomics, transcriptomics, and metabolomic data along with the bioremediation process may help to get a clear-cut picture of the microbial bioremediation process. The integration of all of these with high throughput techniques will enable a step forward in the researches on pesticide bioremediation. Recently, in various studies, the metagenomic approach has been used to investigate the potential biodegradation pathways of persistent pesticides (Fang et al. 2014).

The metagenomic approach was used by Fang et al. (2014) to understand the abundance and diversity of biodegradation genes (BDGs) along with the potential degradation route of DDT, HCH, and atrazine in both freshwater and marine sediments by using 6 datasets. It was found that out of 69 genera identified, *Plesiocystis, Anaerolinea, Jannaschia,* and *Mycobacterium* were found to be potential for pesticide biodegradation in all sediments. Dadhwal et al. (2009) proposed the biostimulation of indigenous HCH degrading microorganisms after analysing the diversity of culturable microorganisms for effective bioremediation of severely HCH-contaminated site in India. Metagenomic analysis carried out by Sangwan et al. (2012) for HCH-contaminated soil samples from India confirms the horizontal transfer of HCH catabolism genes. The study also suggests developing an economically viable bioremediation technology for HCH contaminated sites. Upon degradation, insecticide chlorpyrifos produces a toxic metabolite 3,5,6-trichloro-2-pyridinol (TCP). Gene (tcp3A) encoding a TCP degrading enzyme was cloned from a metagenomic library prepared from cow rumen (Math et al. 2010). TCP was used as a sole source of carbon by recombinant *E. coli* harboring the tcp3A gene. This was a breakthrough in the application of metagenomics approaches for pesticide degradation. Besides, recently a combination of metagenomics and metaproteomics approach has been used for remediation of emerging contaminants (acetaminophen and sulfonamides) and this approach can also be extended for bioremediation of pesticides (Chang et al. 2018).

8. Possibilities, practical hindrance and future direction

Bioremediation is a wonderful process of treating pesticide contaminants by utilizing the natural survival activities of organisms. In this process, the desired organisms utilize the toxic pesticide contaminant as a source of carbon and energy for their growth and formation of new cells/ reproduction. Bioremediation involves both phytoremediation and microbial remediation. Though both methods are effective, however microbial bioremediation is more effective, highly applicable, and economic as compared to phytoremediation. The concept of bioremediation is relatively new, and the first application was for removing oil spills from the coastal area of Santa Barbara, California, USA in 1960. Thereafter, the science of bioremediation has made tremendous improvements and presently bioremediation can be applied for a wide range of contaminants including pesticides, heavy metals, polychlorinated biphenyls (PCBs), polyaromatic hydrocarbons (PAHs), synthetic dyes, pharmaceuticals, etc. In this context, it is important to draw attention to the fact that biodegradation and bioremediation are different. Though both covers the formation of non-toxic components/ metabolites from toxic contaminants, biodegradation is a naturally occurring phenomenon, whereas bioremediation is a technology. Microbial bioremediation holds immense potential to treat pesticide contamination. As discussed earlier, soil acts as a major sink of pesticide contamination, and such contamination can occur in the surface layer, vadose zone, and groundwater. Phytoremediation is possible for surface soil, but it is difficult for treating vadose zone and groundwater. Microbes offer the best possible way to treat pesticides in all three zones, i.e., aerobic and anaerobic layers of soil. However, certain factors like (i) availability of pesticides (considering the concentration and toxic nature) to the microbes (ii) physiological conditions of microbes including growth, metabolism, enzymatic activity, population dynamics, etc. (iii) survival of desired microbes in the soil environment after facing competition, predation and antagonism from native microbes (iv) environmental factors like moisture content, temperature, pH, nutrients, aeration status, etc. affect the performance of microbes on the available forms of toxicant/contaminants. Bioremediation techniques offer several advantages over the traditional methods of remediation like:

- Bioremediation involves natural microbes which immobilise/metabolise/convert toxic pesticide residue into non-toxic molecules or complete degradation to carbon dioxide and water, whereas conventional treatment methods offer primary removal of toxic pesticide residue from one environmental part and have a serious concern of secondary disposals like landfill and incineration. This secondary disposal may lead to indirect contamination of other environmental parts.
- Bioremediation offers a scope of large scale application without displacement of the contaminated media whereby the natural activity of the medium is maintained, whereas traditional methods require physical transport to bring the contaminated medium to the treatment plant. Further, the quantity/scale of treatment for traditional methods is small as compared to the bioremediation methods. Hence, traditional methods require a transport cost also.
- Bioremediation is easy and less effort is required for on-field applications. Labour involvement is also less in bioremediation as compared to conventional methods. Bringing the contaminated medium to the treatment plant and reallocation of the medium after treatment makes a traditional/ conventional treatment method cost and labour intensive.
- The energy requirement (either chemical/thermal) for traditional methods is high, whereas bioremediation techniques do not require any external energy. Microbes degrade contaminants and in each step, it releases carbon and energy which promote natural growth and population of the desired microbes. Hence, the bioremediation technique is a sustainable process.

- There is no requirement for hazardous chemicals in the bioremediation technique. Sometimes fertilizers are used as nutrient supplements for the microbes.
- Microbial bioremediation techniques allow treating areas like vadose zone which is difficult by traditional treatment methods.

The process of bioremediation involves either degradation or transformation or a combination of both pathways. Several bioremediation techniques utilize microbes like bioreactors, landfarming, soil washing with microbes, and bioaugmentation. Biostimulation, bioattenuation, solid-phase bioremediation, biopiles, bioventing, etc. have been developed and applied in the laboratory-scale study, and some have been tried for field studies. However, under field conditions success of a microbial bioremediation technique depends on (i) detailed knowledge of all naturally occurring biological processes at the site/field (ii) detailed scientific information on the biodegradation of contaminant/pesticide generated in the laboratory, and (iii) on-site monitoring of biodegradation process. The success of a microbial bioremediation technique depends primarily on the nature of the microbe. In recent years, thorough investigations have been carried out to understand the potential microbes and their molecular genetics. Several studies for microbial degradation of contaminants like DDTs, endosulfans, carbaryl, monocrotophos, atrazine, lindane, etc. (Lal et al. 2006, Kumar et al. 2007, Barraga et al. 2008) have been carried out at genetic levels. Findings indicated that Ese and Esd genes for endosulfans (Kumar et al. 2007), linA genes (linA1 and linA2) for lindane (Lal et al. 2006), atzA gene for atrazine (Neumann et al. 2004), etc. have a significant role in determining the biodegradation potentiality of desired microbial species. The biodegrading microbes utilise the contaminant/pesticide residue as source of energy and carbon under certain environmental conditions and detailed information is generated in lab studies. Hence, for a successful field application of a microbial bioremediation technique, desired conditions are needed to be created. In many cases, it has been observed that a potential microbe (under laboratory condition) gives a poor performance in the field because of a higher level of contaminant concentration which may be toxic, resulting in low survival of degrading microbes under field condition. Therefore, to promote the growth of the contaminant degrading microbes, a detailed study of the site/field is required. Moreover, these degrading microbes do not always strive for toxicants/contaminants. Easy availability of other sources for carbon and energy may result in a lower affinity for the contaminant/pesticide residues, resulting in poor field performance of a potential microbe. These limitations can be overcome by either applying fertilizers to supply nutrients or bioventing to supply oxygen or a combination of both is required to stimulate microbial growth. The introduction of a genetically modified native or non-native microbe for bioremediation is a promising aspect but the future effects of the genetically modified microbe on native soil population are not fully understood. Further, microbial bioremediation of air contaminants/pollutants is limited and inefficient, considering the volumes of polluted air generated from industries.

Microbial bioremediation of pesticides is a technique of the future. It is a 'green' technique as compared to the traditional methods of remediation. However, more research in this field is required to overcome the limitations discussed earlier. Frontier research in a better understanding of microbes at an ecological, biological, and genetic level will provide a basis for the selection and utilization of microbial population for bioremediations. Omics-based technologies and data information have opened a new horizon for a better understanding of bioremediation. Further, researches in innovative engineering technologies to supply the desired stimulant like nutrients or oxygen to the specific microbial population is also an underexplored area. New techniques to promote contaminant's availability to the degrading microbes are another new area of research.

9. Conclusion

In post-green revolution era, the use of pesticides has been more frequent due to the heavy load of pest and disease problems in field crops. So is the regular occurrence of weeds in the field.

Use of insecticides, fungicides, and herbicides has been a common practice as a prophylactic and also as a remedial measures to check the problem to the best extent. Undoubtedly, these chemicals are effective to handle the pest-diseases and weeds to offset the undesirable damages to get the desirable yield. That is why they became popular and the number of existing and new generation pesticides are getting promoted among farmers. But the problems arise in other ways, when residues of pesticides remain in the system due to recurrent applications, sometimes even after years of application, and pollute the soil and nearby water bodies. Moreover, these pesticide residues can disturb the soil's ecological health and have every possibility of bio-magnification. Here lies the relevance of degradation of pesticide residues employing microbial intervention. It is an emerging researchable issue and a good number of promising outcomes have been reported all over the world. If a single species of microbes or a formulation of microbial consortia can effectively degrade the pesticide residues, that may be the best eco-friendly and cost-effective approach to save the pollution load. Scientists are working in this line in multi-disciplinary mode and hopefully, shortly, we can get suitable remedial measures.

References

Agarry, S.E., Olu-Arotiowa, O.A., Aremu, M.O. and Jimoda L.A. 2013. Biodegradation ofdichlorovos (organophosphate pesticide) in soil by bacterial isolates. Biodegradation 3(8): 11–16.

Allison, D., Kallman,B.J., Cope, O.B. and Van Valin, C.C. 1964. Some Chronic Effects of DDT on Cutthroat Trout, Research Report 64. US Department of Interior, Bureau of Sport Fisheries and Wildlife, Washington DC. pp. 30.

Azubuike, C.C., Chikere, C.B. and Okpokwasili, G.C. 2016. Bioremediation techniques–classification based on site of application: principles, advantages, limitations and prospects. World J. Microbiol. Biotechnol. 32(11): 180.

Barragan-Huerta, B.E., Costa-Perez, C. Peralta-Cruz, J., Esparza-García, F. and Barrera-Corte, J. 2007. Biodegration of organochlorine pesticides by bacteria grown in microniches of the porus structure of green bean coffee. Int. Biodeter. Biodegr. 59: 239–44.

Bhaduri, D., Chatterjee, D., Chakraborty, K., Chatterjee, S. and Saha, A. 2018. Bioindicators of degraded soils. pp. 231–257. In: Sustainable Agriculture Reviews. 33. Springer, Cham.

Bhalerao, T.S. and Puranik, P.R. 2007. Biodegradation of organochlorine pesticide, endosulfan, by a fungal soil isolate, *Aspergillus niger*. Int. Biodeter. Biodegr. 59(4): 315–321.

Bi, Y.F., Miao, S.S., Lu, Y.C., Qiu, C.B., Zhou, Y. and Yang, H. 2012. Phytotoxicity, bioaccumulation and degradation of isoproturon in green algae. J. Hazard. Mater. 243: 242–249.

Bielská, L., Šmídová, K. and Hofman, J. 2013. Supercritical fluid extraction of persistent organic pollutants from natural and artificial soils and comparison with bioaccumulation in earthworms. Environ. Pollut. 176: 48–54.

Briceno, G., Fuentes, M.S., Palma, G., Jorquera, M.A., Amoroso, M.J. and Diez, M.C. 2012. Chlorpyrifos biodegradation and 3, 5, 6-trichloro-2-pyridinol production by actinobacteria isolated from soil. Int. Biodeter. Biodegr. 73: 1–7.

Burnett, M. and Welford, R. 2007. Case study: coca-cola and water in India: episode 2. Crop Soc. Responsib. Environ. Mgmt. 14: 298–304.

Cao, X., Yang, C., Liu, R., Li, Q., Zhang, W., Liu, J., Song, Qiao, C. and Mulchandani, A. 2013. Simultaneous degradation of organophosphate and organochlorine pesticides by *Sphingobium japonicum* UT26 with surface-displayed organophosphorus hydrolase. Biodegradation 24(2): 295–303.

Carvalho, F.P., Montenegro-Guillén, S.,Villeneuve, J.P., Cattini, C., Tolosa, I., Bartocci, J., Lacayo-Romero, M. and Cruz-Granja, A. 2003. Toxaphene residues from cotton fields in soils and in the coastal environment of Nicaragua. Chemosphere 53: 627–636.

Castelo-Grande, T., Augusto, P.A. and Barbos, D. 2005. Removal of pesticides from soil by supercritical extraction—a preliminary study. Chem. Eng. J. 111: 167–171.

Castelo-Grande, T., Augusto, P.A., Monteiro, P., Estevez, A.M. and Barbosa, D. 2010. Remediation of soils contaminated with pesticides: a review. Int. J. Environ. Anal. Chem. 90: 438–467.

Chang, B.V., Fan, S.N., Tsai, Y.C., Chung, Y.L., Tu, P.X. and Yang, C.W. 2018. Removal of emerging contaminants using spent mushroom compost. Sci. Total Environ. 634: 922–933.

Chaurasia, A.K., Adhya, T.K. and Apte, S.K. 2013.Engineering bacteria for bioremediation of persistent organochlorine pesticide lindane (γ-hexachlorocyclohexane). Bioresour. Technol. 149: 439–45.

Chaussonnerie, S., Saaidi, P.L., Ugarte, E., Barbance, A., Fossey, A., Barbe, V., Gyapay, G., Brüls, T., Chevallier, M., Couturat, L., Fouteau, S., Muselet, D., Pateau, E., Cohen, G.N., Fonknechten, N., Weissenbach, J. and Le Paslier, D. 2016. Microbial degradation of a recalcitrant pesticide: Chlordecone. Front. Microbiol. 7: 2025. doi: 10.3389/fmicb.2016.02025.

Chourasiya S., Khillare, P.S. and Jyethi, D.S. 2015. Health risk assessment of organochlorine pesticide exposure through dietary intake of vegetables grown in the periurban sites of Delhi, India. Environ. Sci. Pollut. Res. Int. 22: 5793–806.

Cotillas, S., Sáez, C., Cañizares, P., Cretescu, I. and Rodrigo, M.A. 2018. Removal of 2, 4-D herbicide in soils using a combined process based on washing and adsorption electrochemically assisted. Sep. Puri. Technol. 194: 19–25.

Dadhwal, M., Jit, S., Kumari, H. and Lal, R. 2009. *Sphingobium chinhatense* sp. nov., a hexachlorocyclohexane (HCH)-degrading bacterium isolated from an HCH dumpsite. Int. J. Syst. Evol. Microbiol. 59: 3140–3144.

Deribe, E., Rosseland, B.O., Borgstrom, R., Salbu, B., Gebremariam, Z., Dadebo, E., Skipperud, L. and Eklo, O.M. 2013. Biomagnification of DDT and its metabolites in four fish species of a tropical lake. Ecotox. Environ. Safe. 95:.10–18.

Deutsch, C.A., Tewksbury, J.J., Tigchelaar, M., Battisti, D.S., Merrill, S.C., Huey, R.B. and Naylor, R.L. 2018. Increase in crop losses to insect pests in a warming climate. Science 361(6405): 916–919.

Dillon, A.J.P., Camassola, M., Henriques, J.A.P., Fungaro, M.H.P., Azevedo, A.C.S., Velho, T.A.F. and Laguna, S.E. 2008. Generation of recombinants strains to cellulases production by protoplast fusion between *Penicillium echinulatum* and *Trichoderma harzianum*. Enzyme Microb. Technol. 43(6): 403–409.

Dimitrov, S.D., Dimitorova, N.C., Walker, J.D., Veith, G.D. and Mekenyan, O.G. 2002. Predicting bioconcentration factors of highly hydrophobic chemicals. Effects of molecular size. Pure Appl. Chem. 74: 1823–1830.

Dos Santos, E.V., Sáez, C., Martínez-Huitle, C.A., Cañizares, P. and Rodrigo, M.A. 2015a. Combined soil washing and CDEO for the removal of atrazine from soils. J. Hazar. Mater. 300: 129–134.

Dos Santos, E.V., Saez, C., Martínez-Huitle, C.A., Canizares, P. and Rodrigo, M.A. 2015b. The role of particle size on the conductive diamond electrochemical oxidation of soil-washing effluent polluted with atrazine. Electrochem. Commun. 55: 26–29.

Dos Santos, E.V., Fernanda, S., Saez, C., Cañizares, P., Lanza, M.R.V., Martínez-Huitle, C.A. and Rodrigo, M.A. 2016. Application of electrokinetic soil flushing to four herbicides: a comparison. Chemosphere. 153: 205–211.

Eapen, S., Singh, S. and D'souza, S.F. 2007. Advances in development of transgenic plants for remediation of xenobiotic pollutants. Biotech. Advan. 25(5): 442–451.

El-Temsah, Y.S., Sevcu, A., Bobcikova, K., Cerni, M. and Joner, E.J. 2016. DDT degradation efficiency and ecotoxicological effects of two types of nano-sized zero-valent iron (nZVI) in water and soil. Chemosphere 144: 2221–28.

Evans, M.S., Noguchi, G.E. and Rice, C.P. 1991. The biomagnification of polychlorinated biphenyls, toxaphene, and DDT compounds in a Lake Michigan offshore food web. Arch.Environ. Con. Tox. 20: 87–93.

Environmental Protection Agency (EPA). 2002. Technology news and trends, full-scale bioremediationof organic explosive contaminated soil. EPA 542-N-02-003.

Fang, H., Cai, L., Yang, Y., Ju, F., Li, X., Yu, Y. and Zhang,T. 2014.Metagenomic analysis reveals potential biodegradation pathways of persistent pesticides in freshwater and marine sediments. Sci. Total Environ. 470–471: 983–92.

FAOSTAT. 2017. Pesticides use, Available online at http://www.fao.org/faostat/en /#data/RP. Accessed Aug 2018.

Feng, L., Xiong, M., Cheng, X., Hou, N. and Li, C. 2013. Construction and analysis of an intergeneric fusant able to degrade bensulfuronmethyl and butachlor. Biodegradation 24: 47–56.

Fisher,S.W., Hwang,H., Atanasoff, M. and Landrum, P.F. 1999. Lethal body residues for pentachlorophenol in Zebra mussels (*Dreissena polymorpha*) under varying condition of temperature and pH. Ecotox. Environ. Safe. 43: 274–283.

Frampton, G.K., Jansch, S., Scott-Fordsmand, J.J., Römbke, J. and Van den Brink, P.J. 2006. Effects of pesticides on soil invertebrates in laboratory studies: a review and analysis using species sensitivity distributions. Environ. Toxicol. Chem. 25(9): 2480–89.

Garbarino, J.R., Snyder-Conn, E., Leiker, T.J. and Hoffman, G.L. 2002. Contaminants in arctic snow collected over northwest Alaskan sea ice. Water Air Soil Pollut. 139: 183–214.

Ghosh, R.K. and Singh, N. 2013.Adsorption-desorption of metolachlor and atrazine in Indian soils: effect of fly ash amendment. Environ. Monit. Assess. 185: 1833–45.

Giménez, J., Bayarri, B., González, O., Malato, S., Peral, J. and Esplugas, S. 2015. Advanced oxidation processes at laboratory scale: environmental and economic impacts. ACSSustain. Chem. Eng. 3(12): 3188–96.

Gong, T., Liu, R., Che, Y., Xu, X., Zhao, F., Yu, H., Song, C., Liu, Y. and Yang, C.2016. Engineering *Pseudomonas putida* KT 2440 for simultaneous degradation of carbofuran and chlorpyrifos. Microbiol. Biotech. 9(6): 792–800.

Gotthard, G., Hiblot, J., Gonzalez, D., Elias, M. and Chabriere, E. 2013. Structural and enzymatic characterization of the phosphotriesterase OPHC2 from *Pseudomonas pseudoalcaligenes*. PLoS One. 8(11): e77995.

Goulson, D. 2014. Ecology: pesticides linked to bird declines. Nature. 511: 295–6.

Gu, L.F., He, J., Huang, X., JiaK, Z. and Li, S.P. 2006. Construction of a versatile degrading bacteria *Pseudomonas putida* KT2440-DOP and its degrading characteristics. Wei. Sheng. Wu. Xue. Bao. 46(5): 763–66.

Hanazato, T. 2001. Pesticide effects on freshwater zooplankton: An ecological perspective. Environ. Pollut. 112(1): 1–10.

Hang, B.J., Hong, Q., Xie, X.T., Huang, X., Wang, C.H., He, J. and Li, S.P. 2012. SulE, a sulfonylurea herbicide de-esterification esterase from *Hansschlegelia zhihuaiae* S113. Appl. Environ. Microbiol. 78: 1962–1968.

Ikram, N. and Shoaib, N. 2018. Effects of pesticides on photosynthesis of marine phytoplankton. Bangladesh J. Bot. 47(4): 1007–11.

Javaid,M.K., Ashiq, M. and Tahir, M. 2016. Potential of biological agents in decontamination of agricultural soil. Scientifica. DOI: 10.1155/2016/1598325.

Jayashree, R. and Vasudevan, N. 2007. Effect of tween 80 added to the soil on the degradation of endosulfan by *Pseudomonas aeruginosa*. Int. J. Environ. Sci.Technol. 4(2): 203–210.

Jin, Z. P., Luo, K., Zhang, S., Zheng, Q. and Yang, H. 2012. Bioaccumulation and catabolism of prometryne in green algae. Chemosphere. 87(3): 278–84.

Jinzhong, W.A.N., Linna, W.U., Mao, Y.E., Zhang, S., Jiang, X., Tao, L., Yusuo, L. Zhou, Y.,Xin, J., Yusuo, L. and Xiaohu, L. 2017. Remediation of organochlorine pesticides contaminated soils by surfactants enhanced washing combined with activated carbon selective adsorption. Pedosphere 29(3): 400–408.

John, E.M., Varghese, E.M., Krishnasree, N. and Jisha, M.S. 2018. In situ bioremediation of Chlorpyrifos by *Klebsiella* sp. isolated from pesticide contaminated agricultural soil. Int. J. Curr. Microbiol. Appl. Sci. 7(3): 1418–29.

Kay, S.H. and Vicksburg, M.S. 1984. Potential for biomagnification of contaminants within marine and freshwater food webs. Technical Report D-84-7, U.S. Army Corps of Engineers Waterways Experiment Station.

Khan, M.Z. and Law, F.C.P. 2005. Adverse Effects of pesticides and related chemicals onenzyme and hormone systems of fish, amphibians andreptiles: A review. Proc. Pakistan Acad. Sci. 42(4): 315–323.

Kim, I.S., Ryu, J.Y., Hur, H.G., Gu, M.B., Kim, S.D. and Shim, J.H. 2004. *Sphingomonas* sp. strain SB5 degrades carbofuran to a new metabolite by hydrolysis at the furanyl ring. J. Agr. Food Chem. 52(8): 2309–2314.

Komtchou, S., Dirany, A., Drogui, P., Robert, D. and Lafrance, P. 2017. Removal of atrazine and its by-products from water using electrochemical advanced oxidation processes. Water Res. 125: 91–103.

Kumar, M., Lakshmi, C.V. and Khanna, S. 2008. Biodegradation and bioremediation of endosulfan contaminated soil. Bioresour. Technol. 99(8): 3116–3122.

Kumar, S., Dagar, V.K., Khasa, Y.P. and Kuhad, R.C. 2013. Genetically modified microorganisms (GMOS) for bioremediation in Biotechnology for environmental management and resource recovery. pp. 191–218. *In*: Kuhad, R. and Singh, A. (eds.). Biotechnology for Environmental Management and Resource Recovery. Springer, India.

Kumar, K., Devi, S.S., Krishnamurthi, K., Kanade, G.S. and Chakrabarti, T. 2007. Enrichment and isolation of endosulfan degrading and detoxifying bacteria. Chemosphere. 68: 317–22.

Lal, R., Dogra, C., Malhotra, S., Sharma, P. and Pal, R. 2006. Diversity, Distribution and Divergence of lin genes in hexachlorocyclohexane degrading sphingomonads. Trend. Biotech. 24: 121–29.

Li, R. and Jin, J. 2013. Modelling of temporal patterns and sources of atmospherically transported and deposited pesticides in ecosystems of concern: a case study of toxaphene in the Great Lakes. J. Geophys. Res. Atmos. 118: 11863–11874.

Li, X., He, J. and Li, S. 2007. Isolation of a chlorpyrifos-degrading bacterium, *Sphingomonas* sp. strain Dsp-2, and cloning of the mpd gene. Res. Microbiol. 158: 143–149.

Liang, Y., Zhou, S., Hu, L., Li, L., Zhao, M. and Liu, H. 2010. Class-specific immune affinity monolith for efficient on-line clean-up of pyrethroids followed by high-performance liquid chromatography analysis. J. Chromatogr. B, 878: 278–82.

Liu, Z., Hong, Q., Xu, J.H., Jun, W. and Li, S.P. 2006. Construction of a genetically engineered microorganism for degrading organophosphate and carbamate pesticides. Int. Biodeter. Biodegr. 58: 65–69.

Lu, P., Hong, Y.F., Hong, Q., Jiang, X. and Li, S.P. 2008. Construction of a stable genetically engineered microorganism for degrading HCH & methyl parathion and its characteristics. Huanjing Kexue. 29: 1973–76.

Madhuban, G., Debashis, D., Jha, S.K., Shobhita, K., Saumya, B. and Das, S.K. 2011. Biodegradation of imidacloprid and metribuzin by *Burkholderia cepacia* strain CH9. Pestic. Res. J. 23: 36–40.

Malghani, S., Chatterjee, N., Yu, H.X. and Luo, Z. 2009. Isolation and identification of profenofos degrading bacteria. Braz. J. Microbiol. 40: 893–900.

Math, R.K., Islam, S.M., Cho, K.M., Hong, S.J., Kim, J.M., Yun, M.G., Cho, J.J., Heo, J.Y., Lee, Y.H., Kim, H. and Yun, H.D. 2010. Isolation of a novel gene encoding a 3,5,6-trichloro-2-pyridinol degrading enzyme from a cow rumen metagenomic library. Biodegradation. 21: 565–73.

Mostafalou, S. and Abdollahi, M. 2013. Pesticides and human chronic diseases: evidences, mechanisms, and perspectives. Toxicol. Appl. Pharmacol. 268: 157–77.

Nag, S.K. and Raikwar, M.K. 2011. Persistent organochlorine pesticides residues in animal feed. Environ. Monit. Assess. 174: 327–35.

Nawaz, S. and Kirk, K.L. 1995. Temperature effects on bioconcentration of DDE by Daphnia. Freshw. Biol. 34: 173–78.

Neumann, G., Teras,R., Monson,L., Kivisaar,M., Schauer, F. and Heipieper, H.J. 2004. Simultaneous degradation of atrazine and phenol by Pseudomonas sp. strain ADP: effects of toxicity and adaptation. Appl. Environ. Microbiol. 70: 1907–12.

Niethammer, K.R., White, D.H., Baskett, T.S. and Sayre, M.W. 1984. Presence and biomagnification of organochlorine chemical residues in oxbow lakes of northeastern Louisiana. Arch. Environ. Contam. Toxicol. 13: 63–74.

Oerke, E.C. 2006. Crop losses to pests. J. Agri. Sci. 144: 31–43.

Orris, P., Chary, L.K., Perry, K. and Asbury, J. 2000. Persistent organic pollutants (POPs) and human health. A Publication of the World Federation of Public Health Association's Persistent Organic Pollutant Project. WFPHA, Washington, DC. pp. 1–46.

Osman, K.A. 2011. Pesticides and human health. pp. 206–30. *In*: Stoytcheva, M. (ed.). Pesticides in the Modern World – Effects of Pesticides Exposure. InTech Open.

Ozdal, M., Ozdal, O.G. and Algur, O.F. 2016. Isolation and characterization of α-endosulfan degrading bacteria from the microflora of cockroaches. Pol. J. Microbial. 65(1): 63–68.

Pal, R., Bala, S., Dadhwal, M., Kumar, M., Dhingra, G., Prakashet, O., Prabagaran, S.R., Shivaji, S., Cullum, J., Holliger, C. and Lal, R. 2005. Hexachlorocyclohexane-degrading bacterial strains *Sphingomonas paucimobilis* B90A, UT26 and Sp+, having similar line genes, represent three distinct species, *Sphingobium indicum* sp. nov., *Sphingobium japonicum* sp. nov. and *Sphingobium francense* sp. nov., and reclassification of [*Sphingomonas*] *chungbukensis* as *Sphingobium chungbukense* comb. nov. Int. J. Syst. Evol. Microbial. 55: 1965–72.

Parker, A.M., Lester, Y., Spangler, E.K., VonGunten, U. and Linden, K.G. 2017. UV/H_2O_2 advanced oxidation for abatement of organophosphorous pesticides and the effects on various toxicity screening assays. Chemosphere 182: 477–82.

Parte, S.G., Mohekar, A.D. and Kharat, A.S. 2017. Microbial degradation of pesticide: a review. Afr. J. Microbiol. Res. 11(24): 992–1012.

Paul, D., Pandey, G. and Jain, R.K. 2005. Suicidal genetically engineered microorganisms for bioremediation: need and perspectives. Bioessays 27(5): 563–73.

Pavel, L.V. and Gavrilescu, M. 2008. Overview of *ex-situ* decontamination technologies for soil clean-up. Environ. Eng. Manage. J. 7: 815–34.

Phugare, S.S. and Jadhav, J.P. 2015. Biodegradation of acetamiprid by isolated bacterial strain *Rhodococcus* sp. BCH2 and toxicological analysis of its metabolites in silkworm (*Bombax mori*). Clean Soil Air Water. 43: 296–304.

Pimentel, D. 2009. Pesticides and pest control. pp. 83–87. Peshin, R. and Dhawan, A.K. (eds.). In Integrated pest management: innovation-development process. Springer, Dordrecht.

Rayu, S., Nielsen, U.N., Nazaries, L. and Singh, B.K. 2017. Isolation and molecular characterization of novel chrloropryifos and 3, 5, 6-trichloro-2-pyridinol-degrading bacteria from sugarcane farm soils. Front. Microbiol. 8: 518.

Robinson, T., Ali, U., Mahmood, A., Chaudhry, M.J., Li, J. and Zhang, G. 2015. Concentrations and patterns of organochlorines (OCs) in various fish species from the Indus River, Pakistan: a human health risk assessment. Sci. Total Environ. 541: 1232–42.

Rousidou, K., Chanika, E., Georgiadou, D., Soueref, E., Katsarou, D., Kolovos, P. and Karpouzas, D.G. 2016. Isolation of oxamyl-degrading bacteria and identification of cehA as a novel oxamyl hydrolase gene. Front. Microbiol. 7: 616.

Sagarkar, S., Nousiainen, A., Shaligram, S., Björklöf, K., Lindström, K., Jorgensen, K.S. and Kapley, A. 2013. Soil mesocosm studies on atrazine bioremediation. J. Environ. Manage. 139: 208–16.

Saha, A., Bhaduri, D., Pipariya, A. Jain, N.K. and Basak, B.B. 2015. Behaviour of pendimethalin and oxyfluorfen in peanut field soil: effects on soil biological and biochemical activities. Chem. Ecol. 31: 550–66.

Saha, A., Bhaduri, D., Pipariya, A. and Jain, N.K. 2016a. Influence of imazethapyr and quizalofop-p-ethyl application on microbial biomass and enzymatic activity in peanut grown soil. Environ. Sci. Pollut. Res. 23: 23758–71.

Saha, A., Pipariya, A. and Bhaduri, D. 2016b. Enzymatic activities and microbial biomass in peanut field soil as affected by the foliar application of tebuconazole. Environ. Earth Sci. 75: 558.

Sangwan, N., Lata, P., Dwivedi, V., Singh, A. and Niharika, N. 2012. Comparative metagenomic analysis of soil microbial communities across three hexachlorocyclohexane contamination levels. Plos One. 7(9): e46219.

Satapanajaru, T., Patrick, J.S., Steve, D.C. and Yul, R. 2003.Green rust and iron oxide formation influences metolachlor dechlorination during zerovalent iron treatment. Environ. Sci. Technol. 37: 5219–27.

Sayles, G.D., You,G., Wang,M. and Kupferle, M.J. 1997. DDT, DDD, and DDE dechlorination by zero-valent iron. Environ. Sci. Technol. 31: 3448–54.

Saylor, G.L. and Kupferle, M.J. 2019. The impact of chloride or bromide ions on the advanced oxidation of atrazine by combined electrolysis and ozonation. J. of Environ. Chem. Eng. 7: 103–105.

Schäfer, R.B., Von der Ohe, P.C., Rasmussen, J., Kefford, B.J.,Beketov, M.A., Schulz, R. and Liess, M. 2012. Thresholds for the effects of pesticides on invertebrate communities and leaf breakdown in stream ecosystems. Environ. Sci. Technol. 46: 5134–42.

Sethunathan, N., Megharaj, M., Chen, Z.L., Williams, B.D., Lewis, G. and Naidu, R. 2004. Algal degradation of a known endocrine disrupting insecticide, α-endosulfan, and its metabolite, endosulfansulfate, in liquid medium and soil. J. Agri. Food. Chem. 52: 3030–35.

Sharma, A., Gangola, S., Khati, P., Kumar, G. and Srivastava, A. 2016. Novel pathway of cypermethrin biodegradation in a *Bacillus* sp. strain SG2 isolated from cypermethrin-contaminated agriculture field. Biotech. 6(1): 45.

Shi, H., Pei, L., Gu, S., Zhu, S.,Wang, Y., Zhang, Y. and Li,B. 2012. Glutathione S-transferase (GST) genes in the red flour beetle, *Tribolium castaneum*, and comparative analysis with five additional insects. Genomics 100: 327–335.

Shimizu, H. 2002. Metabolic engineering-integrating methodologies of molecular breeding and bioprocess systems engineering. J. Biosci. Bioeng. 94: 563–73.

Strong, L.C., McTavish, H., Sadowsky, M.J. and Wackett, L.P. 2000. Field-scale remediation of atrazine-contaminated soil using recombinant Escherichia coli expressing atrazine chlorohydrolase. Environ. Microbiol. 2: 91–98.

Swackhamer, D.L. and Skoglund, R.S. 1993. Bioaccumulation of PCBs by algae: Kinetics versus equilibrium. Environ. Toxicol. Chem. 12: 831–38.

Tewari, L., Saini, J. and Arti. 2012. Bioremediation of pesticides by microorganisms: General aspectsand recent advances. pp. 25–48. *In*: Maheshwari, D.K. and Dubey, R.C. (eds.). Bioremediation of Pollutants I.K. Int. Publishing House Pvt. Ltd. India.

Top, M.E., Daele, V.P., DeSaeyer, N. and Forney, L.J. 1998. Enhancement of 2, 4-dichlorophenoxyacetic acid (2, 4-D) degradation in soil by dissemination of catabolic plasmids. Anton. Leeuw. 73: 87–94.

Trivedi, V.D., Jangir, P.K., Sharma, R. and Phale, P.S. 2016. Erratum: Insights into functional and evolutionary analysis of carbaryl metabolic pathway from Pseudomonas sp. strain C5pp. Sci. Rep. 6: 38430.

U.N. 2015. United Nations, Department of Economic and Social Affairs, Population Division.

Wang, G., Zhu, D., Xiong, M., Zhang, H. and Liu, Y. 2016 Construction and analysis of an intergeneric fusion from *Pigmentiphaga* sp. strain AAP-1 and *Pseudomonas* sp. CTN-4 for degrading acetamiprid and chlorothalonil. Environ. Sci. Pollut. Res. 23: 13235–44.

Wasilkowski, D., Swedziol, Z. and Mrozik, A. 2012. The applicability of genetically modified microorganisms in bioremediation of contaminated environments. Chemik. 66: 817–826.

Witczak, A. and Abdel-Gawad, H. 2014. Assessment of health risk from organochlorine pesticides residues in high-fat spreadable foods produced in Poland. J. Environ. Sci. Health Biol. 49: 917–28.

Xiao, P., Mori, T., Kamei, I. and Kondo, R. 2011. Metabolism of organochlorine pesticide heptachlor and its metabolite heptachlor epoxide by white rot fungi, belonging to genus *Phlebia*. FEMS Microbiol. Lett. 314(2): 140–146.

Yang, J., Liu, R., Song, W., Yang, Y., Cui, F. and Qiao, C. 2012. Construction of a genetically engineered microorganism that simultaneously degrades organochlorine and organophosphate pesticides. Appl. Biochem. Biotech. 166(3): 590–598.

Zhang, S., Qiu, C.B., Zhou, Y., Jin, Z.P. and Yang H. 2011. Bioaccumulation and degradation of pesticide fluroxypyr are associated with toxic tolerance in green alga *Chlamydomonas reinhardtii*. Ecotoxicol. 20(2): 337–347.

Zhou, S., Pan, Y., Zhang, L., Xue, B., Zhang, A. and Jin, M. 2018. Biomagnification and enantiomeric profiles of organochlorine pesticides in food web components from Zhoushan Fishing Ground, China. Mar. Pollut. Bull. A. 131: 602–61.

13

Compost-assisted Bioremediation of Polycyclic Aromatic Hydrocarbons

N.S. Bolan,[1,*] *Y. Yan,*[2] *Q. Li*[3] *and M.B. Kirkham*[4]

1. Introduction

Polycyclic aromatic hydrocarbons (PAHs) reach the environment from incomplete combustion of organic substances, and heavy industries and transport, which use fossil fuel, are the main anthropogenic sources for contamination by PAHs in terrestrial and aquatic environments (Ghosal et al. 2016). These PAH organic compounds consist of carbon (C) and hydrogen (H) atoms with two or more fused benzene rings. Although a large number of different PAH compounds exist in soil and aquatic environments, only around 16 of these compounds are considered as priority pollutants (Table 1) (Keith 2015, Lukić et al. 2017b). The specific characteristics of PAH compounds are their high hydrophobicity and low water solubility. These characteristics impact their active adsorption to various soil components, thereby affecting their bioavailability and subsequent degradation. PAH contamination can cause various health hazards to humans and other living organisms. For example, some of the PAH compounds, including benzo-anthracene, benzo-fluoranthene, benzo-pyrene, chrysene, dibenzo-anthracene, and indeno-pyrene, are considered as potential human carcinogens. Additionally, chrysene and benzo-pyrene may cause genetic disorders and impair fertility. Similarly, naphthalene, benzo-anthracene, and benzo-pyrene have been shown to cause teratogenicity or embryo toxicity in animals. Bioremediation of PAH contaminated soil is an attractive technology because of lower capital investments, a limited interruption of contaminated site activity, and a green-based environmentally friendly approach compared to other chemically-based remediation treatment technologies (Table 2). Composting of organic contaminants promotes the bioremediation of these compounds.

2. Sources of PAHs

PAHs are distributed extensively in the environment, and a range of PAH compounds have been identified in waste materials, landfill leachates, soils, sediments, groundwater, and in the atmosphere

---

[1] Global Centre for Environmental Remediation (GCER), Advanced Technology Centre, Facultyof Science, The University of Newcastle, Callaghan, Newcastle, NSW, Australia.
[2] School of Chemistry and Chemical Engineering, Huaiyin Normal University, Huai'an 223300, China.
[3] Jiangsu Key Laboratory of Chemical Pollution Control and Resources Reuse, Nanjing University of Science and Technology, Nanjing 210094, China.
[4] Department of Agronomy, 2004 Throckmorton Plant Sciences Center, Kansas State University, Manhattan, KS, United States.
* Corresponding author: nanthi.bolan@newcastle.edu.au

Table 1. Structure and physico-chemical properties of major priority polycyclic aromatic hydrocarbons (PAH) pollutants (modified from Keith 2015, Lukić et al. 2017).

No.	PAH compound	Melting point (°C)	log K_{ow}	Water solubility at 25°C (µgL^{-1})	Vapour pressure (Pa at 25°C)
1	Naphthalene ($C_{10}H_8$)	81	3.00–4.00	3.17×10^4	10.9
2	Acenaphthylene ($C_{12}H_8$)	95	3.70	/	5.96×10^{-1}
3	Acenaphthene ($C_{12}H_{10}$)	96.2	3.92–5.07	3.93×10^3	5.96×10^{-1}
4	Fluorene ($C_{13}H_{10}$)	115–116	4.18	1.98×10^3	8.86×10^{-2}
5	Anthracene ($C_{14}H_{10}$)	218	4.46–4.76	73	2.0×10^{-4}
6	Phenanthrene	100.5	4.45	1.29×10^3	1.8×10^{-2}
7	Fluoranthene ($C_{14}H_{10}$)	108.8	4.90	260	2.54×10^{-1}
8	Pyrene ($C_{16}H_{10}$)	150.4	4.90	135	8.86×10^{-4}
9	Benzo[a]anthracene ($C_{18}H_{20}$)	160.7	5.61–5.70	14	7.3×10^{-6}
10	Chrysene ($C_{18}H_{20}$)	253.8	5.61	2	5.7×10^{-7}
11	Benzo[b]fluoranthene ($C_{20}H_{12}$)	168.3	6.57	1.2^{11} (20°C)	/
12	Benzo[k]fluoranthene ($C_{20}H_{12}$)	215.7	6.84	0.76	/
13	Benzo[a]pyrene ($C_{20}H_{12}$)	178.1	6.04	3.8	8.4×10^{-7}
14	Dibenzo[a,h]anthracene ($C_{22}H_{14}$)	266.6	5.80–6.50	0.5 (27°C)	3.7×10^{-10}
15	Indeno[1,2,3-cd]pyrene ($C_{22}H_{12}$)	163	7.66	62	/
16	Benzo[g,h,i]perylene ($C_{22}H_{12}$)	278.3	7.23	0.26	6×10^{-8}

(Li et al. 2019). Both natural processes and anthropogenic activities contribute to PAH input to the environment. Natural processes such as bush fires and volcanic eruptions contribute to PAH input to the environment. The major anthropogenic activities contributing to PAH input to the environment include partial combustion of fossil fuels in transport, disposal of petroleum hydrocarbon products, and waste incineration. Most of the industries using fossil fuels in their production system, such as petroleum refining and coal gasification, generate PAHs. The concentration of PAHs in contaminated media including soils and sediments depends mainly on the anthropogenic sources of contamination (Soroji et al. 2007).

3. Interactions of PAH

PAHs undergo a number of reactions including adsorption, volatilization, photolysis, and redox reactions, although microbial transformation is the major natural attenuation process of PAH-contaminated sites. Low molecular weight (LMW) PAH compounds, consisting of 2 or 3 rings, have been reported to cause acute toxicity but are not carcinogenic, while high molecular weight (HMW) PAHs, consisting of 4 to 7 rings, are relatively less toxic but may be carcinogenic, mutagenic, or teratogenic (Bauer et al. 2018, Keith 2015, Ghosal et al. 2016). The most important physical and chemical properties of the 16 priority PAH pollutants are reported in Table 1 (Lukić et al. 2016). These properties control the interactions of PAH with soil components and subsequent bioavailability and biodegradation processes. For example, water solubility of PAHs is likely to decrease with increasing number of fused benzene rings, indicating that HMWPAHs are more slowly mobilised from solid substrates and dissolved into water than LMW PAHs, and, therefore, are less subjected to biodegradation (Yamada et al. 2003). Similarly, partitioning based on octanol and water (K_{ow}) is generally used to predict the affinity of an organic pollutant to be retained onto organic substrates. Higher K_{ow} of an organic compound indicates its lower biodegradability and higher potential for its bioaccumulation (Jonker and Vanderheijden 2007). In soils, the partition value of PAHs based on soil organic matter (K_{oc}) indicates the extent of sorption and mobility of these compounds, and

Table 2. Remediation technologies for contaminated soil and sediments (Modified from Lukić et al. 2017).

Remediation treatment	Major contaminants	Primary environmental media	Remediation process
Bioremediation	PAH, TPH, Pesticides	Soil and sediments	Microorganism mineralize the contaminants to a less toxic, environmentally acceptable form.
Phytoremediation	Heavy metals, PAH, TPH, Pesticides	Soil, sediments and groundwater	Higher plants are used to extract, accumulate, sequester, and detoxify contaminants.
Natural attenuation (NA), Monitored natural attenuation (MNA), Enginerred monitored natural attenuation (EMNA)	PAH, TPH, Pesticides	Soil, sediments and groundwater	Native microorganisms are stimulated (i.e., biostimulation) to facilitate the degradation of mostly organic contaminants
Chemical	PAH, PFAS, TPH, Pesticides	Soil, sediments and groundwater	Chemical reactions involving oxidising/reducing agents destroy, decompose, or neutralize contaminants.
Thermal	PAH, PFAS, TPH, Pesticides	Soil and sediments	Heat is employed to destroy contaminants through incineration, gasification, and pyrolysis.
Physical	Heavy metals, PFAS, PAH, TPH, Pesticides	Soil, sediments and groundwater	Contaminated soil is removed to a landfill site or contained at the contaminated site.
Chemical immobilization	Heavy metals, PAH, PFAS, TPH, Pesticides	Soil, sediments and groundwater	Chemical amendments are used to immobilize, thereby reducing the bioavailability of contaminants.
Permeable reactive barrier	Heavy metals, PAH, TPH, Pesticides	Groundwater	Applied mainly to 'filter' contaminants in aquatic media such as surface and groundwater sources.
Solidification/vitrification	Heavy metals, PAH, TPH, Pesticides	Soil and sediments	Solidification refers to the encapsulation of contaminants within a monolithic solid of high structural integrity. Vitrification involves the use of high temperatures using plasma to fuse contaminated material.
Integrated remediation techniques	Heavy metals, PAH, PFAS, TPH, Pesticides	Soil, sediments and groundwater	Multiple remediation technologies can be applied to the degradation of contaminants.

PAH = polycyclic aromatic hydrocarbons; PFAS = Poly- and perfluoroalkyl substances; TPH = total petroleum hydrocarbon.

the higher the K_{oc} the stronger the partition onto soil organic matter rather than mobilization in the aqueous phase (Zhang et al. 2009, Appert-Collin et al. 1999). Furthermore, the electrochemical stability, resistance toward biodegradation, persistency in environmental media, and carcinogenic index of PAHs increase with increasing number of aromatic rings and hydrophobicity, while volatility of PAHs is likely to decrease with increasing molecular weight (Kanaly and Hariyama 2000, Ghosal et al. 2016).

Most of the PAHs are hydrophobic, leading to a high adsorption onto to soil organic matter and consequent persistency in soils and sediments (Zhang et al. 2017). The major pathways of PAH exposure include ingestion, inhalation, and dermal contact, thereby impacting human and animal health. The bioavailability of PAHs in environmental substrates, such as soils and sediments, is influenced by both the extent of adsorption and also the contact period between substrates and contaminants. The residence period in soils (i.e., ageing) allows the diffusion of contaminants into soil micropores, leading to their ready incorporation into stable phases and decreasing the mobility and bioavailability of contaminants. Bioaccumulation refers to the tendency of PAHs to accumulate in the tissue of organisms resulting from the exposure to a contaminated medium, or by ingestion of contaminated food sources. The bioaccumulation factor (BAF), which refers to the ratio of contaminant concentration in an organism to that in the ambient environmental substrates such as soil and sediment, is usually used to predict the potential for uptake and accumulation and subsequent monitoring of contaminant hazard to human and ecosystem health (Chapman et al. 1996, Feijtel et al. 1997, Sample et al. 1999).

4. Bioremediation process in soil

Bioremediation refers to a microbial process applied to remediate contaminated media, including water, soil, and sediments, by optimising environmental conditions to stimulate the growth of microorganisms, thereby degrading the target pollutants. Bioremediation of a contaminant can be achieved directly through enzymatic reactions by stimulating and bioagumenting the microorganism or indirectly through non-enzymatic reactions by the microbially-induced changes in soil properties impacting contaminant degradation.

The direct enzymatic bioremediation process covers a number of reactions including mineralisation, co-metabolism, polymerisation, and bioaccumulation (Luo et al. 2014). The in-direct non-enzymatic reactions include microbially-induced changes in pH and redox potential that impact the abiotic and biotic degradation processes. In the mineralization process, the contaminants can serve as a substrate for microbial growth, thereby facilitating their degradation. Mostly, the carbon in the contaminants is used as a source of energy and released as carbon dioxide (CO_2). In the co-metabolism process, the contaminants do not serve as a direct source of energy but are transformed by metabolic reactions by the microorganisms. It is the most prevalent form of degradation of organic contaminants. Polymerisation is a microbially mediated, oxidative coupling reaction by which a contaminant or its intermediate compound combines with itself or with other residues and naturally occurring compounds to form a larger molecular polymer. In the bioaccumulation process, the contaminants are incorporated into microorganisms by both active uptake and physical absorption. Bioaccumulation can readily be used to remove contaminants in aquatic environments.

In non-enzymatic reactions, the microbes alter the environmental parameters, such as pH, redox potential, and salt concentration, which promote the secondary or non-enzymatic transformations of contaminants. For example, acidic pH is created by bacteria during the nitrification reaction (equation 1) and elemental sulphur oxidation (equation 2) reaction in soil (Bolan et al. 2003).

$$2NH_4 + 3O_2 \rightarrow 2NO_3 + 8H^+ \tag{1}$$

$$2S^0 + 3O_2 + 2H_2O \rightarrow 2H_2SO_4 \tag{2}$$

Therefore, when these elements (i.e., N and S) are added to stimulate microbial activity and functions, the change in pH is likely to impact the reactions of contaminants and their subsequent bioavailability and degradation. For example, after a period of exposure to organic contaminants, such as pesticides, certain groups of microorganisms proliferate rapidly and the pesticide is metabolised. The increase in the proportion of the functional-specific biochemically active species within the total microbial population is known as the enrichment effect. The characteristic of enrichment effect in soil is the accelerated degradation of subsequent applications of pesticides without a lag period (i.e., enhanced degradation). Under natural conditions, mobilization of organic contaminants from soil and sediments, and their subsequent biodegradation, are dependent on the activity of indigenous microorganisms. Biodegradation is primarily influenced by the metabolic activity of microorganisms that degrade organic contaminants into non-toxic substances, which are then assimilated into natural biogeochemical cycles (Ghattas et al. 2017).

5. Bioremediation of PAHs through composting

Composting is a biological process used to decompose organic solid substrates and to convert them into a beneficial soil amendment rich in bioavailable carbon (humic substances) and nutrients. The composting process has generated interest because it can be used in the bioremediation of soils and sediments as a means to accelerate the biodegradation and subsequent removal of organic contaminants including PAHs (Lukić et al. 2016). Composting is a biochemical process based on the ability of microorganisms to decompose organic substrates, resulting in the release of heat, CO_2, and water, along with biologically stable material (i.e., compost) (Hubble et al. 2010). A number of processes are involved in the removal of organic contaminants including PAHs during the co-composting of contaminated substrates with organic solid materials (Figure 1) (Lukić et al. 2016). These include: (i) an increase in temperature during the composting process; (ii) the addition of carbon and nutrient sources (biostimulation effect); and (iii) the introduction of a wide number of microorganisms capable of degrading organic compounds (bioaugmentation effect). The elevated temperature during the composting process can accelerate the biochemical kinetics involved in the biodegradation process, including the solubility and mass transfer rate of contaminants, thus facilitating the ability of the contaminants to become more accessible and bioavailable to microorganisms for their metabolism and subsequent degradation of contaminants (Haritash and Kaushik 2009).

Composting is also used as a biostimulation and bioaugmentation strategy for PAH contaminated soils. Bioremediation of PAH contaminated soils and sediments by indigenous microorganisms can be stimulated by incorporating organic materials (Table 3) (Mattei et al. 2016). The new microorganisms in the compost also enhance the bioaugmentation process of bioremediation. Composting of organic waste using the biostimulation and bioaugmentation strategy would improve microbial diversity, introduce new microorganisms, enrich the activity of microorganisms in contaminated environments, and increase the supply of moisture and nutrients (Lukić et al. 2017a, b).

Biostimulation of microbiological process in contaminated soils is achieved by promoting optimum environmental conditions, including using the proper pH for microbial growth and adding nutrients to the soil. Microorganisms effectively promote the metabolization of contaminants under favourable environmental conditions for their growth. Biostimulation has been applied successfully

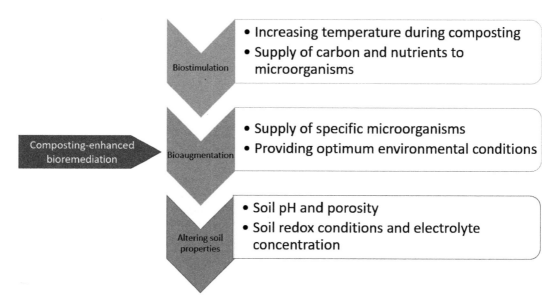

Figure 1. Processes involved in compost-assisted bioremediation of organic contaminants.

Table 3. Selected references on the bioremediation of polycyclic aromatic hydrocarbons (PAHs) with the addition of organic amendments (modified from Lukić et al. 2017).

Contaminants	Organic waste	Soil	Bioremediation process details	Reference
Pyrene	Composted sewage sludge	Spiked soil	Moisture content 40%, temperature was raised from 20°C to 60°C by 5°C day^{-1}, 21 days	Adenuga et al. 1992, Antizar-Ladislao et al. 2004
Naphthalene, acenaphthene, fluorene, phenanthrene, anthracene, fluoranthene, pyrene, benzo[a]anthracene, chrysene	Municipal solid wastes and fertilizer	Creosote-Contaminated soil	Constant temperature at 45°C, 15 days	Antizar-Ladislao et al. 2004, Civilini 1994, Semple et al. 2001
Diesel oil	Biowaste – vegetable, fruit, and garden waste	Diesel oil-spiked soil	Soil-to-biowaste ratio 1:10, 12 weeks	Van Gestel et al. 2003
16 PAHs	Green tree waste and manure	Tar residues contaminated soil	Field scale, green tree waste (< 5 days old), manure:soil ratio 15:5:80, moisture 60%–80% field capacity, temperature reached maximum at 42°C after 35 days, 224 days	Antizar-Ladislao et al. 2004, Guerin 2000
Anthracene, phenanthrene	Sewage sludge, sterilized sludge, sludge to maintain pH, glucose plus N and P source	Soil of the former lake Texcoco	Temperature 22°C, 112 days	Fernández-Luqueño et al. 2008
19 PAHs	Cow manure, modified fertilizer, and activated sewage sludge	Reilly soil*– creosote manufacturing and wood preserving	C:N:P 100:5:1, corn cobs bulking agent, moisture 30%–35%, temperature 41°C to 53°C in the first 15 days and subsequently decreased to ambient temperature, 84 days	Antizar-Ladislao et al. 2004, Potter et al. 1999
16 PAHs	Green waste	Aged coal tar contaminated soil	Constant temperature at 38°C, 55°C, and 70°C separately, and comparative studies using a temperature profile, moisture content of 40%, 60%, and 80% field capacity, soil-to-green waste ratio 0.6:1, 0.7:1, 0.8:1, and 0.9:1, 8 weeks	Antizar-Ladislao et al. 2005, Sayara et al. 2010
PAHs	Tall fescue, arbuscular mycorrhizal fungus and epigeic earthworms	Former gasworks soil	Temperature 28/21°C (day/night), N:P$_2$O$_5$:K$_2$O = 1:0.35:0.8, 60% of field water holding capacity, 120 days	Lu and Lu 2015

Table 3 Contd. ...

...Table 3 Contd.

Contaminants	Organic waste	Soil	Bioremediation process details	Reference
Anthracene, benzo[a]pyrene	Compost	Industrial soil	Temperature 25°C, dark room, air humidity 70%, 274 days	Baldantoni et al. 2017
PAHs	Compost, biochar	Spiked soil	Temperature 20°C, dark, 50% of the soil's maximum water holding capacity, 120 days	Sigmund et al. 2018
16 PAHs	Green compost, meat compost	Spiked soil, coal tar contaminated soil, coal ash contaminated soil	Ratio of compost to soil 250 and 750 tha^{-1}, incubation 3, 6 and 8 months	Wu et al. 2014
PAHs	Sewage sludge, compost	Uncultivated agricultural soil	Temperature 21/18°C (day/night), air humidity = 75–80%, day-night cycle 16 h_{day}/8 h_{night}	Włóka et al. 2017
PAHs	Sewage sludge	Heavily contaminated soil with creosote (> 310,000 mg kg^{-1})	Field scale, moisture content 70% field capacity, aeration every 2 weeks, C:N:P ratio 25:1:1, 10 months	Atagana 2004

* *The Reilly series consists of very deep, excessively drained, rapidly permeable soils that formed in stratified alluvial deposits of mixed origin. These soils are in river valleys.*

to promote the degradation and subsequent removal of PAHs from soils and sediments (Table 3) (Murrieta et al. 2016). For example, Straube et al. (2003) reported PAH removal efficiency of 86% with biostimulation using ground rice husks as a bulking agent and dried blood as a slow-release nutrient source. Similarly, Atagana (2004) demonstrated that co-composting PAH contaminated soil with poultry manure resulted in a rapid degradation of a number of PAH compounds. For example, in the presence of poultry manure, the 2- and 3-ring PAHs (naphthalene, anthracene, phenanthrene, and pyrrole) were removed below the remediation target of 1 mg/kg within four months, whereas in the absence of poultry manure it took more than 16 months to reach the remediation target (Figure 2). In the case of 4- and 5-ring PAHs (pyrene, chrysene, fluoranthene, and benzo(a)pyrene), while degradation continued very slowly in the control, degradation in the compost system increased rapidly.

Composting of contaminants, including PAHs in the organic residues, can also facilitate bioaugmentation. Bioaugmentation involves the introduction of specific but benign microorganisms that are efficient at degrading contaminants from environmental media including soil, sediments, and water (Lukić et al. 2016). In soils with high PAH concentrations, the indigenous microorganisms are generally low, and, hence, bioaugmentation with microorganisms through composting can promote and accelerate the degradation process. The performance of bioaugmentation in promoting biodegradation of contaminants in soils depends on a number of factors, including the viability of introduced microorganisms, the environmental conditions of the polluted soil promoting the functional activity of the microorganisms, and the bioavailability of contaminants (Mrozik and Piotrowska-Seget 2010). The viability of the introduced microorganisms can be impacted from the competition with indigenous microorganisms and also the presence of co-contaminants in the composting substrate that could be toxic to added strains. Therefore, it is important to use indigenous

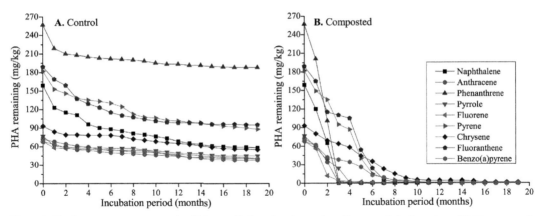

Figure 2. Poultry manure compost-assisted bioremediation of various polycyclic aromatic hydrocarbons (PAH) compounds (Atagana 2004).

microorganisms isolated from the contaminated soil and sediments, which are tolerant to the target contaminant being degraded (Abatenh et al. 2017).

The compost-integrated biodegradation efficiency of PAHs in soils and sediments depends on the physical, chemical, and biological properties of the contaminated matrix, organic compost substrate characteristics, and environmental conditions (Lukić et al. 2016, Poluszyńska et al. 2017). Additionally, the composting substrate enriches the contaminated soil with microorganisms derived from the composting substrate, and it also increases the moisture retention capacity of soil, thereby facilitating the bioremediation of PAHs (Lukić et al. 2016). The compost-integrated bioremediation approach also facilitates the eco-friendly disposal of organic waste used as a composting substrate, since the waste is simultaneously decomposed (Potter et al. 1999, Lukić et al. 2016). The organic amendments used as a compost substrate are likely to improve the soil structure and oxygen transfer and provide an additional nutrient and carbon source for the microorganisms. A number of studies have demonstrated that addition of an organic compost substrate increases the capability of microorganisms for degrading PAHs (Table 3).

In the compost-integrated bioremediation process, most of the PAH degradation has been shown to occur during the active thermo-composting phase and very little degradation occurs during the final curing phase of the composting process (Kästner and Miltner 2016, Lukić et al. 2016). For example, Antizar-Ladislao et al. (2004) have demonstrated that PAH degradation in soil during composting is more efficient with the addition of fresh organic substrates compared to the addition of mature compost, which may be attributed to the active composting of the fresh organic substrates, thereby facilitating the degradation of the contaminant.

6. Conclusions

This chapter provides an overview into the application of composting as a suitable technology to improve the biodegradation of PAHs in soil and sediments. Composting leads to both stimulation and augmentation of the microbial community, thereby facilitating the enzymatic and non-enzymatic degradation of organic contaminants including PAHs. Enzymatic degradation occurs through mineralisation, co-metabolism, polymerisation, and bioaccumulation. In non-enzymatic reactions, the microbes alter the environmental conditions, such as pH, redox potential, and salt concentration, which are conducive to the secondary non-enzymatic transformations of contaminants. Under field conditions, the characteristics of contaminated and organic compost substrates, including co-contaminants, and indigenous microbial communities are critical factors that affect the performance of compost-integrated bioremediation of PAH compounds. Chemical recalcitrance of some of the HMW PAHs to biodegradation can limit the implementation of compost-facilitated bioremediation technology. The concentrations of some of the HMW PAHs in the contaminated substrate may increase after the compost-integrated bioremediation process, especially when a considerable amount of the solid organic compost matrix is decomposed while the recalcitrant HMW PAHs are not degraded by the microorganisms. This could hamper the application of compost-integrated bioremediation of HMW recalcitrant PAHs.

References

Abatenh, E., Gizaw, B., Tsegaye, Z. and Wassie, A. 2017. Application of microorganisms in bioremediation—review. J. of Environ. Microbiol. 1: 2–9.

Adenuga, A.O., Johnson Jr., J.H., Cannon, J.N. and Wan, L. 1992. Bioremediation of PAH contaminated soil via in-vessel composting. Water Science and Technol. 26: 2331–2334.

Antizar-Ladislao, B., Lopez-Real, J.M. and Beck, A.J. 2004. Bioremediation of polycyclic aromatic hydrocarbon (PAH)-contaminated waste using composting approaches. Critical Reviews in Environmental Science and Technol. 34: 249–289.

Antizar-Ladislao, B., Lopez-Real, J. and Beck, A.J. 2005. In-vessel composting-bioremediation of aged coal tar soil: Effect of temperature and soil/green waste amendment ratio. Environment Int. 31: 173–178.

Appert-Collin, J.C., Dridi-Dhaouadi, S., Simonnot, M.O. and Sardin, M. 1999. Nonlinear sorption of naphtalene and phenanthrene during saturated transport in natural porous media. Physics and Chemistry of the Earth, Part B: Hydrology, Oceans and Atmosphere 24: 543–548.

Atagana, H.I. 2004. Co-composting of PAH-contaminated soil with poultry manure. Letters in Appl. Microbiology 39: 163–168.

Bauer, A.K., Velmurugan, K., Plöttner, S., Siegrist, K.J., Romo, D., Welge, P., Brüning, T., Xiong, K. and Käfferlein, H.U. 2018. Environ. prevalent polycyclic aromatic hydrocarbons can elicit co-carcinogenic properties in an *in vitro* murine lung epithelial cell model. Arch. of Toxicol. 92: 1311–1322.

Bolan, N.S., Adriano, D.C. and Curtin, D. 2003. Soil acidification and liming interactions with nutrient and heavy metal transformation and bioavailability. Adva. in Agronomy 78: 215–272.

Chapman, P.M., Allen, H.E., Godtfredsen, K. and Z'Graggen, M.N. 1996. Evaluation of bioaccumulation factors in regulating metals. Environmental Science & Technol. 30: 448–452.

Cheruiyot, N.K., Lee, W., Mwangi, J.K., Wang, L., Lin, N., Lin, Y., Cao, J., Zhang, R. and Chang-Chien, G. 2015. An overview: polycyclic aromatic hydrocarbon emissions from the stationary and mobile sources and in the ambient air. Aerosol and Air Quality Res. 15: 2730–2762.

Civilini, M. 1994. Fate of creosote compounds during composting. Microbiol. Europ. 2: 16–24.

Feijtel, T., Kloepper-Sams, P., den Haan, K., van Egmond, R., Comber, M., Heusel, R., Wierich, P., Ten Berge, W., Gard, A., de Wolf, W. and Niessen, H. 1997. Integration of bioaccumulation in an environment of risk assessment. Chemosphere 34: 2337–2350.

Fernández-Luqueño, F., Marsch, R., Espinosa-Victoria, D., Thalasso, F., Hidalgo Lara, M.E., Munive, A., Luna-Guido, M.L. and Dendooven, L. 2008. Remediation of PAHs in a saline alkaline soil amended with wastewater sludge and the effect on dynamics of C and N. Sci. of The Total Environ. 402: 18–28.

Ghattas, A.K., Fischer, F., Wick, A. and Ternes, T.A. 2017. Anaerobic biodegradation of (emerging) organic contaminants in the aquatic environment. Water Res. 116: 268–295.

Ghosal, D., Ghosh, S., Dutta, T.K. and Ahn, Y. 2016. Current state of knowledge in microbial degradation of polycyclic aromatic hydrocarbons (PAHs): A review. Front. in Microbiol. 7: 1369.

Guerin, T.F. 2000. The differential removal of aged polycyclic aromatic hydrocarbons from soil during bioremediation. Environ. Sci. and Poll. Res. 7: 19–26.

Haritash, A.K. and Kaushik, C.P. 2009. Biodegradation aspects of polycyclic aromatic hydrocarbons (PAHs): a review. J. of Hazardous Materials 169: 1–15.

Jonker, M.T.O. and Vanderheijden, S.A. 2007. Bioconcentration factor hydrophobicity cutoff: An artificial phenomenon reconstructed. Environ. Sci. & Technol. 41: 7363–7369.

Juwarkar, A.A., Singh, S.K. and Mudhoo, A. 2010. A comprehensive overview of elements in bioremediation. Reviews in Environ. Sci. and Bio-Technol. 9: 215–288.

Kanaly, R.A. and Harayama, S. 2000. Biodegradation of high-molecular-weight polycyclic aromatic hydrocarbons by bacteria. J. Bacteriology 182: 2059–2067.

Kästner, M. and Miltner, A. 2016. Application of compost for effective bioremediation of organic contaminants and pollutants in soil. Appl. Microbiology and Biotechnol. 100: 3433–3449.

Keith, L.H. 2015. The source of U.S. EPA's Sixteen PAH priority pollutants. Polycyclic Aromatic Compounds 35: 2–4.

Li, C., Zhang, X., Gao, X., Qi, S. and Wang, Y. 2019. The potential environmental impact of PAHs on soil and water resources in air deposited coal refuse sites in Niangziguan Karst Catchment, Northern China. Int. J. of Environ. Res. and Public Health 16: 1368.

Lukić, B., Huguenot, D., Panico, A., van Hullebusch, E.D. and Esposito, G. 2017a. Influence of activated sewage sludge amendment on PAH removal efficiency from a naturally contaminated soil: application of the landfarming treatment. Environmental Technol. 38: 2988–2998.

Lukić, B., Panico, A., Huguenot, D., Fabbricino, M., vanHullebusch, E.D. and Esposito, G. 2017b. A review on the efficiency of landfarming integrated with composting as a soil remediation treatment. Environmental Technol. Reviews 94–116.

Luo, W., Zhu, X., Chen, W., Duan, Z., Wang, L. and Zhou, Y. 2014. Mechanisms and strategies of microbial cometabolism in the degradation of organic compounds—chlorinated ethylenes as the model. Water Science and Technol. 69: 1971–1983.

Mattei, P., Cincinelli, A., Martellini, T., Natalini, R., Pascale, E. and Renella, G. 2016. Reclamation of river dredged sediments polluted by PAHs by co-composting with green waste. Science of The Total Environment 566/567: 567–574.

Mrozik, A. and Piotrowska-Seget, Z. 2010. Bioaugmentation as a strategy for cleaning up of soils contaminated with aromatic compounds. Microbiological Res. 165: 363–375.

Poluszyńska, J., Jarosz-Krzemińska, E. and Helios-Rybicka, E. 2017. Studying the effects of two various methods of composting on the degradation levels of polycyclic aromatic hydrocarbons (PAHs) in sewage sludge. Water, Air, & Soil Poll. 228: 305.

Potter, C.L., Glaser, J.A., Chang, L.W., Meier, J.R., Dosani, M.A. and Herrmann, R.F. 1999. Degradation of polynuclear aromatic hydrocarbons under bench-scale compost conditions. Environmental Science & Technol. 33: 1717–1725.

Sample, B.E., Beauchamp, J.J., Efroymson R. and Suter, G.W. 1999. Literature-derived bioaccumulation models for earthworms: Development and validation. Environ. Toxicol. and Chemistry 18: 2110–2120.

Sayara, T., Sarrà, M. and Sánchez, A. 2010. Effects of compost stability and contaminant concentration on the bioremediation of PAHs-contaminated soil through composting. J. of Hazardous Materials 179: 999–1006.

Sayara, T., Sarrà, M. and Sánchez, A. 2010. Optimization and enhancement of soil bioremediation by composting using the experimental design technique. Biodegradation 21: 345–356.

Van Gestel, K., Mergaert, J., Swings, J., Coosemans, J. and Ryckeboer, J. 2003. Bioremediation of diesel oil-contaminated soil by composting with biowaste. Environ. Poll. 125: 361–368.

Yamada, M., Takada, H., Toyoda, K., Yoshida, A., Shibata, A., Nomura, H., Wada, M., Nishimura, M., Okamoto, K. and Ohwada, K. 2003. Study on the fate of petroleum-derived polycyclic aromatic hydrocarbons (PAHs) and the effect of chemical dispersant using an enclosed ecosystem, mesocosm. Marine Pollution Bulletin 47: 105–113.

Zhang, J., Zeng, J. and He, M. 2009. Effects of temperature and surfactants on naphthalene and phenanthrene sorption by soil. Journal of Environ. Sci. 21: 667–674.

Zhang, S., Yao, H., Lu, Y., Yu, X., Wang, J., Sun, S., Liu, M., Li, D., Li, Y. and Zhang, D. 2017. Uptake and translocation of polycyclic aromatic hydrocarbons (PAHs) and heavy metals by maize from soil irrigated with wastewater. Sci. Reports 7: 12165.

14

Petroleum Hydrocarbon-Contaminated Soils
Scaling Up Bioremediation Strategies from the Laboratory to the Field

José A. Siles

1. Introduction

Petroleum (crude oil) and its derivatives are still the main source of energy for most of the countries worldwide and is commonly used as a raw material in industrial processes (Kotowicz et al. 2010). The importance of this industry is such that the global crude oil production in 2018 was 100.71 million barrels per day and it is estimated that 102.22 million barrels of oil per day will be needed in 2020 (ttps://www.eia.gov/outlooks/steo/marketreview/crude.php).

Petroleum is mainly a mixture of hydrocarbons, although it also contains other elements such as sulfur, oxygen, and nitrogen. The compounds found in petroleum can be classified into four different categories: (i) saturates (aliphatics), (ii) aromatics, (iii) resins, and (iv) asphaltenes (Figure 1). Saturates are defined as hydrocarbons without double bonds and represent the highest percentage of crude oil constituents. They can further be classified as straight-chain (i.e., n-alkanes), branched-chain, and cyclic compounds. Aromatics are hydrocarbon molecules that have one or several aromatic rings that may be substituted with different alkyl groups and are divided into: (i) monocyclic aromatic hydrocarbons (i.e., benzene, toluene, ethylbenzene, and xylene (BTEX)), and (ii) polycyclic aromatic hydrocarbons (PAHs) (e.g., anthracene, naphthalene, phenanthrene, or pyrene (Figure 1)) (Varjani 2017). Both resins and asphaltenes contain non-hydrocarbon polar compounds and have very complex and mostly unknown carbon (C) structures (Figure 1). Resins consist of compounds containing nitrogen, sulfur, and oxygen that are dissolved in oil, while asphaltenes are large and complex molecules that are colloidally dispersed in oil (Nelyubov et al. 2017). They both constitute around 10% of crude oil composition (Koshlaf and Ball 2017) (Figure 1). The exact chemical composition of crude oil varies depending on the place where it is obtained from (Chandra et al. 2013). Depending on the origin, most of the crude oil components are biodegradable and only a small part is recognized as recalcitrant or not biodegradable (dos Santos and Maranho 2018). PHC biodegradability tends to decrease in the following order: n-alkanes > branched-chain alkanes > branched alkenes > low-molecular-weight n-alkyl aromatics > monoaromatics > cyclic alkanes > PAHs > asphaltenes and resins (Gkorezis et al. 2016).

Department of Plant & Microbial Biology, University of California at Berkeley, Berkeley, CA, USA.
Email: josesimartos@gmail.com

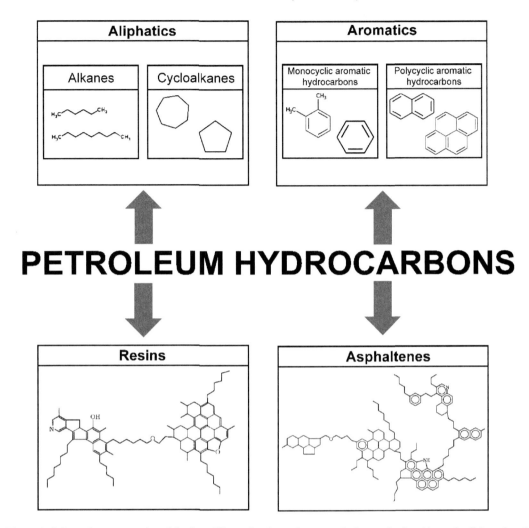

Figure 1. Schematic representation of the four different fractions of compounds that can be found in crude oil (petroleum).

The extensive use of petroleum and its derivatives (mainly gasoline, creosote, and diesel) is negatively impacting the environment (Hussain et al. 2018). Many places worldwide are being affected by PHC contamination as a consequence of industrial runoffs, effluent releases and accidents' spills that occur during the following petroleum industry activities: (i) works of exploration and production; (ii) labors of refining and marketing; and (iii) works of transporting (Brassington et al. 2007, Yavari et al. 2015, Gkorezis et al. 2016). PHC contamination is happening in environmental compartments such as soils (Arias Espana et al. 2018, dos Santos and Maranho 2018, Siles and Margesin 2018, Chaudhary and Kim 2019), groundwaters (Logeshwaran et al. 2018), marshes (Jackson and Pardue 1997, Reddy et al. 2002), shores (Mearns 1997, Wang et al. 2007), and oceans (Cripps 1992, Paniagua-Michel and Fathepure 2018). PHCs are considered as the most widespread class of organic contaminants worldwide. As an example, it has been estimated that approximately $0.8 \pm 0.4\%$ of the entire world production of petroleum eventually reaches oceans (Freitas et al. 2016). Additionally, several compounds derived from crude oil are classified as priority pollutants by the United States Environmental Protection Agency (Siles and Margesin 2018). In humans, chronic PHC exposure can produce damages on the central nervous system, alterations in the endocrine and respiratory systems, and even cancer (Gkorezis et al. 2016).

Soil is one of the ecosystems where PHC contamination can have more severe consequences. After an oil product spill, the behavior of PHCs through the soil profile is different according to their nature (chemical composition) and quantity (volume of the spill), the soil structure, the content and composition of soil organic matter as well as the chemical processes that may occur (Cocârță et al. 2017). Soluble PHCs impact soil properties to a lower extent since they are more susceptible to changes due to degradation, volatilization, and filtration and usually reach groundwater rapidly (Mena et al. 2016). Instead, the PHCs that are more absorbed and adsorbed by the soil organic matter and other soil particles are the most resistant to losses or alterations by other processes as well as those that affect soil properties to a higher extent (Streche et al. 2018). From a physical point of view, it has been reported that the daily maximum surface temperature of PHC-contaminated soils is often higher than that of adjacent uncontaminated soils (Aislabie et al. 2004). Additionally, PHCs cause soil anaerobic conditions by smothering soil particles and blocking air diffusion in the soil pores (Wang et al. 2013) as well as soil hydrophobicity and water repellency, which adversely affect hydrological and ecological soil functions by decreasing infiltration and increasing surface runoff and erosion (Gordon et al. 2018). PHC-induced hydrophobicity is highly persistent, with soils remaining hydrophobic many years after contamination. It has been reported that soils without visible indication of contamination (such as color or odor) and with low PHC contents exhibit severe water repellency and hydrophobicity (Adams et al. 2008). From a chemical point of view, soil PHC-pollution provokes an increase in the content of C, unbalancing C:N (carbon:nitrogen) and C:N:P (carbon:nitrogen:phosphorus) ratios, and alterations in pH as well as in other soil chemical properties (Wang et al. 2010, Devatha et al. 2019). From a biological perspective, PHCs directly or indirectly (as a consequence of the shifts in soil physico-chemical properties) impact soil microorganisms and animals in terms of abundance, diversity and taxonomic composition (Siles and García-Sánchez 2018). In the case of plants, increased PHC contents in soil can produce alterations in their physiology in terms of stem height, photosynthetic rate, overall biomass and germination, leading ultimately to their death (Borah 2018). In the most extreme cases, pollution with PHCs causes infertile soils (Hussain et al. 2018). Soil PHC contamination supposes a great problem in absolute terms. For instance, about 40,000 barrels of crude oil are yearly spilled as a result of pipeline failures and more than 200,000 from underground storage tanks in the United States. Oil spill alone accounts for ~ 15% of all pollution incidents in England with about nine incidents per day resulting in one million tons of oil spillage into terrestrial ecosystems every year in this country (Hazra et al. 2012).

Since humankind is greatly dependent on services provided by soils and PHC contamination causes a great impact on quality services of these ecosystems, the need of finding remediation strategies is thus urgent. Nevertheless, the issue of PHC-contaminated soils and their remediation are among the most challenging tasks in the field of environmental protection from both financial and organizational perspectives (Streche et al. 2018).

Physical (e.g., centrifugation, soil washing, soil flushing, soil vapor extraction, wet classification, encapsulation, or photocatalysis), chemical (e.g., chemical oxidation or electrokinetics), and thermal (incineration, thermal desorption, wet oxidation, vitrification, or supercritical oxidation) treatments have been developed and applied (with varying success rates) to recover PHC-contaminated soils (Kumar and Yadav 2018). Nevertheless, these approaches involve important physical disturbances of the site to be decontaminated and most of them involve changing the pollution from its original phase to another one, creating new ways of contamination such as air pollution (Kumar and Yadav 2018). As an alternative, bioremediation has shown to be an efficient and environmentally sound technology for cleaning up PHC-polluted soils (Varjani 2017, Wu et al. 2017). In the present chapter, I aim at giving an overview of the main approaches used for bioremediation of PHC-contaminated soils and discuss how most of our current knowledge on soil bioremediation is based on laboratory-scale studies. Finally, I review the state of the art of bioremediation of PHC-contaminated soils at field scale.

2. Bioremediation

Soil bioremediation can be defined as "the use of biologically mediated processes to detoxify, degrade or transform pollutants to an innocuous state". This approach is based on the capacity of many microorganisms to use hydrocarbons as a source of C and energy (biodegradation), transforming or mineralizing these pollutants into less harmful or nonhazardous substances, which are then incorporated into natural biogeochemical cycles. Zobell (1946) was the first who demonstrated the microbial ability of using PHCs as C source and the wide distribution of these microorganisms in nature. The advantages of bioremediation with respect to other approaches are that: (i) it is cost-effective, versatile, and easy to carry out; (ii) it uses biological inputs such as microbes (making this technique ecofriendly); and (iii) there are no side effects (Kumar and Yadav 2018).

Each soil ecosystem has an intrinsic PHC-biodegradation capacity, which is known as natural attenuation (combined result of chemical oxidation, photo-oxidation, evaporation, and microbial mineralization). This means that bioremediation can occur on its own via natural attenuation in any soil, but it would take a long time in most of the cases (Polyak et al. 2018). Three groups of factors limit soil PHC-biodegradation: (i) the characteristics of the indigenous microbial communities (in terms of taxonomy, gene regulation and expression, metabolic diversity, tolerance to metals and other toxic xenobiotics, substrate uptake or adherence mechanisms, chemotaxis, and biofilm formation); (ii) environmental conditions (i.e., pH, temperature, pressure, moisture, salinity, and availability of nutrients and terminal electron acceptor groups); and (iii) chemical and physico-chemical properties of the PHCs (i.e., solubility, concentration, hydrophobicity, volatility, and molecular mass) (Gkorezis et al. 2016). Bioremediation seeks the way to stimulate and/or improve the biodegradation potential of a soil. Examples of approaches developed for the improvement of this potential are: phytoremediation, bioaugmentation, and biostimulation. Bioremediation of PHC-contaminated soils has greatly attracted the interest of environmental scientists and engineers as demonstrated by the high number of scientific articles published in the last 20 years (period 1998–2018) on this topic (Figure 2).

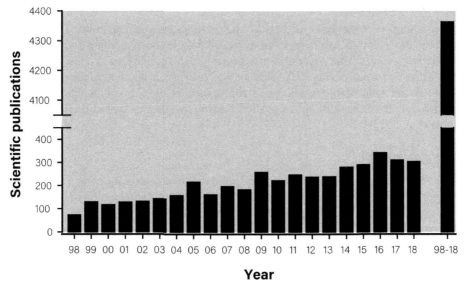

Figure 2. Number of publications found on Scopus database using the searching terms "soil AND bioremediation AND hydrocarbon" in the period 1998–2018.

2.1 Phytoremediation

Phytoremediation consists in planting and growing living plants in PHC-contaminated soils. A useful plant for bioremediation purposes should meet as many as possible of the following requirements: (i) a fibrous root system; (ii) its biomass above the ground shouldn't be consumed by animals; (iii) tolerant to high PHC concentrations; (iv) fast growing rate; (v) versatile to growth under different environmental conditions; (vi) quick capacity to reach the required level of PHC degradation; and (vii) resistance toward diseases and pests (Kumar and Yadav 2018). Some examples of plants showing to be useful for phytoremediation purposes include: (i) native/wild grasses (e.g., Italian ryegrass (*Lolium multiflorum*), forage grass (*Brachiaria brizantha*), tall fescue (*Festuca arundinacea*), or smooth flatsedge (*Cyperus laevigatus*)); (ii) legumes (e.g., white clover (*Trifolium repens*), broad bean (*Vicia faba*), medik (*Melilotus albus*), or alfalfa (*Medicago sativa*)); (iii) agronomic crops (e.g., rice (*Oryza sativa*), wheat (*Triticum aestivum*), maize (*Zea mays*), or Chinese cabbage (*Brassica chinensis*)); (iv) ornamental plants (e.g., blanket flower (*Gaillardia aristata*), devil's beggarticks (*Bidens frondosa*), or purple coneflower (*Echinacea purpurea*)); and (v) shrubs and trees (e.g., flame tree (*Delonix regia*), poplar (*Populus deltoides* x *nigra*), lebbeck (*Albizia lebbeck*), or cassod tree (*Cassia siamea*)) (Correa-García et al. 2018, Hussain et al. 2018, Hunt et al. 2019). Phytoremediation is a cost-effective technology driven by solar energy and results in a natural cleaning of the environment that can be conducted either *in situ* or *ex situ*. Plants can clean up a soil using five different mechanisms (Fernández-Luqueño et al. 2017):

- Phytoextraction: plants transport PHCs from the soil via roots and accumulate them into the shoots.
- Phytodegradation: PHCs are taken up by the plants and degraded in their tissues. Plants can also synthesize and release enzymes into the soil, which degrade PHCs.
- Phytovolatilization: plants take PHCs up, convert them into nontoxic volatile substances, and subsequently release them into the atmosphere.
- Phytostabilization: bioavailability of the PHCs gets reduced because they are bound by the plants and immobilized.
- Phytostimulation: abundance, diversity and PHC-degrading capability of microorganisms living in the plant rhizosphere is stimulated through the exudation by roots of different organic compounds. This process is also known as rhizoremediation. Rhizosphere also plays other important roles during PHC-bioremediation since it has been recognized as a hotspot for horizontal gene transfer and some root exudates have been shown to detach organic contaminants from the soil organic matter, making them more available to microbes (Correa-García et al. 2018).

2.2 Bioaugmentation

Bioaugmentation seeks to enhance the biodegradation capability of a soil by inoculating it with exogenous, endogenous, or constructed (genetically engineered) hydrocarbon-degrading microorganisms (Masy et al. 2016, Wu et al. 2016, Siles and Margesin 2018). This approach has been recognized as the most suitable strategy to recover soils with poor microbial communities in PHC degraders in order to improve their functionality (Kuppusamy et al. 2017). Microorganisms used for bioaugmentation should: (i) be fast-growing and easy to culture; (ii) be effective in degrading as many as possible different PHCs; (iii) be able to adapt to the environmental conditions and stresses prevailing in the soil to be treated; and (iv) neither be pathogenic agents, nor attack native microorganisms (Nur Zaida and Piakong 2018). Since each strain suitable to be used for bioaugmentation may have different PHC-biodegradation capabilities, the selection of the most adequate microorganisms for the process is a key step. The previous characterization of the PHCs present in the contaminated soil may help to select the most appropriate microbial inoculum.

Inocula used for bioaugmentation can consist of either a single strain or a microbial consortium. If the inoculum consists of only one microbial species, it is important to select microorganisms with the ability to use different PHCs as source of energy and growth. The use of a microbial consortium can help to generate an inoculum with wider biodegradation capabilities (Polyak et al. 2018). In this way, consortia of bacteria, bacteria and fungi, or bacteria and algae have been developed and experimentally used in the last years. The use of inocula containing bacteria and algae is especially interesting as microalgae release organic molecules such as organic acids, polypeptides, and polysaccharides that serve as microbial growth substrates, which, in turn, enhance the PHC-degradation potential of bacteria. Likewise, microalgae supply oxygen for enhancing aerobic degradation of contaminants (Kuppusamy et al. 2017). On the other hand, the formulation of the inocula with microorganisms with the capability of naturally producing biosurfactants help to make PHCs more accessible to biodegradation and obtain better bioremediation results (see below for more information on biosurfactants) (Gentry et al. 2004). Bioaugmentation products are typically sold either in liquid slurries or freeze-dried forms and may contain inorganic nutrients (Nwankwegu and Onwosi 2017).

Bacteria belonging to genera such as *Achromobacter*, *Acinetobacter*, *Alcaligenes*, *Burkholderia*, *Collimonas*, *Flavobacterium*, *Pseudomonas*, *Ralstonia*, *Sphingomonas*, or *Variovorax*, among Gram-negatives, and *Arthrobacter*, *Bacillus*, *Corynebacterium*, *Dietzia*, *Gordonia*, *Micrococcus*, *Mycobacterium*, *Nocardia*, *Nocardioides*, or *Rhodococcus*, among Gram-positive, have been shown to be effective in treatments of soil bioaugmentation. Effective fungi have been taxonomically affiliated to genera such as *Absidia*, *Acremonium*, *Aspergillus*, *Candida*, *Cunninghamella*, *Fusarium*, *Phanerochaete*, *Penicillium*, *Mucor*, *Sporobolomyces*, *Rhodotorula*, or *Trichoderma* (Chikere et al. 2011).

The use of genetically engineered microorganisms (GEMs) (microbes modified or supplemented to present improved PHC-degradation capabilities) is another bioaugmentation option. GEMs are obtained by: (i) improving the specificity and affinity of their PHC-degrading enzymes and their metabolic pathway designs; (ii) expanding the range of substrates they can use; (iii) preventing the production of toxic intermediates that inhibit the path by redirection of carbon flux; (iv) enhancing genetic stability of their catabolic activities; and (v) identifying genetically modified bacteria in PHC-polluted soils by marker gene (Paul et al. 2005, Baniasadi and Mousavi 2018). However, this approach faces important ecological and environmental concerns as well as regulatory constraints that prohibit the release of GEMs into the environment. A current hot topic in this field is developing methods to reduce the potential risks of using GEMs under real field conditions. One example is using some genetic barriers that restrict the survival of the modified bacteria as well as the gene transfer in the environment (Baniasadi and Mousavi 2018).

In general, when a bioaugmentation treatment fails, it is a consequence of the death of the introduced microorganisms since they have not been able (i) to adapt to the prevailing environmental conditions (fluctuations or extremes in temperature, water content, pH and/or nutrient availability), (ii) to tolerate co-contaminants like heavy metals, (iii) to compete with indigenous microbiota in cases of limited nutrients, and/or (iv) to survive the predators (protozoa and bacteriophages) or antibiotics produced by competing microorganisms (Juhanson et al. 2009, Siles and García-Sánchez 2018). To overcome these limitations, the use of microorganisms isolated from the soil to be bioremediated (autochthonous bioaugmentation) has proved to yield inocula with better adaptive potentials (Zawierucha and Malina 2011). On the other hand, the immobilization of inoculating microbes on some type of carrier protects them from predatory and sub-optimal environmental conditions and may also help to obtain improved bioaugmentation results (Nwankwegu and Onwosi 2017). For immobilization, both natural (e.g., dextran, agar, agarose, alginate, chitosan polyacrylamides, k-carrageenan, etc.) and synthetic (e.g., poly(carbamoyl)sulphonate, polyacrylamide, polyvinyl alcohol, etc.) materials can be used (Zawierucha and Malina 2011).

2.3 Biostimulation

Biostimulation consists in providing to the indigenous soil microbial community the optimum resources to stimulate its growth and degradation capacity (Mohan et al. 2006, Zawierucha and Malina 2011). Soil biostimulation treatments may involve: (i) the adjustment of moisture (irrigation), temperature (heating) or any other soil physical property; and/or (ii) the addition of inorganic nutrients (e.g., inorganic fertilizers rich in N, P (phosphorus), and K (potassium)), organic materials (e.g., animal manure, domestic sewage, rice straw biochar, crop residues, or different types of composts varying in composition and degree of stabilization), surfactants (synthetic (SDS (sodium dodecyl sulfate), Brij 30, Tween 80, etc.) surfactants produced by microorganisms (rhamnolipids, sophorose, surfactin, etc.)), and/or electron acceptors (e.g., O_2, chelated Fe (III), nitrates, or sulfate).

As previously commented, the contamination of a soil with an oil-derived product provokes an imbalance in the C:N:P ratio that limits the availability of essential nutrients for microbial growth and activity. Consequently, lack of a balanced C:N:P ratio constrains most of the bioremediation processes that involve native microbes (Safdari et al. 2018). The application of fertilizers rich in N and P (e.g., $(NH_4)_2SO_4$, Na_2HPO_4, CH_4N_2O, NH_4NO_3, KH_2PO_4, etc.) seeks to equilibrate such ratio. Theoretically, 150 mg of N and 30 mg of P are required to degrade 1 g of PHCs. Hence, a molar C:N:P ratio of 100:10:1 (equivalent to a weight ratio of 100:12:3) has been regarded as optimum for the biostimulation of PHC-contaminated soils (Zawierucha and Malina 2011). However, the effectiveness of nutrient sources tends to be affected by soil physico-chemical properties and, as a consequence, other C:N:P ratios such as 100:9:2, 100:10:1, 100:10:5, 100:21:16, 120:10:1, and 250:10:3 have also been identified as optimum (Simarro et al. 2011, Gkorezis et al. 2016, Gupta et al. 2016). The optimal C:N:P ratio should thus be identified for each contaminated soil. The addition of the appropriate level of nutrients is important as excessive fertilization can lead to ammonia toxicity, soil eutrophication, and overgrowth of r-strategist microorganisms, which may not be PHC degraders (Yavari et al. 2015, Brzeszcz et al. 2016).

The amendment of a PHC-contaminated soil with organic materials not only enriches it with fresh nutrients, but also with bulking agents, which facilitate soil aeration (Chen et al. 2015, Covino et al. 2016). Soil amendment with compost for biostimulation purposes also involves the enrichment of soil with the microorganisms inhabiting compost, which may be PHC degraders (Kastner and Miltner 2016).

Biosurfactants are amphipathic compounds produced by microorganisms that act in interfaces (Twigg et al. 2018). These compounds reduce surface and interfacial tension, making hydrophobic substances such as PHCs more accessible to microbial action and potentially increasing their mineralization. Biosurfactants have also been shown to: (i) promote foaming; (ii) induct flocculating action; (iii) increase wetting, spreading, and penetrating action(s); (iv) enhance microbial growth and metal sequestration; (v) quickly biodegrade; (vi) have lower toxicity; and (vii) be an environmentally friendly option unlike synthetic surfactants (Hazra et al. 2012, Lopes et al. 2018). Biosurfactants have hydrophilic structures (amino acids, peptides, mono-/disaccharides, and/or polysaccharides) and hydrophobic structures (saturated and/or unsaturated fatty acids) in their molecules. Biosurfactants are classified according to their chemical structure, which depends on the producing microorganisms, the substrate, and the conditions of the fermentation process. A well-known group of bacteria-synthesized biosurfactants is glycolipids, which can be further classified as rhamnolipids, trehalolipids, and sophorolipids. Among these, rhamnolipids (glycolipids containing fatty acid groups linked to a rhamnose) are the best known and the most commonly used in biostimulation treatments (Lopes et al. 2018). Bacteria (determined species belonging to the genera *Azotobacter*, *Bacillus*, *Brevibacterium*, *Burkholderia*, *Corynebacterium*, *Flavobacterium*, *Lactobacillus*, *Micrococcus*, *Nocardiopsis*, *Pseudomonas*, *Pseudoxanthomonas*, and *Rhodococcus*), fungi (*Aspergillus*, *Candida*, *Penicillium*, and *Trichoderma*) and algae (*Chlorella*) have been reported to produce biosurfactants using agroindustrial wastes as source (Radmann et al. 2015, Lopes et al. 2018, Sena et al. 2018).

Based on biostimulation treatments, different *in situ* technologies have been developed. For example, bioventing and biosparging are based on the introduction of air or oxygen into saturated and unsaturated PHC-contaminated soils, respectively (Kumar and Kaur 2018).

3. Laboratory-scale experiments dominate research on bioremediation of petroleum hydrocarbon-contaminated soils

Of the 4363 articles published on bioremediation of PHC-polluted soils on the Scopus database over the last 20 years (Figure 2), only ~ 5% of them contain the terms "field" and "experiment", "assay", "field", or "study". This clearly evidences that most of our current knowledge on soil bioremediation is based on laboratory-based experiments (in this chapter, greenhouse experiments are also considered as laboratory experiments). Laboratory trials are key tools to understand the factors and mechanisms involved in bioremediation processes and to get insights into the potential effectiveness of a specific bioremediation treatment. These assays, ranging in volume from flasks containing a few grams of soil to pans/bioreactors with several kilograms, are based on soils collected from real contaminated sites or soils artificially contaminated with PHC-derived products in the laboratory. Laboratory-scale studies, generally comparing treatments of phytoremediation, bioaugmentation, and/or biostimulation, consist of monitoring in the soil to be decontaminated: (i) contents of the different PHC fractions; (ii) an ample array of physical and chemical properties; (iii) microbial enzyme activities and respiration; and (iv) microbial communities in terms of abundance, taxonomic diversity and composition as well as functionality. In the last few years, the introduction of high-throughput techniques has revolutionized the field of bioremediation since the survival of a microbial inoculum can be examined with detail as well as the characteristics of the microbial community mediating soil recovery (metataxonomics) and the microbial genes involved in PHC degradation (metagenomics) (Breitwieser et al. 2017). Results from these studies, in general terms, have demonstrated that the efficacy of phytoremediation, bioaugmentation and biostimulation is case-specific and varies with the kind of plants, microorganisms and biostimulation treatments applied, respectively. Even inconsistent results among works have been shown (Wu et al. 2016). Among the soils more recalcitrant to be bioremediated are those with high contents in C and clay (carbon-rich clayey soils) (Masy et al. 2016). In successful soil bioremediation treatments, final rates of PHC removal ranging from 70% to 90% have been reported (Siles and García-Sánchez 2018).

Although laboratory surveys on soil bioremediation of PHCs are valuable, they should be recognized only as the first step in developing successful bioremediation technologies and every positive result obtained in the laboratory should be confirmed at field scale. In this way, when a laboratory-scale experiment yields promising results using a determined bioremediation approach, the following question should be addressed: is this approach applicable to real situations in which bioremediation processes are needed on a large scale under real environmental conditions? (Tortella et al. 2015). Nevertheless, promising bioremediation studies at laboratory scale are not always scaled-up to the field level. The reasons for this can be explained by budget-related matters. Unlike other sectors, remediation does not result in the production of value-added products. Consequently, the volume of venture capital invested in the field of bioremediation has been lower in comparison with other industrial sectors, negatively affecting R&D (research and development) activity (Kuppusamy et al. 2017). Additionally, some current cutting-edge technologies in bioremediation are directly not economically viable at field extent (e.g., production and purification of some biosurfactants at industrial level to be used for bioremediation purposes (Effendi et al. 2018)). On the other hand, other logistic and operational difficulties also influence the limited scaling up of bioremediation approaches (Körschens 2006).

In general, there are inherent limitations associated with laboratory studies. These surveys are carried out in flask, column, pan, "bucket", or respirometer volumes, far from what a real soil is. In this way, the spatial and temporal heterogeneity of a soil under natural conditions is rarely captured in a laboratory micro- or mesocosm experiment even using appropriate sampling strategies. Trials in the laboratory are focused on the short-term, lasting from a few days to several months under optimum conditions. Consequently, laboratory studies tend to overestimate the rate of PHC degradation observed in real field conditions. Some studies working with laboratory observations and simulation models have reported that field-scale trials take, on average, three times as long to reach the PHC-biodegradation rates observed at laboratory scale (Diplock et al. 2009). Nevertheless, the results from these models are reliable only if all the bioremediation controlling factors are integrated. Unfortunately, the mechanisms that drive bioremediation under field conditions are not fully understood. Bioremediation studies at field scale to generate mathematical models to predict temporal evolution of PHCs in soil are especially valuable in this context. Ortega et al. (2019), in a field study bioremediating via biostimulation two types of soils contaminated with different doses of diesel oil, demonstrated that soil TPH (total petroleum hydrocarbon) loss fitted an exponential model, following a first-order kinetic equation.

The number of large-scale field trials lasting for years is scarce, but the few ones existing provide very useful information regarding the effectiveness of bioremediation under real conditions. The objective of the following section is providing an overview of key field studies dealing with the decontamination of PHC-contaminated soils using bioremediation approaches.

4. Bioremediation studies of PHC-contaminated soils at field scale: current state of the art

Soil bioremediation experiments at field scale can be performed *in situ* or *ex situ*. While *in situ* approaches occur at the place of contamination directly on the intact polluted soil, *ex situ* techniques involve moving or excavating the soil to be treated at the site of contamination or somewhere else (Baniasadi and Mousavi 2018). Since *ex situ* techniques are operationally more expensive and have a higher environmental impact, this section is reviewing only *in situ* field experiments. Main findings of key bioremediation studies decontaminating PHC-polluted soils by using phytoremediation, bioaugmentation and/or biostimulation have been summarized in Table 1.

A typical remediation scheme for a contaminated soil should include these three phases: (i) site reconnaissance and risk assessment, (ii) assessment of remedial options, and (iii) remediation and monitoring (Ashraf et al. 2014). Once the bioremediation approach has been selected, the subsequent monitoring program should require a multidisciplinary approach integrating the work of microbiologists, geologists, engineers, hydrogeologist, and soil scientists (Kuppusamy et al. 2017). Nevertheless, these multidisciplinary works are uncommon. Most of the field-based studies on bioremediation of PHC-polluted soils are conducted from a microbiological point of view, as shown in the works summarized in Table 1. Additionally, official criteria for evaluating the success or failure of a specific bioremediation strategy have not been established (Kuppusamy et al. 2017). A bioremediation treatment is thus regarded as successful when the reduction in the contents of TPHs is significant in relation to the amount of time invested. As expected, rates of TPH and PAH removal varied between the works listed in Table 1 as a consequence of the different bioremediation approaches applied and the different climatic conditions and physico-chemical properties characterizing each of the soils studied. Interestingly, surveys studying simultaneously two contamination loads showed similar final rates of TPH loss (Ortega et al. 2018, 2019). None of the studies listed in Table 1 obtained dissipation rates of TPHs or PAHs higher than 90%, independent of the approach used and the bioremediation time considered. Even Polyak et al. (2018) did not achieve total recovery of a crude-oil contaminated soil after nine years of bioremedation. These maximum degradation rates concur with those obtained at laboratory scale (Siles and García-Sánchez 2018). The PHC residual fraction of 10% can be explained by two conceptual models. The contaminant sequestration model assumes that a certain fraction of PHCs is "locked up" in small soil pores within soil particles or aggregates and is inaccessible to soil microbes for degradation. According to the inherent recalcitrance model, certain PHCs are inherently recalcitrant to biodegradation or

Table 1. Key studies using bioremediation to decontaminate petroleum hydrocarbon-contaminated soils at field scale.

Geographical location	Source of contamination	Initial petroleum hydrocarbon concentration (mg kg^{-1})[1]	Most effective bioremediation approach/es	Rates of hydrocarbon degradation	Reference
Saskatchewan, Canada	Flare-pit	TPHs: ~ 7,000	Phytoremediation (Altai wild rye, *Elymus angustus* Trin.) + biostimulation (inorganic fertilizers, organic amendments)	55% after 2 years	Phillips et al. (2009)
University of Nigeria at Nsukka, Nigeria	Not specified	TPHs: 203	Phytoremediation (Cowpea, *Vigna Unguiculata*) + biostimulation (chicken manure, cassava peels and NPK Fertilizer)	90% after 2 years	Jidere et al. (2012)
Akala-olu, Nigeria	Crude oil spill	PAHs: 42	Phytoremediation (*Fimbristylis Littoralis*) + biostimulation (NPK fertilizer and poultry dung)	92% after 90 days	Nwaichi et al. (2015)
Chicago, Illinois, USA	Illegal waste dumping	Benzo(a)pyrene: between 1.8 and 0.43	Phytoremediation (yellow coneflower, *Echinacea paradoxa*) + biostimulation (compost)	47% after 3 years	Reddy and Amaya-Santos (2017)
North-east of Strasbourg, France	Artificially contaminated with fuel oil	TPHs: ~ 34,500	Phytoremediation (Goosegrass, *Eleusine indica*)	83% after 5 months	Matsodoum Nguemté et al. (2018)
Pisa, Italy	Illegal waste dumping	TPHs: 1,287	Phytoremediation (*Paulownia tomentosa*) + biostimulation (horse manure)	60% after 2.5 years	Macci et al. (2016)
St. Petersburg, Russia	Artificially contaminated with crude oil (10 L m^{-2})	TPHs: ~ 50,000	Natural attenuation, biostimulation (NPK fertilization) and biostimulation + bioaugmentation (commercial product) did not significantly differ after 9 years	90% after 9 years	Polyak et al. (2018)
Puyang, Henan Province, China	Abandoned petroleum well	TPHs: 6,320	Biostimulation (wheat straw) + bioaugmentation (*Enterobacter cloacae* and *Cunninghamella echinulata*)	64% after 45 days	Zhang et al. (2008)
Poznan, Poland	Artificially contaminated with diesel oil	TPHs: ~ 8,300	Bioaugmentation (consortium containing 8 different bacteria). Addition of rhamnolipids did not have a significant effect.	88% after 1 year	Szulc et al. (2014)

Table 1 Contd. ...

...Table 1 Contd.

Geographical location	Source of contamination	Initial petroleum hydrocarbon concentration (mg kg^{-1})[1]	Most effective bioremediation approach/es	Rates of hydrocarbon degradation	Reference
Coastline of Peninsular Malaysia	Oil and gas facility	TPHs: 10,000	Bioaugmentation (2 different consortia showed similar results)	> 80% after 3 months	Suja et al. (2014)
Spain	Artificially contaminated with diesel fuel	TPHs: 16,000 (low dose) TPHs: 32,000 (high dose)	Low dose: biostimulation (slow-release NPK fertilizer). High dose: biostimulation (slow-release NPK fertilizer + biosurfactant)	Low dose: 86% after 328 days High dose: 76% after 328 days	Ortega et al. (2019)
Not specified	Artificially contaminated with crude fuel	TPHs: 20,000 (low dose) TPHs: 50,000 (high dose)	Low dose: biostimulation (NPK fertilizer + irrigation + aeration). High dose: biostimulation (NPK fertilizer + surfactant + irrigation + aeration)	Low dose: 62% after 486 days High dose: 63% after 486 days	Ortega et al. (2018)

[1] TPHs: total petroleum hydrocarbons; PAHs: polycyclic aromatic hydrocarbons.

are only extremely slowly degradable even under optimal conditions (Huesemann 1997). In this context, some authors have argued that complete recovery of PHC-polluted soils may need the joint action of physical or chemical treatments and biological ones (Kuppusamy et al. 2017). In this context, electrokinetics, consisting in applying direct current within subsurface porous media to induce specific transport phenomena, in conjunction with bioremediation-based technologies has been regarded as a promising approach to obtain PHC-dissipation rates higher than 90% (Gill et al. 2014). On the other hand, further studies are needed in order to elucidate the effects of that residual TPH fraction on soil functionality.

In the case of phytoremediation, it is a common practice to supplement soil with inorganic and organic amendments to promote plant growth and establishment. This soil biostimulation helps plants to grow by providing nutrients, facilitates plant establishment since amendments act as bulking agents, and promotes bacterial activity and abundance in concomitance with decreasing PHC contents (Siles and García-Sánchez 2018). It has been reported that direct transplanting of plants without soil fertilization or amendment may result in the death of the plants in the short term (Matsodoum Nguemté et al. 2018). In the works presented in Table 1, the most effective phytoremediation treatments contained only one plant species. Therefore, using a mix of plants for phytoremediation treatments not always yields better decontamination results (Phillips et al. 2009). However, more studies are needed in this regard. As there are evidences that, for example, grasses and legumes differ in the types of TPHs and PAHs that they degrade, surveys using different mixes of plant species and monitoring degradation rates of both TPHs and PAHs are required.

Although bioaugmentation has been shown to yield improved rates of TPH loss in comparison with natural attenuation, especially in short term (Szulc et al. 2014), phytoremediation and biostimulation seem to be more effective (in terms of final PHC loss) than bioaugmentation at field scale. As happens with phytoremediation approaches, soil biostimulation with inorganic or organic fertilizers improves degradation rates of bioaugmentation treatments independently of the taxonomic features of the inoculum used (Zhang et al. 2008). Nevertheless, according to the field study of Szulc et al. (2014), soil biostimulation with biosurfactants (rhamnolipids, 150 mg kg^{-1} soil) did not improve rates of TPH loss after soil bioaugmentation nor improved degradative capacity of autochthonous microbes. These findings were explained by the facts that rhamnolipids were rapidly degraded in soil by microbes and that the study was conducted on a freshly contaminated soil. It has been argued that rhamnolipids-mediated biodegradation is more evident in long-term contaminated soils (Szulc et al. 2014). Finally, it is worth noting that biostimulation has been shown to not only help treatments of phytoremediation and bioaugmentation to be more effective, but also to yield effective decontamination rates for itself (Ortega et al. 2019). These findings are in line with previous reports showing that the easiest, cheapest and most convenient way to decontaminate a PHC-polluted soil may be through biostimulation by applying appropriate rates of cheap inorganic fertilizers and/or organic amendments and subsequent mixing of soil to facilitate aeration (Paudyn et al. 2008).

5. Conclusions and final remarks

Most of what we know about the effectiveness and the processes mediating the bioremediation of PHC-polluted soils is based on laboratory-scale studies. Since laboratory experiments may not reflect real field conditions, surveys scaling up bioremediation approaches from the laboratory to the field should be a priority. The scarce number of surveys bioremediating PHC-contaminated soils at field scale provides thus very valuable information. They show how bioremediation, via phytoremediation, bioaugmentation, and/or biostimulation approaches, yields decontamination rates of up to 90% in reasonable time frames. In general, faster and improved rates of TPH and PAH loss are obtained when different bioremediation approaches are simultaneously applied. Most of the existing field-scale studies have been conducted from a microbiological perspective and final conclusions have been drawn by considering only final rates of PHC dissipation. However, in the future, further integrative field-scale studies on bioremediation of PHC-contaminated soils integrating the work of geologists, engineers, hydrogeologists, and soil scientists are needed. This should be the priority of funding agencies in this field.

References

Adams, R., Osorio, F.G. and Cruz, J.Z. 2008. Water repellency in oil contaminated sandy and clayey soils. Int. J. Environ. Sci. Te. 5: 445–454.

Aislabie, J.M., Balks, M.R., Foght, J.M. and Waterhouse, E.J. 2004. Hydrocarbon spills on Antarctic soils: Effects and management. Environ. Sci. Technol. 38: 1265–1274.

Arias Espana, V.A., Rodriguez Pinilla, A.R., Bardos, P. and Naidu, R. 2018. Contaminated land in Colombia: A critical review of current status and future approach for the management of contaminated sites. Sci. Total Environ. 618: 199–209.

Baniasadi, M. and Mousavi, S.M. 2018. A comprehensive review on the bioremediation of oil spills. pp. 223–254. *In*: Kumar, V., Kumar, M. and Prasad, R. (eds.). Microbial Action on Hydrocarbons. Springer Singapore, Singapore.

Brassington, K.J., Hough, R.L., Paton, G.I., Semple, K.T., Risdon, G.C., Crossley, J., Hay, I., Askari, K. and Pollard, S.J.T. 2007. Weathered hydrocarbon wastes: A risk management primer. Crit. Rev. Env. Sci. Tec. 37: 199–232.

Breitwieser, F.P., Lu, J. and Salzberg, S.L. 2017. A review of methods and databases for metagenomic classification and assembly. Brief Bioinform. 2017: 1–15.

Brzeszcz, J., Steliga, T., Kapusta, P., Turkiewicz, A. and Kaszycki, P. 2016. R-strategist versus K-strategist for the application in bioremediation of hydrocarbon-contaminated soils. Int. Biodeter. Biodegr. 106: 41–52.

Cocârță, D.M., Stoian, M.A. and Karademir, A. 2017. Crude oil contaminated sites: Evaluation by using risk assessment approach. Sustainability 9: 1365.

Correa-García, S., Pande, P., Séguin, A., St-Arnaud, M. and Yergeau, E. 2018. Rhizoremediation of petroleum hydrocarbons: A model system for plant microbiome manipulation. Microb. Biotechnol. 11: 819–832.

Covino, S., Stella, T., D'Annibale, A., Lladó, S., Baldrian, P., Čvančarová, M., Cajthaml, T. and Petruccioli, M. 2016. Comparative assessment of fungal augmentation treatments of a fine-textured and historically oil-contaminated soil. Sci. Total Environ. 566-567: 250–259.

Cripps, G.C. 1992. The extent of hydrocarbon contamination in the marine environment from a research station in the Antarctic. Mar. Pollut. Bull. 25: 288–292.

Chandra, S., Sharma, R., Singh, K. and Sharma, A. 2013. Application of bioremediation technology in the environment contaminated with petroleum hydrocarbon. Ann. Microbiol. 63: 417–431.

Chaudhary, D.K. and Kim, J. 2019. New insights into bioremediation strategies for oil-contaminated soil in cold environments. Int. Biodeter. Biodegr. 142: 58–72.

Chen, M., Xu, P., Zeng, G., Yang, C., Huang, D. and Zhang, J. 2015. Bioremediation of soils contaminated with polycyclic aromatic hydrocarbons, petroleum, pesticides, chlorophenols and heavy metals by composting: Applications, microbes and future research needs. Biotechnol. Adv. 33: 745–755.

Chikere, C.B., Okpokwasili, G.C. and Chikere, B.O. 2011. Monitoring of microbial hydrocarbon remediation in the soil. 3 Biotech. 1: 117–138.

Devatha, C.P., Vishnu Vishal, A. and Purna Chandra Rao, J. 2019. Investigation of physical and chemical characteristics on soil due to crude oil contamination and its remediation. Appl. Water Sci. 9: 89.

Diplock, E.E., Mardlin, D.P., Killham, K.S. and Paton, G.I. 2009. Predicting bioremediation of hydrocarbons: Laboratory to field scale. Environ. Pollut. 157: 1831–1840.

dos Santos, J.J. and Maranho, L.T. 2018. Rhizospheric microorganisms as a solution for the recovery of soils contaminated by petroleum: A review. J. Environ. Manage. 210: 104–113.

Effendi, A.J., Kardena, E. and Helmy, Q. 2018. Biosurfactant-enhanced petroleum oil bioremediation. pp. 143–179. In: Kumar, V., Kumar, M. and Prasad, R. (eds.). Microbial Action on Hydrocarbons. Springer Singapore, Singapore.

Fernández-Luqueño, F., López-Valdez, F., Sarabia-Castillo, C.R., García-Mayagoitia, S. and Pérez-Ríos, S.R. 2017. Bioremediation of polycyclic aromatic hydrocarbons-polluted soils at laboratory and field scale: A Review of the literature on plants and microorganisms. pp. 43–64. In: Anjum, N.A., Gill, S.S. and Tuteja, N. (eds.). Enhancing Cleanup of Environmental Pollutants-Volume 1: Biological Approaches. Springer, Cham.

Freitas, B.G., Brito, J.G., Brasileiro, P.P., Rufino, R.D., Luna, J.M., Santos, V.A. and Sarubbo, L.A. 2016. Formulation of a commercial biosurfactant for application as a dispersant of petroleum and by-products spilled in oceans. Front. Microbiol. 7: 1646.

Gentry, T., Rensing, C. and Pepper, I. 2004. New approaches for bioaugmentation as a remediation technology. Crit. Rev. Env. Sci. Tec. 34: 447–494.

Gill, R.T., Harbottle, M.J., Smith, J.W.N. and Thornton, S.F. 2014. Electrokinetic-enhanced bioremediation of organic contaminants: A review of processes and environmental applications. Chemosphere 107: 31–42.

Gkorezis, P., Daghio, M., Franzetti, A., Van Hamme, J.D., Sillen, W. and Vangronsveld, J. 2016. The interaction between plants and bacteria in the remediation of petroleum hydrocarbons: An environmental perspective. Front. Microbiol. 7: 1836.

Gordon, G., Stavi, I., Shavit, U. and Rosenzweig, R. 2018. Oil spill effects on soil hydrophobicity and related properties in a hyper-arid region. Geoderma 312: 114–120.

Gupta, G., Kumar, V. and Pal, A.K. 2016. Biodegradation of polycyclic aromatic hydrocarbons by microbial consortium: a distinctive approach for decontamination of soil. Soil Sediment Contam. 25: 597–623.

Hazra, C., Kundu, D. and Chaudhari, A. 2012. Biosurfactant-assisted bioaugmentation in bioremediation. pp. 631–664. In: Satyanarayana, T. and Johri, B.N. (eds.). Environmental Management: Microbes and Environment. Springer Netherlands, Dordrecht.

Huesemann, M.H. 1997. Incomplete hydrocarbon biodegradation in contaminated soils: limitations in bioavailability or inherent recalcitrance? Bioremediat. J. 1: 27–39.

Hunt, L.J., Duca, D., Dan, T. and Knopper, L.D. 2019. Petroleum hydrocarbon (PHC) uptake in plants: A literature review. Environ. Pollut. 245: 472–484.

Hussain, I., Puschenreiter, M., Gerhard, S., Schöftner, P., Yousaf, S., Wang, A., Syed, J.H. and Reichenauer, T.G. 2018. Rhizoremediation of petroleum hydrocarbon-contaminated soils: Improvement opportunities and field applications. Environ. Exp. Bot. 147: 202–219.

Jackson, A. and Pardue, J.H. 1997. Seasonal variability of crude oil respiration potential in salt and fresh marshes. J. Environ. Qual. 26: 1140–1146.

Jidere, C.M., Akamigbo, F.O.R. and Ugwuanyi, J.O. 2012. Phytoremediation potentials of cowpea (*Vigina unguiculata*) and maize (*Zea mays*) for hydrocarbon degradation in organic and inorganic Manure-Amended tropical typic paleustults. Int. J. Phytoremediat. 14: 362–373.

Juhanson, J., Truu, J., Heinaru, E. and Heinaru, A. 2009. Survival and catabolic performance of introduced Pseudomonas strains during phytoremediation and bioaugmentation field experiment. FEMS Microbiol. Ecol. 70: 446–455.

Kastner, M. and Miltner, A. 2016. Application of compost for effective bioremediation of organic contaminants and pollutants in soil. Appl. Microbiol. Biot. 100: 3433–3449.

Körschens, M. 2006. The importance of long-term field experiments for soil science and environmental research—A review. Plant Soil Envir. 52: 1–8.

Koshlaf, E. and Ball, A.S. 2017. Soil bioremediation approaches for petroleum hydrocarbon polluted environments. AIMS Microbiol. 3: 25–49.

Kotowicz, J., Chmielniak, T. and Janusz-Szymańska, K. 2010. The influence of membrane CO_2 separation on the efficiency of a coal-fired power plant. Energy 35: 841–850.

Kumar, R. and Kaur, A. 2018. Oil spill removal by mycoremediation. pp. 505–526. In: Kumar, V., Kumar, M. and Prasad, R. (eds.). Microbial Action on Hydrocarbons. Springer Singapore, Singapore.

Kumar, R. and Yadav, P. 2018. Novel and cost-effective technologies for hydrocarbon bioremediation. pp. 543–565. In: Kumar, V., Kumar, M. and Prasad, R. (eds.). Microbial Action on Hydrocarbons. Springer Singapore, Singapore.

Kuppusamy, S., Thavamani, P., Venkateswarlu, K., Lee, Y.B., Naidu, R. and Megharaj, M. 2017. Remediation approaches for polycyclic aromatic hydrocarbons (PAHs) contaminated soils: Technological constraints, emerging trends and future directions. Chemosphere 168: 944–968.

Logeshwaran, P., Megharaj, M., Chadalavada, S., Bowman, M. and Naidu, R. 2018. Petroleum hydrocarbons (PH) in groundwater aquifers: An overview of environmental fate, toxicity, microbial degradation and risk-based remediation approaches. Environ. Technol. Inno. 10: 175–193.

Lopes, P.R.M., Montagnolli, R.N., Cruz, J.M., Claro, E.M.T. and Bidoia, E.D. 2018. Biosurfactants in improving bioremediation effectiveness in environmental contamination by hydrocarbons. pp. 21–34. *In*: Kumar, V., Kumar, M. and Prasad, R. (eds.). Microbial Action on Hydrocarbons. Springer Singapore, Singapore.

Macci, C., Peruzzi, E., Doni, S., Poggio, G. and Masciandaro, G. 2016. The phytoremediation of an organic and inorganic polluted soil: A real scale experience. Int. J. Phytoremediat. 18: 378–386.

Masy, T., Demanèche, S., Tromme, O., Thonart, P., Jacques, P., Hiligsmann, S. and Vogel, T.M. 2016. Hydrocarbon biostimulation and bioaugmentation in organic carbon and clay-rich soils. Soil Biol. Biochem. 99: 66–74.

Matsodoum Nguemté, P., DjumyomWafo, G.V., Djocgoue, P.F., KengneNoumsi, I.M. and WankoNgnien, A. 2018. Potentialities of six plant species on phytoremediation attempts of fuel oil-contaminated soils. Water Air Soil Poll. 229: 88.

Mearns, A.J. 1997. Cleaning oiled shores: Putting bioremediation to the test. Spill Sci. Technol. Bull. 4: 209–217.

Mena, E., Villaseñor, J., Rodrigo, M.A. and Cañizares, P. 2016. Electrokinetic remediation of soil polluted with insoluble organics using biological permeable reactive barriers: Effect of periodic polarity reversal and voltage gradient. Chem. Eng. J. 299: 30–36.

Mohan, S.V., Kisa, T., Ohkuma, T., Kanaly, R.A. and Shimizu, Y. 2006. Bioremediation technologies for treatment of PAH-contaminated soil and strategies to enhance process efficiency. Rev. Environ. Sci. Bio. 5: 347–374.

Nelyubov, D.V., Semikhina, L.P., Vazhenin, D.A. and Merkul'ev, I.A. 2017. Influence of resins and asphaltenes on the structural and rheological properties of petroleum disperse systems. Petrol. Chem. 57: 203–208.

Nur Zaida, Z. and Piakong, M.T. 2018. Bioaugmentation of petroleum hydrocarbon in contaminated soil: A review. pp. 415–439. *In*: Kumar, V., Kumar, M. and Prasad, R. (eds.). Microbial Action on Hydrocarbons. Springer Singapore, Singapore.

Nwaichi, E.O., Frac, M., Nwoha, P.A. and Eragbor, P. 2015. Enhanced phytoremediation of crude oil-polluted soil by four plant species: effect of inorganic and organic bioauugmentation. Int. J. Phytoremediat. 17: 1253–1261.

Nwankwegu, A.S. and Onwosi, C.O. 2017. Microbial cell immobilization: A renaissance to bioaugmentation inadequacies. A review. Environ. Technol. Rev. 6: 186–198.

Ortega, M.F., Guerrero, D.E., García-Martínez, M.J., Bolonio, D., Llamas, J.F., Canoira, L.and Gallego, J.L.R. 2018. Optimization of landfarming amendments based on soil texture and crude oil concentration. Water Air Soil Poll. 229: 234.

Ortega, M.F., García-Martínez, M.J., Bolonio, D., Canoira, L. and Llamas, J.F. 2019. Weighted linear models for simulation and prediction of biodegradation in diesel polluted soils. Sci. Total Environ. 686: 580–589.

Paniagua-Michel, J. and Fathepure, B.Z. 2018. Microbial consortia and biodegradation of petroleum hydrocarbons in marine environments. pp. 1–20. *In*: Kumar, V., Kumar, M. and Prasad, R. (eds.). Microbial Action on Hydrocarbons. Springer Singapore, Singapore.

Paudyn, K., Rutter, A., Kerry Rowe, R. and Poland, J.S. 2008. Remediation of hydrocarbon contaminated soils in the Canadian Arctic by landfarming. Cold Reg. Sci. Technol. 53: 102–114.

Paul, D., Pandey, G. and Jain, R.K. 2005. Suicidal genetically engineered microorganisms for bioremediation: Need and perspectives. Bio Essays 27: 563–573.

Phillips, L.A., Greer, C.R., Farrell, R.E. and Germida, J.J. 2009. Field-scale assessment of weathered hydrocarbon degradation by mixed and single plant treatments. Appl. Soil Ecol. 42: 9–17.

Polyak, Y.M., Bakina, L.G., Chugunova, M.V., Mayachkina, N.V., Gerasimov, A.O. and Bure, V.M. 2018. Effect of remediation strategies on biological activity of oil-contaminated soil—A field study. Int. Biodeter. Biodegr. 126: 57–68.

Radmann, E.M., Morais, E., Cibele, F., Zanfonato, K. and Costa, J. 2015. Microalgae cultivation for biosurfactant production. Afr. J. Microbiol. Res. 9: 2283–2289.

Reddy, C.M., Eglinton, T.I., Hounshell, A., White, H.K., Xu, L., Gaines, R.B. and Frysinger, G.S. 2002. The west falmouth oil spill after thirty years: The persistence of petroleum hydrocarbons in marsh sediments. Environ. Sci. Technol. 36: 4754–4760.

Reddy, K.R. and Amaya-Santos, G. 2017. Effects of variable site conditions on phytoremediation of mixed contaminants: Field-scale investigation at big marsh site. J. Environ. Eng. 143: 04017057.

Safdari, M.S., Kariminia, H.R., Rahmati, M., Fazlollahi, F., Polasko, A., Mahendra, S., Wilding, W.V. and Fletcher, T.H. 2018. Development of bioreactors for comparative study of natural attenuation, biostimulation, and bioaugmentation of petroleum-hydrocarbon contaminated soil. J. Hazard. Mater. 342: 270–278.

Sena, H.H., Sanches, M.A., Rocha, D.F.S., Segundo Filho, W.O.P., de Souza, É.S. and de Souza, J.V.B. 2018. Production of biosurfactants by soil fungi isolated from the Amazon forest. Int. J. Microbiol. 2018: 5684261.

Siles, J.A. and García-Sánchez, M. 2018. Microbial dynamics during the bioremediation of petroleum hydrocarbon-contaminated soils through biostimulation: An overview. pp. 115–134. *In*: Prasad, R. and Aranda, E. (eds.). Approaches in Bioremediation: The New Era of Environmental Microbiology and Nanobiotechnology. Springer, Cham.

Siles, J.A. and Margesin, R. 2018. Insights into microbial communities mediating the bioremediation of hydrocarbon-contaminated soil from an Alpine former military site. Appl. Microbiol. Biot. 102: 4409–4421.

Simarro, R., González, N., Bautista, L.F., Sanz, R. and Molina, M.C. 2011. Optimisation of key abiotic factors of PAH (naphthalene, phenanthrene and anthracene) biodegradation process by a bacterial consortium. Water Air Soil Poll. 217: 365–374.

Streche, C., Cocârță, D.M., Istrate, I.-A. and Badea, A.A. 2018. Decontamination of petroleum-contaminated soils using the electrochemical technique: Remediation degree and energy consumption. Sci. Rep. 8: 3272.

Suja, F., Rahim, F., Taha, M.R., Hambali, N., Rizal Razali, M., Khalid, A. and Hamzah, A. 2014. Effects of local microbial bioaugmentation and biostimulation on the bioremediation of total petroleum hydrocarbons (TPH) in crude oil contaminated soil based on laboratory and field observations. Int. Biodeter. Biodegr. 90: 115–122.

Szulc, A., Ambrożewicz, D., Sydow, M., Ławniczak, Ł., Piotrowska-Cyplik, A., Marecik, R. and Chrzanowski, Ł. 2014. The influence of bioaugmentation and biosurfactant addition on bioremediation efficiency of diesel-oil contaminated soil: Feasibility during field studies. J. Environ. Manage. 132: 121–128.

Tortella, G., Durán, N., Rubilar, O., Parada, M. and Diez, M.C. 2015. Are white-rot fungi a real biotechnological option for the improvement of environmental health? Crit. Rev. Biotechnol. 35: 165–172.

Twigg, M.S., Tripathi, L., Zompra, A., Salek, K., Irorere, V.U., Gutierrez, T., Spyroulias, G.A., Marchant, R. and Banat, I.M. 2018. Identification and characterisation of short chain rhamnolipid production in a previously uninvestigated, non-pathogenic marine pseudomonad. Appl. Microbiol. Biot. 102: 8537–8549.

Varjani, S.J. 2017. Microbial degradation of petroleum hydrocarbons. Bioresource Technol. 223: 277–286.

Wang, L., Lee, F.S., Wang, X., Yin, Y. and Li, J. 2007. Chemical characteristics and source implications of petroleum hydrocarbon contaminants in the sediments near major drainage outfalls along the coastal of Laizhou Bay, Bohai Sea, China. Environ. Monit. Assess. 125: 229–237.

Wang, X., Feng, J. and Zhao, J. 2010. Effects of crude oil residuals on soil chemical properties in oil sites, Momoge Wetland, China. Environ. Monit. Assess. 161: 271–280.

Wang, Y., Feng, J., Lin, Q., Lyu, X., Wang, X. and Wang, G. 2013. Effects of crude oil contamination on soil physical and chemical properties in Momoge wetland of China. Chinese Geogr. Sci. 23: 708–715.

Wu, M., Dick, W.A., Li, W., Wang, X., Yang, Q., Wang, T., Xu, L., Zhang, M. and Chen, L. 2016. Bioaugmentation and biostimulation of hydrocarbon degradation and the microbial community in a petroleum-contaminated soil. Int. Biodeter. Biodegr. 107: 158–164.

Wu, M., Ye, X., Chen, K., Li, W., Yuan, J. and Jiang, X. 2017. Bacterial community shift and hydrocarbon transformation during bioremediation of short-term petroleum-contaminated soil. Environ. Pollut. 223: 657–664.

Yavari, S., Malakahmad, A. and Sapari, N.B. 2015. A review on phytoremediation of crude oil spills. Water Air Soil Pollut. 226: 279.

Zawierucha, I. and Malina, G. 2011. Bioremediation of contaminated soils: Effects of bioaugmentation and biostimulation on enhancing biodegradation of oil hydrocarbons. pp. 187–201. *In*: Kumar, V., Kumar, M. and Prasad, R. (eds.). Microbial Action on Hydrocarbons. Springer Singapore, Singapore.

Zhang, K., Hua, X.F., Han, H.L., Wang, J., Miao, C.C., Xu, Y.Y., Huang, Z.D., Zhang, H., Yang, J.M., Jin, W.B., Liu, Y.M. and Liu, Z. 2008. Enhanced bioaugmentation of petroleum- and salt-contaminated soil using wheat straw. Chemosphere 73: 1387–1392.

Zobell, C.E. 1946. Action of microorganisms on hydrocarbons. Bacteriol. Rev. 10: 1–49.

15

Heavy Metal Pollution in Agricultural Soils
Consequences and Bioremediation Approaches

Abdul Majeed

1. Introduction

Metals and metalloids which have densities higher than water (generally having atomic density > 4 g/cm^3) are regarded as heavy metals (Nagajyoti et al. 2010), although many other physical and chemical properties of metals are also taken into consideration while defining the term 'heavy metals' (Appenroth 2010). Naturally, the main sources of heavy metals are in the Earth's crust, rocks, volcanoes, and sea-water, where they are found as oxides, sulfides, phosphates and in several other forms (Masindi and Muedi 2018). Weathering processes of mineralized rocks, volcanic eruptions and flooding from oceanic water bodies, rivers and lakes containing heavy metals are some of the natural drivers which add metal contaminants to soils (Alloway 2012, Majeed et al. 2019). Anthropogenic origins of soil's contamination with heavy metals are largely linked with industries, mining and processing (Islam et al. 2018). Fertilizers and pesticides used in agriculture for dealing with nutrient deficiencies in soil and insects and pests atrocities are also considered as main anthropogenic agents which have increased heavy metal concentrations in soils.

Plants' growth and development are affected by several factors among which the chemistry and compositional aspects of soil are important. Availability of necessary nutrients and water and stable communities of microbes in soil govern physiological growth and developmental features of plants. Discrepancies in soil's nutritional and microbial components decline the health and fertility of soils that correspond to the overall disturbance of plants. The presence of heavy metals in soils beyond threshold levels induces stress and toxic environment which significantly influence plants' potential to absorb nutrient and water leading to reduced physiological and metabolic activities in them. Negative interactions of heavy metals with soil's beneficial microbes can contribute to reduced soil fertility and impaired plants-microbe relations. Consequently negative effects of heavy metal stress on morphology, development and yield attributes of plants depends on the type of heavy metals, their concentrations and the responding plant species. Some heavy metals like lead (Pb), cadmium (Cd), mercury (Hg), and arsenic (As) that occur in many physical and valiancy forms are not required by plants for their growth or other physiological activities and hence they are highly

---

Department of Botany, Government Degree College Pabbi (Nowshera), Khyber Pakhtunkhwa 24210, Pakistan.
Email: majeedpsh@gmail.com

toxic in very limited concentrations (Saxena et al. 2019). The occurrence of these metals in soils affect plants' germination, growth, and development due to poor water and nutrient uptake, and metabolic abnormalities.

Several techniques have been used as cleanup strategies for the reclamation of polluted soils with variable outcomes and cost efficiency (Kong and Glick 2017). Physical, chemical and thermal methods are employed to remove heavy metal-pollutants from soil and water; however, the costs, technical feasibility, health hazards, and the emergence of secondary pollution make these strategies less effective (Lone et al. 2008, Ashraf et al. 2017, Kong and Glick 2017). Instead, bioremediation techniques (the use of microorganisms and plants) for extraction of soil contaminants has been well received during the last few years because of their potential sustainability, low costs and risks they pose to the environment and human health (Ma et al. 2011). Although plants mediated remediation (phytoremediation) alone has been documented to yield promising results, microbial assistance of plants in the phytoremediation processes may further accelerate the extraction of pollutants and can enhance the efficiency of bioremediation (Ma et al. 2011, Ashraf et al. 2017).

Key components in a successful bioremediation approach involve the identification of bio-remediating agents, their tolerance level to heavy metals, and their suitability in the polluted environment. Further exploration of metal-tolerant species of bacterial genera and their use as co-inoculants with phyto-extractant plants would lead to reduced reliance on traditional approaches used for soil reclamation. In this chapter, sources of heavy metal pollution in soils, the effect of heavy metal stress on plants, and potential bioremediation approaches for reclamation of heavy metal polluted soils are addressed.

2. Heavy metals, sources of contamination in soils and effects on plants

Heavy metals (and metalloids) are those which are denser than water, have greater atomic weight, and show biological toxicity (Tchounwou et al. 2012, Li et al. 2019). The most important heavy metals from toxicological perspectives (and some for nutritional essentiality) are mercury (Hg), copper (Cu), nickel (Ni), lead (Pb), cadmium (Cd), chromium (Cr), strontium (Sn), zinc (Zn), iron (Fe), manganese (Mn), molybdenum (Mn), arsenic (As), aluminum (Al), cobalt (Co), and silver (Ag). Some heavy metals like Zn, Fe, Mn, Mo, Ni, Co, and Cu are essential micronutrients required for the proper functioning of both plants and animals but in lesser amounts (Zwolak et al. 2019). The essential heavy metals in trace amounts (depending on living beings) play major roles in the regulation of structural and functional diversity of enzymes, vitamins, photosynthetic pigments, and many other metabolic activities in both humans and plants. In plants, their role in cellular metabolism, enzymatic activities, electron transporters, synthesis of biological molecules and membranes, and photosynthetic pigments has been acknowledged in different studies and reviews (Arif et al. 2016, Oves et al. 2016). However, excess amounts of even essential heavy metals may cause several physiological and metabolic abnormalities in plants and humans leading to specific hyper-concentration-symptoms in them. Non-essential heavy metals, particularly Pb, Hg, Cd, Cr and As, are highly toxic to living organisms. Since their biological role is unknown, they are regarded as 'hazards' by many environmental protection agencies among which the United States Environmental Protection Agency considers them as highly toxic and carcinogenic (Tchounwou et al. 2012).

Soil pollution emerges when heavy metals and several other organic and inorganic contaminants are introduced to the soil (Cachada et al. 2018). Among the inorganic pollutants, heavy metals contaminations in soils are of significant concern for living beings because of their toxic effects on flora, fauna and microbes. Leading sources of soil contamination are human activities which include industrial processes, the use of agrochemicals, and waste disposal (Shankar 2017). Natural agents and anthropogenic activities are the leading sources of the incursion of heavy metals to soils. Naturally, the main sources of heavy metals are in the Earth's crust, rocks, volcanoes, and sea-water, where they are found as oxides, sulfides, phosphates and in several other forms (Masindi and Muedi

2018). Weathering processes of mineralized rocks, volcanic eruptions and flooding from oceanic water bodies, rivers and lakes containing heavy metals are some of the natural drivers which add metal contaminants to soils (Alloway 2012, Majeed et al. 2019). Anthropogenic origins of soil's contamination with heavy metals are largely linked with industries, mining and processing (Islam et al. 2018). Fertilizers and pesticides used in agriculture for dealing with nutrient deficiencies in soil and insects and pests atrocities are also considered as main anthropogenic agents which have increased heavy metal concentrations in soils. The distribution and occurrence of heavy metals in soils depend on the site, nature of industrial operations and mining activities.

Soils contaminated with heavy metals adversely affect the soil's microbial communities, plants, animals and humans and the whole ecological system (Edwards 2002, Padmavathiamma and Li 2007, Khan et al. 2009). In a managed ecosystem, polluted soils offer problems of diverse nature and intensity to crop growers as well as to consumers. Generally, polluted soils can directly challenge the germination and subsequent growth of cultivated crops by altering their metabolic activities, reducing water and nutrient absorption, and root injury (Khan et al. 2008). The rendered functionality of important physiological processes such as photosynthesis and respiration in response to metal and pollution stress would result in deficit accumulation of photosynthate and subsequent breakdown of the respiratory substrate in different plant organs. This can lead to stunted growth and a reduction in yield outputs of the crops. Other drastic consequences of polluted soils are the negative impacts on or even elimination of microbial communities (He et al. 2005, Wang et al. 2007), which are particularly alarming in the case of soil beneficial microbes because of their contributions toward plant growth promotion. Several studies have outlined the adverse effects of both nonessential heavy metals and essential heavy metals (in exceeding concentrations) on germination, growth, physiology and developmental aspects of plants. Inhibited germination and impaired seedling growth in *Arabidopsis thaliana* were observed in response to Cd, Pb and Hg stress (Li et al. 2005). In many other works, negative consequences of both essential heavy metals (Ni, Cu, Co, Al, etc.) at higher concentration and non-essential heavy metals (Pb, Hg, Cd, As, Cr, etc.) have been documented in different plant species in terms of abnormalities in their nutrient and mineral uptake, physiology, biochemistry, and photosynthetic activities (Table 1). Thus, reclamation of soils polluted with heavy metals in agro-farming systems is vitally important for mitigation of the adverse effects on soil microbes, growth and yield of crops.

3. Bioremediation methods of heavy metal contaminated soils

It is estimated that nearly 20 million hectares of soils are contaminated with different heavy metals, both originating from natural sources as well as anthropogenic activities (Liu et al. 2018). Many approaches are employed to reduce heavy metal pollution in soils which include physical, chemical, biological techniques and integrated approaches. Physical approaches of remediation rely mainly on removal of contaminated soil, applying electro-kinetic approaches, and verification while chemical methods include soil washing, immobilization and encapsulation; in biological methods, different organisms (plants and microbes) are used to extract, stabilize, volatilize and remove heavy metals from polluted soils (Khalid et al. 2017). In integrated methods, physical, chemical, or biological approaches are implied in combination. Particularly employed techniques in the remediation processes have their own merits and demerits. Liu et al. (2018) advocated *in situ* and landfilling approaches as more effective than the other methods. Aruliah et al. (2019) asserted that an integrated approach using chemical and biological techniques is more effective and environmentally sound. A combination of nanotechnology with the use of microorganisms is also considered as a promising method to reclaim contaminated soils (Cao et al. 2019). In many recent reviews, employment of bioremediation techniques has been favored for reclamation of heavy metal contaminated soils because of their cost-effectiveness, feasibility and eco-friendliness (Ashraf et al. 2019, Fu et al. 2019, Yang et al. 2019, Muthusaravanan et al. 2020).

Table 1. Heavy metals and their consequences on germination, growth, physiological, metabolical and yield attributes of different plants.

Plants under heavy metal stress	Heavy metals	Nature of heavy metals	Effects on plants	References
Phaseolus vulgaris	As	Non-essential, toxic	At concentration 5 mg dm^{-3}, As stress significantly reduced growth, physicochemical activities and protein contents	Stoeva et al. (2005)
Hordeum vulgare	Ni	Essential	100 µM concentration of Ni caused necrosis, chlorosis and reduced growth and mineral distribution	Rahman et al. (2005)
Sesbania drummondii	Pb	Non-essential, toxic	Poor seedling growth, photosynthesis and anti-oxidative responses	Israr et al. (2006)
Brassica pekinensis	Cu	Essential	At higher concentration, growth and nitrogen metabolism reduced	Xiong et al. (2006)
Hordeum vulgare	Cu, Al, Cd	Essential and non-essential	Poor growth and metabolic activities at higher concentrations	Guo et al. (2007)
Elsholtzia argyi	Pb	Non-essential, toxic	Photosynthesis, leaf growth and membranes' abnormalities at 200 µM	Islam et al. (2008)
Leucaena leucocephala	Cd, Pb	Non-essential	75 ppm concentration of Pb while 50 ppm of Cd reduced seedling root and shoot growth and dry weight	Shafiq et al. (2008)
Vetiveria zizanioides	Cd	Non-essential, toxic	Protein, chlorophyll contents, reduced water uptake and poor root activity	Aibibu et al. (2010)
pakchoi and mustard	Cd	Non-essential	Reduced growth and photosynthesis, shoot and root weight decreased	Chen et al. (2011)
Linum usitatissimum	Cd, Cr	Non-essential	Decreased photosynthesis and plant growth attributes	Ali et al. (2015)
Eichhornia crassipes	Pb	Non-essential	Plant growth and chlorophyll content drastically decreased	Malar et al. (2016)
Eruca sativa	Cu, Ni, Zn, Hg, Cr, Pb	Essential/non-essential	Decreased germination and seedling growth	Zhi et al. (2015)
Brassica juncea	Cr, Cd, Pb, Hg	Non-essential	Increased level of heavy metals decreased photosynthesis, biomass and chlorophyll content	Sheetal et al. (2016)
Zea mays	Ni	Essential	At higher concentrations, Ni stress induced low mineral contents, photosynthesis, growth and biomass	Rehman et al. (2016)
Pisum sativum	Co, Pb, Cd	Essential/Non-essential	Abnormal growth and biomass at concentration 750–1250 ppm	Majeed et al. (2019)
Spinacia oleracea	As, Hg	Non-essential	Abnormal and reduced plant growth, and metal distribution	Zubair et al. (2019)
Triticum aestivum	Zn, Cu	Essential	Germination parameters and seedling growth attributes responded negatively to higher concentration of metals	Wang et al. (2019)
T. aestivum	Cu, Pb	Essential/non-essential	Reduced plant growth and enzymatic activities	Jiang et al. (2019)
Desmodesmus sp.	Cu	Essential	Negative effect on growth and filament structure	Buayam et al. (2019)
Zea mays	Cu	Essential	At higher concentrations, plant height and weight reduced while mineral uptake was disturbed	Reckova et al. (2019)

Table 1 Contd. ...

...Table 1 Contd.

Plants under heavy metal stress	Heavy metals	Nature of heavy metals	Effects on plants	References
Petunia hybrid	Cd, Cr, Cu, Ni and Pb	Essential/non-essential	Morphological and biochemical abnormalities	Khan et al. (2019)
Urginea maritime	Al, Cr, Cd, Pb	Essential/non-essential	Decline in photosynthesis and chlorophyll pigments	Houri et al. (2020)
Nicotiana alata	Cd, Cr, Co, Ni, Pb	Essential/non-essential	Stress imposition, negative effects on physiology and photosynthesis	Khan et al. (2020)

Bioremediation in wider terms implies the use of living organisms for minimizing the concentration and adverse effects of heavy metals in contaminated soils. Generally, plants (herbs, shrubs, trees and even algae) are used as remediating agents in most of the polluted sites and the technique is referred to as "phytoremediation". Microorganisms such as bacteria and some fungi may also be used in remediation processes and this approach is termed as "microbial remediation". Phytoremediation coupled with the microbial application (bacteria, microfungi, and microalgae) is termed as "microbial assisted phytoremediation" (Dotaniya et al. 2018). Plants used wither alone or in combination with microbes, and employ different strategies such as chelation, accumulation, extraction, volatilization and stabilization of heavy metals (Figure 1).

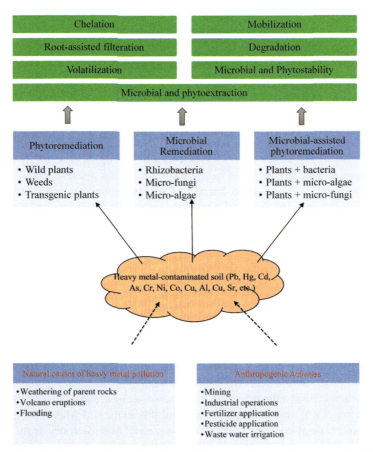

Figure 1. An illustration of the heavy metal input into soils and bioremediation strategies employed by plants and microbes.

3.1 Plant mediated, microbial and microbial-assisted phytoremediation

A comprehensive approach towards bioremediation of heavy metals is the utilization of different plant species (both wild and transgenic), microorganisms or both of them in combination. Different plants have diverse eco-physiological and growth characteristics that suit them to fit in different environments. Considering the problem of heavy metal pollution, some plants are sensitive while some show a certain degree of tolerance to heavy metal stress. The later offer opportunities for their utilization in phytoremediation processes. However, the only ability of plants to tolerate heavy metal stress does not make them ideal candidates for remediation, instead, their ability to extract heavy metals in larger quantities and biotransformation are some of the important features which can make the phytoremediation technique more suitable. Moreover, contaminated soil, types and concentration of heavy metals and plant species to be used in the remediation approach are influencing factors for a targeted phytoremediation method. In many empirical studies and reviews, the major steps involved in phytoremediation have been outlined below.

3.1.1 Phytoextraction

In phytoextraction, heavy metals are absorbed by plants, retained in their roots or translocated to above-ground parts (Gupta et al. 2019). Heavy metal absorption by roots and subsequent transport to other parts is influenced by many factors among which the role of specific transporters, enzymes, hormones, gene expression and metabolic modulation are significant (Dal Corso et al. 2019). Ideally, phytoextractant plants should possess extensive root-system, broad leaves, and hyper-accumulation potentials and should be short-lived. Several plants such as *Brassicrapa*, *Cannabis sativa*, *Helianthus annuus* and *Zea mays* (Meers et al. 2005), *Vertiveriazizanioides, Dianthus chinensis*, and *Rumex* spp. (Zhuang et al. 2007), *Amaranthushypochondriacus*, and *Averrhoacarambola* (Li et al. 2009, 2012), *Rosa multiflora* and *Sidahermaphrodita* (Antonkiewicz et al. 2017), *Cichoriumintybus, Ricinuscommunis*, and *Sesuviumportulacastrum* (Ayyappan et al. 2016, Bursztyn Fuentes et al. 2018), and *Salix* spp. (Ishikawa et al. 2018) and several other plants have been successfully utilized in phytoextraction methods.

Like plants, microbial agents are also capable of extraction of heavy metals. Unlike plants, microbes have lower biomass and lack translocation organs, but they still carry out a significant proportion of heavy metals extraction from polluted soils. When used in combination with plants, they augment the phytoextraction process by facilitating roots to absorb efficiently, mobilize heavy metals and synergistically stimulate plant species to traffic them up. Microorganism whether bacteria, microalgae or fungi should be tolerant of heavy metal stress, and possess symbiotic potentials with plants. Rajkumar et al. (2010) reviewed the role of siderophore producing bacteria and concluded that these microbes facilitate heavy metal binding to plants' roots. Ma et al. (2016b) suggested that endophytic bacteria could help plants to tolerate heavy metal stress and efficiently remove heavy metals from soil. Khan and Banu (2019) highlighted that microbes in polluted soil detoxify the toxic heavy metals, make them mobile and available to plants for uptake, facilitate solubilization and by releasing chelating agents.

3.1.2 Phytostability

Phytostability defined as chemical stabilization of heavy metals in soil through complex formation, roots attachment, amendments application, allelopathic interactions, and phytochelation (Radziemska et al. 2017, Shackira and Puthur 2019). Many plants and microorganisms have the potential to release allelochemicals, and exudates that can interact with heavy metals, leading to the formation of phytochemical-metal complexes. The complex formation depends on the nature of released phyto-microbial chemicals and the reactivity potential of heavy metals. The complex formation may either reduce the toxicity of heavy metals or their mobility in the soil which will correspond to their stabilization. Besides metal-complex formation, plant roots and microbes sequester heavy metals by accumulation and adsorption, which reduces their bioavailability and results in the heavy metals' stability (Yang et al. 2014). Several plants and microbes have been identified to phytostabilize heavy metal load in polluted soils by adsorption, accumulation, and through the release of phytochemicals into the rhizosphere. Dary et al. (2010) revealed that *Lupinusluteus* assisted with plant growth-promoting bacteria (*Bradyrhizobium* sp. 750, *Pseudomonas* sp. and *Ochrobactrumcytisi*) proved effective remediation potentials by phytostabilizing heavy metals Cu, Cd and Pb. Dary et al. (2010) revealed that plant growth-promoting bacteria (*Bradyrhizobium* sp. 750, *Pseudomonas* sp. and *Ochrobactrumcytisi*) promotes plant growth (*Lupinus luteus*) by phytostabilizing heavy metals such as Cu, Cd and Pb. Similarly, Yang et al. (2014) documented some shrubs and grasses as good phytostabilizing agents for heavy metals (Pb, Zn, Mn, Cd) in mine tailing. Javaid (2011) highlighted the role of arbuscular mycorrhizal fungi in heavy metal stabilization referring to their potential of releasing polyphosphate granules, chelation and adsorption. A significant reduction in the bioavailability of Pb and its stability was observed when Pb-polluted agricultural waste was treated with the compost of *Phanerochaetechrysosporium* (Huang et al. 2017).

3.1.3 Phytovolatilization

Another key component of the phytoremediation of heavy metals is volatilization carried out by plants and microbial communities. It is an effective mechanism to convert more toxic heavy metals to less toxic volatile compounds and released by plants or microbes to the atmosphere in modified forms (Tangahu et al. 2011). Taken up by roots, heavy metals are transported to shoots and leaves where they can interact with available biomolecules and may be transformed into less toxic complexes. During the process of transpiration, the transformed chemicals may escape to the atmosphere thus reducing their loads inside plant tissues and in soils. Once transformed, the toxicity of heavy metals may be greatly reduced and in the air, their presence may not pose any serious problems. Microorganisms may not directly participate in phytovolatilization processes; they, however, can enhance their transport to roots and the ability of plants to take them up effectively. The nature of heavy metals plays a key role in phytovolatilization. Generally, mercury, selenium and arsenic are effectively volatilized by *Pterisvittata* and Arabidopsis thaliana (Sakakibara et al. 2010, Kumar et al. 2017). Limmer and Burken (2016) stated that plants can either directly volatilize pollutants through their stems and leaves or indirectly via root activities in the soil. Aromatic plants and those which have high evaporation potential may be regarded as ideal candidates for phytovolatilization. Muthusaravanan et al. (2020), while referring to Pilon-Smits et al. (1999), listed many plants as potential volatilizers among which *Canna glauca, Colocasiaesculenta, Cyperus papyrus, Azollacaroliniana, Arundodonax, Pterisvittata, Brassica juncea, Lupinus* sp., *Liriodendron tulipifera* and *Typhaangustifolia* effectively volatilized selenium. Anarado et al. (2019) reported that *Murrayakoenigii, Ocimumgratissimum, Amaranthushybridus, Capsicum annuum* and *Moringaoleifera* showed some degree of volatilization of heavy metals from the contaminated environment; however, since these vegetables are consumed by humans and could offer health issues.

Besides phytoextraction, phyto-stability and phytovolatilization of heavy metals by plants, microbes and plants-microbes, some other strategies such as rhizofiltration, phytodegradation, phyto-accumulation/hyper-accumulation, plants or microbes-mediated detoxification of heavy metals and phytomining are also widely acknowledged strategies in cleaning up of polluted soils (Tangahu et al. 2011, Suman et al. 2018). Ma et al. (2016a) discussed the role of plants and microbes in phytoremediation processes. They outlined that plants either alone or assisted by microbes carry out detoxification, bioaccumulation, bioleaching, bioexclusion, mobilization and immobilization, and biotransformation, which effectively contributes to minimizing heavy metal concentration in soil. Mosa et al. (2016) also provided a comprehensive mechanism of heavy metal removal by different microorganisms. They described that microbes could reduce heavy metal concentration in the contaminated environment by biosorption and bioaccumulation, siderophore formation, and production of biosurfactants.

Studies have revealed promising results for whether the bioremediation agents (plants and microbes) were applied alone or in combination. Strains of *Microbacterium* (G16) and *Pseudomonas fluorescens* (G10) were found to promote the growth and lead uptake in *Brassicanapus* under Pb-contaminated soils (Sheng et al. 2008), while *Achromobacterxylosoxidans* (Ax10) contributed to growth and biomass increase in *B. juncea* with an enhanced potency of Cu uptake in the host plant (Ma et al. 2008). In other reports, bacterial species *Flavobacterium* sp., *Rhodococcus* sp., *Variovoraxparadoxus*, *Chrysiobacteriumhumi*, *Ralstoniaeutropha*, *Microbacterium*, *Pseudomonas fluorescens*, and *P. aeruginosa* were identified as effective synergists to stimulate extraction of heavy metals (Cd, Cr, Zn, Pb, Co, Ni) and pesticide residues from the polluted soils with augmentative effects on the growth of different plants (Belimov et al. 2005, Rajkumar and Freitas 2008, Ahemad and Khan 2011, Sobariu et al. 2017).

Similarly, different plant species such as *Tithoniadiversifolia* and *Helianthus annuus* (Adesodun et al. 2010), *Pterisvittata, P. cretica, Boehmerianivea,* and *Miscanthusfloridulu* (Sun et al. 2014), *Piper marginathum* and *Stecherusbifidus, Jatropacurcas* and *Capsicum annuum* (Marrugo-Negrete et al. 2016), microbial assisted-*Ricinuscommunis* (Annapurna et al. 2016), *Lemna minor* (Bokhari et al. 2016), compost and biochar assisted-*Moringaoleifera* (Ogundiran et al. 2018), *Lemna minor* and *Azollafiliculoides* (Amare et al. 2018), *Brassica campestris, Rorippapalustris, Sinapisarvensis* and *Thlaspiarvens* (Drozdova et al. 2019), *Coronopusdidymus* (Sidhu et al. 2020), and *Corchoruscapsularis* (Saleem et al. 2020) have been reported for varying degrees of tolerance to heavy metals and phytoremediation potentials of different heavy metals in different environmental conditions.

4. Limitations and prospects in bioremediation approaches

The process of bioremediation offers excellent opportunities for utilizing plants, microorganisms or plants assisted by microorganisms to clean up the polluted environment with heavy metals and other organic and inorganic pollutants. However, bioremediation strategies have some limitations in application. The major problem concerned with the processes is the identification of suitable plants and microbial candidates for phytoremediation. Since the majority of the plant species and microorganisms exhibit limited tolerance to heavy metal stress, using of non-tolerant bioremedial agents would not work efficiently in cleaning up the polluted environment. Microbial and plant species which are tolerant to heavy metal stress can work better to remediate the contaminants from the given environment. They also show variable efficacy because some species may have long life period, low biomass, low uptake potential for heavy metals, and not well-developed rooting system. Ideally, plants with hyperaccumulation potential, desired biomass, broad leaves, and extensive roots can promote the process of phytoremediation. Further, the most widely known strategies in phytoremediation such as phytoextraction, phytostability, phytoaccumulation, and phytovolatilization may not be exhibited by a single plant species. Similarly, microbes have low biomass and limited mechanisms of remediation due to lack of roots, and their variable responses to heavy metal stress. Consumable plants are not ideal for phytoremediation because they may cause food chain contamination.

Besides the limitations, bioremediation has several advantages over traditional clean-up methods. Physical, chemical, thermal and other methods have labor and cost burdens in addition to their lack of feasibility in different sites, possible secondary pollution they may cause, and problems concerned with the disposition (Sharma et al. 2018). Bioremediation, in contrast, seems more effective, eco-friendly, cost-effective, relatively feasible, and less hazardous to the environment (Awa and Hadibarata 2020, Haq et al. 2020). After identification of plants and microbes which have ideal characters of heavy metal tolerance, high potential of metal uptake and translocation, rhizofilteration, accumulation, volatilization, biodegradation and transformation of heavy metals could be used as alternative agents to other methods. To prevent food chain contamination, weeds, and wild or cultivated but non-edible plants can be applied as phytoremediation agents. To further augment their efficacy, suitable bacterial, algal and fungal strains may be utilized to assist them in bioremediation of heavy metals. Selvi et al. (2019) favored an integrative approach to maximize the efficiency of bioremediation by employing amendments and upgraded traditional strategies. Utilization of transgenic plants that are modified for phytoremediation purposes may also be considered as potential bioremedial agents in the reclamation of heavy metal polluted soils (Gunarathne et al. 2019).

5. Conclusions

Heavy metal pollution (Hg, Cu, Ni, Pb, Cd, Cr, Sn, Zn, Fe, Mn, Mo, As, Al, Co, Ag) in land and aquatic environment is one of the major abiotic constraints affecting microorganisms, plants, animals and humans. In agricultural land, the presence of heavy metals is of particular concern because they drastically affect cultivated plants and may lead to food chain contamination. Natural sources of heavy metal pollution are weathering phenomena, the release of volcanoes and flooding of land with water containing heavy metals while anthropogenically, industrial operations, wastewater irrigation, use of agrochemicals and mining activities are adding substantial quantities of heavy metals to soils. Except for essential heavy metals in smaller quantities, most of the heavy metals are toxic to flora and fauna and trigger a number of abnormalities in them. Plants challenged with heavy metal stress may experience altered germination, growth, physiology, development and yield. To prevent domestic plants and subsequent food chain contamination, cleanup remediation of the heavy metal polluted environment is necessary. Bioremediation which employs various plants and microorganisms is an attractive cleanup approach for reclamation of polluted soils due to cost-effectivity, feasibility and eco-friendliness as compared to traditional methods. Plants may be used either alone or they may be assisted by different microorganisms to extract, accumulate, volatilize, biodegrade, bio-transform and bio-stabilize heavy metals present in the soil. Besides its potential benefits and evident outcomes, bioremediation approaches need more rigorous research and attention for a wide-scale utilization in the reclamation of heavy metal pollution.

References

Adesodun, J.K., Atayese, M.O., Agbaje, T.A., Osadiaye, B.A., Mafe, O.F. and Soretire, A.A. 2010. Phytoremediation potentials of sunflowers (*Tithonia diversifolia* and *Helianthus annuus*) for metals in soils contaminated with zinc and lead nitrates. Water Air Soil Poll. 207(1-4): 195–201.

Aibibu, N., Liu, Y., Zeng, G., Wang, X., Chen, B., Song, H. and Xu, L. 2010. Cadmium accumulation in *Vetiveria zizanioides* and its effects on growth, physiological and biochemical characters. Bioresour. Technol. 101(16): 6297–6303.

Ali, N., Masood, S., Mukhtar, T., Kamran, M.A., Rafique, M., Munis, M.F.H. and Chaudhary, H.J. 2015. Differential effects of cadmium and chromium on growth, photosynthetic activity, and metal uptake of *Linum usitatissimum* in association with *Glomus intraradices*. Environ. Monit. Assess. 187(6): 311.

Alloway, B.J. (ed.). 2012. Heavy Metals in Soils: Trace Metals and Metalloids in Soils and Their Bioavailability (Vol. 22). Springer Sci. & Business Media.

Amare, E., Kebede, F., Berihu, T. and Mulat, W. 2018. Field-based investigation on phytoremediation potentials of *Lemna minor* and *Azolla filiculoides* in tropical, semiarid regions: Case of Ethiopia. Int. J. of Phytoremediat. 20(10): 965–972.

Anarado, C.E., Anarado, C.J.O., Okeke, M.O., Ezeh, C.E., Umedum, N.L. and Okafor, P.C. 2019. Leafy vegetables as potential pathways to heavy metal hazards. J. Agric. Chem. Environ. 8(01): 23.

Annapurna, D., Rajkumar, M. and Prasad, M.N.V. 2016. Potential of Castor bean (*Ricinus communis* L.) for phytoremediation of metalliferous waste assisted by plant growth-promoting bacteria: possible cogeneration of economic products. pp. 149–175. *In*: Prasad, M.N.V. (ed.). Bioremediation and Bioeconomy. Elsevier.

Antonkiewicz, J., Kołodziej, B. and Bielińska, E.J. 2017. Phytoextraction of heavy metals from municipal sewage sludge by *Rosa multiflora* and *Sida hermaphrodita*. Int. J. Phytoremediat. 19(4): 309–318.

Appenroth, K.J. 2010. Definition of "heavy metals" and their role in biological systems. pp. 19–29. *In*: Sherameti, I. and Varma, A. (eds.). Soil Heavy Metals. Springer, Berlin, Heidelberg.

Arif, N., Yadav, V., Singh, S., Singh, S., Ahmad, P., Mishra, R.K. and Chauhan, D.K. 2016. Influence of high and low levels of plant-beneficial heavy metal ions on plant growth and development. Front. Environ. Sci. 4: 69.

Aruliah, R., Selvi, A., Theertagiri, J., Ananthaselvam, A., Kumar, K.S., Madhavan, J. and Rahman, P. 2019. Integrated remediation processes towards heavy metal removal/recovery from various environments—a review. Front. Environ. Sci. 7: 66.

Ashraf, M.A., Hussain, I., Rasheed, R., Iqbal, M., Riaz, M. and Arif, M.S. 2017. Advances in microbe-assisted reclamation of heavy metal contaminated soils over the last decade: a review. J. Environ. Manag. 198: 132–143.

Ashraf, S., Ali, Q., Zahir, Z.A., Ashraf, S. and Asghar, H.N. 2019. Phytoremediation: Environmentally sustainable way for reclamation of heavy metal polluted soils. Ecotoxicol. Environ. Saf. 174: 714–727.

Awa, S.H. and Hadibarata, T. 2020. Removal of heavy metals in contaminated soil by phytoremediation mechanism: a review. Water Air Soil Poll. 231(2): 47.

Ayyappan, D., Sathiyaraj, G. and Ravindran, K.C. 2016. Phytoextraction of heavy metals by *Sesuvium portulacastrum* L. a salt marsh halophyte from tannery effluent. Int. J. Phytoremediat. 18(5): 453–459.

Bokhari, S.H., Ahmad, I., Mahmood-Ul-Hassan, M. and Mohammad, A. 2016. Phytoremediation potential of *Lemna minor* L. for heavy metals. Int. J. Phytoremediat. 18(1): 25–32.

Buayam, N., Davey, M.P., Smith, A.G. and Pumas, C. 2019. Effects of copper and pH on the growth and physiology of *Desmodesmus* sp. AARLG074. Metabolites 9(5): 84.

Bursztyn Fuentes, A.L., José, C., de Los Ríos, A., do Carmo, L.I., de Iorio, A.F. and Rendina, A.E. 2018. Phytoextraction of heavy metals from a multiply contaminated dredged sediment by chicory (*Cichorium intybus* L.) and castor bean (*Ricinuscommunis* L.) enhanced with EDTA, NTA, and citric acid application. Int. J. Phytoremediat. 20(13): 1354–1361.

Cao, X., Alabresm, A., Chen, Y.P., Decho, A.W. and Lead, J. 2019. Improved metal remediation using a combined bacterial and nanoscience approach. Sci. Total Environ. 135378.

Chen, X., Wang, J., Shi, Y., Zhao, M.Q. and Chi, G.Y. 2011. Effects of cadmium on growth and photosynthetic activities in pakchoi and mustard. Bot. Stud. 52(1).

DalCorso, G., Fasani, E., Manara, A., Visioli, G. and Furini, A. 2019. Heavy metal pollutions: state of the art and innovation in phytoremediation. Int. J. Mol. Sci. 20(14): 3412.

Dary, M., Chamber-Pérez, M.A., Palomares, A.J. and Pajuelo, E. 2010. "*In situ*" phytostabilisation of heavy metal polluted soils using *Lupinusluteus* inoculated with metal resistant plant-growth promoting rhizobacteria. J. of Hazardous Materials 177(1-3): 323–330.

Dotaniya, M.L., Rajendiran, S., Dotaniya, C.K., Solanki, P., Meena, V.D., Saha, J.K. and Patra, A.K. 2018. Microbial assisted phytoremediation for heavy metal contaminated soils. pp. 295–317. *In*: Kumar, V., Kumar, M. and Prasad, R. (eds.). Phytobiont and Ecosystem Restitution. Springer, Singapore.

Drozdova, I., Alekseeva-Popova, N., Dorofeyev, V., Bech, J., Belyaeva, A. and Roca, N. 2019. A comparative study of the accumulation of trace elements in Brassicaceae plant species with phytoremediation potential. Appl. Geochem. 108: 104377.

Fu, J.T., Yu, D.M., Chen, X., Su, Y., Li, C.H. and Wei, Y.P. 2019. Recent research progress in geochemical properties and restoration of heavy metals in contaminated soil by phytoremediation. J. Mt. Sci. 16(9): 2079–2095.

Gunarathne, V., Mayakaduwa, S., Ashiq, A., Weerakoon, S.R., Biswas, J.K. and Vithanage, M. 2019. Transgenic plants: benefits, applications, and potential risks in phytoremediation. pp. 89–102. *In*: Prasad, M.N.V. (ed.). Transgenic Plant Technol. for Remediation of Toxic Metals and Metalloids. Academic Press.

Guo, T.R., Zhang, G.P. and Zhang, Y.H. 2007. Physiological changes in barley plants under combined toxicity of aluminum, copper and cadmium. Colloids Surf. B Biointerfaces 57(2): 182–188.

Gupta, P., Rani, R., Usmani, Z., Chandra, A. and Kumar, V. 2019. The role of plant-associated bacteria in phytoremediation of trace metals in contaminated soils. pp. 69–76. *In*: Singh, J.S. and Singh, D.P. (eds.). New and Future Developments in Microbial Biotechnology and Bioengineering. Elsevier.

Haq, S., Bhatti, A.A., Dar, Z.A. and Bhat, S.A. 2020. Phytoremediation of heavy metals: an eco-friendly and sustainable approach. pp. 215–231. *In*: Hakeem, K.R., Bhat, R.A. and Qadri, H. (eds.). Bioremediation and Biotechnology. Springer, Cham.

Houri, T., Khairallah, Y., Al Zahab, A., Osta, B., Romanos, D. and Haddad, G. 2020. Heavy metals accumulation effects on the photosynthetic performance of geophytes in Mediterranean reserve. J. King Saud Univ. Sci. 32(1): 874–880.

Huang, C., Zeng, G., Huang, D., Lai, C., Xu, P., Zhang, C., Cheng, M., Wan, J., Hu, L. and Zhang, Y. 2017. Effect of Phanerochaete chrysosporium inoculation on bacterial community and metal stabilization in lead-contaminated agricultural waste composting. Bioresour. Technol. 243: 294–303.

Ishikawa, Y., Yabuki, S., Nagasawa, S., Sugimoto, H., Aoki, Y., Satoh, S. Haykawa, A. and Takashi, T. 2018. Study on phytoextraction of heavy metal contaminated soil by fast-growth willow (Salix spp.). J. of Arid Land Stud. 28(S): 193–196.

Islam, E., Liu, D., Li, T., Yang, X., Jin, X., Mahmood, Q.,Tian, S. and Li, J. 2008. Effect of Pb toxicity on leaf growth, physiology and ultrastructure in the two ecotypes of Elsholtzia argyi. J. Hazard. Mater. 154(1-3): 914–926.

Islam, M.M., Karim, M., Zheng, X. and Li, X. 2018. Heavy metal and metalloid pollution of soil, water and foods in Bangladesh: A critical review. Int. J. Environ. Res. Public Health 15(12): 2825.

Israr, M., Sahi, S., Datta, R. and Sarkar, D. 2006. Bioaccumulation and physiological effects of mercury in *Sesbania drummondii*. Chemosphere 65(4): 591–598.

Javaid, A. 2011. Importance of arbuscular mycorrhizal fungi in phytoremediation of heavy metal contaminated soils. pp. 125–141. *In*: Khan, M.S., Zaidi, A., Goel, R. and Musarrat, J. (eds.). Biomanagement of Metal-Contaminated Soils. Springer, Dordrecht.

Jiang, K., Wu, B., Wang, C. and Ran, Q. 2019. Ecotoxicological effects of metals with different concentrations and types on the morphological and physiological performance of wheat. Ecotoxicol. Environ. Saf. 167: 345–353.

Khalid, S., Shahid, M., Niazi, N.K., Murtaza, B., Bibi, I. and Dumat, C. 2017. A comparison of technologies for remediation of heavy metal contaminated soils. J. Geochem. Explor. 182: 247–268.

Khan, A.H.A., Butt, T.A., Mirza, C.R., Yousaf, S., Nawaz, I. and Iqbal, M. 2019. Combined application of selected heavy metals and EDTA reduced the growth of *Petunia hybrida* L. Sci. Rep. 9(1): 1–12.

Khan, A.H.A., Nawaz, I., Qu, Z., Butt, T.A., Yousaf, S. and Iqbal, M. 2020. Reduced growth response of ornamental plant *Nicotianaalata* L. upon selected heavy metals uptake, with co-application of ethylenediaminetetraacetic acid. Chemosphere 241: 125006.

Kumar, B., Smita, K. and Flores, L.C. 2017. Plant mediated detoxification of mercury and lead. Arab. J. Chem. 10: S2335–S2342.

Li, C., Zhou, K., Qin, W., Tian, C., Qi, M., Yan, X. and Han, W. 2019. A review on heavy metals contamination in soil: effects, sources, and remediation techniques. Soil and Sediment Contamination: An International Journal 28(4): 380–394.

Li, J.T., Liao, B., Dai, Z.Y., Zhu, R. and Shu, W.S. 2009. Phytoextraction of Cd-contaminated soil by carambola (*Averrhoa carambola*) in field trials. Chemosphere 76(9): 1233–1239.

Li, N.Y., Fu, Q.L., Zhuang, P., Guo, B., Zou, B. and Li, Z.A. 2012. Effect of fertilizers on Cd uptake of *Amaranthushypochondriacus*, a high biomass, fast growing and easily cultivated potential Cd hyperaccumulator. Int. J. Phytoremediat. 14(2): 162–173.

Li, W., Khan, M.A., Yamaguchi, S. and Kamiya, Y. 2005. Effects of heavy metals on seed germination and early seedling growth of Arabidopsis thaliana. Plant Growth Regulation 46(1): 45–50.

Limmer, M. and Burken, J. 2016. Phytovolatilization of organic contaminants. Environ. Sci. Technol. 50(13): 6632–6643.

Liu, L., Li, W., Song, W. and Guo, M. 2018. Remediation techniques for heavy metal-contaminated soils: principles and applicability. Sci. Total Environ. 633: 206–219.

Ma, Y., Rajkumar, M. and Freitas, H. 2009. Inoculation of plant growth promoting bacterium Achromobacter xylosoxidans strain Ax10 for the improvement of copper phytoextraction by *Brassica juncea*. J. Environ. Manag. 90(2): 831–837.

Ma, Y., Prasad, M.N.V., Rajkumar, M. and Freitas, H. 2011. Plant growth promoting rhizobacteria and endophytes accelerate phytoremediation of metalliferous soils. Biotechnol. Adv. 29(2): 248–258.

Ma, Y., Oliveira, R.S., Freitas, H. and Zhang, C. 2016a. Biochemical and molecular mechanisms of plant-microbe-metal interactions: relevance for phytoremediation. Front. Plant Sci. 7: 918.

Ma, Y., Rajkumar, M., Zhang, C. and Freitas, H. 2016b. Beneficial role of bacterial endophytes in heavy metal phytoremediation. J. Environ. Manag. 174: 14–25.

Majeed, A., Muhammad, Z. and Siyar, S. 2019. Assessment of heavy metal induced stress responses in pea (*Pisum sativum* L.). Acta Ecol. Sin. 39(4): 284–288.

Malar, S., Vikram, S.S., Favas, P.J. and Perumal, V. 2016. Lead heavy metal toxicity induced changes on growth and antioxidative enzymes level in water hyacinths [*Eichhornia crassipes* (Mart.)]. Bot. Stud. 55(1): 1–11.

Marrugo-Negrete, J., Marrugo-Madrid, S., Pinedo-Hernández, J., Durango-Hernández, J. and Díez, S. 2016. Screening of native plant species for phytoremediation potential at a Hg-contaminated mining site. Sci. Total Environ. 542: 809–816.

Masindi, V. and Muedi, K.L. 2018. Environmental contamination by heavy metals. Heavy Metals; Intech Open: Aglan, France 115–133.

Meers, E., Ruttens, A., Hopgood, M., Lesage, E. and Tack, F.M.G. 2005. Potential of *Brassicrapa, Cannabis sativa, Helianthus annuus* and *Zea mays* for phytoextraction of heavy metals from calcareous dredged sediment derived soils. Chemosphere 61(4): 561–572.

Mosa, K.A., Saadoun, I., Kumar, K., Helmy, M. and Dhankher, O.P. 2016. Potential biotechnological strategies for the cleanup of heavy metals and metalloids. Front. Plant Sci. 7: 303.

Muthusaravanan, S., Sivarajasekar, N., Vivek, J.S., Priyadharshini, S.V., Paramasivan, T., Dhakal, N. and Naushad, M. 2020. Research updates on heavy metal phytoremediation: enhancements, efficient post-harvesting strategies and economic opportunities. pp. 191–222. *In*: Naushad, M. and Lichtfouse, E. (eds.). Green Materials for Wastewater Treatment. Springer, Cham.

Muthusaravanan, S., Sivarajasekar, N., Vivek, J.S., Priyadharshini, S.V., Paramasivan, T., Dhakal, N. and Naushad, M. 2020. Research updates on heavy metal phytoremediation: enhancements, efficient post-harvesting strategies and economic opportunities. pp. 191–222. *In:* Green Materials for Wastewater Treatment. Springer, Cham.

Nagajyoti, P.C., Lee, K.D. and Sreekanth, T.V.M. 2010. Heavy metals, occurrence and toxicity for plants: a review. Environ. Chem. Lett. 8(3): 199–216.

Ogundiran, M.B., Mekwunyei, N.S. and Adejumo, S.A. 2018. Compost and biochar assisted phytoremediation potentials of *Moringaoleifera* for remediation of lead contaminated soil. J. Environ. Chem. Eng. 6(2): 2206–2213.

Oves, M., Khan, S., Qari, H., Felemban, N. and Almeelbi, T. 2016. Heavy metals: biological importance and detoxification strategies. J. Bioremed. Biodegrad. 7: 334. doi: 10.4172/2155-6199.1000334.

Pilon-Smits, E.A.H., De Souza, M.P., Hong, G., Amini, A., Bravo, R.C., Payabyab, S.T. and Terry, N. 1999. Selenium volatilization and accumulation by twenty aquatic plant species. J. Environ. Qual. 28(3): 1011–1018.

Radziemska, M., Vaverková, M.D. and Baryła, A. 2017. Phytostabilization—management strategy for stabilizing trace elements in contaminated soils. Int. J. Environ. Res. Public Health 14(9): 958.

Rahman, H., Sabreen, S., Alam, S. and Kawai, S. 2005. Effects of nickel on growth and composition of metal micronutrients in barley plants grown in nutrient solution. J. Plant Nut. 28(3): 393–404.

Rajkumar, M. and Freitas, H. 2008. Influence of metal resistant-plant growth-promoting bacteria on the growth of Ricinus communis in soil contaminated with heavy metals. Chemosphere 71(5): 834–842.

Rajkumar, M., Ae, N., Prasad, M.N.V. and Freitas, H. 2010. Potential of siderophore-producing bacteria for improving heavy metal phytoextraction. Trends Biotechnol. 28(3): 142–149.

Reckova, S., Tuma, J., Dobrev, P. and Vankova, R. 2019. Influence of copper on hormone content and selected morphological, physiological and biochemical parameters of hydroponically grown *Zea mays* plants. Plant Growth Regulat. 89(2): 191–201.

Rehman, M.Z.U., Rizwan, M., Ali, S., Fatima, N., Yousaf, B., Naeem, A., Sabir, M., Ahmad, H.R. and Ok, Y.S. 2016. Contrasting effects of biochar, compost and farm manure on alleviation of nickel toxicity in maize (*Zea mays* L.) in relation to plant growth, photosynthesis and metal uptake. Ecotoxicol. Environ. Saf. 133: 218–225.

Sakakibara, M., Watanabe, A., Inoue, M., Sano, S. and Kaise, T. 2010 January. Phytoextraction and phytovolatilization of arsenic from As-contaminated soils by *Pterisvittata*. In Proceedings of the Annual Int. Conf. on Soils, Sediments, Water and Energy 12(1): 26.

Saleem, M.H., Fahad, S., Khan, S.U., Ahmar, S., Khan, M.H.U., Rehman, M., Maqbool, Z. and Liu, L. 2020. Morpho-physiological traits, gaseous exchange attributes, and phytoremediation potential of jute (*Corchorus capsularis* L.) grown in different concentrations of copper-contaminated soil. Ecotoxicol. Environ. Saf. 189: 109915.

Saxena, G., Purchase, D., Mulla, S.I., Saratale, G.D. and Bharagava, R.N. 2019. Phytoremediation of heavy metal-contaminated sites: eco-environmental concerns, field studies, sustainability issues, and future prospects. Rev. Environ. Contaminat. Toxicol. 249: 71–131.

Selvi, A., Aruliah, R., Theertagiri, J., Ananthaselvam, A., Kumar, K.S., Madhavan, J. and Rahman, P. 2019. Integrated remediation processes towards heavy metal removal/recovery from various environments—a review. Front. Environ. Sci. 7: 66.

Shackira, A.M. and Puthur, J.T. 2019. Phytostabilization of heavy metals: understanding of principles and practices. pp. 263–282. *In*: Srivastava, S., Srivastava, A.K. and Suprasanna, P. (eds.). Plant-Metal Interactions. Springer, Cham.

Shafiq, M., Iqbal, M.Z. and Mohammad, A. 2008. Effect of lead and cadmium on germination and seedling growth of *Leucaena leucocephala*. J. Appl. Sci. Environ. Manag. 12(3).

Sharma, S., Tiwari, S., Hasan, A., Saxena, V. and Pandey, L.M. 2018. Recent advances in conventional and contemporary methods for remediation of heavy metal-contaminated soils. 3 Biotech 8(4): 216.

Sheetal, K.R., Singh, S.D., Anand, A. and Prasad, S. 2016. Heavy metal accumulation and effects on growth, biomass and physiological processes in mustard. Indian J. Plant Physiol. 21(2): 219–223.

Sidhu, G.P.S., Bali, A.S., Singh, H.P., Batish, D.R. and Kohli, R.K. 2020. Insights into the tolerance and phytoremediation potential of *Coronopus didymus* L. (Sm) grown under zinc stress. Chemosphere 244: 125350.

Stoeva, N., Berova, M. and Zlatev, Z. 2005. Effect of arsenic on some physiological parameters in bean plants. Biologia Plantarum 49(2): 293–296.

Suman, J., Uhlik, O., Viktorova, J. and Macek, T. 2018. Phytoextraction of heavy metals: a promising tool for clean-up of polluted environment?. Front. Plant Sci. 9: 1476.

Sun, L., Liao, X., Yan, X., Zhu, G. and Ma, D. 2014. Evaluation of heavy metal and polycyclic aromatic hydrocarbons accumulation in plants from typical industrial sites: potential candidate in phytoremediation for co-contamination. Environ. Sci. Pollut. Res. 21(21): 12494–12504.

Tangahu, B.V., Abdullah, S., Rozaimah, S., Basri, H., Idris, M., Anuar, N. and Mukhlisin, M. 2011. A review on heavy metals (As, Pb, and Hg) uptake by plants through phytoremediation. Int. J. Chem. Eng. 2011.

Tchounwou, P.B., Yedjou, C.G., Patlolla, A.K. and Sutton, D.J. 2012. Heavy metal toxicity and the environment. pp. 133–164. *In*: Luch, A. (ed.). Molecular, Clinical and Environmental Toxicology. Springer, Basel.

Wang, R., Fu, W., Wang, J., Zhu, L., Wang, L., Wang, J. and Ahmad, Z. 2019. Application of rice grain husk derived biochar in ameliorating toxicity impacts of Cu and Zn on growth, physiology and enzymatic functioning of wheat seedlings. Bullet. Environ. Contaminat. Toxicol. 103(4): 636–641.

Wang, Y., Shi, J., Wang, H., Lin, Q., Chen, X. and Chen, Y. 2007. The influence of soil heavy metals pollution on soil microbial biomass, enzyme activity, and community composition near a copper smelter. Ecotoxicol. Environ. Saf. 67(1): 75–81.

Xiong, Z.T., Liu, C. and Geng, B. 2006. Phytotoxic effects of copper on nitrogen metabolism and plant growth in *Brassica pekinensis* Rupr. Ecotoxicol. and Environ. Safety 64(3): 273–280.

Yang, J., You, S. and Zheng, J. 2019, March. Review in strengthening technology for phytoremediation of soil contaminated by heavy metals. In IOP Conference Series: Earth Environ. Sci. 242(5): 052003. IOP Publishing.

Yang, S., Liang, S., Yi, L., Xu, B., Cao, J., Guo, Y. and Zhou, Y. 2014. Heavy metal accumulation and phytostabilization potential of dominant plant species growing on manganese mine tailings. Front. Environ. Sci. Eng. 8(3): 394–404.

Zhi, Y., Deng, Z., Luo, M., Ding, W., Hu, Y., Deng, J. and Huang, B. 2015. Influence of heavy metals on seed germination and early seedling growth in Eruca sativa Mill. American J. Plant Sci. 6(05): 582.

Zhuang, P., Yang, Q.W., Wang, H.B. and Shu, W.S. 2007. Phytoextraction of heavy metals by eight plant species in the field. Water Air Soil Pollut. 184(1-4): 235–242.

Zubair, M., Khan, Q.U., Mirza, N., Sarwar, R., Khan, A.A., Baloch, M.S. and Shah, A.N. 2019. Physiological response of spinach to toxic heavy metal stress. Environ. Sci. Pollut. Res. 26(31): 31667–31674.

Zwolak, A., Sarzyńska, M., Szpyrka, E. and Stawarczyk, K. 2019. Sources of soil pollution by heavy metals and their accumulation in vegetables: a review. Water Air Soil Pollut. 230(7): 164.

16

Arsenic Toxicity in Water-Soil-Plant System
An Alarming Scenario and Possibility of Bioremediation

Ganesh Chandra Banik, Shovik Deb,* Surajit Khalko, Ashok Chaudhury, Parimal Panda and *Anarul Hoque*

1. Introduction

Arsenic is present in soil and water by natural means like volcanic eruption, weathering, leaching and by anthropogenic activities like smelting of ores, mining, burning of coal, and manufacturing of herbicides and pesticides (Cavalca et al. 2013). Arsenic was first isolated by Albert Magnus, a German Alchemist, in 1250 AD via heating of soap with orpiment (arsenic trisulphide). By the 18th century, arsenic was well known as a unique toxic metalloid and was used as poison. It was named as 'Poison of Kings' for its use to kill several kings in 17th and 18th centuries. In agriculture, arsenic was mainly used to manufacture insecticides by Chinese. Arsenic containing insecticide such as sandarach (realgar, AsS) was effective for protection of grapes. Widespread use of another arsenic containing insecticide (Paris Green) was done to control Colorado potato beetle. This was followed by the use of London Purple (a mixture of calcium arsenate and arsenite with some organic matter) as an insecticide.

Arsenic is not abundant in the Earth's crust (0.0001%) and its normal concentration in soils is considered as < 15 mg kg^{-1} (Oreml and Stolz 2003). It is present in soil in inorganic forms combined with other elements but without carbon and also in organic forms containing carbon. Inorganic arsenic compounds are more toxic and soluble in soil system than its organic forms. Inorganic arsenic compounds are present in the Earth's crust mostly as metal arsenic, arsenic sulphide and arsenides and also as insoluble sulphide compounds and sulfosalts such as arsenopyrite (FeAsS), lollingite (FeAs$_2$), realgar (As$_4$S$_4$), orpiment (As$_2$S$_3$) and enargite (Cu$_3$AsS$_4$), etc. Arsenic usually exists in soil in four oxidation states, viz. As$^{(-III)}$ (arsine), As^0 (elemental arsenic), As$^{(III)}$ (arsenite) and As$^{(V)}$ (arsenate) of which arsenite is the most toxic and mobile but arsenate is the most abundant species in groundwater. Toxicity of arsenite predominantly occurs due to its binding ability to the sulfhydryl groups, present in proteins (Gebel 2002). On the contrary, arsenate is present in the environment as a toxic analog of inorganic phosphate. It is generally introduced into the living cell by mimicking the same phosphate transport mechanism through the cell membrane and also

---

Uttar Banga Krishi Viswavidyalaya, Cooch Behar 736165, India.
* Corresponding author: shovik@ubkv.ac.in

interferes in phosphorylating metabolisms occurring in living systems (Shrestha et al. 2008). Arsenic is also present in atmosphere as minute particles and can stay there for a long time as a mixture of both arsenite and arsenate. Arsenic compounds are not biodegradable, get converted from organic to inorganic form or vice-versa and easily react with other chemicals.

Large doses of arsenic have been known to be a human poison for centuries. However, it is creating alarming environmental concern in relatively low doses also in recent times. The United States Environmental Protection Agency (USEPA) declared arsenic a *Class A* carcinogen. Presently, high toxicity of arsenic and its increased appearance in the biological system has sparked public and political concern. Presence of arsenic in groundwater, above the World Health Organization (WHO) recommended maximum permissible limit for drinking purpose, has been found in several affected zones of the world. About 296 million people, spread in more than 100 countries, are at risk from drinking groundwater contaminated with arsenic above the permissible limit (Chakrabarty et al. 2018). South and South-East Asian countries are facing the worst scenario with more than 187 million people exposed to arsenic contamination. Groundwater lifted from shallow tube wells, especially in lean period (December to April), is used as the major source of irrigation for growing rice and vegetables in these countries. This irrigation water with severe arsenic contamination may lead to the accumulation of arsenic in soil and plant system. Eventually, high accumulation of arsenic in the top soil causes substantial build-up of arsenic in crops as well as its entry in food chain with potential biomagnifications. Magnified arsenic concentration in the biological food chain through crops or fodder is considered as more dangerous for humans than arsenic intake through drinking water. The chronic exposure to arsenic causes severe toxic effects on human health, which is defined by the generic term 'arsenicosis'. The common symptoms of arsenic toxicity are weight loss, loss of appetite, weakness, skin itching, skin cancer, vomiting, diarrhoea, dryness and burning of mouth and throat, dysphasia, fatigue, moderate to severe anaemia, chronic respiratory disorder, and gastrointestinal disorders (Banerjee et al. 2011). Considering its harmful effects, remediation of arsenic toxicity in water-soil-plant system has become a great concern.

Several physical, chemical and biological approaches have been attempted for arsenic removal by researchers worldwide. The physical approaches like adsorption, excavation, coagulation, filtration, solidification, stabilization, solid liquid separation, etc. and chemical approaches like acid washing, absorption, oxidation, precipitation, ion-exchange, chelation, reverse osmosis, coagulation with metal salts, iron filing, photo oxidation, etc. were not found too effective to necessarily reduce the risk and are quite expensive. Therefore, the biological approaches are now being tested as potential arsenic remediation measures. Use of plants and microorganisms for arsenic remediation is considered as a stable, natural and economical solution (Hadis 2011). The interactions between micro-organisms and plant roots are also considered as effective biological mechanism for removal of arsenic. The present chapter overviews the sources and distribution of arsenic in soil-water-plant system and explores the bioremediation measures to reduce its toxicity in food-chain.

2. Origin of arsenic contamination in soil and aquatic environment

The reason of arsenic contamination in soil is still a debated issue. It is believed that arsenic has entered into the environment either through geogenic (geological) or anthropogenic (human activities) sources. In addition, small amount of arsenic also enters in the soil and water from arsenic-rich biological (biogenic) sources. The detailed sources and mechanisms of arsenic release are presented in Figure 1.

2.1 Geogenic origin

The geogenic arsenic contamination in soil and aquatic systems has primarily been derived from volcanic eruption and by release of arsenic containing minerals. Arsenic is present in almost all the geological substances with varying degree of concentration. The average concentration of arsenic is 1.5 to 2.0 mg kg^{-1} in the continental Earth's crust, with presence of 245 arsenic bearing

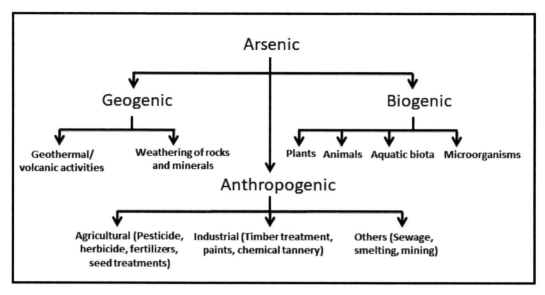

Figure 1. Sources of arsenic in soil and aquatic system (adopted from Mahimairaja et al. 2005, Hasanuzzaman et al. 2015).

minerals. These minerals are mostly sulphide-containing ores of copper, nickel, lead, cobalt, zinc, gold or other base metals. The most predominant sulphide ores of arsenic include arsenopyrite (FeAsS), realgar (AS_4S_4) and orpiment (As_2S_3). Arsenopyrite (FeAsS) is the most abundant arsenic containing mineral, which exists in anaerobic environments as well as in the crystal lattice structure of various sulphide minerals, as the substitute of sulphur. Arsenolite (As_2O_3) is the oxidized form of arsenic present in soil while common reduced forms are orpiment (As_2S_3) and realgar (As_4S_4). The dissolution of arsenic containing minerals was observed to follow the order: native-arsenic > arsenolite > orpiment > realgar > arsenopyrite > tennantite. The release of arsenic in soil and groundwater in deltaic and alluvial plains is mainly attributed to dissolution and desorption of naturally occurring arsenic bearing minerals followed by subsequent leaching and runoff. The geogenic arsenic contamination in soil and groundwater has been observed in different parts of the globe, and the Ganga-Brahmaputra-Meghna fluvial plains of West Bengal, India and adjacent Bangladesh are the typical examples of it.

Based on arsenic geochemistry, three natural phenomena have been believed to be responsible for geogenic arsenic release in groundwater. They are (i) oxidation of arsenic-bearing primary minerals, (ii) dissolution of sorbed arsenic from iron oxyhydroxides and (iii) release of arsenic by competitive ion exchange with phosphates (Bose and Sharma 2002, Hasanuzzaman et al. 2015). (i) Oxidation of arsenic-bearing pyrite minerals is believed to mobilize arsenic in soil. When exposed to atmosphere, insoluble arsenopyrite (FeAsS) rapidly oxidises and releases soluble arsenite, sulphate and divalent ferrous iron (Fe^{II}). The oxidation of arsenopyrite, however, depends on oxygen availability and oxidation of sulphide (S^{2-}) to sulphate (SO_4^{2-}). The released arsenite is further oxidised to arsenate by microbially mediated reactions. (ii) In oxidized condition, the iron arsenic may be sorbed by goethite (FeOOH) coatings, present on the surface of soil particles. Dissolution of this goethite coating, in waterlogged reduced condition, liberates arsenic. The process is further driven by fermentation of peat in the subsurface layer, releasing organic acids such as acetic acid (CH_3COOH), resulting in release of arsenite and arsenate from FeOOH coatings. (iii) The third geogenic mechanism of arsenic release in soil is competitive ion exchange with phosphates ($H_2PO_4^-$), added through the application of phosphatic fertilizers in the cropped fields. Migration of these phosphates' ions to the subsurface aquifers leads to release of sorbed arsenic from aquifer minerals. Among these three geogenic processes, dissolution of FeOOH under reduced conditions is considered as the most probable cause for arsenic accumulation in groundwater.

2.2 Anthropogenic and biogenic origin

Substantial amount of arsenic is added in soil and in aquatic systems by various human (anthropogenic) activities. The major anthropogenic activities responsible for arsenic release are ore dressing, smelting of non-ferrous metals, mining, electronic industries, chemical industries, dye industries, tanning industries, burning of fossils fuels, glass and mirror manufacture, wood processing and preservation, production and use of arsenic containing pesticides, paints, pigments, cosmetics, fungicides, insecticides, etc. The arsenic released from these various anthropogenic activities differs greatly in chemical nature and bioavailability (Mahimairaja et al. 2005). Anthropogenic activity is considered as the predominant sources of arsenic contamination in industrial zones of developed countries as well as in the mega-cities of the developing countries (Cullen and Reimer 1989). A study in the landfill area near Kolkata, India exhibited high arsenic load due to continuous reception of urban and industrial waste (Chakraborty et al. 2017). Industrial wastewater is also responsible for arsenic pollution in urban agglomerations (Deb and Dutta 2017).

Arsenic has widely been used in agriculture since long. However, rapid increase of fertilizers and pesticide application after the green revolution accelerated the use of arsenic. The field application of arsenic containing pesticides causes contamination in soil and water bodies. Sodium arsenite was a known fungicide for protection of grapevines from excoriosis and was in use until 2001. Several arsenic containing compounds such as zinc arsenite, zinc arsenate, lead arsenate, calcium arsenate, magnesium arsenate, Paris Green, etc. are still being used as pesticides in several parts of the world. Disposal from pesticide manufacturing industries is also responsible for arsenic contamination in soil and water system. Herbicides are also reported to release arsenic in soil. Monosodium methanearsonate and disodium methanearsonate are two examples of such herbicides which were used since mid-70s (Smith et al. 1998). Sodium arsenite was used to control aquatic weeds and was reported to be responsible for contamination of lakes and small fish ponds in many areas in United States of America (USA) (Adriano 2001). Fortunately, there has been a gradual decline in the use of arsenicals in agriculture. On the contrary, emission of arsenic oxide through burning of arsenic-rich fossil fuels is considered as one of the important anthropogenic reasons for atmospheric arsenic contamination. Mining and smelting of metals, like gold, copper, lead and zinc, etc. also releases a large amount of arsenic because it is widely present in the sulphide ores of these metals.

A small amount of arsenic can be added in the environment by some plants and microorganisms. However, biological activities affect the redistribution of arsenic by several mechanisms like bioaccumulation (e.g., biosorption), biotransformation (e.g., biomethylation) and transfer (e.g., volatilization). The bioaccumulated arsenic can be transferred from soil to plant, plant to animal and then ultimately to the human through food-chain. The arsenic in the food-chain gets biomagnified in concentration several times than that in water and affects animal and human health. Organisms living in aquatic environment also accumulate higher concentration of arsenic than the water in which they live and subsequently become a source of contamination.

3. Guideline value of maximum arsenic contamination

Considering the widespread and verified negative health effects on humans, WHO recommended provisional health-based guideline value of total arsenic concentration of 10 µg L^{-1} in drinking water as 'safe' limit since 1993 (WHO 1993). Considering the difficulties in practical quantification and removal of arsenic in range of 1 to 10 µg L^{-1} and below from drinking water, especially from small supplies, the WHO guideline value of 10 µg L^{-1} is designated as provisional. However, there are evidences of arsenic related health hazards even from low concentration on long-term exposure. Many countries, viz. Japan, Jordan, Laos, Mongolia, Namibia, Syria, USA and the European Union have adopted this 'safe limit' as their national standard for drinking water. However, several other counties like India and Bangladesh have accepted 50 µg arsenic L^{-1} in drinking water as National Standard based on 1971 WHO advice. Some other countries such as Bahrain, Bolivia, China, Egypt, Indonesia, Oman, Philippines, Saudi Arabia, Sri Lanka, Vietnam, Zimbabwe, etc. also did not update their drinking water standards as per recent WHO recommendation and retain the old WHO guideline of 50 µg L^{-1} (Johnston et al. 2001). Australia has set most stringent national standard for acceptable arsenic concentration in drinking water (7 µg L^{-1}). The joint Food and Agriculture Organization (FAO)-WHO Expert Committee on Food Additives has set 0.002 mg arsenic kg^{-1} of body weight as provisional maximum tolerable daily intake limit of inorganic arsenic in 1983.

4. Arsenic distribution in the world

Due to the worldwide distribution of arsenic contamination and its severe impact on health, arsenic problem is frequently illustrated by the term 'mass poisoning'. United Nations considered arsenic poisoning as the second most important drinking water related health-threat (Johnston et al. 2001). The arsenic-induced health effects were first reported from Cordoba province of Argentina in 1917 and was popularly known as "Bell Ville Disease" (Chakraborti et al. 2018). Earlier (before 2000), only a few countries worldwide were considered as arsenic contaminated, viz. Bangladesh, China, India (West Bengal), Taiwan and Thailand. In the first decade of this century, arsenic related groundwater problems emerged in different countries across the globe. Currently, 107 countries in five continents are considered as affected with arsenic contamination in groundwater (Figure 2). In all these countries, around 296 million peoples are reported to be at potential risk of toxicity due to arsenic consumption through drinking water and also from food.

In India, arsenic contamination in groundwater in the range of 50 to 3700 µg L^{-1} has been reported from several states, namely Punjab, Haryana, Uttar Pradesh, Madhya Pradesh, Jharkhand, Chhattisgarh, Assam, Manipur, Tripura, Nagaland, Arunachal Pradesh, Bihar and West Bengal (Figure 3). Among these, Bihar and West Bengal are the most severely affected. More than 40 percent of the people of Bihar and West Bengal are affected with arsenic contamination in groundwater and are suffering from serious health hazards. The presence of arsenic in groundwater of West Bengal, exceeding the acceptable limit (50 µg L^{-1}), was first detected in 1978 and the first cases of arsenicosis

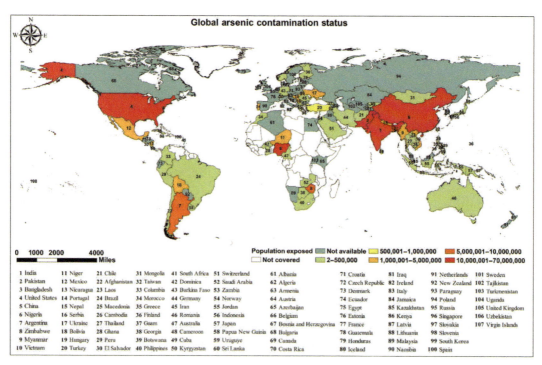

Figure 2. Current global groundwater arsenic contamination situation and potentially exposed population (Source: Chakraborti et al. 2018).

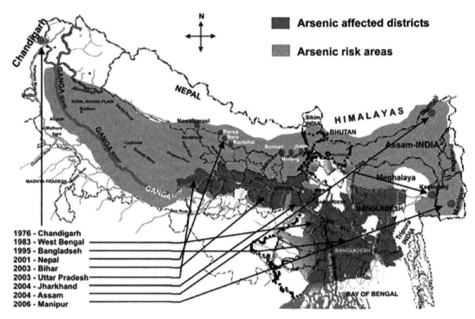

Figure 3. Arsenic affected stretches of Ganga-Meghna-Brahmaputra plains of North and North-Eastern India (source: Datta 2015).

were identified from one village of 24 Parganas district (Garai et al. 1984). In next few years, scientists identified arsenic contamination in several other adjoining districts. At present, approximately 111 blocks spreading over 12 districts along with the Kolkata city are considered as arsenic affected. In the northern states, high arsenic contamination has been reported from 17 districts of Punjab and 14 districts of Haryana. In North-Eastern part of the country, arsenic contamination in groundwater has been reported from several districts of Assam, Manipur, Nagaland, Tripura and Arunachal Pradesh (Mukherjee et al. 2006, Singh 2007, Chakraborti et al. 2008, Banerjee et al. 2011).

Arsenic contamination in groundwater of Bangladesh is considered as one of the greatest environmental disasters in the world. In Bangladesh, the groundwater of 61 out 64 districts is severely contaminated with arsenic. Over 35 million people are exposed to arsenic contamination in drinking water. In addition, around 57 million people are at the potential risk of arsenic exposure, especially in areas where groundwater is used for drinking, cooking and other household activities. In some parts of Bangladesh, arsenic in drinking water is considered as major cause of chronic health problem and at least 1 million people are feared to be affected with arsenicosis (Hasanuzzaman et al. 2015).

In Americas (North and South America), 21 countries are affected with arsenic. Among those, the problem is acute in Mexico, USA, Chile and Argentina. In Latin America, over 4 million people are affected with arsenic contamination in drinking water. In USA, 30 million peoples are in danger of arsenic poisoning through drinking water and food-chain. In USA, arsenic problem is mostly present in states like Alaska, Arizona, California, Hawaii, Idaho, New Hampshire, Nevada, Utah, Oregon and Washington.

As per the current status, 34 European countries are facing arsenic exposure, threatening over 8 million people. Portugal is the most severely affected country with 2.8 million people at potential risk followed by Serbia with 2 million affected people. The other alarming arsenic affected European countries are Hungary, Romania and Croatia. A study of groundwater quality reported that drinking water of around 400 towns and villages situated in the Great Hungarian Plain have arsenic contamination several times higher than the WHO guidelines (Csalagovits 1999).

About 15 African countries are found to be affected with groundwater arsenic contamination with 24.3 million potentially exposed population. In Africa, high concentration of arsenic has been noticed predominantly in Nigeria, Ethiopia, Ghana, Morocco, Togo, Zimbabwe, and in some areas of Tanzania while the coastal countries are the least affected (Ahoulé et al. 2015).

5. Entry of arsenic in the food chain

Arsenic contamination in groundwater is often considered as a grave environmental issue on global scale. However, it is not the only intake source of arsenic for humans. Cereals, pulses, vegetables and other crops grown on soils of the arsenic-contaminated zones can also act as a potential source of arsenic (Williams et al. 2005). Therefore, the total dietary exposure to arsenic by humans as well as animals should be calculated considering arsenic intake through both drinking water and foodstuffs (Arslan et al. 2017). Since 85–90% groundwater is extensively used for purpose of irrigation, particularly in the arsenic affected belts of India and Bangladesh, the possibility of arsenic accumulation in agricultural soils and products is anticipated. Irrigation with arsenic laden groundwater can enhance the arsenic level in agricultural soils up to five-fold than the normal unaffected soils (Ahsan et al. 2009). It was reported that the soil arsenic levels could be raised by 1 µg g^{-1} per annum due to irrigation with groundwater contaminated with arsenic (Meharg and Rahman 2003).

Entry of arsenic in the food-chain also results in diffused arsenic poisoning. On absorption by the agricultural crops, arsenic may substantially be added to the dietary intake through bio-magnification in food-chain. Consumption of these foods can affect a larger population even beyond the geographically affected zones. However, the uptake of arsenic in different crops varies widely. The concentration of arsenic in the edible parts of agricultural crops is reported to vary from 0.007 to 7.5 mg kg^{-1} (Dahal et al. 2008). Some hyper-accumulator plants can uptake huge amount of arsenic and also can translocate it from root to shoot. Among the cereals, rice accumulates good quantity of mobile and reduced form of arsenic (arsenite) as rice is grown in flooded submerged field condition (Srivastava et al. 2012). Rice grains are susceptible to the enrichment of arsenic in the affected belts of India and Bangladesh. The arsenic translocation factor of the rice plant (0.8) is also high (Xu et al. 2008). Leafy vegetables, grown in arsenic affected soils and irrigated by arsenic contaminated water, can also accumulate elevated amount of arsenic (Deb and Dutta 2017).

6. Bioremediation of arsenic toxicity in water-soil-plant system

Following the worldwide severe arsenic pollution scenario, remediation or treatments to remove arsenic from environment by physical, chemical and biological techniques are major challenges. The physical and chemical arsenic removal techniques attempt to convert mobile arsenite to less mobile arsenate to arrest it in soil and to restrict its entry in the water-plant systems. Adsorption, precipitation, coagulation, and filtration are some such physical arsenic removal processes. Physical adsorption can be performed using different types of clays like kaolinite and illite, which can adsorb and increase the oxidation of arsenite to arsenate (Manning and Goldberg 1997). Few other adsorbents like humic acid, activated alumina, granulated ferric hydroxide, etc. have also been used to adsorb arsenic. Chemical oxidation and precipitation reactions are also being used as the effective key to reduce arsenic toxicity. The oxidizing agents like potassium permanganate, chlorine, ozone, hydrogen peroxide or manganese oxides play key role in reducing arsenic toxicity. Besides these, some agronomic management practices such as use of surface water (e.g., harvested rainwater) for irrigation during the lean period have also been suggested to reduce accumulation of arsenic in food crops. However, these physical and chemical approaches are proved to be expensive and labour-intensive. Therefore, scientists started to focus on biological remediation or bioremediation processes in last decades using biological agents like green plants, bacteria, fungi, etc. to remove or neutralize arsenic contamination in environment. The following section details a few such methods.

6.1 Phytoremediation

Phytoremediation of arsenic is the process of using green plants to clear hazardous arsenic from the contaminated soil and groundwater. This is a low-cost green technology and proper implementation makes it eco-friendly. This does not require luxurious equipment or highly-specialized human resources and is simple to implement. Among phytoremediation technologies, phytoextraction and phytostabilization are the two major approaches in remediation of soils contaminated with arsenic. The plants used for arsenic phytoremediation should be highly tolerant to arsenic and efficient in accumulation of arsenic into aboveground biomass.

Chinese brake fern (*Pteris vittata* L.) is a very efficient arsenic hyper-accumulator and can be used for phytoextracting arsenic from contaminated soils (Ma et al. 2001). This plant is a fast growing vegetation of tropical and subtropical areas. The mechanism of arsenic removal from soil by *P. vittata* is principally credited to fast arsenic mobilization in rhizosphere, efficient uptake by the roots and unique metabolism characteristic of the plant (Mathews et al. 2010). As per study, a period of 7–8 years is required for *P. vittata* to totally remove arsenic from top 15 cm soil (Ma et al. 2001). Root exudates and bacteria associated with *P. vittata* enhance arsenic solubilisation in the rhizosphere. Further, arsenic-resistant bacteria (like *Pseudomonas* sp., *Comamonas* sp. and *Stenotrophomonas* sp.) present in the *P. vittata* rhizosphere exhibit incredible ability to boost arsenic concentration in the uptake solution by solubilizing insoluble $FeAsO_4$ and $AlAsO_4$ (Ghosh et al. 2011). The pyochelin-type siderophores, produced by arsenic-resistant bacteria in rhizosphere, also play a role in arsenic solubilisation. There are also other arsenic hyperaccumulators like *Pityrogramma calomelanosis*, some species of *Pteridaceae* family, etc. The translocation and accumulation of arsenic is higher in younger plants than the older plants, possibly due to higher metabolic activity of younger plants (Gonzaga et al. 2007). This signifies the requirement of using young plants in phytoextraction and harvesting of the shoots before plant-death to minimize the return of arsenic to soil. Proper agricultural practices and plant management like application of specific fertilizers and rhizosphere manipulation also help to acquire efficient phytoextraction. For instance, application of phosphorus in soil can enhance arsenic uptake by increasing soil arsenic bioavailability (Fayiga and Ma 2006). Rhizosphere manipulation through application of mycorrhiza can also change arsenic accumulation dynamics and the species of arbuscular mycorrhizal fungi plays an important role here (Trotta et al. 2006).

Phytostabilization is a method where plant roots and associated microbes limit the contamination of arsenic in soil through restricting its mobility and bioavailability. In this process, the amount of water percolating through soil matrix gets reduced by the plants and this acts as a hurdle towards direct contact of plants with arsenic. Plant species like *Populusand salix* is capable of this process. Such plants generally have high adaptability and tolerance to multiple metals and metalloids. In arsenic contaminated areas with poor soil nutritional status, adoption of leguminous plants is also a good option as their nitrogen fixing ability helps in revegetation. In addition, legume plants do not have the ability of shoot translocation of metals and thus results in low exposure risk of arsenic to animals and human. The white lupin (*Lupinusalbus*) is a good nitrogen fixer plant for arsenic contaminated soil.

Uptake of arsenic from the soil, transforming it into volatile forms and releasing it to the atmosphere through transpiration is known as phytovolatilization. Here, the plants obtain the soil arsenic through water, passes it through the xylem towards leaves and convert it into non-toxic forms. These non-toxic forms (methylated and volatile arsenic compounds) are finally volatilized in the atmosphere (Heaton et al. 1998). The *P. vittata* is considered as an efficient arsenic volatilizer.

6.2 Microbial bioremediation of arsenic

Use of microorganisms for bioremediation of arsenic polluted soils has an excellent potential and is an environment friendly approach. Several microorganisms showed potential mechanisms to live in arsenic-enriched environments and fight against the toxicity of arsenic. Besides, association of these microorganisms with arsenic for a considerable period helps those (microorganisms) to adapt in such ecology (Banerjee et al. 2013). Certain bacteria and fungi also have arsenic detoxification mechanisms that can neutralize its toxic effects (Kruger et al. 2013). A range of microorganisms assimilate arsenic, utilize arsenate as the electron receiver in anaerobic soil or generate energy by oxidation of arsenite (Kruger et al. 2013). These microorganisms are ideal for management of arsenic contaminated soils. Microorganisms can also methylate arsenite into volatile compounds through bioaccumulation and biotransformation (Yang and Rosen 2016).

A number of rhizospheric bacteria (like *Rhizobium, Frankia, Klebsiella, Clostridium, Bacillus, Pseudomonas, Azotobactor*) are partially resistant to arsenic and also have plant growth promoting abilities and thus have considerable role in bioremediation of arsenic (Gupta et al. 2015). On the contrary, some of the bacterial genera like *Aeromonas* and *Exiguobacterium* can directly interact with the soil arsenic. Many other bacteria are able to oxidise arsenite into the less toxic arsenate. The oxidation of arsenite is catalysed by arsenite oxidase enzyme (Krugar et al. 2013). A common site for arsenite oxidation is the periplasm of the bacteria (Silver and Phung 2005). Few genera of bacteria are also able to reduce arsenate to arsenite using arsenate reductase enzyme. The reduced arsenite is either released from the cell or accumulates into the intracellular compartments, in the form of free arsenite or gets combined with glutathione or other thiols. In general, gram-positive bacteria has been found as more efficient in using higher concentrations of arsenic compared to gram-negative bacteria. Few bacterial genera like *Alcligenes, Pseudomonas* and *Mycobacterium* have been found to have role in arsenic methylation (Hughes 2002).

There is scope to use fungi (like *Aspergillus flavus, Alcaligenes* sp., *Thiomonas* sp. and *Trichoderma* sp.) in remediation of arsenic contaminated soil and water (Duquesne et al. 2007). Fungi are omnipresent in the soil, particularly in acidic conditions. The ability of fungi to solubilize, transform and uptake arsenic plays a crucial role in its remediation. Fungi uses a numbers of mechanisms like efficient transformation, intracellular precipitation and rapid uptake to tolerate, detoxify and absorb arsenic. Several fungal species (like *Aspergillus, Mucor, Fusarium, Paecilomyces*) are also capable of transforming inorganic arsenic compounds by biomethylation (Pickett et al. 1981). In this process, arsenite and arsenate are transformed into less toxic monomethylarsonic acid, dimethylarsinic acid, trimethylarsine and trimethylarsine-oxide. Such transformation of arsenic into less toxic forms plays a vital role in reducing arsenic pollution. Alternatively, *Trichoderma* can transform inorganic arsenic to organic form and then to volatile compounds (Tripathi et al. 2017). A number of *Aspergillus* species, isolated from different environments, have been found capable of removing considerable amount of arsenic by bioaccumulation and biovolatilization (Urik et al. 2007, Vala 2010). Similarly, *Penicillium gladioli, Penicillium glabrum, Penicillium purpurogenum,* and *Penicillium chrysogenum* showed ability to remove substantial quantity of arsenic from environment (Urik et al. 2007). The filamentous fungi (like *Fomitopsis pinicola, Penicillium gladioli, Fusarium oxysporummeloni* and *Scopulariopsis koningii*) are more resistant to arsenic pollution and have higher accumulating and volatilizing capacity of arsenic, as compared to unicellular yeasts. Arbuscular mycorrhizal fungi benefit their hosts by increasing the uptake of competitive nutrient (phosphorus), thus curbing arsenic toxicity in plants (Long et al. 2010). However, the capacity of arbuscular mycorrhizal fungi to reduce arsenic toxicity changes with the host plant (Jankong and Visoottiviseth 2008). A few species of *Glomus* also play a role in arsenic accumulation and thereby can reduce soil arsenic toxicity.

Some algae (like *Anabena, Nostoc, Chlorella, Dunaliella, Desmodesmus*) can also be effectively used to accumulate excess amount of arsenic from water and soil. Different genera of arsenic hyperaccumulating algae (like *Oscillatoria, Nitzschia, Gloeotrichia, Navicula*) are suggested to be grown in the rice field to curb the arsenic uptake by rice (Huq et al. 2007). Bioremediation capacity has also been reported in some other species of algae like *Ankistrodesmus convolutes, Euglena gracilis, Fucusgardneri, Lessonianigrescens, Spirulina platensis,* etc. Some marine algae can also perform oxidation and reduction of arsenic species, arsenic methylation and transformation of arsenic into arseno-sugars or arseno-lipids by their metabolism (Duncan et al. 2015, Wang et al. 2015).

7. Future perspective and conclusion

Arsenic contamination in groundwater and food-web is a recognized serious global environmental threat, which causes severe health damage to millions of people. Therefore, scientists around the world are looking into different kinds of remediation techniques. However, most of these are still in laboratory test condition and hard to implement due to its complexity and cost. At present, use of cheap and available adsorbents like activated carbon, activated alumina, clays, hydroxyl apatite, zeolites and iron oxide is considered as the best physical option to remove arsenic from drinking water. However, the large-scale use of these sorbents at the village level in contaminated areas is financially and practically not feasible, particularly in developing countries. Further, these adsorbents have limitation in terms of arsenic adsorption capacity. Separation of arsenic from these adsorbents or dumping of these after use also may result in return of arsenic in soil and water. Likewise, use of chemicals for oxidation and precipitation of arsenic is not suitable on a large scale.

In this context, bioremediation, using plant and microbes, has emerged as a potential tool to remove arsenic from the environment. Bioremediation has the merits for being cost-effective and environmentally friendly. These techniques can also be coupled with other physical or chemical treatments for better and more efficient result. However, different bioremediation techniques also have their own limitations. Most of the hyperaccumulating plants are slow growers and have low biomass. Phytoremediation processes like phytoextraction, rhizofiltration, phytostabilization, etc. depend on many environmental factors, such as suitability of soil and weather condition for plant growth, depth and extent of the contamination, depth of plant root system etc. In addition, more researches are required for better understanding of the physiology, biochemistry, nutrition and other biological parameters of these plants. Dumping of the hyperaccumulating plants after arsenic phytoextraction is also an issue to be considered seriously. Proper post-harvest management needs to be developed for this. Significant research gap still exists in selection and popularization of arsenic tolerant plant species except *P. vittata*. Likewise, microbial bioremediation has established itself as a potential tool for arsenic removal but is yet to be proved sufficient on global scale. Developing genetically engineered microorganisms may be a potential option for more effective and large scale bioremediation. More knowledge about the metabolism and detoxification pathway of arsenic by different types of microorganisms will be helpful in this context. Recent developments in genomics have identified and characterized arsenic-tolerant genes and proteins. These genes could be engineered into other microorganisms, which can adapt well in certain environmental conditions for improved bioremediation efficiency. More studies should be carried out to enhance the use of transgenic microorganisms and to understand their metabolic pathways for arsenic detoxification. Selection of potentially efficient arsenic oxidizing/reducing microorganisms from indigenous microbial population is also essential for successful bioremediation. The meticulous use of indigenous and genetically engineered microorganisms with cutting-edge biotechnical knowledge can protect the ecosystems from the arsenic menace.

References

Adriano, D.C. 2001. Trace Elements in Terrestrial Environments: Biogeochemistry, Bioavailability and Risks of Metals. Second Edn. Springer, New York.

Ahouléet, D.G., Lalanne, F., Mendret, J., Brosillon, S. and Maïga, A.H. 2015. Arsenic in African waters: a review. Water Air Soil Poll. 226: 302.

Ahsan, D.A., DelValls, T.A. and Blasco, J. 2009. Distribution of arsenic and trace metals in the floodplain agricultural soil of Bangladesh. B. Environ. Contam. Tox. 82: 11–15.

Arslan, B., Djamgoz, M.B.A. and Akun, E. 2017. Arsenic: a review on exposure pathways, accumulation, mobility and transmission into the human food chain. Rev. Environ. Contam. T. 243: 27–51.

Banerjee, S., Datta, S., Chattyopadhyay, D. and Sarkar, P. 2011. Arsenic accumulating and transforming bacteria isolated from contaminated soil for potential use in bioremediation. J. Environ. Sci. Heal. A. 46: 1736–1747.

Banerjee, S., Majumdar, J., Samal, A.C., Bhattachariya, P. and Santra, S.C. 2013. Biotransformation and bioaccumulation of arsenic by Brevibacillus brevis isolated from arsenic contaminated region of West Bengal. IOSR J. Environ. Sci. Toxicol. Food Technol. 3: 1–10.

Bose, P. and Sharma, A. 2002. Role of iron in controlling speciation and mobilization of arsenic in subsurface environment. Wat. Res. 36: 4916–4926

Cavalca, L., Corsini, A., Zaccheo, P., Andreoni, V. and Muyzer, G. 2013. Microbial transformations of arsenic: perspectives for biological removal of arsenic from water. Future Microbiol. 8: 753–68.

Chakraborti, D., Singh, E.J., Das, B., Shah, B.A., Hossain, M.A., Nayak, B., Ahamed, S. and Singh, N.R. 2008. Ground water arsenic contamination in Manipur, one of the seven north-eastern hill states of India: a future danger. Environ. Geol. 56: 381–390.

Chakraborti, D., Singh, S.K., Rashid, M.H. and Rahman, M.M. 2018. Arsenic: occurrence in groundwater. pp. 153–168. In: Nriagu, J. (ed.). Encyclopedia of Environmental Health. 2nd Edition. Elsevier.

Chakraborty, S., Weindorf, D.C., Deb, S., Li, B., Paul, S., Choudhury, A. and Ray, D.P. 2017. Rapid assessment of regional soil arsenic pollution risk via diffuse reflectance spectroscopy. Geoderma 289: 72–81.

Csalagovits, I. 1999. Arsenic–bearing artesian waters of Hungary. Annual Report of the Geological Institute of Hungary. 1992-1993/II. pp. 85–92.

Dahal, B.M., Fuerhacker, M., Mentler, A., Karki, K.B., Shrestha, R.R. and Blum, W.E.H. 2008. Arsenic contamination of soils and agricultural plants through irrigation water in Nepal. Environ. Pollut. 155: 157–163.

Datta, S. 2015. Hydrological aspects of arsenic contamination of groundwater in eastern India. Adv. Agron. 132: 75–137.

Deb, S. and Dutta, P. 2017. Wastewater in agriculture: possibilities and limitations. pp. 215–225. In: Rakshit, A., Abhilash, P.C., Singh, H.B. and Ghosh, S. (eds.). Adaptive Soil Management: From Theory to Practices. Springer Nature, Singapore.

Duncan, E.G., Maher, W.A. and Foster, S.D. 2015. Contribution of arsenic species in unicellular algae to the cycling of arsenic in marine ecosystems. Environ. Sci. Technol. 49: 33–50.

Fayiga, A.O. and Ma, L.Q. 2006. Using phosphate rock to immobilize metals in soil and increase arsenic uptake by hyperaccumulator *Pteris vittata*. Sci. Total Environ. 359: 17–25.

Garai, R., Chakraborti, A.K., Dey, S.B. and Saha, K.C. 1984. Chronic arsenic poisoning from tube well water. J. Indian. Med. Assoc. 82: 34–5.

Gebel, T.W. 2002. Arsenic methylation is a process of detoxification through accelerated excretion. Int. J. Hyg. Envir. Heal. 205: 505–508.

Ghosh, P., Rathinasabapathi, B. and Ma, L.Q. 2011. Arsenic-resistant bacteria solubilized arsenic in the growth media and increased growth of arsenic hyperaccumulator *Pteris vittata* L. Bioresour. Technol. 102: 8756–8761.

Gonzaga, M.I.S., Ma, L.Q. and Santos, J.A.G. 2007. Effects of plant age on arsenic hyperaccumulation by *Pteris vittata* L. Water Air Soil Poll. 186: 289–295.

Gupta, G., Parihar, S.S., Ahirwar, N.K., Snehi, S.K. and Singh, V. 2015. Plant growth promoting rhizobacteria (PGPR): current and future prospects for development of sustainable agriculture. J. Microb. Biochem. Technol. 7: 96–102.

Hadis, G. 2011. Investigation of bioremediation of arsenic by bacteria isolated from contaminated soil. Afr. J. Microbiol. Res. 5: 5889–5895.

Hasanuzzaman, M., Nahar, K., Hakeem, K.R., Ozturkand, M. and Fujita, M. 2015. Arsenic toxicity in plants and possible remediation. pp. 433–501. In: Hakeem, K.R., Sabir, M., Ozturk, M. and Murmut, A. (eds.). Soil Remediation and Plants. Academic press, Elsevier.

Heaton, A.C.P., Rugh, C.L., Wang, N. and Meagher, R.B. 1998. Phytoremediation of mercury and methyl mercury-polluted soils using genetically engineered plants. J. Soil Contam. 7: 497–510.

Hughes, M.F. 2002. Arsenic toxicity and potential mechanisms of action. Toxicol. Lett. 133: 1–16.

Huq, S.M.I., Abdullah, M.B. and Joardar, J.C. 2007. Bioremediation of arsenic toxicity by algae in rice culture. Land Contam. Reclamat. 15: 327–333.

Jankong, P. and Visoottiviseth, P. 2008. Effects of arbuscular mycorrhizal inoculation on plants growing on arsenic contaminated soil. Chemosphere 72: 1092–1097.

Johnston, R., Heijnen, H. and Wurzel, P. 2001. Safe water technology, Chapter 6. United Nations Synthesis Report on Arsenic in Drinking Water, World Health Organization, pp. 210–316.

Kruger, M.C., Bertin, P.N., Heipieper, H.J. and Arsène-Ploetze, F. 2013. Bacterial metabolism of environmental arsenic mechanisms and biotechnological applications. Appl. Microbiol. Biot. 97: 3827–3841.

Long, L.K., Yao, Q., Guo, J., Yang, R.H., Huang, Y.H. and Zhu, H.H. 2010. Molecular community analysis of arbuscular mycorrhizal fungi associated with five selected plant species form heavy metal polluted soils. Eur. J. Soil Biol. 46: 288–294.

Ma, L.Q., Komar, K.M., Tu, C., Zhang, W. and Cai, Y. 2001. A fern that hyperaccumulates arsenic. Nature 409: 579.

Mahimairaja, S., Bolan, N.S., Adriano, D.C. and Robinson, B. 2005. Arsenic contamination and its risk management in complex environmental settings. Adv. Agron. 86: 1–82.

Manning, B.A. and Goldberg, S. 1997. Adsorption and stability of arsenic (III) at the clay mineral-water interface. Environ. Sci. Technol. 31: 2005–2011.

Mathews, S., Ma, L.Q., Rathinasabapathi, B., Natarajan, S. and Saha, U.K. 2010. Arsenic transformation in the growth media and biomass of hyperaccumulator *Pteris vittata* L. Bioresource Technol. 101: 8024–8030.

Meharg, A.A. and Rahman, M.M. 2003. Arsenic contamination of Bangladesh paddy field soils: implications for rice contribution to arsenic consumption. Environ. Sci. Technol. 37: 229–234.

Mukherjee, A., Sengupta, M.K., Hossain, M.A., Ahamed, S., Das, B., Nayak, B., Lodh, D., Rahman, M.M. and Chakraborti, D. 2006. Arsenic contamination in groundwater: A global perspective with emphasis on the Asian scenario. J. Health. Popul. Nutr. 24: 142–163.

Pickett, A.W., McBride, B.C., Cullen, W.R. and Manji, H. 1981. The reduction of trimethylarsine oxide by *Candida humicola*. Can. J. Microbiol. 27: 773–778.

Shrestha, R.A., Lama, B., Joshi, J. and Sillanpaa, M. 2008. Effect of Mn (II) and Fe (II) on microbial removal of arsenic (III). Environ. Sci. Pollut. Res. 15: 303–307.

Silver, S. and Phung, L.T. 2005. Genes and enzymes involved in bacterial oxidation and reduction of inorganic arsenic. Appl. Environ. Microb. 71: 599–608.

Singh, A.K. 2007. Approaches for removal of arsenic from groundwater of north-eastern India. Curr. Sci. 92: 1506–1515.

Smith, A.H., Goycolea, M., Haque, R. and Biggs, M.L. 1998. Marked increase in bladder and lung cancer mortality in a region of northern Chile due to arsenic in drinking water. Am. J. Epidemiol. 147: 660–669.

Srivastava, S., Suprasanna, P. and D'Souza, S.F. 2012. Mechanisms of arsenic tolerance and detoxification in plants and their application in transgenic technology: a critical appraisal. Int. J. Phytoremediat. 14: 506–517.

Tripathi, P., Singh, P.C., Mishra, A., Srivastava, S., Chauhan, R. and Awasthi, S. 2017. Arsenic tolerant *Trichoderma* sp. reduces arsenic induced stress in chickpea (*Cicer arietinum*). Environ. Pollut. 223: 137–145.

Trotta, A., Falaschi, P., Cornara, L., Minganti, V., Fusconi, A., Drava, G. and Berta, G. 2006. Arbuscular mycorrhizae increase the arsenic translocation factor in the As hyper accumulating fern *Pteris vittata* L. Chemosphere 65: 74–81.

Urik, M., Cernansky, S., Sevc, J., Simonovicova, A. and Littera, P. 2007. Biovolatilization of arsenic by different fungal strains. Water Air Soil Poll. 186: 337–342.

Vala, A.K. 2010. Tolerance and removal of arsenic by a facultative marine fungus *Aspergillus candidus*. Bioresource Technol. 101: 2565–2567.

Wang, Y., Wang, S., Xu, P., Liu, C., Liu, M. and Wang, Y. 2015. Review of arsenic speciation, toxicity and metabolism in microalgae. Rev. Environ. Sci. Bio. 14: 427–451.

WHO. 1993. Guidelines for Drinking Water Quality. 3rd edition, Geneva, Switzerland.

Williams, P.N., Price, A.H., Raab, A., Hossain, S.A., Feldmann, J. and Meharg, A.A. 2005. Variation in arsenic speciation and concentration in paddy rice related to dietary exposure. Envir. Sci. Tech. Lib. 39: 5531–5540.

Xu, X.Y., McGrath, S.P., Meharg, A.A. and Zhao, F.J. 2008. Growing rice aerobically markedly decreases arsenic accumulation. Environ. Sci. Technol. 42: 5574–5579.

Yang, H.C. and Rosen, B.P. 2016. New mechanisms of bacterial arsenic resistance. Biomed. J. 39: 5–13.

17
Bioremediation of Fluoride and Nitrate Contamination in Soil and Groundwater

Lal Chand Malav,[1] *Gopal Tiwari,*[1,]* *Abhishek Jangir*[1] and *Manoj Parihar*[2]

1. Introduction

In most of the Asian countries, groundwater is mainly used for drinking and irrigation purpose (Al-Hatim et al. 2015, Raj and Shaji 2017). As a result, groundwater demand is increasing continuously due to its over-exploitation (Gleeson et al. 2012, Gupta et al. 2013, Alhababy and Al-Rajab 2015). Fluoride (F^-) is naturally found in soil and groundwater, but some anthropogenic activities such as iron, steel, glass and aluminum industries or agricultural practices by using phosphate fertilizers raise the concentration of F^- in groundwater (Gupta et al. 2015, 2019, Srivastav et al. 2018, Maurya et al. 2019).

Fluoride is an all-encompassing, 13th most abundant electronegative element in the crust of the Earth (Ibrahim et al. 2011). Fluorine is an important trace element that maintains normal animal and human physiological function while excessive or inadequate intake may affect the health of living beings (Liu et al. 2014). Lack of fluoride may create teeth problems, whereas excessive fluorine has a detrimental effect on environmental quality and causes health disorders, including dental fluorosis, skeletal fluorosis, impaired thyroid function and decreased childhood intelligence. With rapid economic growth and widespread human activity, environmental pollution from non-point agricultural and point industrial sources is becoming more serious, resulting in air, water and soil contamination (Wu and Sun 2016). Fluoride (F) exhibits non-biodegradability by biological enrichment and food chain, with apparent toxicity to many species. Thus, a small amount of F pollution can pose harm to living beings (Szostek and Ciecko 2017).

The fluorosis epidemic is all embracing, and nearly 25 nations of the world are under its terrible fate. Many countries, notably India, Sri Lanka and China, the countries of the Rift Valley in East Africa, Turkey and parts of South Africa have reported the occurrence of high fluoride concentrations in groundwater (Singh and Gothalwal 2016b). In India, 15 states are endemic to fluorosis and about 62 million people are affected by fluorosis. Fluoride can act as double edged sword in drinking water (Biswas et al. 2016). While the concentration of fluoride is even less than 0.5 mg L^{-1} of drinking water, it has negative impact on human's health (Fawell et al. 2006). The optimal level of fluoride for drinking water is 0.6–1.2 mg L^{-1}, whereas the discharge cap for industrial wastewater is 2 mg L^{-1} (BIS 2012).

---

[1] ICAR-National Bureau of Soil Survey and Land Use Planning, Nagpur 440033, India.
[2] ICAR-Vivekananda Parvatiya Krishi Anusandhan Sansthan, Almora, Uttarakhand 263601, India.
* Corresponding author: gopalmpkv@gmail.com

Nitrogen, on the other hand, is an essential element for all living things. The atmosphere of Earth is rich in nitrogen, and it is naturally extracted from atmosphere. Nitrogen is a crucial building block in the synthesis of proteins, contributing 13% of the mass. Nitrogen acts as nutrient or bio-stimulant when it is present in water or soil. Nitrogen chemistry is complex as it occurs in eight different oxidation states, and it is noteworthy that living organisms can alter the state. Excess nitrogen may pose a serious environmental problem, including cultural eutrophication, methemoglobinemia, and other health problems. The pollution of nitrates in water and soil has thus become a growing concern for the environment. The maximum level of nitrate contamination, according to USEPA guidelines (1996), is 45 mg L^{-1}, and BIS adopts the same standard. The adverse health and ecosystem effects of nitrate and concerns about decreasing water quality have increased interest in technologies for nitrate removal. The use of any mechanism in the nitrogen cycle that serves as a reservoir for nitrogen in groundwater is advantageous from a water quality point of view. Denitrification is the final step in the production of nitrogen. If oxygen is in shortage, bacteria in the soil and water use nitrate as an electron acceptor. Natural denitrification is a long term process and the rate of denitrification is increased in the biological denitrification technique by continuously supplying the carbon source and maintaining constant process parameters.

Chemical process such as precipitation, weathering, absorption and exchange of ions may influence concentrations of F$^-$ and nitrate in groundwater. Various modern methods of defluoridation and nitrate removal are already being used but these traditional methods have certain drawbacks, including high cost and energy consumption, secondary contamination after treatment and lower efficiency. In this context, bioremediation could be a cost-effective and environmentally friendly way to remove fluoride and nitrate compounds from soil and groundwater.

2. Contamination of F$^-$ in the Asian countries

Groundwater varies from region to region based on temperature, precipitation, rock composition, and terrain. In most of the Asian countries, F$^-$ concentration reached 1.5 mg L^{-1} of groundwater. Several studies in the Province of Afghanistan recorded the highest level of F$^-$ up to 79.2 mg L^{-1}. The F$^-$ content in different regions of South Korea ranged from 20 to 79.2 mg L^{-1} (Chae et al. 2007). The reason behind high F$^-$ content could be due to the geological activities and the presence of F$^-$ containing minerals in groundwater (Msonda et al. 2007). In the Muteh region of Iran, F$^-$ concentration in groundwater ranged from 0.02 to 9.2 mg L^{-1} (Keshavarzi et al. 2010), while it was reported up to 14.10 mg L^{-1} in Yuncheng Basin of China (Li et al. 2015). In many areas, such as Anuradhapura in Sri Lanka (Chandrajith et al. 2012), East Java in Indonesia (Heikens et al. 2005), and some regions of Anatolia in Turkey (Oruc 2008), the F$^-$ content in groundwater ranged from 10 to 20 mg L^{-1}.

3. Indian scenario of fluoride and nitrate contamination

India is one of the countries most affected by F$^-$ contamination in groundwater. Nearly 12 million tons of F$^-$ deposits on the Earth's crust are found in our country, which might be the reason of high F$^-$ occurrence in groundwater. Arsenic and F$^-$ are known as two major pollutants in terms of the number of people affected and the distribution area (Sahu 2019). In many areas, such as Birbhum in West Bengal, the F$^-$ content in groundwater is ranged from 10 to 20 mg L^{-1} (Batabyal and Gupta 2017, Das et al. 2016a). Nevertheless, for irrigation, this water can be used (Sahu et al. 2017). F$^-$ contamination has been recognized as one of the most important natural groundwater quality issues affecting India's most of arid and semi-arid regions. For example, high F$^-$ concentrations were found in Andhra Pradesh's Nalgonda area. Several parts of the country were then found with high concentrations of F$^-$ among which Andhra Pradesh, Rajasthan, Haryana, Punjab, Gujarat and Assam were affected severely (Kumar et al. 2016). For areas near river, the groundwater flow rate is higher with lower residence period which provides lesser contact time between water and F$^-$ bearing rocks. This has resulted in lower groundwater F$^-$ concentrations. Groundwater F$^-$

concentrations below 5 mg L^{-1} were recorded in Indian states of Bihar, Chhattisgarh, Jammu and Kashmir, Jharkhand, Karnataka, Maharashtra, Tamil Nadu, and Uttar Pradesh. Concentrations of 5 to 10 mg L^{-1} were registered in Andhra Pradesh, Delhi, Gujarat, Kerala, Orissa, and West Bengal F$^-$. Fluoride concentrations $\geq$ 10–20 mg L^{-1} was recorded in areas of Madhya Pradesh, Punjab, Assam and Haryana. It has been confirmed that the maximum F$^-$ concentration exceeded 20 mg L^{-1} in many parts of Rajasthan (CGWB 2014). Rajasthan and Gujarat (80–100 percent) were stated to be the highest F$^-$ affected regions, followed by Andhra Pradesh and Punjab (60–80 percent). F$^-$ contamination affects about 21–40% of areas in Madhya Pradesh, Maharashtra, Kerala, Tamil Nadu, Jharkhand, West Bengal and Sikkim.

Nitrate is one of the most common groundwater pollutants in rural areas and is recorded from several areas of Tamil Nadu, Orissa, Karnataka, Maharashtra, Bihar, Gujarat, Madhya Pradesh and Rajasthan. The permissible acceptable level of nitrate concentration in groundwater is 45 mg L^{-1} according to the BIS standard for drinking water. Approximately 37.7 million Indians are expected to be affected by waterborne diseases annually, 1.5 million children reportedly die from diarrhea alone with the loss of 73 million working days each year due to waterborne disease (WHO 2005).

4. Sources of fluoride contamination

Water has two main sources of contamination, i.e., geogenic (contamination that occurs naturally) and anthropogenic (human activity related). There are a wide range of rock types and soil-water interactions resulting in a variety of geogenic contamination in India.

4.1 Geogenic sources

Fluoride is naturally present in rocks, soil and water. The chances of fluoride contamination in groundwater increase due to the presence of the fluoride bearing rocks like fluorspar or calcium fluoride (CaF_2), apatite or rock phosphate [$Ca_3F(PO_4)_3$], cryolite (Na_3AlF_6) and magnesium fluoride (MgF_2) (Biswas et al. 2009). The groundwater contamination of fluoride may also rise as a result of drainage and gradual dissolution from the weathering of the rocks (Das et al. 1998, Koteswar and Metre 2014). The other causes are chemical waste from different industries, contributing to groundwater contamination. Southern India's rocks are rich in fluoride, which is the main cause of pollution of fluoride in groundwater (Brindha and Elango 2011) and granites in the Andhra Pradesh district of Nalgonda contain far higher levels of fluoride than the global average of 810 mgkg^{-1} (Tikki 2014).

Study by Koteswar and Metre (2014) revealed that F$^-$ release in groundwater can occur through fluoride-bearing rocks such as fluorspar and fluorapatite, which further depends on parameters such as mineral solubility, availability, concentration, pH and flowing water velocity. Such rocks' weathering processes could be one of the reason for water fluoride contamination. The F$^-$ and OH$^-$ ions are negatively charged ions and F$^-$ can easily replace OH$^-$ ions in the rock during chemical reaction and increase their concentration in rocks and minerals. Whenever carbonate and bicarbonate-rich water passes through such rock types, F$^-$ are released as a result of chemical reactions and percolate to groundwater (Saxena and Ahmed 2001).

$CaF_2 + Na_2CO_3 \rightarrow CaCO_3 + 2F^- + 2Na^+$

$CaF_2 + 2NaHCO_3 \rightarrow CaCO_3 + 2Na^+ + 2F^- + H_2O + CO_2$

4.2 Anthropogenic sources

Fluoride can be absorbed into soil and water through many point and diffused sources, i.e., industrial effluents discharged from aluminum factories, phosphate industries, brick kilns, coal power plants and from phosphatic fertilized agricultural fields. The processing industries of aluminum, utilize the fluoride compound cryolite as a catalyst in the process of bauxite mining and the resulting gaseous fluoride is discharged directly into the atmosphere or in the waste streams. The average F⁻ content in drainage sources for aluminum reduction plants is 107–145 mg L⁻¹. Concentrations of F⁻ ranging from 1000 to 3000 mg L⁻¹ were reported from the production of glass (Sun et al. 1998). The other non-point groundwater sources also lead to fluoride contamination. Modern agricultural activities such as chemical fertilizers (rich in phosphate) and fluoride containing pesticides increase fluoride levels in groundwater. The appreciable amount of fluoride is also reported from some food items (e.g., cereals and pulses), cosmetics and medicines. The use of NaF drugs for osteoporosis and dental carriers is very common; different toothpaste brands use excessive amounts of F⁻ through which the fluoride is introduced into the oral cavity and reaches the body of humans (Meenakshi and Maheshwari 2006).

5. Sources of nitrate contamination

In the world, there are multiple sources that contribute to the overall nitrate content of natural waters such as geological characteristics, man-made sources, atmospheric nitrogen fixation and soil nitrogen. Wastewater in the upper soil layer can infiltrate the groundwater aquifer from the disposal ponds. The lack of a sewage system encourages these forms of nitrate pollution. Nitrate in groundwater and soil can be derived from natural or point sources, such as water distribution systems and livestock infrastructure by percolation induces contamination of surface water, groundwater and wells. One of the anthropogenic causes of groundwater nitrate pollution is waste materials. The use of nitrogen (N) fertilizer in agriculture has significantly increased over the past 30 years to meet the food requirements of the speedily growing population. Therefore, the use of nitrate fertilizers causes the foremost predicament in groundwater contamination. Nitrate is a key contaminant present in effluent wastewater of various industries including fertilizer, metal ores, explosives, and paper mills. Generally, nitrate level in the contaminated water varies from 200 to 500 mgL⁻¹ based on the nature of source as listed in Table 1. Low-level waste from nuclear industries contains as high as 50000 mg NO_3^- L⁻¹. Waste disposal site drainage and municipal waste are liable to get oxidized to nitrate that also augments pollution of groundwater with nitrate. Apart from chemical industries, nitrate waste is also formed in electronic, mining, and petroleum industries. All these wastewaters, when disposed into the natural water bodies, cause severe health and environmental concerns.

Table 1. Nitrate levels of various wastewater sources as reported in the literature (Rajmohan et al. 2018).

Sr. No.	Wastewater source	Nitrate level (mg l⁻¹)	Reference
1.	Domestic wastewaters	70–85	Oladoja and Ademoroti (2006), Wu et al. (2007)
2.	Fertilizer, diaries, metal finishing industries	200	Peyton et al. (2001)
3.	Brackish water	1000	Dorante et al. (2008)
4.	Tannery	222	Munz et al. (2008)
5.	Glasshouses waste	325	Park et al. (2009)
6.	Explosives factory	3600	Shen et al. (2009)
7.	Nuclear plant	50,000	Francis and Hatcher (1980)

6. Remediation technologies available for fluoride and nitrate removal

Traditionally, various methods are being used for removal of fluoride (Figure 1) and nitrate from contaminated water, among them liming (Harrison 2005), ion-exchange or precipitation (Tressaud 2006), activated alumina (Ghorai and Pant 2005), alum sludge and calcium, reverse osmosis and electro-coagulation are popular in developing nations (Hu et al. 2003, Sehn 2008). Nalgonda technique is most popular in our country but major drawback found in this technology is the presence of high residual aluminum concentration (0.2 mg L⁻¹), even higher than WHO standards (Ayoob

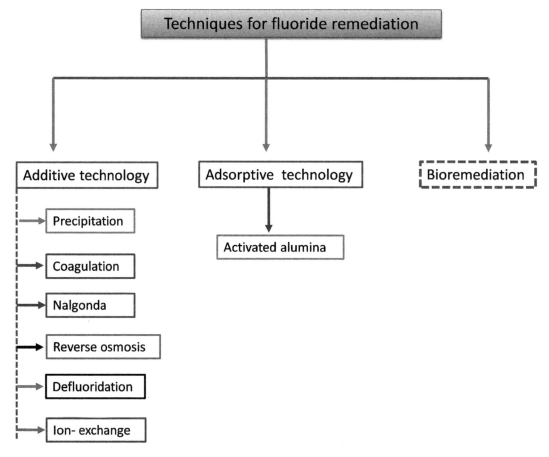

Figure 1. Flow chart showing various technologies for remediation of fluoride.

et al. 2008). Among other methods, adsorption is the reliable technique because it is easy to operate, and provides wider choice for adsorbents' selection (Mohapatra et al. 2009).

Conventional methods of fluoride and nitrate removal from contaminated water (Piddennavar and Krishnappa 2013) have some limitations as they produce chemical waste in the environment system, have high cost and energy involved, and presence of secondary pollutants in wastewater (Gentili and Fick 2016). Therefore, microbial methods can be a viable alternative when they have the capability of developing resistance to different pollutants through bioaccumulation, biotransformation and biosorption (Chouhan et al. 2012). Cost-effectiveness, operational simplicity and less sludge production are major advantages of biological approach of remediation. However, in many studies phytoremediation and bioremediation are found as two of the most promising techniques for removal of pollutants from ground water and soil (Cho et al. 2011, Hu et al. 2000, Kushwaha et al. 2014, Sindelar et al. 2015).

6.1 Bioremediation processes of fluoride removal

A natural process of mineralization of organic and inorganic compounds through various microbes is called biodegradation. These microbes are bacteria, fungi and algae (Rutkowska et al. 2002). An essential step in the biodegradation is dehalogenation of organohalogen compounds catalyzed by different dehalogenases (haloaciddehalogenases, halohydrindehalogenases and haloalkanedehalogenase), which have the capability to dehalogenate the aliphatic and aromatic compounds (Janssen et al. 2005). The fluoroacetate degrading bacteria are *Acinetobacter*,

Arthrobacter, Aureobacterium, Bacillus, Pseudomonas, Streptomyces and *Weeksella* (Kumar and Haripriya 2013) with *Synergistetes* (a single ruminal bacterial phylum) which shows the ability to degrade sodium fluoroacetate (Leong et al. 2010, Davis et al. 2011) (Table 2).

The fluoroacetate is defluorinated by fluoroacetate dehalogenase and is found in *Pseudomonas* sp. The enzyme fluoroacetatedehalogenase is well known for its ability to break the highly stable carbon–fluorine bond (Donnelly and Murphy 2009) and is isolated and purified from various microbial species like *Burkholderia* sp. FA1, *Fusariumsolani*, *Moraxella* sp. B, *Pseudomonas fluorescens* DSM8341 and a soil *Pseudomonas* sp. (Leong et al. 2017). Fluorobenzene is also used by microbial consortiums as the sole source of carbon and energy (Chen et al. 2015).

The monofluorobenzoates are generally degradable under aerobic conditions and fluorophenols and fluorobenzoates under anaerobic conditions (Kuntze et al. 2011). Transformations of isomers like luorobenzoate have been studied already with the strict anaerobic *Syntrophusaciditrophicus* (Mouttaki et al. 2009). Under aerobic conditions, there are two possible pathways for the transformation of 2-fluorobenzoate (2-FB), both involve dioxygenation. The *Pseudomonas aerugenosa* and *Acinetobacter* RH5 are fluoride-resistant bacteria (Chouhan et al. 2012, Mukherjee et al. 2015). Such understanding will be useful for designing strategies for removing micro pollutants from drinking water in engineered systems. The role of protozoans and phages in shaping prokaryotic degraders and manipulating *in situ* degradation is totally ignored. Only few studies have confirmed the availability of bacteriophages in groundwater with lack of information to consider its influence on degrader (Meckenstock et al. 2015). However, some factors limit the bioremediation process, i.e., slow rate of degradation, adverse climatic conditions such as temperature, moisture, pH, ionic strength or redox status, limited supply of nutrients, limited amounts of electron acceptors and high concentrations of pollutants (Kumar et al. 2017).

Table 2. List of fluoride compound degrading microorganism (Singh and Gothalwal 2016a).

Sr. No.	Fluoride compound	Microorganism	Reference
1.	3-Fluorobenzoate.	Sphingomonas sp.	Boersma et al. (2004)
2.	Fluorobenzene	Sphingobacterium/Flavobacterium and Alcaligenes	Carvalho et al. (2002)
3.	Fluoranthene	M. vanbaalenii PYR-1	Kweon et al. (2007)
4.	4-Fluorobenzyl	Alcohol Pseudomonas spp.	Maeda et al. (2007)
5.	Fluoranthene	Mycobacterium sp. JS14	Lee et al. (2007)
6.	2-Fluorophenol	Rhodococcus sp. FP1	Duque et al. (2012)
7.	a,a,a-Tri fluoroacetophenone	Gordonia sp. SH2	Hasan (2010)
8.	4-fluorocinnamic acid	Consortium of Arthrobacter sp. strain G1 and Ralstonia sp. strain H1	Hasan et al. (2011)
9.	Fluoroacetate	Synergistetes	Davis et al. (2012)
10.	Sodium fluoride	Aspergillus sp. and Rhizopus sp.	Tamilvani et al. (2015)
11.	Sodium fluoride	Acinetobacter sp. RH5	Mukherjee et al. (2015)

6.1.1 Bioadsorbents for defluoridation

Biosorption is a promising method for treatment of water having abundantly available biomaterials. Various bioadsorbents have been developed for fluoride removal, and among those chitin and chitosan-derivatives have gained wide attention as effective bioadsorbents due to their low cost and high contents of amino and hydroxyl functional groups, which show significant adsorption potential for the removal of various aquatic pollutants. Algal biomass pretreated with Ca^{2+} was also evaluated for the biosorption of fluoride from polluted waters. The phenomenon like fluoride sorption on fungal bio-adsorbent was called as chemical type of interaction. Collagen fiber, which is profuse natural biomass with abundant functional groups, reacts with various metal ions and can be utilized as a carrier for metal ions (Jagtap et al. 2011). There are several materials that could be used as bioabsorbants such as:

6.1.1.1 Agricultural wastes as sorbents

Agricultural waste materials are cost-effective and ecological significant due to their unique chemical composition, availability in large quantity and biodegradable nature. Low-cost adsorbents from different agricultural waste materials such as coconut shell, coconut shell fibers, and rice husk were developed and employed for the removal of various pollutants in industrial wastewater including fluoride (Alagumuthu and Rajan 2007).

6.1.1.2 Industrial waste as sorbents

Widespread industrial activities generate huge amount of solid waste materials as by-products and these byproducts such as fly ash, carbon slurry, original waste mud, red mud, acid-activated and precipitated waste mud, solid waste from edible oil processing industry, sludge and spent bleaching earth can be used as defluoridating agents with some treatment (Kemer et al. 2009).

6.1.1.3 Nanosorbents

In the past decade, nanotechnology has come out as a promising technology in various fields and use of nanoparticles as sorbents for water treatment is also gaining wide attention in recent years. Aligned carbon nanotubes, prepared by catalytic decomposition of xylene using ferrocene as catalyst, proved their good performance for fluoride removal from water. Both the surface and inner cavities of aligned carbon nanotubes were found to be readily accessible for fluoride sorption (Sarkar et al. 2007). A variety of nano-sized inorganic oxides prepared through thermolysis of a polymeric-based aqueous precursor were capable of giving the solution of the desired inorganic ions (Sarkar et al. 2007). Similarly, various efforts in this field have been made to take the advantage of nanoparticles (Ayoob and Gupta 2009).

6.1.1.4 Low-cost adsorbents for fluoride removal

Different low-cost adsorbent materials are available for effective removal of fluoride from water. The naturally available adsorbents, chalk powder, pineapple powder, orange peel powder, *multani mitti* (special type of clay soil), activated neem and babul leaves, rice husk, etc. are some of the different materials investigated as adsorptive agents for the removal of fluoride from water.

- a) **Chalk Powder:** Chalk powder, due to certain porosity, adsorbs fluoride from aqueous solution. The adsorption of fluoride by chalk powder is 86%. The main component of chalk is calcium carbonate ($CaCO_3$), a form of limestone.

- b) **Pineapple peel powder:** Pineapple (*Ananascosmosus*) is a tropical fruit which grows in countries which are situated in the tropical and subtropical regions. Hence, it is an agricultural waste which is cheap for modern communities and easily available. The fluoride adsorption efficiency of pineapple peel powder is 86%.

- c) **Orange peel powder:** Orange peel is rich in flavonoids with some trace elements, ascorbic acid, carotenoids, dietary fibers, and polyphenols, and the antiradical efficiency was assessed in the dried peels. Due to certain porosity, orange peel powder adsorbs fluoride from aqueous solution.

- d) **Multani mitti:** *Multani mitti* or fuller's earth is clay material that is popularly used as a skin care ingredient. It is rich in magnesium chloride and has great surface area with excellent bonding and sealing properties. It contains fine sand particle with complex multicenter crystalline structures of oxides and hydroxide of aluminum, zinc, magnesium and silicon. It has the tendency to filter, decolorize, and clarifying properties without any chemical treatment. The efficient removal of fluoride can be achieved to 56% using *multani mitti*.

e) **Babul (*Acacia nilotica*) leaves:** Babul is well known for its amazing use in treating gum problems and its leaf, bark, seeds and gum are used for medicinal purposes in India. Leaves bark, and flowers contain tannin and other polyphenolic compounds while fruits contain gallic acid. Babul tree has many medicinal properties with antiviral and antifungal characteristics. It also has antidiarrheal, antioxidant, antibacterial, antimalarial, anthelmintic and anti-inflammatory properties. Babul leaves are cheap adsorbents for fluoride removal with the removal efficiency of 59.2%.

f) **Rice husk:** Rice husk is the by-product of rice milling. This husk contains about 75% organic volatile matter and remaining 25% of the weight of this husk is converted into ash during the firing process (gasification) known as rice husk ash (RHA). RHA is rich source of floristic fiber, protein, and some functional groups such ascarboxyl, hydroxyl and amidogen, which makes adsorption processes possible. There is marginal variation in fluoride removal by rice husk over pH range of 2 to 10 from 83% to 84% removal efficiency.

g) **Neem (Azadiractaindica) leaves:** Neem leaves can be used by rural communities as it is the most common and easily available tree. With the help of neem leaves, fluoride can be removed up to 58%.

6.2 Bioremediation processes of nitrate removal

Treatment of industrial waste polluted with the nitrate needs interdisciplinary approach, as nitrate is highly stable at low concentrations and does not undergo co-precipitation. Different techniques are used for nitrate remediation such as ion exchange phenomena, reverse osmosis (RO) and chemical denitrification. The removal of nitrate (NO_3^-) using microorganisms, known as denitrification/biological remediation of nitrate, is an promising approach method as it has high separation efficiency. Biological denitrification provides economic and social advantages with minimal toxic substances and polluted effluent production.

6.2.1 Denitrification

Denitrification is a very important process of nitrate removal from water and soil by transforming it into atmospheric nitrogen. It is a biological process which is observed under aerobic and anaerobic environmental conditions. Under aerobic conditions, denitrification is carried out by sulfur bacteria, i.e., *Paracoccuspantotrophus*, which is not dependent on cytoplasmic membrane nitrate transport. Generally, two types of denitrification is found in nature, viz. assimilatory and dissimilatory denitrification (Figure 2). In assimilatory denitrification, nitrate is transformed into ammonia (NH_3) by microbes, which serves as a nitrogen (N_2) source for cell and protein synthesis but in case of dissimilatory denitrification, facultative microorganisms will utilize nitrate as terminal electron acceptor (e^-) in energy metabolism and convert into nitrogen gas principally with different gaseous end products [e.g., nitrous oxide (N_2O) and nitric oxide (NO)]. Facultative anaerobes (bacteria and fungi) can grow in anoxic environment by using oxygen containing compounds like nitrate in their metabolic activities to extract oxygen. An environment in which very little or no free dissolved oxygen (dissolved O_2 concentration < 0.2 mg L^{-1}) is present is known as anoxic environment. On the basis of microbes involved in denitrification process, dissimilatory denitrification is subdivided into heterotrophic and autotrophic denitrification.

6.2.2 Heterotrophic and autotrophic denitrification

The nitrate compound generated from autotrophic nitrification is converted into nitrogen gas during the heterotrophic denitrification process and substrates used in this mechanism were methane (CH_4), carbon monoxide (CO), methanol (CH_3OH), ethanol (C_2H_5OH), acetic acid (CH_3COOH), lactate and glucose. In autotrophic denitrification, microbes can utilize hydrogen, sulfide, sulfur and

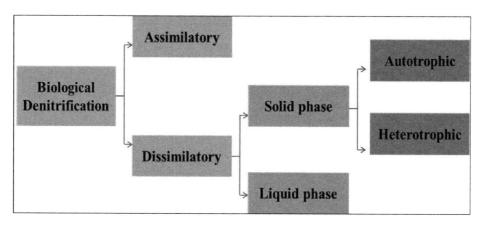

Figure 2. Flow chart showing biological denitrification process.

Table 3. Comparison of autotrophic and heterotrophic denitrification (Rajmohan et al. 2018).

Parameters	Heterotrophic denitrification	Autotrophic denitrification
Environmental condition	Anaerobic	Aerobic
Requirement	Require organic carbon source	No requirement of any organic carbon source
Responsible microorganism	*Pseudomonos, Bacillus, Micrococcus, Methanomonas* and *Spirillum*	*Paracoccus, Thiobacillus* and *Thiosphaera*
Energy and carbon source	Methane (CH_4), carbon monoxide (CO), methanol (CH_3OH), ethanol (C_2H_5OH), lactate, glucose acetic acid (CH_3COOH), acetone, etc.	Sulfur (S), hydrogen (H_2) CO_2 and HCO_3, iron (Fe), or alloyed nanoparticles and uranium
Complexity	Less complex	More complex
Microbial growth rate	High	Slow
Efficiency of nitrate remediation	High	Less
Alkalinity	Increase alkalinity	Reduce alkalinity
Reactors used	Activated sludge, packed bed reactor, fluidized-bed reactor and biodent tube reactor	Hydrogenotrophic denitrification reactor
End product	Harmless nitrogen (N_2), NO and N_2O	Harmless N_2
Secondary pollution	Yes	No
Cost	Less expensive	More expensive

thiosulfates compounds as substrates or electron donors for removing nitrogen from wastewater. Both denitrification processes are different with respect to requirement of oxygen. Autotrophic and heterotrophic denitrification requires aerobic and anaerobic conditions, respectively. A comparison between heterotrophic and autotrophic denitrification is shown in Table 3. Mostly, heterotrophic denitrification is prominent and shows higher efficiency in nitrate remediation from wastewater than the autotrophic denitrification.

6.2.3 Suspended growth and fixed film process

In several instances, the practices of activated sewage and sludge were used (Isaka et al. 2007). In a suspended-growth process, such as activated sludge method, the waste flows through and around free-floating microorganisms, accumulating into biological flocks that settle out of the wastewater

(Breisha and Winter 2010). The biomass appears to be washed out in the case of suspended biofilms and contributes to the development of hyper-concentrated cultures. Attached growth systems, on the other hand, use a medium for retaining and growing microorganisms. Immobilization can occur naturally or artificially (Fierro et al. 2008, Hill and Khan 2008) and gel trapping method is used on a suitable support for artificial immobilization that is non-toxic, photo transparent and mechanically stable and has the ability to retain cellular viability. Attached growth processes also provide an incentive for industrial applications to extend laboratory processes (Yan and Hu 2009). The required power plant size is significantly reduced, shows resistance to temperature variations and toxic pollutants and thus allows for long-term uninterrupted operation (Morita and Tojo 2007, Yan and Hu 2009).

6.2.4 Microbiology of bioremediation

Microorganisms are widely distributed in the ecosystem due to their highly impressive metabolic capacity and can easily grow in a diverse range of environmental conditions. Microorganism's dietary flexibility can also be used for pollutant biodegradation. It is continued on the basis of the ability of certain microorganisms to absorb, alter and use toxic pollutants for the production of energy and biomass in the phase. Denitrification is a typical respiratory process that requires an energy source, or oxidizable substrate. The genera of bacteria containing denitrifying species are *Pseudomonas, Spirillum, Thiobacillus, Moraxella, Methanomonas, Paracoccus, Propionibacterium* and *Xanthomonas,* most of which are optional organisms using nitrate as electron acceptors under anaerobic conditions. *Nitrosomonas, Nitrosovibrio* and *Nitrosolobus, Nitrosococcus* and *Nitrosopir* are known to oxidize atmospheric ammonia to nitrite (Uemoto and Saiki 2000). Mostly, denitrifiers are the Gram-negative bacteria. Several Gram-positive bacteria such as *Bacillus* and some halophilicarchaeal microorganisms like *Haloferaxdenitrificans* also showed ability to denitrify (Kim et al. 2005). *Thiomicrospiradenitrifiers* and *thiobacillusdenitrifiers* are also referred to as autotrophic denitrifiers (Breisha and Winter 2010).

7. Challenges and future perspective

Several fluoride (F^-) remediation issues and challenges that exist in our country include:

- Scarcity of attentiveness and knowledge of F^- contamination;
- Low understanding about the hazardous effects of F^- on health;
- Lack of clear policies and guidelines on groundwater F^- management;
- Inadequacy of groundwater F^- monitoring and regulation systems;
- Lack of clear enforcement of laws and regulations of F^- contaminated groundwater;
- Undefined water rights in developing countries;
- Costly treatment of remedial technologies on a large scale;
- Deficiency of coordination between groundwater management agencies and local populations;
- Shortage of effective, cheap and region-based F^- remedial approaches;
- People being unconscious about future climate change that can affect F^- contamination in groundwater;
- Inadequate medical facilities in the contaminated areas and no routine check-ups facilities;
- Lack of proper guidelines and appropriate demarcation of F^- affected risk zones.

In future, better groundwater resource utilization coupled with economical, sustainable and adoptable measures should be employed. These solutions are listed below:

- Implementing the appropriate policy to manage groundwater F^- contaminations;
- Appropriate monitoring F^- concentrations in groundwater by building new institutions;

- Obtaining actual data on groundwater resources by strengthening data base management, adopting effective techniques for recharge assessment and mapping aquifers effectively;
- Forceful execution of policies and regulations to manage F^- concentrations in groundwater;
- Government needs to recommend guidelines for F^- concentration in drinking water;
- Strict penalties need to be levied for violations of rules concerning pollution and contamination of groundwater resources;
- Community based awareness campaigns to educate the population in affected areas;
- Adoption of community-level regulatory options;
- Establishment of medical facilities and regular health check-ups;
- Formation of regular F^--monitoring facilities in affected areas;
- Provision of basic facilities for poorer to overcome health issues; and
- Government should develop specific criteria for labeling better groundwater resource utilization having F^- contaminated water (e.g., low risk, medium risk, high risk, and severe risk).

Fluoride contamination of drinking water is currently one of the most vital global problems. The main focus of scientific researches in this field has immense potential for the development of more cost-effective, sustainable and eco-friendly bioremediation technology. After field trials, the most competent fluoride-degrading microbes should be multiplied at large scale to degrade the fluoride at commercial level. Technology for heterotrophic denitrification is well developed in comparison with autotrophic denitrification. Inhibition effects of reaction intermediates or by-products need to be investigated. Recent advancements in denitrification has focused on evaluating the negative impact of dissolved oxygen in the bioreactor, limiting its presence, evaluating the hydrodynamic behavior of the reactor system, and effect of variation of nitrate concentration in the raw wastewater being treated. Further development of *in situ* treatment including introducing substrates and nutrients into the aquifers and their long-term impact on ecosystem needs to be studied. Optimization of reaction conditions based on the nutrient, substrate, microorganism, pH, temperature, and dissolved oxygen needs to be studied for every wastewater source. Combination of biological denitrification with other technologies such as ion exchange, electrochemical methods, filtration, and electro-dialysis will pave a new way to overcome the challenges associated with the individual methods.

This lack of information is required to understand the behavior of F^- in natural water and how that impacts local hydrogeological settings, climatic conditions, and industrial and agricultural practices. Remediation of groundwater fluoride contamination is very difficult and on a large scale, F^- treatment is very cost effective; therefore, the local governments should pay more attention to region specific F^- treatment programs. Local population should be made aware and educated to aid in providing suitable and cheap remedial technology.

8. Conclusion

Screening of efficient fluoride degrading microbes or consortium from the all available sources is still needed. The characterization of efficient fluoride degrading microbes is still not available, which helps in commercializing the fluoride biodegradation. The nitrate concentration range in the wastewater may vary from low to high, which needs a suitable method for their remediation. Selection of a particular method depends upon the nitrate concentration in wastewater, volume of wastewater to be processed, land and power availability, environmental policy, and economic concerns. Biological denitrification can be employed to treat the nitrate-contaminated wastewater. Nitrite accumulation is the concern in bio-denitrification which can be addressed by maintaining optimum pH, temperature, and hydraulic residence time. For higher concentrations (> 1000 mg NO_3^- L^{-1}), bio-denitrification can be combined with other technologies such as electrochemical reduction. Combination of traditional and innovative methods for denitrification for nitrate needs to be applied properly after careful evaluation. In particular, the challenges in biological denitrification can be addressed by combining it with other treatment methods to conserve our natural resources like water and soil.

Defluoridation can be categorized into three different strategies, specifically (a) using alternative water sources; (b) better nutrition and (c) defluoridation of water. Lack of alternative water sources and better nutrition for people living in our country are not always possible, so the defluoridation of water may be the best strategy. Bioremediation of pollutant is a potential approach against the conventional techniques. In this area, the main focus of scientific research must be on to develop more cost-effective, sustainable and environmentally friendly bioremediation technology.

References

Alagumuthu, G. and Rajan, M. 2007. Equilibrium and kinetics of adsorption of fluoride onto zincromium impregenated cashewnut shell carbon. Chem. Eng. J. 158: 451–457.

Alhababy, A.M. and Al-Rajab, A.J. 2015. Groundwater quality assessment in Jazan region, Saudi Arabia. Curr. World Environ. 10: 22–28.

Al-Hatim, H.Y., Alrajhi, D. and Al-Rajab, A.J. 2015. Detection of pesticide residue in dams and well water in Jazanarea, Saudi Arabia. Am. J. Environ. Sci. 11: 358–365.

Ayoob, S., Gupta, A.K. and Venugopal, T.B. 2008. A conceptual overview on sustainable technologies for the defluoridation of drinking water. Crit. Rev. Environ. Sci. Technol. 38: 401–470.

Ayoob, S. and Gupta, A.K. 2009. Performance evaluation of alumina cement granules in removing fluoride from natural and synthetic waters. Chem. Eng. J. 150: 485–491.

Batabyal, A.K. and Gupta, S. 2017. Fluoride-contaminated groundwater of Birbhum district, West Bengal, India: interpretation of drinking and irrigation suitability and major geochemical processes using principal component analysis. Environ. Monit. Assess. 189: 369.

Biswas, G., Dutta, M., Dutta, S. and Adhikari, K. 2016. A comparative study of removal of fluoride from contaminated water using shale collected from different coal mines in India. Environ. Sci. Pollut. Res. 23: 9418–9431.

Biswas, K., Gupta, K. and Ghosh, U.G. 2009. Adsorbtion of fluoride by hydrous iron (III)-tin (IV) bimetaloxide from the aqueous solution. Chem. Eng. J. 149: 196–206.

Boersma, F.G., McRoberts, W.C., Cobb, S.L. and Murphy, C.D. 2004. A ^{19}F NMR study of fluorobenzoate biodegradation by *Sphingomonas* sp. HB-1. FEMS Microbiol. Lett. 237: 355–61.

Breisha, G.Z. and Winter, J. 2010. Bio-removal of nitrogen from wastewaters—a review. J. Am. Sci. 6(12): 508–528.

Brindha, K. and Elango, L. 2011. Fluoride in groundwater: causes, implications and mitigation measures. pp. 111–136. *In*: Monroy, S.D. (ed.). Fluoride Properties, Applications and Environmental Management.

BIS, 2012 (IS-10500) Indian standard drinking water-specification (second revision), Bureau of Indian Standards, New Delhi.

Camboim, E.K.A., Tadra-Sfeir, M.Z. and Souza, E.M. 2012. Defluorination of sodium fluoroacetate by bacteria from soil and plants in Brazil. Sci. World J. 149–152.

Carvalho, M.F., Alves, C.C.T., Ferreira, M.I.M., Marco, P.D. and Castro, P.M.L. 2002. Isolation and initial characterization of a bacterial consortium able to mineralize fluorobenzene. Appl. Environ. Microbiol. 68: 102–105.

Central Ground Water Board (CGWB). 2014. Ground Water Year Book 2012–13 Rajasthan State. Government of India, Ministry of Water Resources, Regional Office Data Centre Western Region Rajasthan.

Chae, G.T., Yun, S.T., Mayer, B., Kim, K.H., Kim, S.Y., Kwon, J.S., kim, K. and Koh, Y.K. 2007. Fluorine geochemistry in bedrock groundwater of South Korea. Sci. Total Environ. 385: 272–283.

Chandrajith, R., Padmasiri, J.P., Dissanayake, C.B. and Prematilaka, K.M. 2012. Spatial distribution of fluoride in groundwater of Sri Lanka. J. Natl. Sci. Found. 40: 303–309.

Chen, C.S., Wu, T.W., Wang, H.L., Wu, S.H. and Tien, C.J. 2015. The ability of immobilized bacterial consortia and strains from river biofilms to degrade the carbamate pesticide methomyl. Int. J. Environ. Sci. Technol. 12: 2857–2866.

Cho, S., Luong, T.T., Lee, D., Oh, Y.K. and Lee, T. 2011. Reuse of effluent water from a municipal wastewater treatment plant in microalgae cultivation for biofuel production. Bioresour. Technol. 102: 8639–8645.

Chouhan, S., Tuteja, U. and Flora, S.J.S. 2012. Isolation, identification and characterization of fluoride resistant bacteria: possible role in bioremediation. Appl. Biochem. Microbiol. 48: 43–50.

Das, N., Deka, J.P., Shim, J., Patel, A.K., Kumar, A., Sarma, K.P. and Kumar, M. 2016a. Effect of river proximity on the arsenic and fluoride distribution in the aquifers of the Brahmaputra Flood plains, Assam, Northeast India. Groundwater Sustain. Dev. 2-3: 130–142.

Das, S., Mehta, B.C., Das, P.K., Srivastava, S.K. and Samantha, S.K. 1998. Source of high fluoride in groundwater around Angul, Dhenkanal District, Orissa. Pollut. Res. 17: 385–392.

Davis, C.K., Denman, S.E., Sly, L.I. and McSweeney, C.S. 2011. Development of a colorimetric colony-screening assay for detection of defluorination by micro-organisms. Lett. Appl. Microbiol. 53: 417–423.

Davis, C.K., Webb, R.I., Sly, L.I., Denman, S.E. and McSweeney, C.S. 2012. Isolation and survey of novel fluoroacetate degrading bacteria belonging to the phylum synergistetes. FEMS Microbiol. Ecol. 80: 671–684.

Donnelly, C. and Murphy, C.D. 2009. Purification and properties of fluoroacetate dehalogenase from *Pseudomonas fluorescens* DSM8341. Biotechnol. Lett. 31: 245–250.

Dorante, T., Lammel, J., Kuhlmann, H., Witzke, T. and Olfs, H.W. 2008. Capacity, selectivity, and reversibility for nitrate exchange of a layered double-hydroxide (LDH) mineral in simulated soil solutions and in soil. J. Plant Nutr. Soil Sci. 171: 777–784.

Duque, A.F., Hasan, S.A., Bessa, V.S., Carvalho, M.F., Samin, G., Janssen, D.B. and Castro, P.M. 2012. Isolation and characterization of a *Rhodococcus* strain able to degrade 2-fluorophenol. Appl. Microbiol. Biotechnol. 95: 511–520.

Fawell, J., Bailey, K., Chilton, J., Dahi, E., Fewtrell, L. and Magara, Y. 2006. World Health Organization. IWA publishing, London UK: 1–144.

Fierro, S., delPilar Sánchez-Saavedra, M. and Copalcua, C. 2008. Nitrate and phosphate removal by chitosan immobilized scenedesmus. Biores. Technol. 99(5): 1274–1279.

Francis, C.W. and Hatcher, C.W. 1980. Biological denitrification of high nitrate wastes generated in the nuclear industry. Bio Fluid Bed Treat Water and Waste 1: 235–250.

Gentili, F.G. and Fick, J. 2016. Algal cultivation in urban wastewater: an efficient way to reduce pharmaceutical pollutants. J. Appl. Phycol. http://dx.doi.org/10.1007/s10811-016-0950-0.

Ghorai, S. and Pant, K.K. 2005. Equilibrium, kinetics and breakthrough studies for adsorption of fluoride on activated alumina. Sep. Purif. Technol. 42: 265–271.

Gleeson, T., Wada, Y., Bierkens, M.F. and van Beek, L.P. 2012. Water balance of global aquifers revealed by groundwater footprint. Nature 488: 197–200.

Gupta, N., Yadav, K.K., Kumar, V. and Singh, D. 2013. Assessment of physic-chemical properties of Yamuna river in Agra city. Int. J. Chem. Res. 5(1): 528–531.

Gupta, N., Yadav, K.K. and Kumar, V. 2015. A review on current status of municipal solid waste management in India. J. Environ. Sci. 37: 206–217.

Gupta, N., Yadav, K.K., Kumar, V., Kumar, S., Chadd, R.P. and Kumar, A. 2019. Trace elements in soil-vegetables interface: translocation, bioaccumulation, toxicity and amelioration—a review. Sci. Total Environ. 651: 2927–2942.

Harrison, P.T.C. 2005. Fluoride in water: A UK perspective. J. Fluorine. Che. 26: 1448–1456.

Hasan, S.A. 2010. Biodegradation of Fluorinated Environmental Pollutants Under Aerobic Conditions. PhD Thesis, University of Groningen, Netherlands.

Hasan, S.A., Ferreira, M.I.M., Koetsier, M.J., Arif, M.I. and Janssen, D.B. 2011. Complete biodegradation of 4-fluorocinnamic acid by a consortium comprising *Arthrobacter* sp. strain G1 and *Ralstonia* sp. strain H1. Appl. Environ. Microbiol. 77: 572–579.

Heikens, A., Sumarti, S., van Bergen, M., Widianarko, B., Fokkert, L., van Leeuwen, K. and Seinen, W. 2005. The impact of the hyperacid Ijen Crater lake: risks of excess fluoride to human health. Sci. Total Environ. 346(1-3): 56–69.

Hill, C. and Khan, E. 2008. A comparative study of immobilized nitrifying and co-immobilized nitrifying and denitrifying bacteria for ammonia removal from sludge digester supernatant. Water Air Soil Pollut. 195(1-4): 23–33.

Hu, C.Y., Lo, S.L. and Kuan, W.H. 2003. Effect of co-exiting anions on fluoride removal in electrocoagulation process using aluminium electrodes. Water Res. 37: 4513–4523.

Hu, Q., Weterhoff, P. and Vermaas, W. 2000. Removal of nitrate from groundwater by cyanobacteria quantitative assessment of factors influencing nitrate uptake. Appl. Env. Microbiol. 133–139.

Ibrahim, M., Amirasheed, M., Sumalatha, M. and Prabhakar, P. 2011. Effect of fluoride contents in groundwater: a review. Int. J. Pharm. Appl. 2: 128–134.

Isaka, K., Sumino, T. and Tsuneda, S. 2007. High nitrogen removal performance at moderately low temperature utilizing anaerobic ammonium oxidation reactions. J. Biosci. Bioeng. 103(5): 486–490.

Jagtap, S., Yenkie, M.R., Das, S. and Rayalu, S. 2011. Synthesis and characteristics of lathanum impregnated chitosan flakes for fluoride removal in water. Desalination 273: 267–275.

Janssen, D.B., Dinkla, I.J.T., Poelarends, G.J. and Terpstra, P. 2005. Bacterial degradation of xenobiotic compounds: evolution and distribution of novel enzyme activities. Environ. Microbiol. 7: 1868–1882.

Kemer, B., Ozdes, D., Gondogdu, A., Bulut, V.N., Duran, C. and Soylak, M. 2009. Removal of fluoride ions from aqueous solution by waste mud. J. Hazard Mater. 168: 888–894.

Keshavarzi, B., Moore, F., Esmaeili, A. and Rastmanesh, F. 2010. The source of fluoride toxicity in Muteh area, Isfahan, Iran. Environ. Earth Sci. 61: 777–786.

Kim, J., Park, K., Cho, K., Nam, S., Park, T. and Bajpai, R. 2005. Aerobic nitrification–denitrification by heterotrophic *Bacillus* strains. Biores. Technol. 96(17): 1897–1906.

KoteswarRao, M. and Metre, M. 2014. Effective low cost adsorbents for removal of fluoride from water. Int. J. Sci. Res. 3(6): 120–124.

Kumar, M., Das, A., Das, N., Goswami, R. and Singh, U.K. 2016. Co-occurrence perspective of arsenic and fluoride in the groundwater of Diphu, Assam, Northeastern India. Chemosphere 150: 227–238.

Kumar, S., Sangam, P. and Gaur, R. 2017. Heavy metals contamination, inauspicious wallop on microbial diversity and their possible remediation for environmental restoration. J. Bacteriol. Mycol. 5: 00124.

Kumar, V.P. and Hari Priya, V.R. 2013. Molecular characterization of fluorine degrading bacteria from soil samples for its industrial exploitation. Int. J. Adv. Life Sci. 6: 351–355.

Kuntze, K., Kiefer, P., Baumann, S., Seifert, J., Bergen, M.V., Vorholt, J.A. and Boll, M. 2011. Enzymes involved in the anaerobic degradation of meta-substituted halobenzoates. Mol. Biol. 82: 758–769.

Kushwaha, D., Saha, S. and Dutta, S. 2014. Enhanced biomass recovery during phycoremediation of Cr(VI) using cyanobacteria and prospect of biofuel production. Ind. Eng. Chem. Res. 53: 19754–19764.

Kweon, O., Kim, S.J., Jones, R.C., Freeman, J.P., Adjei, M.D., Edmondson, R.D. and Cerniglia, C.E. 2007. A polyomic approach to elucidate the fluoranthene degradative pathway in *Mycobacterium vanbaalenii* PYR-1. J. Bacteriol. 189: 4635–4647.

Lee, S.E., Seo, J.S., Keum, Y.S., Lee, K.J. and Li, Q.X. 2007. Fluoranthene metabolism and associated proteins in *Mycobacterium* sp. JS14. Proteomics 7: 2059–2069.

Leong, L.E.X., Denman, S.E., Davis, C.K., Huber, T. and McSweeney, C.S. 2010. Peptide utilization of the novel fluoroacetate-degrading ruminal bacterium. Proc. Aust. Soc. Anim. Prod. 28: 64.

Leong, L.E.X., Khan, S., Davis, C.K., Denman, S.E. and McSweeney, C.S. 2017. Fluoroacetate in plants—A review of its distribution, toxicity to livestock and microbial detoxification. J. Anim. Sci. Biotechnol. 8: 1–11.

Li, C., Gao, X. and Wang, Y. 2015. Hydrogeochemistry of high-fluoride groundwater at Yuncheng Basin, Northern China. Sci. Total Environ. 508: 155–165.

Liu, X.J., Wang, B.B. and Zheng, B.S. 2014. Geochemical process of fluorine in soil. Chin. J. Geochem. 33: 277–9.

Maeda, T., Okamura, D., Yokoo, M., Yamashita, E. and Ogawa, H.I. 2007. Fluorine elimination from 4-Fluorobenzyl alcohol by *Pseudomonas* spp. J. Environ. Biotechnol. 7: 45–53.

Maurya, P.K., Malik, D.S., Yadav, K.K., Kumar, A., Kumar, S. and Kamyab, H. 2019. Bioaccumulation and potential sources of heavy metal contamination in fish species in River Ganga basin: possible human health risks evaluation. Toxicol. Rep. 6: 472–481.

Meckenstock, R.U., Elsner, M., Griebler, C., Lueders, T., Stumpp, C., Aamand, J., Agathos, S.N., Albrechtsen, H.J., Bastiaens, L., Bjerg, P.L. and Boon, N. 2015. Biodegradation: Updating the concepts of control for microbial cleanup in contaminated aquifers. Environ. Sci. Technol. 49: 7073–7081.

Meenakshi and Maheswari, R.C. 2006. Fluoride in drinking water and its removal. J. Hazard Matter. 137: 526–527.

Mohapatra, M., Anand, A., Mishra, B.K., Dion, E., Giles, D.E. and Singh, P. 2009. Review of fluoride removal from drinking water. J. Environ. Manage. 91: 67–77.

Morita, Y. and Tojo, M. 2007. Modifications of PARP medium using fluazinam, miconazole, and nystatin for detection of *Pythium* spp. in soil. Plant Disease 91(12): 1591–1599.

Mouttaki, H., Nanny, M.A. and McInerney, M.J. 2009. Metabolism of hydroxylated and fluorinated benzoates by *Syntrophusaciditrophicus* and detection of a fluorodiene metabolite. Appl. Environ. Microbiol. 75: 998–1004.

Msonda, K.W.M., Masamba, W.R.L. and Fabiano, E. 2007. A study of fluoride groundwater occurrence in Nathenje, Lilongwe, Malawi. Phys. Chem. Earth J. 32: 1178–1184.

Mukherjee, S., Yadav, V., Mondal, M., Banerjee, S. and Halder, G. 2015. Characterization of a fluoride resistant bacterium *Acinetobacter* sp. RH5 towards assessment of its water defluoridation capability. Appl. Water Sci. 7: 1923–1930.

Munz, G., Gori, R., Cammilli, L. and Lubello, C. 2008. Characterization of tannery wastewater and biomass in a membrane bioreactor using respirometric analysis. Biores. Technol. 99(18): 8612–8618.

Oladoja, N.A. and Ademoroti, C.M.A. 2006. The use of fortified soil-clay as on-site system for domestic wastewater purification. Water Res. 40(3): 613–620.

Oruc, N. 2008. Occurrence and problems of high fluoride waters in Turkey: an overview. Environ. Geochem. Health 30: 315–323.

Park, J.B.K., Craggs, R.J. and Sukias, J.P.S. 2009. Removal of nitrate and phosphorus from hydroponic wastewater using a hybrid denitrification filter (HDF). Biores. Technol. 100(13): 3175–3179.

Peyton, B.M., Mormile, M.R. and Petersen, J.N. 2001. Nitrate reduction with *Halomonascampisalis*: kinetics of denitrification at pH 9 and 12.5% NaCl. Water Res. 35(17): 4237–4242.

Piddennavar, R. and Krishnappa, P. 2013. Review on defluoridation techniques of water. Int. J. Eng. Sci. 2: 86–94.

Raj, D. and Shaji, E. 2017. Fluoride contamination in groundwater resources of Alleppey, Southern India. Geosci. Front. 8: 117–124.

Rajmohan, K.S., Gopinath, M. and Chetty, R. 2018. Bioremediation of nitrate-contaminated wastewater and soil. pp. 387–409. *In*: Varjani, S., Agarwal, A., Gnansounou, E. and Gurunathan, B. (eds.). Bioremediation: Applications for Environmental Protection and Management. Energy, Environment, and Sustainability. Springer, Singapore.

Rutkowska, M., Heimowska, A., Krasowska, K. and Janik, H. 2002. Biodegradability of polyethylene starch blends in sea water. Pol. J. Environ. Stud. 11: 267–274.

Sahu, B.L., Banjare, G.R., Ramteke, S., Patel, K.S. and Matini, L. 2017. Fluoride contamination of groundwater and toxicities in Dongargaon block, Chhattisgarh, India. Expo. Health 9: 143–156.

Sahu, P. 2019. Fluoride pollution in groundwater. pp. 329–350. *In*: Sahu, P. (ed.). Groundwater Development and Management. Springer, Cham.

Sarkar, S., Blaney, L.M., Gupta, A., Ghosh, D. and Gupta, A.K.S. 2007. Use of ArsenXnp, a hybrid anion exchanger for arsenic exchanger for arsenic removal in remote villages in the Indian subcontinent. React. Funct. Polym. 67(12): 1599–1611.

Saxena, V.K. and Ahmed, S. 2001. Dissolution of fluoride in groundwater: a water rock interaction study. Environ. Geol. 40: 1084–1087.

Sehn, P. 2008. Fluoride removal with extra low energy reverse osmosis membranes: three years of large scale field experience in Finland. Desalination 223: 73–84.

Shen, J., He, R., Han, W., Sun, X., Li, J. and Wang, L. 2009. Biological denitrification of high-nitrate wastewater in a modified anoxic/oxic-membrane bioreactor (A/O-MBR). J. Hazard Mater. 172(2): 595–600.

Sindelar, H.R., Yap, J.N., Boyer, T.H. and Brown, M.T. 2015. Algae scrubbers for phosphorus removal in impaired waters. Eco. Eng. 85: 144–158.

Singh, A. and Gothalwal, R. 2016a. A reappraisal on biodegradation of fluoride compounds: role of microbes. Water and Environment Journal 32(3): 481–7.

Singh, A. and Gothalwal, R. 2016b. Fluoro database: An open access database related to unveiling of fluoride. Eur. J. Pharm. Med. Res. 3: 476–478.

Srivastav, A., Yadav, K.K., Yadav, S., Gupta, N., Singh, J.K., Katiyar, R. and Kumar, V. 2018. Nano-phytoremediation of pollutants from contaminated soil environment: current scenario and future prospects. pp. 383–401. *In*: Ansari, A., Gill, S., Gill, R., Lanza, R.G. and Newman, L. (eds.). Phytoremediation. Springer.

Sun, Z., Cheng, Y., Zhou, J. and Wei, R. 1998. Research on effect of fluoride pollution in atmosphere near an aluminium electrolysis plant on regional fall wheat growth. *In*: Proceedings of air and waste management association 91st annual meeting, San Diego, TPE 09/P1-TPE09/P7.

Szostek, R. and Ciecko, Z. 2017. Effect of soil contamination with fluorine on the yield and content of nitrogen forms in the biomass of crops. Environ. Sci. Pollut. R. 1–14.

Tamilvani, T., King Solomon, E. and Rajesh Kannan, V. 2015. Biodegradation on fluoride contaminated soil and water in Dharmapuri district of Tamil nadu India. CIBTech. J. Microbiol. 4: 78–84.

Tiedje, J.M., Sexstone, A.J., Myrold, D.D. and Robinson, J.A. 1983. Denitrification: ecological niches, competition and survival. Antonie Van Leeuwenhoek 48(6): 569–583.

Tikki, M.A. 2014. Fluoride removal from water—A review. Int. J. Sci. Eng. Res. 5: 515–519.

Tressaud, A. 2006. Fluorine and the environment: Agrochemicals, archaeology, green chemistry and water. pp. 1–296. *In*: Advances in Fluorine Science (ed. 1). Vol. 2. Elsevier, Amsterdam, Boston.

Uemoto, H. and Saiki, H. 2000. Nitrogen removal reactor using packed gel envelopes containing *Nitrosomonaseuropaea* and *Paracoccusdenitrificans*. Biotechnol. Bioeng. 67(1): 80–86.

USEPA. 1996. R.E.D. Facts, Cryolite, EPA-738-F-96-016. United States Environmental Protection Agency.

World Health Organization (WHO). 2005. Naturally occurring hazards. Fluoride, Geneva, Switzerland.

Wu, C., Chen, Z., Liu, X. and Peng, Y. 2007. Nitrification–denitrification via nitrite in SBR using real-time control strategy when treating domestic wastewater. Biochem. Eng. J. 36(2): 87–92.

Wu, J. and Sun, Z. 2016. Evaluation of shallow groundwater contamination and associated human health risk in an alluvial plain impacted by agricultural and industrial activities, mid-west China. Expo. Health 8(3): 311–29.

Yan, J. and Hu, Y.Y. 2009. Partial nitrification to nitrite for treating ammonium-rich organic wastewater by immobilized biomass system. Biores. Technol. 100(8): 2341–2347.

18

Soil Degradation in Mediterranean Region and Olive Mill Wastes

Victor Kavvadias,[1,*] *Evangelia Vavoulidou*[1] and *Christos Paschalidis*[2]

1. Introduction

Soils in the Mediterranean region present an enormous variability according to different soil taxonomy systems (USDA, FAO, etc.). This soil diversity reflects differences in climate, geological origin, vegetation, land use and historical development of Mediterranean landscapes. In general, soils have medium to poor fertility, with low organic matter contents due to low natural vegetation developed on them and because the active human activity from more than 2000 years of cultivation. In Southern Europe high rates of soil loss are recorded due to surface sealing deriving from the rabid urbanization in the coastal zones and transport infrastructure. In most of the coastal areas, there is high competition for the usage of land, with important consequences to the soil resources and in general to the environment. In fact, competition between different uses of soil, leading to soil contamination and consumption of the soil resource, is becoming more severe in the Mediterranean region, mainly as the result of increasing urbanization and tourism. Contamination is high in restricted areas, in urban areas and in hazardous industrial compounds, due to both diffuse and localized sources. In addition, salinization and soil degradation through forest fires play an important role as well. Furthermore, soil has a determining role in maintaining slope stability. However, soil degradation, soil erosion and soil sealing in combination with intense precipitation can be the cause of various catastrophic events.

Above all, the main environmental problems that the Mediterranean area encounters are scarce water availability for human activity and the soil degradation by soil erosion especially in areas where high quality and easily erodible soils are subject to more intensive agriculture. Soil erosion is mainly due to low vegetation cover, intensity of rainfall, and intensive cropping and over-grazing management. In terms of Mediterranean agricultural systems, farmers practice both extensive rain fed agriculture and intensive irrigated agriculture. Extensive rain-fed agriculture deteriorates the initial soil properties and quality and increases desertification problems, especially when practiced in soils with low organic matter contents and poor quality and where stubble burning and over-grazing practices and lack of organic residue incorporation reduce fertility and biodiversity. Intensive irrigated agriculture with excessive and inefficient water use and chemical rate applications can increase negative environmental impacts associated to soil degradation, by chemical pollution and salinization.

---

[1] Department of Soil Science of Athens, Institute of Soil and Water Resources, Hellenic Agricultural Organization-"DEMETER", Lykovrisi, Attiki, Greece.
[2] University of Peloponnese, School of Agriculture and Food, Department of Agriculture, 24100, Kalamata, Greece.
* Corresponding author: vkavvadias.kal@nagref.gr

Soil degradation in the form of soil carbon loss has increased rapidly during the last 200 years. Agronomic measures related to managing the vegetation cover, soil management (tillage and nutrient supply) and mechanical methods resulting in durable changes to the landscape are among the measures that are taken so as to alleviate soil degradation problems. The application of organic wastes could be a way of solving two problems, the waste disposal and the correction of the low organic matter content of many agricultural soils. Olive Mill Wastes could be applied as an amendment on the agricultural soils under specific conditions. In this context we review the soil degradation in Mediterranean region and how the sustainable use of Olive Mill Wastes in particular liquid wastes can reduce soil degradation.

2. Soil protection

The functions provided by soils are manifold because they are related to processes such as filtering, storage and buffer functions, serving human activities such as plant production, recreation and leisure (Graeber and Blankenburg 2009). Soil protection refers not simply to the physical soil itself, but to the soil as part of a functioning and living ecosystem that provides all the eco-services. In accordance with the European thematic strategy for soil protection (EC 2006a, b), the following soil threats were identified: erosion, organic matter decline, contamination, salinization, compaction, soil biodiversity loss, sealing, landslides and flooding (Payá Pérez and Rodríguez Eugenio 2018).

3. Soil degradation

Soil degradation is a serious threat for an increasing number of areas all over the world (Lal et al. 1989). Soil degradation is defined as a process that causes deterioration of soil productivity and lowers soil utility, as a result of natural or anthropogenic factors (Ayoub 1991, Mashali 1991, UNEP Staff 1992, Wim and El Hadji 2002). It can be the result of one or more factors which are potential threats for soil. Globally, it has been estimated that nearly 2 billion hectares of land are affected by human-induced soil degradation (UN 2000). Soil is subject to a series of human-induced degradation processes, which are, namely, displacement of soil material and internal soil deterioration (Dwivedi 2002).

The main impact of agriculture on soil degradation is erosion, salinization, compaction, reduction of organic matter and non-point source pollution. Loss of organic matter and soil biodiversity are often driven by unsustainable agricultural practices such as overgrazing of pasturelands, over intensive annual cropping, deep ploughing on fragile soils, continuous use of heavy machinery destroying soil structure through compaction and unsustainable irrigation systems contributing to the salinization of cultivated lands (German Advisory Council on Global Change (WBGU) 1994, EEA 1999, Darwish and Kawy 2008). Soil degradation involves physical loss (soil erosion by water or wind) and the reduction in quality of topsoil associated with nutrient decline and contamination (chemical, physical, and biological degradation). Soil erosion is one of the most important sources of soil degradation, particularly in the most extreme cases (arid and sub-humid climate), together with the destruction of vegetation cover (mainly due to frequently repeated forest fires) and structure and the increase of desertification of marginal lands in the Mediterranean basin (Perez-Treto 1994). Current rates of erosion in the Mediterranean countries show that irreversible processes of soil degradation and desertification are already occurring in Mediterranean region. The chemical degradation mainly consists of soil pollution and acidification. Its consequences are mobilization of harmful elements/compounds, salinization and/or sodification, unfavorable changes in the nutrient regime and decrease of natural buffering capacity. Physical deterioration is comprised of surface sealing or crusting of top soil, soil compaction, structure destruction and extreme moisture regime. Biological deterioration includes imbalance of biological activities via loss of soil organic matter and biodiversity. Biodiversity and organic matter can decline due to erosion or pollution, leading to a reduction in soil functions such as control of water and gas flows. Reduced aboveground plant diversity as a result of tillage, overgrazing, pollutants and pesticides decreases the microbial diversity in the soil ecosystem and disturbs its normal functioning (Christensen 1989, Boddy et al. 1988). The degree of soil degradation depends on soil's susceptibility to degradation processes, land use and the duration of degradative land use (UNEP Staff 1991, Darwish and Abdel Kawy 2008). The processes of soil degradation have major implications at the following: (a) global carbon cycle, mainly due to the decrease in soil organic matter and the release of CO_2 to the atmosphere, (b) reduction in soil buffering capacity, that is the capacity of soil to adsorb contaminants, (c) water and air quality, (d) biodiversity, (e) food production, food and feed safety, and (f) human health.

4. Soil degradation in Mediterranean region

Soil degradation differs markedly across Europe due to the fact that the quality of Europe's soils is a result of natural factors (e.g., climate, soil type, vegetation, topography) (EEA-UNEP 2000). In Europe, damage to soils from modern human activities is increasing and leads to irreversible losses mainly due to local and wide-spread contamination and soil erosion (EEA 2000). Land degradation in the Mediterranean region is estimated to threaten over 60% of the land in southern Europe (UNEP 1991). The degradation of soil resources is a major threat in the Mediterranean region due to (1) climate conditions and global warming, (2) topography, (3) soil characteristics, and, (4) changes in land-use (e.g., abandonment of marginal land with very low vegetation cover and increases in the frequency and extension of forest fires), and characteristics of agriculture (EEA 1995, Boydak and Dogru 1997, Cammeraat and Imeson 1998, Zalidis et al. 2002). Severe erosion and other degradation processes are taking place in Mediterranean soils (Martınez-Mena et al. 2002). Land degradation within the southern European Mediterranean is partially due to dramatic land-use changes that occurred during the second half of this century, as well as due to climate change linked to human intervention and their possible adverse influence on the environment (Jeftic et al. 1993, Perez-Trejo 1994, Hill 2003). More particularly, the Mediterranean islands have been recognized as "hot spots" for various forms of soil degradation (EEA 2000).

5. Decline of organic matter

With respect to soil organic matter, it is affected mostly by climate, soil parent material, texture, hydrology (drainage), topography, land use (tillage) and land cover and/or vegetation (grasslands, forests, agricultural crops) (Smith et al. 2005, Hanegraaf 2009). Soil organic matter, especially the humified fraction (Graham et al. 2002), constitutes an important source of nutrients, while it is a key factor in maintaining or improving soil structure (Ilay et al. 2008). In fact, soil is one of most important natural capital providing ecosystem services pivotal to sustain life. Soil is the second biggest reservoir for carbon after the oceans. There is more carbon stored in soil than in the atmosphere and in vegetation combined (Mokany et al. 2006). Soil therefore plays an important role in C cycle and in turning on the global climate system.

The general threshold of SOC is 2%; however, local conditions such as soil texture introduce large variation (ca. 1–3%), indicating the importance to stratify thresholds according to qualitative data such as soil type. In recent decades, SOC content has been depleted down to approximately 1% mainly because of its overexploitation and inappropriate management practices. The decline of soil organic matter caused by changes in land use, the intensification of agricultural practices (e.g., deep plugging, rapid rotation) at the expense of the naturally forested areas of Europe and, possibly, the climate change are among the main threats to soil fertility and quality (Matson et al. 1997, Vleeshouwers and Verhagen 2002, Freibauer et al. 2004, Davidson and Janssens 2006). In European Union, most soils are out of equilibrium as regards soil organic matter contents. Land use and climate change have resulted in soil organic carbon loss at a rate equivalent to 10% of the total fossil fuel emissions for Europe as a whole. Jones et al. (2005) calculated that 0.6% of soil carbon in European terrestrial ecosystems is lost annually. In addition, approximately 45% of soils in Europe have a low or very low SOM (0–2% organic carbon) and 45% have a medium SOM (2–6% organic carbon). Almost half of European soils have low organic matter content, principally in southern Europe, but also in areas of France, the United Kingdom and Germany.

Soils in many Mediterranean areas are generally characterized by low organic matter content, fertility and productivity and are subject to erosion (Albaladejo et al. 1994). Mediterranean area is facing such a low level of SOC concentration because of dry and warm seasons. Low level of SOC hampers most of soil functions causing disservices (e.g., soil erosion, reduction of soil water holding capacity, low crop yield) and triggering increased external inputs. In the Mediterranean region, the loss of organic matter during last decades is estimated at around 50% of the original content (Van Camp et al. 2004). Nearly 75% of the total area in Southern Europe has a low (3.4%) or very low (1.7%) soil organic matter content (Montanarella 2005). Soils with less than 1.7% organic matter are considered to be in pre-desertification stage. The decline of soil organic carbon content is threatening the diversity of organisms in soils. It can also limit the soil's ability to provide nutrients to crops, leading to lower yields and affect food security. Loss of soil organic matter reduces the water infiltration capacity of a soil, leading to increased run-off and erosion and vice versa. More particularly, the loss of SOM in arid and semi-arid areas of Europe is closely linked to the process of soil erosion. Erosion reduces the organic matter content by washing away fertile topsoil, which may lead to desertification under semi-arid conditions (Montanarella et al. 2004).

Desertification is defined by the United Nations Convention to Combat Desertification (UNCCD1994) as "the degradation of the land in arid, semi-arid and dry-sub-humid areas, as a result of several factors, including climatic change and human activities". Once the desertification process has started, physical processes like wind erosion, soil crusting and surface water erosion accelerate the land degradation process. Furthermore, desertification contributes to environmental crises, such as the loss of biodiversity and global warming. The Mediterranean area is identified as sensitive to desertification due to a combination of climate conditions, soil and terrain characteristics, agriculture and exploitation of water resources (Castillo et al. 2004b). According to Montanarella and Toth (2008), the thematic strategy (EC-COM 2006, EC-SEC 2006) emphasizes that soil degradation processes or threats or a combination of some of the threats will lead to desertification. Desertification in Europe occurs everywhere, for example in central and northern Europe, and is not solely linked to poor land management or poverty. In the most extreme cases soil erosion, coupled with other forms of soil degradation, has led to desertification in some areas of the Mediterranean. Soil desertification is a specific phenomenon of arid, semi-arid, and dry sub-humid regions. According to the United Nations Environmental Program's (UNEP) report, desertification is now considered a worldwide phenomenon, affecting about one-fifth of the world population, 70% of all dry-lands and one-quarter of the total land area of the world (Tolba et al. 1992). Vacca et al. (2009) noted that the main factors of soil degradation in the Mediterranean region are essentially human related which consist mainly of accelerated erosion, loss of prime farmland, compaction, salinization, contamination, decrease of organic matter content and adverse alteration of biological processes.

6. Use of olive mill wastes to reduce soil degradation

Total waste production in the EU amounts to 2.5 billion tons every year. The main guideline in EU environmental policy is the 7th Environment Action Program, which promotes the protection, conservation and enhancement of the natural capital of the Member States as one of the main objectives for 2020 (González-Martínez et al. 2019). The most common groups of waste are the following: agricultural, industrial, municipal and nuclear (Alloway 1995).

Agricultural wastes (AW) are the by-products of various agricultural activities such as crop production, harvesting of crops, saw milling, and agro-industrial processing (Garg 2017). Agricultural wastes are characterized by the seasonality of their production and the need of rapid withdrawal from the field, avoiding interferences with other agricultural managements. Agricultural industry generates mainly liquid and solid residues with a high load of organic matter. The seasonal character of this type of industry means that high amounts of residues are generated in a short period of time. Large quantities of AW are produced in Mediterranean region annually. For example, it is estimated that cereal cultivation produces about 5.5–11.0 ton dry matter of residues per ha, residues from woody tree pruning constitute about 1.3–3.0 ton dry matter per ha, while olive mill industry waste generates huge quantities of solid and liquid wastes, more than 30 million m^3 per year (Mechri et al. 2007). Solid wastes are known as pomace (skins, pulp, seeds and stems of the fruit) and liquid wastes are known as Olive Mill Wastewater (OMW). OMW may be used for restoration of degraded soils in Mediterranean region (Brunetti et al. 1995, Cabrera et al. 1996). However, the wastewater is produced over a short period every year (November–April), creating a major environmental problem (Moraetis et al. 2011). One ton of olives produces approximately 0.8 t of OMW which are characterized as acidic (pH 4–5), with an average chemical oxygen demand (COD) and biochemical oxygen demand (BOD) content of 120 and 60 g/L, respectively, high concentration of suspended solids (7–15 g/L) and phenolic compounds up to 24 g/L.

A number of technologies such as physical, physico-chemical, biological and thermal have been developed for OMW treatment (Kappekalis et al. 2008, Chartzoulakis et al. 2010, Mekki et al. 2013), but when used individually suffer from drawbacks, e.g., low efficiency or high cost. However, combined or advanced alternative methods show encouraging results (Ben Sassi et al. 2006, Niaounakis and Halvadakis 2006, Mekki et al. 2007, Komnitsas et al. 2011, Camarsa et al. 2010).

Recycling of AW by land application for plant uptake and crop production is a traditional and proven waste utilization technique. Properly done, it is an environmentally sound method of waste management, also producing economic benefits due to the reduction of commercial fertilizers' use. Although many research studies focused on effective treatment of OMW and its use in agriculture, only limited contradictory evidence is available for the negative effects of OMW on soil properties. OMW in the Mediterranean area can be discharged directly into sewer systems and water streams or concentrated in evaporation ponds, despite the fact that such disposal methods are prohibited in many Mediterranean countries (Kavvadias et al. 2016). Their disposal may contaminate soils, surface- and ground-waters and thus it can be a significant environmental problem in all the Mediterranean countries (Altieri and Esposito 2008, 2010), especially for the polyphenolic substances (Ferri et al. 2002).

OMW spreading on soil has been until now subject to great controversy because of its fertilization properties and the negative effects to its acidity, salinity and phenolic compounds. As olive mill wastes are rich in inorganic nutrients (micro- and macro-elements) and organic matter, the recycling of these types of wastes in agriculture would contribute to:

- Significant reduction of harmful wastes disposed in the environment,
- Recycling of elements and water in agriculture, which in turn, will reduce production cost and contribute to the increase in European products' competiveness and profits,
- Protection of renewable and non-renewable sources (soil, aquatic bodies, phosphoric minerals) by the disposal and also via nutrients' recycling.

Some studies indicated a positive effect of OMW on physical, chemical and microbiological properties of soil (Pagliai et al. 2001) and thus on soil fertility. In fact, soil amendment with OMW increases organic carbon content (Brunetti et al. 1995), improves structure and nutrient levels (Cabrera et al. 1996). Ayoub et al. (2014) and Chaari et al. (2014) reported that OMW spreading had beneficial effects on top soil, such as nutrient availability for plant growth. Chaari et al. (2015) showed that yearly application of three OMW doses (50, 100 and 200 m^3·ha^{-1}) for 9 successive years improved the fertility of a Tunisian sandy soil. Furthermore, Magdich et al. (2012) displayed an improvement of olive yield after the application of OMW on soil for six successive years.

The high content of organic matter and plants' nutrients makes OMW a low-cost fertilizer. Olive mill wastes are rich in nutrients, especially potassium (K), nitrogen (N) and phosphorous (P) (Peri and Proietti 2014). Due to its organic matter content, it improves soil aggregation stability, improving therefore soil porosity and water retention capacity (Mahmoud et al. 2012, Regni et al. 2017). Chaari et al. (2015) studied the effects of olive mill wastewater onto soil physico-chemical characteristics. Three OMW doses of 50, 100 and 200 m^3 ha^{-1} year^{-1} were applied for ten successive years on sandy soil. The findings showed that the pH of the soil, the electrical conductivity and the organic matter, total nitrogen, phosphorus, sodium, and potassium soil contents increased with increased OMW supply. Di Bene et al. (2013) showed that a long-term repeated OMW spreading has no residual effects or negative trends in soil chemical and biochemical changes. Vella et al. (2016) reported that long-term irrigation with OMW waters (2003–2015) with quantities of about 30 m^3 ha^{-1} year^{-1} had little impact on pH, electrical conductivity, organic matter, concentrations of main cations and polyphenolic content. Chartzoulakis et al. (2010) did not find significant differences in organic matter content between control and raw OMW-treated soils. Yasemin and Killi (2008) stated that the significant enhancement in the organic carbon content that was recorded after 1–2 months of increased rates application of olive solid waste was a short term effect.

OMW water can also be a source of nutrients and irrigation water in Mediterranean regions suffering from water scarcity (Caputo et al. 2013). It is a inexpensive source of nutrients that could replace chemical fertilizers, which are extensively employed in agricultural practices of Mediterranean countries. In fact, Chartzoulakis et al. (2010) showed that after 3 years of raw OMW water application, there were no significant differences in pH, electrical conductivity (EC), P, Na and organic rates between the control and OMW water treated soils. Vella et al. (2016) suggested that OMW water, without pre-treatments, can be annually used for crops and tree irrigation.

On the other hand, olive mill wastewater has a very high polluting power, resulting in high levels of COD (chemical oxygen demand), high salinity and strong phenolic compounds, causing environmental pollution (Dakhli et al. 2018). Chaari et al. (2015) concluded that OMW increased the soil electrical conductivity significantly with the increase of doses of olive mill waste at the depth of 0–20 cm. A cumulative effect of soil salinization was evident when the highest dose of 200 m^3·ha^{-1} wastewater was applied, indicating transformation of soil into an unproductive one. In fact, the presence of phenolic compounds in OMW is the most significant limiting factor for the agronomic use of OMW (Sobhi et al. 2004). A significant increase in soil phenolic compounds content was reported immediately or after some months of the application. Mekki et al. (2007) noticed the presence of phenolic compounds at a depth of 1.2 m of soil treated with OMW after four months, while phytotoxic levels of phenolic compounds were detected in surface soil layer one year after OMW application. In contrast, Feria (2000) reported that the residual levels of polyphenols can remain significant even 6 years after OMW spreading. Kavvadias et al. (2014) reported that the long term uncontrolled and elevated disposal of OMW water resulted in a source of pollution mainly on the soil surface, concluding it was necessary to establish soil quality standards in order to identify soils well-suited to OMW spreading. In addition, leaching of phenolics also seemed to be negligible in soils rich in carbonate and clay materials (Sierra et al. 2007, Hanifi and Handrami 2008a).

Mechri et al. (2011) showed the inhibition of the arbuscular mycorrhizal fungal root colonization by phenolic compound fraction on reducing the nutrient uptake of the olive trees. In addition, the phenolic compounds remained high, as compared with the control essentially on top layers (0–40 cm). Wright et al. (1998) demonstrated that the abundance of AM fungi and the root colonization decreased significantly in olive trees irrigated with OMW, as compared to the control. AM fungal colonization stimulated sufficiently the rate of photosynthesis to compensate for the carbon requirement of fungus. Opposite results reported that OMW application favors fungi populations. AM fungi populations were usually increased as a result of soil amendment with organic matter, which is beneficial for plants, improving plant growth (Joner and Jakobsen 1995, Joner 2000, Ryan et al. 1994, Douds et al. 1997).

Kavvadias et al. (2015) assessed the impact of OMW spreading on soil microbiological properties and explored the relationship to soil chemical properties. They proposed the following measures to ameliorate the effects of direct application of OMW on soils:

- OMW should be partially neutralized before their disposal in evaporation ponds or on soil.
- OMW can be spread immediately after a short period of treatment in evaporation ponds. The annual amount of disposed OMW has to be based on the maximum allowable levels of any

toxic elements defined by the relative EC and national regulations and limits for the non-toxic micro- and macro-nutrients. The amount of distributed OMW to soil should also consider the areas with low hydraulic conductivity, and high underground water level. Disposal of OMW on soils with light texture, acid soils, and soils rich in salts should be avoided due to deterioration of soil structure. On the other hand, OMW can be disposed on alkaline soils and those rich in carbonates, thus avoiding short term excessive acidification of soil.

- Land spreading of OMW can preferably take place in spring, when favorable soil humidity and temperature conditions allow quick decomposition of organic load.
- Soil parameters should be periodically monitored and evaluated. Soil quality chemical and biological indicators have to be adopted, in order to define soils affected by the disposal of OMW.

7. Conclusion

Soil organic matter plays a key role in soil quality and fertility, influencing several ecosystem services. However, a major decline in SOM content in many soils is taking place as a consequence of the intensification of agriculture. Mediterranean soils are low on organic matter, which is mainly affected by the current climate change, erosion and intensified cropping systems. Practices that favor the enrichment of soil with organic matter need to be implemented. Application of C rich inputs is among practices than can be used for optimizing the SOM content of agricultural soils. The amendment of agricultural soils with organic wastes represents a way to increase organic matter of the soil and improve soil fertility. Among the organic waste materials produced by agricultural activities, olive mill wastes disposal on soil is a suitable practice. On the other hand, application of this waste material to the soil may have negative environmental implications mainly due to the high content of potentially phytotoxic compounds, such as phenols. In fact, although there is a lot of information on the treatment of OMW and the use of OMW as soil amendment, in several cases contradictory evidence is available for the effects of OMW on soil properties. Studies on the long term effects of OMW disposal on soil properties are limited, as well as the fate of inorganic and organic load. Further research studies have to be conducted with regard to the definition of effective OMW management strategies, which will include recycling and re-use as well as appropriate remediation schemes in case of soil contamination.

References

Alloway, B.J. 1995. Heavy Metals in Soils. Blackie Academic and Professional, Chapman and Hall, London, 368 p.
Altieri, R. and Esposito, A. 2008. Olive orchard amended with two experimental olive mill wastes mixtures: Effects on soil organic carbon, plant growth and yield. Bioresour. Technol. 99: 8390–8393.
Altieri, R. and Esposito, A. 2010. Evaluation of the fertilizing effect of olive mill waste compost in short-term crops. Int. Biodeter. Biodegr. 64: 124–128.
Ayoub, S., Al-Absi, K., Al-Shdiefat, S., Al-Majali, D. and Hijazean, D. 2014. Effect of olive mill wastewater land-spreading on soil properties, olive tree performance and oil quality. Sci. Hortic. 175: 160–166.
Ayoub, A.T. 1991. An assessment of human induced soil degradation in Africa. U.N. environmental program, Second Soil Sci. conf. Cairo Egypt.
Ben, Sassi, A., Boularbah, A., Jaouad, A., Walker, G. and Boussaid, A. 2006. A comparison of Olive oil Mill Wastewaters (OMW) from three different processes in Morocco. Process Biochem. 41(1): 74–78.
Boddy, L., Watling, R. and Lyon, A.J.E. 1988. Fungi and ecological disturbance. Proc. R. Soc. (Edinburg) B, 94.
Boydak, M. and Dogru, M. 1997. The exchange of experience and state of the art in sustainable forest management by ecoregion: Mediterranean forests. pp. 179–199. *In*: Boydak Dog, M. and Dogru, M. (eds.). Proceedings of the XI World Forestry Congress, Vol. 6. Antalya, Turkey, 13–22 October 1997.
Brunetti, G., Miano, T.M. and Senesi, N. 1995. Effetti dell'applicazione di reflui da frantoi oleari variamente trattati su alcune proprietà del suolo. Atti XIII Conv. Naz. SICA, pp. 36–37.
Cabrera, F., Lopez, R., Martinez-Bordiu, A., Dupuy de Lome, E. and Murillo, J.M. 1996. Land treatment of olive oil wastewater. Int. Biodeterior. Biodegrad. 38: 215–225.
Camarsa, G., Gardner, S., Jones, W., Eldridge, J., Hudson, T. and Thorpe, E. 2010. LIFE among the olives: Good practice in improving environmental performance in the olive oil sector. European Commission. Environment Directorate-General. Luxembourg: Office for Official Publications of the European Union, ISBN 978-92-79-14154-6.

Cammeraat, L.H. and Imeson, A.C. 1998. Deriving indicators of soil degradation from soil aggregation studies in southeastern Spain and southern France. Geomorphology 23: 307–321

Caputo, M.C., De, Girolamo, A.M. and Volpe, A. 2013. Soil amendment with olive mill wastes: impact on ground water. J. Environ. Manage. 131: 216–221.

Castillo, V., Arnoldussen, A., Bautista, S., Bazzoffi, P., Crescimanno, G. and Imeson, A. 2004b. Desertification. pp. 275–295. *In*: Van-Camp, L., Bujarrabal, B., Gentile, A.R., Jones, R., Montanarella, L., Olazabal, C. and Selvaradjo, S.-K. (eds.). Reports of the Technical Working Groups established under 'The Thematic Strategy For Soil Protection'. Volume II. Erosion. Task Group 6 on Desertification. Office for Official Publications of the European Communities, Luxembourg.

Chaari, L., Elloumi, N., Mseddi, S., Gargouri, K., Bourouina, B., Mechichi, T., Kallei, M. and Bourouna, B. 2014. Effects of olive mill wastewater on soil nutrients availability. Int. J. Interdiscip. Multidiscip. Stud. 2: 175–183.

Chaari, L., Elloumi, N., Mseddi, S., Gargouri, K., Rouina, B.B., Mechichi, T. and Kallel, M. 2015. Changes in soil macronutrients after a long-term application of olive mill wastewater. J. Agric. Chem. Environ. 4: 1–13.

Chartzoulakis, K., Psarras, G., Moutsopoulou, M. and Stefanoudaki, E. 2010. Application of olive wastewater to cretan olive orchard: effects on soil properties, plant performance and the environment. Agr. Ecosyst. Environ. 138: 293–298.

Cortón, E. and Viale, A. 2006. Solucionando grandes problemas ambientales con la ayuda de pequeños amigos: Lastécnicas de biorremediación. Rev. Ecosistemas 15.

Dakhli, R., Khatteli, H., Ridha, L. and Taamallah, H. 2018. Agronomic application of olive mill waste water: short-term effect on soil chemical properties and barley performance under semiarid mediterranean conditions. EQA 27: 1–17.

Darwish, Kh.M. and Abdel, Kawy, W.A. 2008. Quantitative assessment of soil degradation in some areas North Nile Delta, Egypt. Int. J. Geol. 2: 17–22.

Davidson, E.A. and Janssens, I.A. 2006. Temperature sensitivity of soil carbon decomposition and feedbacks to climate change. Nature 440: 165–173.

Di, Bene, C., Pellegrino, E., Debolini, M., Silvestri, N. and Bonari, E. 2013. Short- and long-term effects of olive mill wastewater land spreading on soil chemical and biological properties. Soil Biol. Biochem. 56: 21–30.

Douds, D.D., Galvez, L., Franke-Snyder, M., Reider, C. and Drinkwater, L.E. 1997. Effect of compost addition and crop rotation point upon VAM fungi. Agric. Ecosyst. Environ. 65: 257–266.

Dwivedi, R.S. 2002. Spatio-temporal characterization of soil degradation. Trop. Ecol. 43(1): 75–90.

EC – Commission of the European Communities. 2006a. Thematic Strategy for Soil Protection, COM(2006)231 final, Brussels.

EC – Commission of the European Communities. 2006b. Impact Assessment of the Thematic Strategy on Soil Protection. SEC(2006)620 Brussels.

EEA (European Environment Agency). 1995. *In*: Stanners, D. and Boureau, P. (eds.). Europe's Environment: The Dobris Assessment. Office for Official Publications of the European Communities, Luxemburg.

EEA (European Environment Agency). 1999. Management of contaminated sites in Western Europe. Christensen, M. 1989. A view of fungal ecology. Mycologia 81: 1–19.

EEA (European Environment Agency). 2000. Down to earth: Soil degradation and sustainable development in Europe. Environmental issues series No 16. European Environment Agency, Copenhagen Denmark, 32 pp.

EEA-UNEP. 2000. Down to earth: Soil degradation and sustainable development in Europe. A challenge for the 21 century. Environmental issues Series No 6. EEA, UNEP, Luxembourg.

FAO and ITPS. 2015. Status of the World's Soil Resources (SWSR) – Main Report. Rome, Italy, Food and Agriculture Organization of the United Nations and Intergovernmental Technical Panel on Soils.

Feria, A.L. 2000. The generated situation by the O.M.W. in Andalusia, Actas/Proceedings-Workshop Improlive, Annex A1, 55–63. FAIR CT96 1420.

Ferri, D., Convertini, G., Montemurro, F., Rinaldi, M. and Rana, G. 2002. Olive Wastes Spreading in Southern Italy: Effects on Crops and Soil 12th ISCO Conference Beijing 2002.

Freibauer, A., Rounsevell, M. and Smith, P. 2004. Carbon sequestration in European agricultural soils. Geoderma 122: 1–23.

Garg, S. 2017. Bioremediation of agricultural, municipal, and industrial wastes. pp. 341–363. *In*: Bhakta, J. (ed.). Handbook of Research on Inventive Bioremediation Techniques. Hershey, PA: IGI Global.

German Advisory Council on Global Change (WBGU). 1994. World in Transition—The Threat to Soils: 1994 Annual Report Huthig Pub Ltd.

González-Martínez, A., de Simón-Martín, M., López, R., Táboas-Fernández, R. and Bernardo-Sánchez, A. 2019. Remediation of potential toxic elements from wastes and soils: analysis and energy prospects. Sustainability 11: 3307.

Graeber, P.W. and Blankenburg, R. 2009. Modelling of soil degradation and restoration under climate change and land use impact. In Proceedings of International Conference of Soil Degradation, Riga 137–138.

Hanegraaf, M.C., Hoffland, E., Kuikman, P.J. and Brussaard, L. 2009. Trends in soil organic matter contents in Dutch grasslands and maize fields on sandy soils. Eur. J. Soil Sci. 60: 213–222. doi:10.1111/j.1365-2389.2008.01115.x.

Hanifi, S. and El, Hadrami, I. 2008a. Olive mill wastewaters fractioned soil-application for safe agronomic reuse in date palm (Phoenix dactylifera L.) fertilization. J. Agron. 7: 63–69.

Henis, Y. 1997. Bioremediation in agriculture: dream or reality? pp. 481–489. *In*: Rosen, D., Tel-Or, E., Hadar, Y. and Chen, Y. (eds.). Modern Agriculture and the Environment. Developments in Plant and Soil Sciences, vol. 71. Springer, Dordrecht.

Hill, J. 2003 Land and Soil Degradation Assessments in Mediterranean Europe – Contribution to the GMES Full Details Report p. 9.

Igiri, B.E., Okoduwa, S.I.R., Idoko, G.O., Akabuogu, E.P., Adeyi, A.O. and Ejiogu, I.K. 2018. Toxicity and bioremediation of heavy metals contaminated ecosystem from Tannery wastewater: A review. J. Toxicol. 2018: 2568038.

Jeftic, L., Milliman, J.D. and Sestini, G. 1993. Climatic Change and the Mediterranean. Edward Arnold, London, 673 pp.

Jelusic, M. and Lestan, D. 2015. Remediation and reclamation of soils heavily contaminated with toxic metals as a substrate for greening with ornamental plants and grasses. Chemosphere 138: 1001–7.

Joner, E.J. and Jakobsen, I. 1995. Growth and extracellular phosphatase activity of arbuscular mycorrhizal hyphae as influenced by soil organic matter. Soil Biol. Biochem. 27: 1153–1159.

Joner, E.J. 2000. The effect of long-term fertilization with organic or inorganic fertilizers on mycorrhiza-mediated P uptake in subterranean clover. Biol. Fertil. Soils 32: 435–440.

Kappekalis, I.E., Tsagarakis, K.P. and Crowth, J.C. 2008. Olive oil history, production by product management. Rev. Environ. Sci. Bio. 7: 1–26.

Kapellakis, I.E., Paranychianakis, N.V., Tsagarakis, K.P. and Angelakis, A.N. 2012. Treatment of olive mill wastewater with constructed wetlands. Water 4: 260–271.

Kavvadias, V., Doula, M.K., Komnitsas, K. and Liakopoulou, N. 2010. Disposal of olive oil mill wastes in evaporation ponds: Effects on soil properties. J. Hazard. Mater. 182(1-3): 144–55.

Kavvadias, V., Doula, M., Papadopoulou, M. and Theocharopoulos, S.I.D. 2015. Long-term application of olive-mill wastewater affects soil chemical and microbial properties. Soil Res. 53: 461–473.

Kavvadias, V., Elaiopoulos, K., Theocharopoulos, S.I.D. and Soupios, P. 2016. Fate of potential contaminants due to disposal of olive mill wastewaters in unprotected evaporation ponds. Bull. Environ. Contam. Toxicol. 98: 323.

Kavvadia, S.V., Doula, M. and Theocharopoulos, S. 2014. Long term effects of olive oil mill wastes disposal on soil. Enviro. Forensics 15(1): 37–51.

Komnitsas, K., Zaharaki, D., Doula, M. and Kavvadias, V. 2011. Origin of recalcitrant heavy metals present in olive mill wastewater evaporation ponds and nearby agricultural soils. Environ. Forensics 12: 1–8.

Kumpiene, J., Lagerkvist, A. and Maurice, C. 2008. Stabilization of As, Cr, Cu, Pb and Zn in soil using amendments—a review. Waste Manag. 28(1): 215–25.

Lal, R., Hall, G.F. and Miller P. 1989. Soil degradation: I. Basic processes. Land Degrad. Rehabil. 1: 51–69.

Lal, R. 1993. Tillage effects on soil degradation, soil resilience, soil quality, and sustainability. Soil Tillage Res. 27: 1–8.

Magdich, S., Jarboui, R., Ben, Rouina, B., Boukhris, M. and Ammar, E. 2012. A yearly spraying of olive mill wastewater on agriculture soil over six successive years: impact of different application rates on olive production, phenolic compounds, phytotoxicity and microbial counts. Sci. Total Environ. 430: 209–216.

Mahmoud, M., Janssen, M., Peth, S., Horn, R. and Lennartz, B. 2012. Long-term impact of irrigation with olive mill wastewater on aggregate properties in the top soil. Soil Tillage Res. 124: 24–31.

Martınez-Mena, M., Alvarez-Rogel, J., Castillo, V. and Albaladejo, J. 2002. Organic carbon and nitrogen losses influenced by vegetation removal in a semiarid Mediterranean soil. Biogeochemistry 6: 309–321.

Mashali, A.M. 1991. Land degradation and desertification in Africa. 2nd African Soil Sci. Soc. Conf.

Matson, P.A., Parton, W.J., Power, A.G. and Swift, M.J. 1997. Agricultural intensification and ecosystem properties. Science 227: 504–509.

Mechri, B., Echbili, A., Issaoui, M., Braham, M., Elhadij, S.B. and Hammami, M. 2007. Short-term effects in soil microbial community following agronomic application of olive mill wastewaters in a field of olive trees. Appl. Soil Ecol. 36: 216–223.

Mechri, B., Cheheb, H., Boussadia, O., Attia, F., Ben, Mariem, F., Braham, M. and Hammami, M. 2011. Effects of agronomic application of olive mill wastewater in a field of olive trees on carbohydrate profiles, chlorophyll a fluorescence and mineral nutrient content. Environ. Exp. Bot. 71: 184–191.

Mekki, A., Dhouib, A. and Sayadi, S. 2006. Changes in microbial and soil properties following amendment with treated and untreated olive mill wastewater. Microbiol Res. 161: 93–101.

Mekki, A., Dhouib, A. and Sayadi, S. 2007. Polyphenols dynamics and phytotoxicity in a soil amended by olive mill wastewaters. J. Environ. Manage. 84: 134–140.

Mekki, A., Dhouib, A. and Sayadi, S. 2013. Review: effects of olive mill wastewater on soil properties and plant growth. Int. J. Recycl. Org. Waste Agric. 2: 15.

Mokany, K., Raison, R.J. and Prokushkin, A.S. 2006. Critical analysis of root: shoot ratios in terrestrial biomes, Glob. Change Biol. 12: 84–96.

Montanarella, L., Olazabal, C. and Selvaradjou, S.K. (eds.). 2004. Reports of the Technical Working Groups established under the Thematic Strategy for Soil Protection. Volume II Erosion. European Commission Joint Research Centre and European Environmental Agency. EUR 21319 EN/2: 275–294.

Montanarella, L. 2005. The state of European soils. pp. 9–20. In: Strategies Science and Low for the Conservation of World Soil Resources. International Workshop, Agricultural University of Iceland Selfoss Iceland, September 14–18, 2005.

Montanarella, L. and Toth, G. 2008. Desertification in Europe. 15th International Congress of the International Soil Conservation (ISCO Congress) Soil and Water Conservation, Climate Change and Environmental Sensitivity. 18 May 2008, Budapest, Hungary.

Moraetis, D., Stamati, F.E., Nikolaidis, N.P. and Kalogerakis, N. 2011. Olive mill wastewater irrigation of maize: impacts on soil and groundwater. Agric. Water Manag. 98: 1125–1132.

Niaounakis, M. and Halvadakis, C.P. 2006. Olive processing waste management. Literature Review and Patent Survey', Second Edition, Elsevier, Amsterdam.

Oyeyiola, A.O., Adeosun, W. and Fabunmi, I.A. 2017. Use of agricultural wastes for the immobilization of metals in polluted soils in Lagos State, Nigeria. J. Health Poll. 7(13): 56–64. doi:10.5696/2156-9614-7-13.56.

Pagliai, M., Pellegrini, S., Vignozzi, N., Papini, R., Mirabella, A., Piovanelli, C., Gamba, N., Miclaus, M., Castaldini, C. and De Simone. 2001. Influenza dei reflui oleari sulla qualita` del suolo. Supplemento a L'Informatore Agrario 50: 13–18.

Payá, Pérez, A. and Rodríguez, Eugenio, N. 2018. Status of local soil contamination in Europe: Revision of the indicator Progress in the management Contaminated Sites in Europe, EUR 29124 EN, Publications Office of the European Union, Luxembourg, 2018, ISBN 978-92-79-80072-6, doi:10.2760/093804, JRC107508.

Perez-Trejo, F. 1994. Desertification and land degradation in the European Mediterranean. Report EUR 14850 En. European Commission, Luxembourg, 63 pp.

Peri, C. and Proietti, P. 2014. Olive mill waste and by-products. pp. 283–302. In: Peri, C. (ed.). The Extra-Virgin Olive Oil Handbook. John Wiley & Sons, Ltd, Chichester, UK.

RED ESPAÑOLA DE COMPOSTAJE. Residuos agrícolas; Ediciones Paraninfo: Madrid, Spain, 2014; ISBN 978-84-8476-698-8. 2. European Commission 7th Environment Action Programme. Available online: http://ec.europa.eu/environment/action-programme/index.htm.

Regni, L., Gigliotti, G., Nasini, L., Agrafioti, E., Galanakis, C.M. and Proietti, P. 2017. Reuse of olive mill waste as soil amendment. pp. 97–117. In: Galanakis, C.M. (ed.). Olive Mill Waste: Recent Advances for Sustainable Management. Elsevier-Academic Press, London, United Kingdom.

Ryan, M.H., Chilvers, G.A. and Dumaresq, D.C. 1994. Colonization of wheat by VA mycorrhizal fungi was found to be higher on a farm managed in organic manure than on a conventional neighbour. Plant Soil 160: 33–40.

Sierra, J., Marti, E., Garau, M.A. and Cruanas, R. 2007. Effects of the agronomic use of olive oil mill wastewater: Field experiment. Sci. Total Environ. 378: 90–94.

Smith, P., Smith, Jo Andrén, O., Perälä, T., Regina, K. and Wesemael, B. 2005. Carbon sequestration potential in European croplands has been overestimated. Glob. Chang Biol. 11: 2153–2163.

Sobhi, B., Isam, S., Ahmad, Y., Jacob, H. and Ahlam, S. 2004. Reducing the environmental impact of olive mill wastewater in Jordan, Palestine and Israel. In: Water for Life in the Middle East. 2nd Israeli-Palestinian International Conference, Turkey 10–14 October 2004.

Tolba, M.K., El-Kholy, O.A., El-Hinnawi, E., Holdgate, M.W., McMichael, D.F. and Munn, R.T. 1992. The World Environment 1972–1992: Two Decades of Challenge. London: Chapman & Hall.

UNCCD. 1994. United Nations Convention to Combat Desertification, Adopted 17 June 1994 in Paris, France, United Nations, New York, NY, USA. http://www.fao.org/desertification/article_html/en/1.htm.

UNEP Staff. 1991. Global assessment of soil degradation. UNEP. UN. GLASOG. Project.

UNEP Staff. 1992. World atlas of decertification. Publ. E. Arnold, London, 69 pp.

Vacca, A., Loddo, S., Serra, G. and Aru, A. 2009. Soil Degradation in Sardinia (Italy): main factors and processes, Options Méditerranéennes, Série A n.50 p. 11.

Van-Camp, L., Bujarrabal, B., Gentile, A.R., Jones, R.J.A., Montanarella, L. and Olazabal, C. 2004. Reports of the Technical Working Groups Established under the Thematic Strategy for Soil Protection. EUR 21319 EN/1, 872 pp. (Office for Official Publications of the European Communities: Luxembourg). Varallyay.

Vella, F., Galli, E., Calandrelli, R., Cautela, D. and Laratta, B. 2016. Effect of olive mill wastewater spreading on soil properties. Bull. Environ. Contam. Toxicol. DOI 10.1007/s00128-016-1830-7.

Venegas, A., Rigol, A. and Vidal, M. 2015. Viability of organic wastes and biochars as amendments for the remediation of heavy metal-contaminated soils. Chemosphere 119: 190–8.

Vleeshouwers, L.M. and Verhagen, A. 2002. Carbon emission and sequestration by agricultural land use: a model study for Europe. Glob. Change Biol. 8: 519–530.

Walker, D.J., Clemente, R., Roig, A. and Bernal, M.P. 2003. The effects of soil amendments on heavy metal bioavailability in two contaminated Mediterranean soils. Environ. Pollut. 122(2): 303–12.

Wim, G. and E.l. Hadji, M. 2002. Causes, general extent and physical consequence of land degradation in arid, semi arid and dry sub humid areas. Forest Conservation and Natural Resources, Forest Dept. FAO, Rome, Italy.

Wright, D.P., Read, D.J. and Scholes, J.D. 1998. Mycorrhizal sink strength influences whole plant carbon balance of Trifolium repens L., Plant Cell Environ. 21: 881–891.

Yasemin, K. and Killi, D. 2008. Influence of olive oil solid waste applications on soil pH, electrical conductivity, soil nitrogen transformations, carbon content and aggregate stability. Bioresour. Technol. 99: 2326–2332.

Zalidis, G., Stamatiadis, S., Takavakoglou, V., Eskridge, K. and Misopolinos, N. 2002. Impacts of agricultural practices on soil and water quality in the Mediterranean region and proposed assessment methodology. Agriculture, Ecosyst. and Environ. 88: 137–146.

19

Membrane Bioreactor for Perchlorate Treatment

Benny Marie B. Ensano, Sivasankar Annamalai and *Yeonghee Ahn**

1. Introduction

Perchlorate (ClO_4^-) is a highly persistent and toxic oxyanion that is continuously being released in our waters by natural and anthropogenic means (e.g., poor industrial practices, military testing, unregulated disposal, and natural deposition). Salts of perchlorate are widely used as the primary ingredient in the manufacture of explosives and solid-state rocket fuel due to their high chemical stability and powerful oxidizing capacity. Since perchlorate is highly soluble in water (i.e., water solubility of $NaClO_4^-$ at 25°C equals 2,100 mg/l) and has poor retention in soils and minerals, it becomes extremely mobile and strongly persists in the aquatic environment (Urbansky and Brown 2003, Cao et al. 2019).

Numerous cases of perchlorate detection in water resources have been reported globally and some even mentioned ClO_4^- concentrations as high as 811 µg/L, 120 mg/L and 3,700 mg/L in drinking water, surface water and groundwater, respectively (Guan et al. 2015). Such concentrations can pose potential health risks to aquatic and terrestrial organisms. Schmidt et al. (2012) investigated the effects of the different perchlorate concentrations on zebrafish and results showed conspicuous alterations on the vertebrate's thyroidal tissue and pituitary gland at concentrations greater than 250 µg/L. Chronic exposure of perchlorate (10 mg/L) to subadult threespine stickleback (*Gasterosteus aculeatus*) was found by Gardell et al. (2017) to cause obesogenic (adipose-inducing) effects. Moreover, perchlorate can also be accumulated in plants in two ways: (1) when water contaminated with perchlorate is used for irrigation, or (2) when perchlorate-containing fertilizer is added to soil (Calderón et al. 2017). The rate of perchlorate uptake by plants mainly depends on several factors, namely, plant species, ClO_4^- concentration and co-existing ions (Yu et al. 2004).

Ingestion of perchlorate-contaminated drinking water or edible plants, such as fruits and vegetables, may affect the thyroid function in humans. Perchlorate is a known competitive inhibitor of sodium/iodide symporter (NIS) that is responsible for the uptake of iodine needed for the production of metabolic and developmental hormones (Leung et al. 2010). Thyroid hormone deficiency during pregnancy or after birth has been linked to problems in neuropsychological functions such as visual, motor, language and memory skills on the fetus or infant (Zoeller and Rovet 2004). Higher doses (6 mg/kg of body wt·d) even result in fatal bone marrow disorders (Coates et al. 2001). The official reference dose (RfD) for perchlorate as set by the United States Environmental Protection Agency

---

Department of Environmental Engineering, Dong-A University, Busan 49315, South Korea.
* Corresponding author: yahn@dau.ac.kr

(U.S. EPA) is 0.7 μg/kg of body wt·d, corresponding to a Drinking Water Equivalent Level (DWEL) of 24.5 μg/L in solution (Yao et al. 2017).

Physicochemical techniques, namely, anion exchange (Gu et al. 2007), membrane filtration (Yoon et al. 2009) and adsorption (Mahmudov and Huang 2010), were found effective for small-scale treatment of perchlorate-laden aqueous solution. However, these methods only separate perchlorate from the bulk liquid and undesirable by-products are usually generated requiring subsequent treatment, thereby increasing the total treatment costs (Wan et al. 2019). In particular, contaminated resins or brine solutions from ion exchange processes comprise 4–6% of the total disposal costs (Boles et al. 2012). Membrane filtration utilizing porous membranes, such as reverse osmosis (RO), nanofiltration (NF) and ultrafiltration (UF) membranes, was also investigated (Yoon et al. 2005a, b, 2009).

Rejection mechanisms of perchlorate by the membranes were found to be governed by size exclusion and electrostatic exclusion, and that the removal was enhanced at increasing solution pH and conductivity (Yoon et al. 2005a, b). RO membrane can be used for direct pollutant treatment since it can retain low-molecular mass compounds and ions due to the very high dense properties of its separating layer (Velizarov et al. 2005). With RO filtration, total retention of ions can be achieved, albeit undesired because non-toxic minerals such as hardness (Ca^{2+}, Mg^{2+}) may drop to very low levels causing corrosion of some metals (e.g., lead, copper, iron, zinc, etc.) (Velizarov et al. 2005). NF membrane is another alternative technology as it selectively separates ions according to their valent forms (i.e., multi-valent or mono-valent ions) via diffusion through the membranes (at low pressure), convection (at high pressure) or repulsion (Donnan exclusion) (Velizarov et al. 2005). However, pH and ionic strength of the source water significantly affect NF membrane and its ability to reject target anions, thereby requiring a proper selection of operating conditions (Yoon et al. 2009). UF membranes, on the other hand, need larger aggregates for effective filtration, especially in the presence of co-existing anions; hence they are seldom used in the direct removal of perchlorate from solution. Nevertheless, they can be combined with other processes (i.e., biodegradation, coagulation, etc.) for solid exclusion and better effluent quality (Ensano et al. 2019).

Biodegradation is an economically sustainable treatment method because of its capability to completely convert perchlorate into harmless end products at low operating and capital costs. Under anaerobic or anoxic condition, ClO_4^- is used as an electron acceptor, while organic (e.g., methane, ethanol, glucose, acetate, etc.) and inorganic (e.g., elemental iron (Fe^o) hydrogen gas (H_2), and elemental sulfur (S^o), etc.) compounds are used as electron donors by heterotrophic and autotrophic perchlorate-reducing bacteria (PRB), respectively (Han et al. 2011, Zhang et al. 2016, Song et al. 2019) (Figure 1). Heterotrophic PRB use organic compounds as the electron donor and carbon source, while autotrophic PRB use CO_2 as the carbon source. Perchlorate is reduced to chlorate (ClO_3^-) and chlorite (ClO_2^-) by perchlorate reductase and ClO_2^- is further degraded to nontoxic Cl^- and O_2 by chlorite dismutase (1).

$$ClO_4^- \to ClO_3^- \to ClO_2^- \to Cl^- + O_2 \qquad (1)$$

The addition of electron donor to the contaminated water must be carefully controlled as insufficient amount may lead to incomplete perchlorate removal, while over dozing, particularly for organic compounds, can stimulate substantial biomass accumulation in water distribution systems as well as the possible formation of toxic disinfection by-products (Gao et al. 2016). This constitutes one of the major drawbacks of the biological method. In lieu of this, scientists around the world have been searching for new and promising approach to improve the biodegradation process of perchlorate for both *in situ* and *ex situ* applications, the latter being more prevalent in practice. One emerging technology is the membrane bioreactor (MBR), which is the combination of membrane technology and biodegradation process. This paper puts an emphasis on the applicability of perchlorate treatment by MBR, its advantages over other treatment methods, limitation for practical implementations and some future challenges.

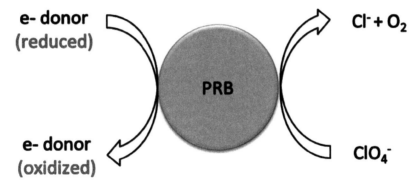

Figure 1. Reduction of perchlorate by PRB. Adapted from Ahn (2019).

2. Membrane bioreactor

2.1 Principles of MBR

Over the years, MBR has been the focus of extensive research in the field of environmental engineering owing to its many advantages over the conventional biological treatment methods. MBR effectively separates the suspended solids and soluble organic matter in the water, particularly the pathogenic bacteria and viruses; thus, additional purification and disinfection treatment may not be required (Ensano et al. 2016). This reduces the operational costs and further improves the effluent quality by the prevention of formation of toxic and hazardous by-products. MBR also has higher biomass concentration yielding higher rate of BOD and COD removal and lower excess sludge production (Ensano et al. 2019).

Most studies regarding MBR mainly focus on the conventional type, widely known as the pressure-driven MBR. In this paper, the authors define MBR as a water treatment technology that combines biological degradation process and membrane technology in a single reactor. Aside from pressure-driven MBR, two other configurations were defined, namely, the ion exchange MBR and the gas-transfer MBR. Each configuration has its own advantages and disadvantages, which are discussed in detail in the succeeding sections.

2.2 Types of MBR

2.2.1 Pressure-driven MBR

In a pressure-driven MBR, the separation of permeate from contaminated water is facilitated by the pressure difference across a porous membrane. The first pressure-driven MBR was built in 1968 via external/side stream configuration, wherein an external UF membrane module was installed after the activated sludge tank as a replacement of the secondary clarifier to overcome sludge settling difficulties (Smith 1968). A new and better configuration was designed by Yamamoto et al. (1989), known as submerged membrane bioreactor, wherein the membrane module was immersed directly inside the aeration tank and a pump was used to recover the permeate (Figure 2A). Internal/submerged MBR is more advantageous than the external/side-stream MBR because it has lower cost, lower energy demand, needs less space and operates at considerably low flux. Moreover, since the sludge retention time is independent from the hydraulic retention time, slow growing microorganisms are maintained inside the bioreactor, which is beneficial to complex micropollutants that require longer degradation periods (Velizarov et al. 2005). Despite these advantages, large scale applications of MBR technology are limited due to the fouling caused by deposition of sludge into the membrane pores. Some of the membrane fouling control techniques successfully implemented by various studies include membrane backwashing, excess aeration, frequent membrane cleaning and the integration with electrochemical processes (Ensano et al. 2019). Pressure-driven anoxic

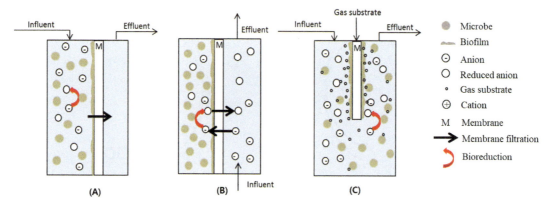

Figure 2. Three types of membrane bioreactor (A) Pressure-driven MBR, (B) Ion exchange MBR, and (C) Gas-transfer MBR.

MBR has been widely used in the treatment of several anions including nitrate (NO_3^-); however, little is known about its application in perchlorate treatment.

2.2.2 Ion exchange MBR

Ion exchange MBR (IEMB) is a tailor made technology that combines physical separation of charged pollutant, using ion exchange membranes, with biodegradation of pollutant by a suitable microbial culture in a bioreactor (Figure 2B). Ion exchange membranes can be either anion or cation exchange membrane depending on the charge of the target contaminants. The orientation of the liquid flow in an IEMB is opposite to that of the conventional MBR. In IEMB, membrane separation precedes biodegradation, such that the ionic micropollutant from the aqueous feedstock (water-compartment) passes through a selective barrier (non-porous ion exchange membrane) towards the opposite compartment (bio-compartment) that is enriched with microorganisms for biodegradation (Figure 2B). The ion transfer across the membrane is governed by Donnan dialysis in which the addition of high concentration of driving counter-ions at the bio-compartment permits the flow of the target ions from the water stream to the bio-compartment, hence maintaining overall electroneutrality between compartments (Crespo et al. 2004). Donnan exclusion effect is also observed, wherein co-ions are rejected by the similar charged membrane (Fox et al. 2014). After the target ion crosses the membrane, it enters the bio-compartment where it is degraded into innocuous products by an appropriate microbial culture. For an effective biodegradation performance, the microorganisms should be fed with carbon and energy source as well as essential nutrients. Moreover, accumulation of the ions can create favorable conditions for biofilm growth at the membrane surface in contact with the bio-compartment. This microbial biofilm increases the ion transmembrane flux and prevents the diffusion of the carbon source from the bio-compartment to the water stream compartment (Fox et al. 2016). However, excess biofilm growth may lead to low transport rate of ionic pollutants.

The first application of IEMB in water treatment was successfully conducted by Fonseca et al. (2000) when nitrate was removed from a synthetic groundwater using IEMB at denitrification rate of 7g N/m^2·d. Portuguese National Patent No. 102385 N and European Patent EP1246778 were awarded to Crespo et al. (1999) and Crespo and Reis (2003), respectively. IEMB was then used extensively for the selective removal of ionic pollutants, such as nitrate, perchlorate and bromate (Matos et al. 2005, 2006a, 2008).

The advantages of IEMB include the following (Fonseca et al. 2000, Crespo et al. 2004): (1) it prevents secondary contamination of the treated water by microbial cells, excess nutrients and/or metabolic by-products since biodegradation occurs in the opposite compartment; (2) the hydraulic residence time in both compartments can be separately adjusted which can aid in the optimization of the pollutant extraction; (3) the selection of appropriate driving counter-ions in the bio-compartment keeps the concentration of the target pollutant in the bio-compartment at low levels; (4) water treatment performance of IEMB mainly depends on the transport rate of the target ions through the membrane and not on the pollutant biodegradation rate; hence, the effluent quality is not affected even if the bio-compartment is operated under extreme conditions (e.g., excessive nutrients and carbon source); and (5) O_2 in the feed water does not enter the bio-compartment, which can otherwise interfere with the biodegradation (i.e., anoxic or anaerobic) process.

2.2.3 Gas-transfer MBR

In gas-transfer MBR, gases such as O_2, H_2, and methane (CH_4) are used as electron donor by microorganisms to degrade contaminants in aqueous solution. It employs gas-permeable membranes, which allow efficient transfer of the gaseous substrate for microbial consumption without bubble formation. Bubble-less gas transfer prevents the accumulation of explosive amount of the gas substrate in the headspace above water, as in the case of highly flammable H_2 (Crespo et al. 2004). Gas-transfer MBR differs significantly from the conventional pressure-driven MBR and ion exchange MBR in terms of the substrate introduction and the microbial orientation. In a gas-transfer MBR, the gaseous substrate is introduced into the membrane module (lumen) and diffuses outside the walls towards the bulk liquid (bio-compartment) (Figure 2C). The microbes can either be suspended in the solution or naturally grown on the membrane surface as biofilm, the latter being designated as the membrane biofilm reactor (MBfR). *In situ* water remediation of contaminated site by gas-transfer MBR can be done by installing the membrane system across the plume while the groundwater flows through it. As the gaseous substrate diffuses from inside the membrane module towards the plume, the biofilm or suspended microorganisms consume it while simultaneously degrading the pollutants.

The concept of MBfR was pioneered by Schaffer et al. (1960) when O_2 was introduced inside the permeable plastic films coated with slimes to oxidize organic pollutants in sewage water. Years passed and the application of O_2-based MBfR (also known as membrane-aerated biofilm reactor, MABR) did not develop into a major water treatment technology since there are so many well-established methods to provide aeration for microorganism consumption (Rittmann 2007). In addition, the major drawback of the MABR technique is the lower oxygen transfer to the biofilm due to the occurrence of gas back diffusion, which affects the efficiency of the wastewater treatment process (Nerenberg 2016).

Most of the researchers focusing on the treatment of wastewater by MBfR utilize H_2-based and CH_4-based gas-transfer MBfR for the effective reduction of micropollutants. Methane-based MBfR uses CH_4 as both carbon and energy sources by heterotrophic bacteria under anoxic condition (Luo et al. 2015). CH_4 is an inexpensive gas that can either be extracted from large fossil reserves or can be produced via anaerobic digestion of biomass. Meanwhile, the H_2-based MBfR has been gathering good reviews from scientific community because of its capability to bio-reduce a variety of oxidized compounds (e.g., NO_3^-, ClO_4^-, BrO_3^-, SeO_4^{2-}, $H_2AsO_4^-$, CrO_4^{2-}, etc.) (Rittmann 2007). H_2 is consumed by autotrophic bacteria, which use inorganic carbon as carbon source, therefore preventing the production of excess biomass in the water distribution system. H_2 is also inexpensive in terms of electron-equivalent that is required for contaminant reduction. Lastly, H_2 is non-toxic to human health and has reliable and safe methods for transport and storage (Rittmann 2007).

3. Perchlorate treatment by MBR

3.1 Ion exchange MBR

The removal of perchlorate by IEMB starts in its physical extraction from the contaminated water stream using an anion exchange membrane (AEM) (Figure 3) (Fox et al. 2014). A driving counter-ion (usually Cl^-) is continually fed to the bio-compartment to favor the transport of ClO_4^- from the feed water to the bio-compartment via Donnan dialysis. Maximized extraction of ClO_4^- can be achieved when the Cl^- is added in excess (Matos et al. 2006a). After the perchlorate passes through

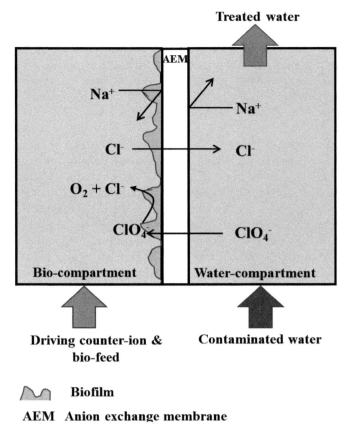

Figure 3. Schematic diagram of perchlorate treatment by IEMB. Modified from Matos et al. (2006b).

the positively charged membrane, it is then consumed either by the biofilm developed on the membrane bio-compartment side or by the microbes suspended in the feed medium. Biodegradation of perchlorate takes place primarily on the membrane-biofilm interface rather than in the bulk of the bio-compartment (Ricardo et al. 2012).

Only few literatures are available for the treatment of perchlorate by IEMB (Table 1). Matos et al. (2005) successfully reduced the amount of oxy-anions (ClO_4^-, NO_3^-, and BrO_3^-) from synthetically prepared polluted water to values below the lowest recommended levels (25 mg/L of NO_3^- by EU, 4 µg/L of ClO_4^- by US EPA and 25 µg/L of BrO_3^- by WHO). In a separate study, Matos et al. (2006b) performed three independent experiments to study the mechanism of perchlorate and nitrate removal via Donnan dialysis, biological reduction and IEMB processes. Since commercial ion exchange membrane typically used in IEMB was highly expensive, a cheaper AEM (Excellion I-200 membrane) was evaluated and compared to the usually employed and costly Neosepta ACS membrane (Matos et al. 2008). Along with perchlorate and nitrate, removal of arsenate and ionic mercury was also studied using various commercial ion exchange membranes (i.e., Neosepta ACS, Ionac MA-3475, Nafion 112, Neosepta CMX) (Velizarov et al. 2005). Ricardo et al. (2012) conducted a kinetic study to assess the effects of the presence of nitrate on the bio-reduction of perchlorate. In the same study, the biofilm growth on the membrane surface was also analyzed for possible biofilm stratification. Above mentioned studies used ethanol as the carbon and energy source of the microbial culture in the bio-compartment. In the studies of Fox et al. (2014, 2016), glycerol was used as carbon substrate instead of ethanol for the biodegradation of high concentrations of perchlorate in synthetic groundwater using IEMB.

Table 1. Summary of the literature on perchlorate treatment using IEMB.

Wastewater	Inoculum	e⁻ donor	Membrane (Manufacturer)	Operational conditions#	Results	References
Synthetic WW (100 µg ClO_4^-/L, 60 mg NO_3^-/L)	Primary inoculum of municipal WWTP	Ethanol	Neosepta ACS (Tokuyama Soda, Japan)	V = 500 mL; T = 23 ± 1°C; HRT_w = 1.4–8.3 h; HRT_b = 5 h; F/A = 3.1–18.5 L/m²h; Driving counter-ion = 5.84 g NaCl/L	• 96.5% ClO_4^- and 99.6% NO_3^- removal efficiencies. • No bacterial and ethanol contamination found in the treated water.	Matos et al. 2006b
Synthetic WW (100 µg ClO_4^-/L, 60 mg NO_3^-/L and 200 µg BrO_3^-/L)*	Enriched microbial culture	Ethanol	Neosepta ACS; Excellion™ I-200 (SnowPure, USA)	T = 23 ± 1°C; HRT_b = 5 h; F/A = 3.1–69.9 L/m²h; Driving counter-ion = 3.55 g NaCl/L	Neosepta ACS • 96% ClO_4^- and 99% NO_3^- removal efficiencies. • No bacterial and ethanol contamination found in the treated water. Excellion™ I-200 • 85% ClO_4^- and 88% NO_3^- removal efficiencies • Allowed transport of sulfate and phosphate ions to water-compartment. • No bacterial and ethanol contamination found in the treated water.	Matos et al. 2008
Synthetic WW (100 µg ClO_4^-/L, 60 mg NO_3^-/L)	Primary inoculum of municipal WWTP	Ethanol	Neosepta ACS	T = 25 ± 1°C; HRT_w = 8.3 h; F/A = 3.1 L/m²h; Driving counter-ion = 5.84 g NaCl/L	• Final anion conc. in treated water: 7.0 ± 0.8 µg ClO_4^-/L and 2.8 ± 0.5 mg NO_3^-/L. • NO_3^- reduction was unaffected but ClO_4^- reduction rate decreased by 10% under ammonia limitation.	Ricardo et al. 2012
Synthetic GW (250 mg ClO_4^-/L)	Enriched culture from soil sample of ClO_4^- contaminated site	Glycerol	Neosepta ACS	F/A_w = 0.66 mL/cm²h; F/A_b = 0.13–0.16 mL/cm²h; HRT_w = 12 h; HRT_b = 100–125 h; T_w = 26–29°C; T_b = 25°C; Driving counter-ion = 5.86 g NaCl/L	• Increase in initial ClO_4^- conc. decreased its removal efficiency in the feed water-compartment. • 99% ClO_4^- removal efficiency in the bio-compartment.	Fox et al. 2014
Synthetic GW (1 mM ClO_4^-, 0.1–4 mM NO_3^-)	Enriched culture from soil sample of ClO_4^- contaminated site	Glycerol	Neosepta ACS	F/A_w = 0.62 mL/cm²h; F/A_b = 0.13–0.19 mL/cm²h; HRT_w = 15 h; HRT_b = 125–83 h; T_w = 26–27°C; T_b = 25°C	• Over 99% biodegradation efficiency for both ClO_4^- and NO_3^- in the bio-compartment.	Fox et al. 2016

All reactors used are in continuous mode; * Bromate was evaluated only in Donnan dialysis studies and not in IEMB (V, reactor vol.; HRT, hydraulic retention time; w, water-compartment; b, bio-compartment; T, temperature; F/A, water flow rate per membrane area); GW, groundwater; WW, wastewater; WWTP, wastewater treatment plant.

3.1.1 Effect of membrane permselectivity

One important consideration for the effective treatment of perchlorate by IEMB system is the permselectivity of the AEM. Some AEM have low permselectivity towards counter-ions of the same valence, hence affecting the quality of the treated water. Low permselectivity exhibits high resistance to ion transport which can be attributed to the membrane's low ion exchange capacity, high water content, and high thickness (Nagarale et al. 2006). In order to enhance permselectivity of the membrane, various modifications can be made including the synthesis of a thin, highly cross-linked surface modifying layer on top of a non-permselective support (Saracco and Zanetti 1994). The product causes a higher steric repulsion on multi-valent anions than the mono-valent anions, and is hence called mono-valent permselective membrane. One of the most commonly employed mono-valent permselective membrane in IEMB application is the Neosepta ACS membrane (Matos et al. 2006b, Velizarov et al. 2008, Ricardo et al. 2012).

Neosepta ACS membrane allows high pollutant fluxes at longer periods of time and can successfully prevent the transfer of microbial cells, carbon sources and metabolic by-products that could otherwise lead to secondary contamination of the treated water (Matos et al. 2008). Consequently, transport of multi-valent ions (e.g., PO_4^{3-}, HPO_4^{2-}, $H_2PO_4^-$) towards the opposite compartment can also be minimized by ACS membrane, thus preserving the concentration of these ions in the feed water and biomedium. However, the major drawback of Neosepta ACS membrane which hinders the large-scale application of IEMB is its expensive cost. This prompted the search for low-cost alternative membranes. Matos et al. (2008) compared the performance of a relatively cheaper Excellion™ I-200 (SnowPure, USA) to that of Neosepta ACS (Tokuyama Soda, Japan) in the removal of NO_3^- and ClO_4^- by IEMB system. Although Excellion I-200 obtained lower removal efficiencies of ClO_4^- (85%) and NO_3^- (88%) than that of the Neosepta ACS (96% ClO_4^- and 99% NO_3^- removal efficiencies) and allowed subsequent transport of sulfate and phosphate ions ($H_2PO_4^-$, HPO_4^{2-}) from the bio-compartment to the treated water stream, there was no observed flux decline of the target anions during the one month operation and no secondary contamination due to organic compounds in the treated water (Matos et al. 2008). The latter characteristics are two important criteria in ion exchange membrane selection. These results showed that Excellion I-200 can be a cheaper alternative to Neosepta ACS for perchlorate removal by IEMB technique.

3.1.2 Effect of biofilm formation

As mentioned earlier, the bioconversion of perchlorate to harmless species in IEMB mainly occurs in the biofilm rather than in the suspended culture. This is because the majority of the active microbial population is found in the biofilm (Fox et al. 2014). The formation of biofilm on the membrane surface in contact with the bio-compartment is due to the ion transport across membranes creating a region with favorable microbial growth conditions (Matos et al. 2006a). Thin biofilm formation facilitates perchlorate reduction and ion transport from water-compartment to bio-compartment by maintaining higher perchlorate flux across the membranes. As demonstrated by Fox et al. (2014), when they compared the fluxes obtained in IEMB with and without biofilm, perchlorate removal efficiency was higher by 27 ± 6.6% in IEMB with biofilm. Fox et al. (2016) also observed a 5% increase in perchlorate transmembrane flux in the presence of biofilm. Additionally, biofilm acts as a secondary barrier against transport of the carbon substrate to the water-compartment, thereby preventing the secondary contamination of the treated water. On the contrary, excess biofilm decreases the IEMB efficiency over time, which can be minimized by controlling the concentration of the ionic pollutants.

3.1.3 Effect of carbon source limitation

Use of heterotrophic PRB for perchlorate treatment by IEMB was reported so far. Ethanol and glycerol have been extensively employed as both the electron donor and carbon source in the effective reduction of ClO_4^- by an enriched heterotrophic culture in IEMB. Ethanol is non-toxic and has low diffusivity through the AEM, that is why a lot of investigations regarding anion treatment by IEMB used this compound as carbon source (Matos et al. 2006b, 2008, Ricardo et al. 2012). However, ethanol is also an expensive carbon source. On the other hand, glycerol, a by-product of bio-diesel production industry, is much cheaper and has 60% lower diffusion coefficient ($6.9 \times 10^{-9} \pm 4.7 \times 10^{-10}$ cm^2/s) than that of ethanol (1.8×10^{-8} cm^2/s), and is hence a suitable alternative. The applicability of glycerol as cheap carbon/energy source in IEMB treatment of perchlorate has just been recently investigated (Fox et al. 2014, 2016). According to Fox et al. (2016), the dissimilatory reduction of perchlorate using glycerol as the electron donor involves two stages: glycerol is first fermented by the suspended particles in the bio-compartment and the fermentation products are then consumed by the biofilm attached to the membrane to reduce perchlorate. The perchlorate bio-reduction reactions with ethanol and glycerol are presented in equations (2) (He et al. 2019) and (3) (Fox et al. 2016), respectively:

$$C_2H_5OH + ClO_4^- \rightarrow Cl^- + 2H_2 + 2CO_2 + H_2O \qquad (2)$$

$$4C_3H_8O_3 + 7ClO_4^- \rightarrow 7Cl^- + 12CO_2 + 16H_2O \qquad (3)$$

The concentration of the carbon source in the bio-compartment must be carefully controlled in order to ensure the production of high-quality treated water. High carbon concentration fed into the bio-compartment was found to cause elevated dissolved organic carbon (DOC) levels in the water-compartment, which then decreased the effluent quality (Fox et al. 2014). This cross-contamination can be attributed to the formation of secondary metabolites from the excess carbon assimilation by the microbial culture in the bio-compartment (Fox et al. 2014). In contrast, insufficient amount of carbon substrate results in accumulated perchlorate ions in the bio-compartment. This, however, does not affect the concentration of perchlorate in the water-compartment since perchlorate transport across membrane is driven by Donnan dialysis principle. In IEMB, the quality of the treated water is the main consideration and not the bioreactor effluent. This is an added advantage of the IEMB process when compared to other techniques, viz. fluidized bed reactor (FBR). In FBR, carbon source limitation results in production of water with higher pollutant concentrations requiring subsequent treatment, which adds to the overall treatment cost.

3.1.4 Effect of co-existing nitrates

It is important to understand the effect of nitrate on perchlorate reduction by IEMB since these two anions often co-exist in contaminated waters. Biodegradation studies involving perchlorate have reported that most PRB can also use NO_3^- as an electron acceptor and that the presence of NO_3^- slows down the degradation of perchlorate (Nerenberg et al. 2008, Gal et al. 2008). A kinetic study was conducted by Ricardo et al. (2012) to investigate the reduction rates of these two oxy-anions in IEMB using a mixed anoxic microbial culture and ethanol as the carbon source. They observed that perchlorate reduction rate was inhibited at the initial stage and only increased after nitrate and nitrite (NO_2^-) were completely depleted. The sequential reduction can be attributed to the presence of low PRB and high denitrifying bacteria population in the mixed culture used (Ricardo et al. 2012). Moreover, both microorganisms have shown more preference to nitrate than perchlorate as primary electron acceptor, which can be seen on their kinetic parameter values: the value of the maximum specific substrate utilization rate (q_{max}) for nitrate (10.79 mg NO_3^-/mg VSS·d) was 35 times higher than perchlorate (0.30 mg ClO_4^-/mg VSS·d). In addition, the lower half saturation constant of nitrate (K_n = 1.05 mg NO_3^-/L) also indicates higher affinity of the mixed culture to nitrate than perchlorate (K_p = 4.97 mg ClO_4^-/L). These values are in agreement with the results of other studies (Halling-Sørensen and Jorgensen 1993, Dudley et al. 2008).

3.1.5 Effect of nutrient limitation

Ricardo et al. (2012) examined the performance of IEMB at excess and limiting nutrient concentrations. Two experiments were carried out feeding the bio-compartment with 9 mg/d and 0.9 mg/d of ammonia. In the presence of excess ammonia, both nitrate and perchlorate were significantly reduced to concentrations below the quantification limits: 1 mg NO_3^-/L and 2 μg ClO_4^-/L. This was due to the development of biofilm on the membrane bio-side where the active biodegradation usually takes place. At limiting ammonia concentration, no biofilm was formed due to the rapid consumption of ammonia by the suspended culture. The nitrate reduction rate was almost unaffected, while the perchlorate reduction rate was decreased by 10%. The same results were obtained by another study (Choi and Silverstein 2008) in which the perchlorate reduction rate was decreased by 70% in the presence of equimolar nitrate at limited acetate condition. This can be attributed to the competition of the microbial cells for the limiting amount of nutrients present (Ricardo et al. 2012). Even if both nitrate and perchlorate accumulated in the bio-compartment due to insufficient bio-nutrients, the final nitrate and perchlorate concentrations in the treated water of IEMB were not affected since the transport of anionic pollutants across the membrane to the bio-compartment by Donnan dialysis primarily dictated the IEMB performance.

3.2 H_2-based gas transfer MBfR

One of the major disadvantages of using heterotrophic microorganisms in the reduction of perchlorate is the secondary contamination of the effluent brought by excessive supplementation of organic carbon and nutrients (Gao et al. 2016). To overcome this problem, autotrophic perchlorate reduction has been recently explored using inorganic electron donor such as H_2 (Zhao et al. 2011) (Table 2). Compared to other inorganic electron donors (i.e., S° and Fe°), H_2 can be readily obtained at a reasonable cost. It can either be purchased from outside suppliers or it can be produced onsite via water electrolysis or methane reforming (Zhou et al. 2019). Furthermore, H_2 is non-toxic and does not produce residue when applied to wastewater treatment. However, the application of H_2 as an effective electron donor for PRB is limited due to its low water solubility and high combustibility.

The idea of using gas-permeable membrane in MBfR prevents the problems related to H_2 off-gassing and combustibility. Such method provides bubble-less transport of H_2 into the reactor, where it is consumed by the autotrophic PRB that forms biofilm on the membrane surface. MBfR can therefore provide high gas utilization efficiency and excellent pollutant removal rates (He et al. 2019). The effectivity of H_2 as electron donor for perchlorate reduction in MBfR was previously demonstrated by Nerenberg et al. (2002), who observed 39% perchlorate removal even without special inoculation required (Table 2). The stoichiometric reaction of perchlorate reduction using H_2 as electron donor is given in the equation (4) below:

$$ClO_4^- + 4H_2 \rightarrow Cl^- + 4H_2O \tag{4}$$

3.2.1 Effect of co-existing electron acceptors

Several electron acceptors (i.e., NO_3^-, NO_2^-, O_2, and SO_4^{2-}) were found to inhibit perchlorate reduction using H_2-based MBfR. Ziv-El and Rittmann (2009) were the first to quantitatively analyze these effects. They found that electron-acceptor surface loading controls the reduction kinetics of oxidized contaminants including perchlorate at limiting H_2 concentration (Table 2). These electron acceptors (NO_3^-, NO_2^-, O_2) compete with perchlorate for the electron availability and the observed hydrogen utilization was in the order of: $O_2 > NO_3^- > NO_2^- > ClO_4^-$.

In the presence of SO_4^{2-} and NO_3^-, Zhao et al. (2013) successfully achieved complete perchlorate reduction by controlling the nitrate surface loadings in a two-stage MBfR (i.e., lead and lag MBfRs) (Table 2). The presence of sulfate in the treated water should be minimized as it may lead to unwanted reaction. For instance, sulfate reacts with H_2 resulting in hydrogen sulfide (H_2S), which can strongly affect the nitrate reduction (Dalsgaard and Bak 1994). Sulfate reduction was prevented without affecting perchlorate reduction by increasing the flow rate, which gave equivalent nitrate surface loading in the lead and lag MBfR equal to 0.94 ± 0.05 g N/m²·d and 0.53 ± 0.03 g N/m²·d, respectively.

Table 2. Summary of the literature on perchlorate treatment using H_2-based MBfR.

Wastewater	Inoculum	Membrane	Operational conditions	Results	References
Synthetic GW (1 mg ClO_4^-/L; 5 mg NO_3^--N/L); GW from well in Main San Gabriel Basin, CA (6, 100, and 50 µg ClO_4^-/L)	Combination of pure culture *Ralstonia eutropha* and tap water	Hydrophobic hollow-fiber membrane (Porous Media Inc., Minneapolis, Minn.)	Flow rate = 10 mL/min; HRT = 44 min; pH = 6.5–8.8; P_{H_2} = 1.5–5.5 psi	• Max. ClO_4^- removal efficiency (35%) was achieved at the highest applied P_{H_2} of 4 psi. • Increasing ClO_4^- load had no effect on nitrate removal but significantly lowered ClO_4^- removal efficiency. • ClO_4^- reduction was pH-sensitive, having an optimum removal at pH 8.	Nerenberg et al. 2002
Purolite brine (170 mg ClO_4^-/L and 2.5 g NO_3^--N/L, ~15% salinity); Synthetic brine (0.5 g ClO_4^-/L; 1 g NO_3^--N/L)	Cultures enriched from salt pond near San Francisco Bay	Hydrophobic hollow-fiber membrane (Model MHF 200TL, Mitsubishi Rayon)	Feed rate = 0.001 mL/min; HRT = 48 h; P_{H_2} = 3, 4, and 5 psi; NaCl conc. = 1, 2, and 4%	• Faster reduction rates were observed for both NO_3^- and ClO_4^- when the purolite brine (~15% salinity) was diluted to 75, 50, 25, or 10%. • NO_3^- and ClO_4^- removal efficiencies were reduced from 88% to 32% and 63% to <1%, respectively, when the salinity was increased from 1% to 4% at 3 psi P_{H_2}. • At 2% salinity, NO_3^- flux increased from 0.085 to 0.091 g/m² and ClO_4^- flux from 0.014 to 0.027 g/m²·d when P_{H_2} was increased from 3 to 5 psi.	Chung et al. 2007
Synthetic IX brine: (8.2–100 g ClO_4^-/L, 23–73 mg NO_3^--N/L, 3–4.5% NaCl)	Sediments taken from 3 saline sources and a freshwater source (backwash sludge from ClO_4^- degrading packed bed reactor)	Hydrophobic Hollow-fiber membrane (Model MHF 200TL.)	HRT = 1 h; NaCl conc. = 0.1, 3, and 4.5 g/L; P_{H_2} = 2 psi and 15 psi	• Max. NO_3^- and ClO_4^- removal fluxes obtained were 5.4 g NO_3^--N/m²·d and 5.0 g ClO_4^-/m^2·d. • ClO_4^- reduction rates were inversely proportional to NO_3^- loadings and salinity. • Average ClO_4^- removal fluxes were 88, 37, and 15 mg ClO_4^-/m^2·d at 0.1, 3, and 4.5% NaCl, respectively.	Van Ginkel et al. 2008
Synthetic GW (98.8 mg NO_3^--N/L); Synthetic IX brine (1800 mg NO_3^--N/L or 20 mg ClO_4^-/L, 1.25% NaCl)	Non-halophilic cultures enriched from denitrifying WW seed	Hydrophobic tubular membranes with perfluoropolymer inside coatings (Compact Membranes Systems, Inc., Wilmington, DE)	40–50 mg NO_3^--N/L; pH = 9–10	• H_2 mass transfer coefficient was twice lower in brine solution than in tap water at high Reynolds number (Re = 30,000). • Average ClO_4^- removal efficiency was 42% within 23 days of batch operation. • Accumulation of effluent nitrite was observed due to pH stress, biofouling and H_2 limitation. • Metabolically active community in the IX brines was composed of *Proteobacteria* (67.5%), *Bacteroidetes* (30.2%), and *Firmicutes* (2.3%).	Sahu et al. 2009

Table 2 Contd....

...*Table 2 Contd.*

Wastewater	Inoculum	Membrane	Operational conditions	Results	References
GW from Sunset Well of the City of Pasadena, CA (12.5 µg ClO_4^-/L; 9.4 mg NO_3^--N/L)	GW from Sunset Well of the City of Pasadena, CA	Hydrophobic Hollow-fiber membrane (Model MHF 200TL)	HRT = 27.5 min; P_{H2} = 0.034 –0.659 atm (0.5–9.5 psi)	• More than 99.5% reduction was achieved for both NO_3^- and ClO_4^- with max. surface loading equal to 0.21 mg NO_3^--N/cm²·d and 3.4 µg ClO_4^-/cm²·d. • At limiting H_2 availability, hydrogen utilization by electron acceptors was in the order of: $O_2 > NO_3^- > NO_2^- > ClO_4^-$.	Ziv-El and Rittmann 2009
Synthetic WW (1 and 10 mg ClO_4^-/L, 2.3, 11.3, and 22.6 mg NO_3^--N/L)	Diluted laboratory sludge enriched with ClO_4^- in a mineral salt medium	Non-porous polypropylene fiber (Teijin, Ltd., Japan)	Recirculation rate = 100 mL/min; Feed rate = 0.25 L/min; P_{H2} = 17.5 psi; pH = 7.5 ± 0.2	• At sufficient H_2, complete removal of NO_3^- and ClO_4^- was achieved. • At insufficient H_2 delivery, NO_3^- out-competed ClO_4^- for electron supply.	Zhao et al. 2011
Synthetic WW (100 µg ClO_4^-/L, 25 mg SO_4^{2-}/L, 65 mg NO_3^-/L)	Flush water from MBfR pilot plant	Non-porous polypropylene fiber (Teijin, Ltd.)	P_{H2} = 17.0 psi; Recirculation rate = 150 mL/min for 24 h; Feed rate = 0.28 mL/min for Phase I and 0.42 mL/min for Phase II; pH = 7.5 ± 0.2	• Optimization of NO_3^- surface loading (0.94 g N/m²·d in the lead MBfR and 0.53 g N/m²·d in the lag MBfR) successfully prevented sulfate reduction without compromising complete ClO_4^- reduction.	Zhao et al. 2013

Note: P_{H2}, H_2 pressure; HRT, hydraulic retention time; WW, wastewater; GW, groundwater; IX, ion exchange.

3.2.2 Effect of salinity

Several studies have investigated the applicability of H_2-based MBfR in reducing perchlorate in high saline environment (Chung et al. 2007, Van Ginkel et al. 2008, Sahu et al. 2009) (Table 2). Chung et al. (2007) used two types of salt solutions: Purolite brine and synthetic high-strength salt medium. Under both saline solutions, perchlorate removal rates, in the presence or absence of co-contaminant nitrate, were observed to increase at decreasing salinity, and vice versa. Faster reduction rates were detected for both nitrate and perchlorate when the purolite brine (~ 15% salinity) was diluted to 75, 50, 25, and 10% (Chung et al. 2007). NO_3^- and ClO_4^- removal efficiencies were reduced from 88% to 32% and 63% to < 1%, respectively, when the salinity was increased from 1% to 4% at constant H_2 pressure (3 psi) (Chung et al. 2007). Similar results were achieved by Van Ginkel et al. (2008): average ClO_4^- removal fluxes were 88, 37 and 15 mg $ClO_4^-/m^2 \cdot d$ when NaCl concentrations were 1, 30 and 45 g/L, respectively. This indicates the negative inhibition of NaCl concentration to the microbial reduction of the two anions. Chung et al. (2007) claimed that high salinity impaired all microorganisms, with perchlorate reduction being more strongly affected. Meanwhile, Sahu et al. (2009) found out that the salinity of the solution reduces the gas mass transport to the biofilm, particularly at higher Reynolds number (Re). The mass transfer coefficients of H_2 in brine solution was the same as that in tap water when Re < 22,000 but was twice lower when Re = 30,000. In freshwater medium, the increase in turbulence helps decrease the thickness of the membrane's liquid boundary layer and hence enhances the gas transfer rates. However, in high saline solutions, the gas transport is counteracted by the high salt transport leading to the decline of the gas transfer coefficients.

3.2.3 Effect of H_2 pressure

Influence of H_2 on simultaneous reduction of perchlorate and nitrate becomes significant when H_2 availability is limiting (Table 2). According to Zhao et al. (2011), insufficient H_2 supply significantly reduced both nitrate and perchlorate removals but it was the latter that has been largely affected. Particularly for NO_3^-, partial denitrification occurs when electron donor is minimal leading to the accumulation of NO_2^- ions in the solution (Sahu et al. 2009). As comprehensively demonstrated by Ziv-El and Rittmann (2009), nitrate and nitrite reduction had higher H_2 utilization priority compared to perchlorate, hence affecting perchlorate reduction. The key to high perchlorate removal therefore lies in sufficient H_2 availability, which can be controlled by the applied H_2 pressure. Increased H_2 pressure delivers more H_2 to the biofilm, allowing continued removal of perchlorate even in the presence of other electron acceptors and at increased surface loading (Ziv-El and Rittmann 2009). In the study of Chung et al. (2007), at 2% salinity, nitrate flux increased from 0.085 to 0.091 $g/m^2 \cdot d$ and perchlorate flux from 0.014 to 0.027 $g/m^2 \cdot d$ when H_2 pressure was increased from 3 to 5 psi.

3.3 Methane (CH_4)-based MBfR

Researchers also investigated the feasibility of CH_4-based MBfR for wastewater treatment applications. Methane can be employed as the sole electron donor and carbon source in the biological reduction of ClO_4^-. Compared to common organics, CH_4 is relatively affordable and widely available. CH_4 is the by-product of anaerobic digestion of biosolids and being a potent greenhouse gas, the use of this gaseous compound for perchlorate treatment may help reduce its impact to climate change (Sun et al. 2017). Furthermore, CH_4 has lower solubility than other organic compounds at standard temperature and pressure, which is advantageous to prevent possible secondary contamination of the treated water (Xie et al. 2018). However, this low solubility also hinders the availability of methane to microbial culture and can result in methane losses to the atmosphere. The use of "bubble-less" membranes in MBfR can therefore reduce the limitations of low gas solubility (Sun et al. 2017).

Methane-based MBfR combines aerobic or anaerobic CH_4 oxidation with ClO_4^- reduction (Chen et al. 2016, Xie et al. 2018, Wu et al. 2019). In aerobic methane oxidation coupled with perchlorate or nitrate reduction (AMO-PR, AMO-D), two distinct bacterial groups are involved:

aerobic methanotrophs (methane oxidizers) and PRB. Methanotrophs oxidize methane using external O_2 in an initial mono-oxygenation step and produce organic intermediates (i.e., methanol, acetate, etc.), which serve as the electron donors for PRB in perchlorate reduction.

On the other hand, anaerobic methane oxidation coupled with perchlorate or nitrate reduction (ANMO-PR, ANMO-D) follows two different pathways, namely, reverse methanogenesis and intra-aerobic pathways (Chen et al. 2016). Reverse methanogenesis pathway employs Archaea and bacteria for synergistic methane oxidation and perchlorate reduction, respectively. Anaerobic methane oxidation is catalyzed by methyl-coenzyme M reductase (MCR) of the anaerobic methanotrophic archaea and the product H_2 can be used by PRB to respire ClO_4^- (Lv et al. 2019). Intra-aerobic pathway, on the other hand, is known to be mediated by only one microorganism. Aerobic methanotrophic bacteria reduce perchlorate to Cl^- and O_2 via the stepwise catalysis of perchlorate reductase and chlorite dismutase. The O_2 released from chlorite dismutation is then used by same microorganism to oxidize methane intracellularly by the action of membrane bound particulate methane mono-oxygenase (MMO) (Lv et al. 2019). Overall, the stoichiometric reactions of perchlorate reduction with methane in AMO-PR and ANMO-PR are given in equations (5) and (6), respectively (Chen et al. 2016):

$$CH_4 + \frac{1}{2}ClO_4^- + O_2 = \frac{1}{2}Cl^- + CO_2 + 2H_2O \tag{5}$$

$$CH_4 + ClO_4^- = Cl^- + CO_2 + 2H_2O \tag{6}$$

In the application of methane-based MBfR for perchlorate treatment, several factors should be considered as these can significantly affect reactor performance. These factors include presence of other ions (NO_3^- or NO_2^-), CH_4 delivery, ClO_4^- loading, hydraulic retention time (HRT), temperature, limited O_2 supply, and salinity. Brief discussions of their effects on perchlorate reduction are provided below. Table 3 shows the summary of literature on perchlorate treatment using CH_4-based MBfR.

3.3.1 *Effect of co-existing anions*

Co-contaminant anions such as NO_3^- and NO_2^- were found by several studies to cause inhibition on perchlorate removal in MBfR system (Luo et al. 2015, Sun et al. 2017, Lv et al. 2020) (Table 3). NO_3^- slows down perchlorate respiration due to the competition for scarce resources such as electron donors and reductase enzymes (Sun et al. 2017). In a study conducted by Luo et al. (2015), perchlorate reduction dropped from 100% to ≤ 5% when NO_3^- surface loading was increased from < 0.32 g N/m²·d to 0.78 g N/m²·d at limiting CH_4 delivery (Table 3). Lv et al. (2020) reported that addition of NO_3^- (15 mg/L) decreased perchlorate reduction rate from 1.66 to 0.64 mmol/m²·d even under sufficient CH_4 supply. However, rate of perchlorate reduction returned back to 1.68 mmol/m²·d when all nitrate was consumed, indicating a reversible inhibition effect of nitrate on perchlorate reduction.

The preference of bacteria to nitrate over perchlorate could be due to the enzymatic specificity. Youngblut et al. (2016) performed biochemical study using perchlorate reductase (PcrAB) purified from *Azospira suillum* PS and nitrate reductase (NarGHI) purified from *Escherichia coli*. The biochemical analysis showed that PcrAB had a higher catalysis rate for NO_3^- (k_{cat} = 51/e·min) than for ClO_4^- (k_{cat} = 27.1/e·min). It also revealed that NarGHI showed lower affinity for ClO_4^- (K_m = 1.1 mM) than for NO_3^- (K_m = 0.2 mM). PcrAB showed substrate inhibition, while NarGHI had no substrate inhibition. Lv et al. (2020) further explained that ClO_4^- conversion to ClO_3^- has higher energy barrier (18 kcal/mol) than NO_3^- to NO_2^- reduction for proton-coupled electron-transfers.

Meanwhile, even at low concentrations, NO_2^- can cause inhibition on both methane oxidation and perchlorate reduction. Luo et al. (2015) observed that perchlorate removal efficiency was complete in the absence of NO_2^- but decreased when NO_2^- surface loading was ≥ 0.10 g N/m²·d. The inhibitory effect of NO_2^- on methanogenesis and methane oxidation was reported previously (King et al. 1994, Kluer et al. 1998).

Table 3. Summary of the literature on perchlorate treatment using CH_4-based MBfR.

Wastewater	Inoculum	Membrane	Operational conditions	Results	References
Synthetic WW (1 and 5 mg ClO_4^-/L, 1.1–11.3 mg NO_3^--N/L).	ANMO-D culture	Hydrophobic microporous PE fiber (Mitsubishi, Ltd., Japan)	V, 65 mL; Influent feeding rate, 0.5 mL/min; HRT, 130 min; P_{CH4}, 10 and 15 psi; T, 29 ± 1°C; pH, 7.0 ± 0.2; Influent DO, ~ 0.2 mg/L	• When CH_4 delivery was limiting, NO_3^- inhibited ClO_4^- reduction due to the competition for the scarce e^- donor. • NO_2^- inhibited ClO_4^- reduction due to its cellular toxicity. • Bacteria dominated in the biofilm rather than Archaea.	Luo et al. 2015
Synthetic WW (~ 2 mg ClO_4^-/L, 5 and 10 mg NO_3^-/L, 2 and 10 g NaCl/L).	Enriched culture from CH_4-based MBfR	Hydrophobic microporous PE fiber (Mitsubishi, Ltd.)	V, 65 mL; Membrane surface area, 90.24 cm²; Influent feeding rate, 0.5 mL/min; HRT, 130 min; P_{CH4}, 10 psi; T, 29 ± 1°C; pH, 7.5 ± 0.2	• At non-saline condition, NO_3^- and ClO_4^- removal rates were 100% and 57.2%, respectively. • At 1% NaCl, ClO_4^- removal was completely inhibited, microbial morphology was dramatically changed, and the microbial population and functional genes were significantly declined.	Zhang et al. 2016
Synthetic WW (1, 5, and 10 mg ClO_4^-/L)	Mixture of freshwater sediments from Liuyang River, anaerobic digester sludge from a piggery and return activated sludge from STP	Hydrophobic microporous PE fiber	V, 1.2 L; Membrane surface area, 1885 cm²; Influent feeding rate, 0.83 mL/min; P_{CH4}, 20, 40, and 60 kPa; pH, 7.4; HRT, 24 h; T, 30 ± 1°C	• Theoretical max. CH_4 flux decreased from 272.7 to 191.8 and then to 90.9 mM/m²·d when P_{CH4} was lowered from 60 kPa to 40 kPa and 20 kPa, respectively. • Average ClO_4^- removal flux increased from 10.68 to 44.46 and 92.75 mg/m²·d when influent ClO_4^- conc. was increased from 1 to 5 and then to 10 mg/L, respectively.	Xie et al. 2018
Synthetic WW (0.21–0.56 mM ClO_4^-)	Enriched culture from CH_4-based MBfR	Composite hollow-fiber membrane (Mitsubishi, Ltd.)	V, 1000 mL; Membrane surface area, ~ 76 cm²; P_{CH4}, 15 psi; T, 35 ± 1°C	• ClO_4^- reduction rates increased from 0.94 to 2.34 mM/m²·d as ClO_4^- conc. increased. • Pyrosequencing analysis of biofilm microbial community showed that PRB, methanotrophic bacteria, and archaeon *Methanosarcina* were dominant.	Lv et al. 2019
Synthetic WW (40–160 mg NO_3-N/L or 4.5–12 mg ClO_4^--Cl$^-$/L)	Enriched culture from N-DAMO/annamox reactor	Composite hollow-fiber membrane (Mitsubishi, Ltd.)	V, 800 mL; pH, 7–8; HRT, 5 and 1 d; T, 22 ± 2°C and 35 ± 2°C; 0 and 10 mg DO/L·d	• ClO_4^- removal efficiency decreased from 90% to 40% when HRT was lowered from 5 d to 3 d but quickly recovered when temperature was increased from 22°C to 35°C. • At limited O_2 supply, VFAs (acetic acid and propionic acid) were generated and were consumed as carbon source for ClO_4^- reduction.	Wu et al. 2019

Note: N/A, Not applicable; STP, sewage treatment plant; ANMO-D, anaerobic methane oxidation coupled to denitrification; N-DAMO, nitrite/nitrate-dependent anoxic methane oxidation; PE, polyethylene; P_{CH4}, CH_4 pressure; V, reactor vol.; HRT, hydraulic retention time; T, temperature; WW, wastewater; IX, ion exchange; VFA, volatile fatty acids.

3.3.2 Effect of CH_4 delivery and perchlorate loading

It is generally known that the availability of the electron donor and acceptor significantly affects microbial community and hence its performance towards pollutant biodegradation. Chen et al. (2016) reported that the microbial community conducting ANMO-PR (ANMO-D) significantly changed after it was inoculated to the anoxic CH_4-based MBfR fed with perchlorate, nitrate, and nitrite. *Anaerolineaceae* sp. and *Ferruginibacter* sp. were dominant in the initial inoculum. However, they were replaced by anoxic methanotrophs (*Methylomonas* sp. and *Methylocystis* sp.) that were enriched and dominated as CH_4 flux was increased in the reactor. The detected *Methylomonas* sp. and *Methylocystis* sp. are phylogenetically close to the known AMO-D. The more CH_4 are available, the more organic intermediates (electron donor) are produced for PRB utilization and, therefore, more perchlorate ions are reduced.

Similarly, increasing perchlorate loading led to the enrichment of PRB in the biofilm (Chen et al. 2016). These results, however, did not take into account the combining effects of methane and perchlorate concentrations on perchlorate removal efficiency. In a simulation study of the biofilm model in a CH_4-based MBfR, the same perchlorate removal efficiency (80%) was achieved at different P_{CH4} and ClO_4^- loading combinations (e.g., 30 kPa and 0.08 g Cl/m^2·d, 120 kPa and 64 g Cl/m^2·d) (Sun et al. 2017). ClO_4^- removal efficiency did not go beyond 80% when P_{CH4} was set from 100 to 120 kPa under perchlorate loading of more than 0.64 g Cl/m^2·d (Sun et al. 2017), suggesting that CH_4 and ClO_4^- concentrations have to be properly controlled at certain critical level to maximize perchlorate reduction. In particular, exceeding amount of methane may lead to high energy loss as well as greenhouse gas emission.

3.3.3 Effect of HRT and temperature

HRT and temperature are two of the most important parameters for effective reduction of perchlorate using methane-based MBfR. Longer HRT allows enough time for PRB to degrade perchlorate, while optimum temperature is conducive for the growth and activity of PRB. This was observed by Wu et al. (2019) when the removal efficiency of perchlorate in a methane-based MBfR at 22°C decreased from 90% to 40% after the HRT was lowered from 5 d to 3 d (Table 3). When the temperature was increased to 35°C, perchlorate removal efficiency quickly recovered to 100% within a day (Wu et al. 2019).

It is important to know the optimum range of temperature within which a certain microbial culture can thrive to be able to maximize its enzymatic activity for biodegradation performance. Prior studies involving PRB reported that the optimum temperature usually ranges from 28 to 37°C (Bruce et al. 1999, Giblin and Frankenberger 2001). Below this range, the affinity of PRB to perchlorate may be reduced due to the stiffening of its membrane lipids, which hinders the ability of transport proteins to sequester perchlorate from the external environment (Nedwell 1999). Above the optimum temperature range, denaturation of key cellular components may occur disrupting the cellular function (Nedwell 1999).

3.3.4 Effect of limiting oxygen supply

Limited oxygen supply can significantly improve perchlorate degradation in CH_4-based MBfR due to the formation of volatile fatty acids (VFAs), which help drive perchlorate reduction even at shorter HRT. VFAs are intermediate products of methane partial oxidation by methanogens (*Methanosarcina*) and fermenters (*Veillonellaceae*). VFAs such as acetic acid and propionic acid can be utilized by PRB (*Denitratisoma* and *Rhodocyclaea*) for heterotrophic perchlorate degradation (Wu et al. 2019). When no oxygen is supplied (anoxic condition), perchlorate reduction can still be achieved, although minimally, via the synergistic interactions of aerobic methanotrophs (*Methylocystaceae* and *Methylococcaceae*) and PRB (*Denitratisoma* and *Rhodocyclaceae*). The aerobic methanotrophs use the oxygen produced from the chlorite dismutation by PRB and in return, methanotrophs oxidize methane and produce intermediates that are utilized by PRB as electron donors in degrading perchlorate. However, the O_2 released by PRB might not be sufficient to aerobic methanotrophs, hence the need for external limiting O_2 supply (Wu et al. 2019).

3.3.5 Effect of salinity

Microorganisms in CH_4-based gas-transfer MBfR showed more sensitivity to salt concentration than those in H_2-based one (Chung et al. 2007, Van Ginkel et al. 2008). Zhang et al. (2016) have studied the effects of salinity on the simultaneous reduction of NO_3^- and ClO_4^- in a CH_4-based MBfR (Table 3). Under non-saline condition, the NO_3^- and ClO_4^- reduction rates were about 100% and 57.2%, respectively. At 1% (W/V) NaCl, the NO_3^- removal efficiency drastically dropped to 50% and ClO_4^- removal was completely inhibited. When the salinity was subsequently removed, to know if the microbial activity can recover after salinity shock, only 28.3% NO_3^- and 64.4% ClO_4^- removal efficiencies were obtained. This poor recovery suggests that high salinity can cause irreversible damage to the microbial activity. A significant decrease in the number of functional genes (*pcrA, nirS,* and *pMMO*) and microbial population (*Methylomonas* and *Denitratisoma*) at 1% NaCl was also observed. The functional genes are encoding enzymes involved in perchlorate and nitrite reduction, and methane oxidation, respectively. *Methylomonas* is a genus of methanotrophic bacteria, while *Denitratisoma* is a genus of denitrifying bacteria.

4. Limitations of MBR for perchlorate removal

In IEMB, the major limitation that hinders its practical application is the expensive cost of the ion exchange membrane and carbon source. Membranes with high selectivity to mono-ions and high ion exchange capacities are highly suitable for IEMB treatment but they also diminish the cost efficiency of the process. High cost of membrane can be resolved by trying cheaper alternatives such as non-monovalent anion permselective membranes. However, this type of membrane also allows transport of some multi-valent anions (e.g., sulfate, bicarbonate, phosphate ions) to the water-compartment, which can reduce the quality of the treated water. This can be minimized by careful optimization of the biomedium composition and the concentration of the driving counter-ions in the bio-compartment.

In gas-transfer MBfR, the challenges that need to be addressed focus mainly on the biofilm development and gas-transfer through membranes. The development of biofilm on the membrane surface occurs naturally depending on the concentration of electron donor and acceptor as well as other nutrients needed for microbial growth. The thickness of biofilm should be controlled because biofilm that is too thin has insufficient biomass to completely reduce perchlorate, while too thick biofilm has limitation of mass transfer into the biofilm. To manage the formation of biofilm on the membrane surface, periodic biofilm detachment can be carried out via recirculation loop of the treated water or by gas sparging (Zhou et al. 2019). This method can also increase the metabolic activity of the non-active biomass. With regard to gas-transfer through membranes, there is a possibility for gas back-diffusion, particularly in the hollow-fiber membranes, due to the presence of other dissolved gases in the bulk liquid and biofilm. This can affect the delivery of H_2 or CH_4 to the biofilm which is needed for microbial growth (Ahmed et al. 2004). A solution to this is to periodically vent the membrane so that other gases will escape. There is a possibility of H_2 or CH_4 loss but it can be reestablished afterwards (Perez-Calleja et al. 2017).

5. Conclusion and future perspectives

Perchlorate is a toxic anion and its presence in the environment poses a serious threat not just to humans but also to other living organisms. The integration of membrane technology with biological degradation proved to be effective in the removal of perchlorate from contaminated waters. It can completely reduce perchlorate to harmless end products and at the same time minimize the occurrence of microbial contamination of the biologically treated water due to the installed membrane.

Perchlorate treatment by MBR is an emerging technology and has to be further explored. Especially, further study on microbial ecology in MBR is necessary to better understand perchlorate biodegradation and improve performance of MBR. Development of new membranes and investigation on the use of other electron sources can be carried out to provide cost-effective options. Optimization studies and cost-analysis would also be helpful for its large-scale application. Finally, a Life Cycle Analysis can be performed to quantify the technology's environmental footprint.

Acknowledgment

This research was supported by the National Research Foundation, Korea (project no. NRF-2017R1A2B4011805).

References

Ahmed, T., Semmens, M.J. and Voss, M.A. 2004. Oxygen transfer characteristics of hollow-fiber, composite membranes. Adv. Environ. Res. 8: 637–646. https://doi.org/10.1016/S1093-0191(03)00036-4.

Ahn, Y. 2019. Removal of perchlorate from salt water using microorganisms. J. Life Sci. 29: 1294–1303. https://doi.org/10.5352/JLS.2019.29.11.1294.

Boles, A.R., Conneely, T., McKeever, R., Nixon, P., Nüsslein, K.R. and Ergas, S.J. 2012. Performance of a pilot-scale packed bed reactor for perchlorate reduction using a sulfur oxidizing bacterial consortium. Biotechnol. Bioeng. 109: 637–646. https://doi.org/10.1002/bit.24354.

Bruce, R.A., Achenbach, L.A. and Coates, J.D. 1999. Reduction of (per)chlorate by a novel organism isolated from paper mill waste. Environ. Microbiol. 1: 319–329. https://doi.org/10.1046/j.1462-2920.1999.00042.x.

Calderón, R., Godoy, F., Escudey, M. and Palma, P. 2017. A review of perchlorate (ClO_4^-) occurrence in fruits and vegetables. Environ. Monity. Assess. 189: 82. https://doi.org/10.1007/s10661-017-5793-x.

Cao, F., Jaunat, J., Sturchio, N., Cancès, B., Morvan, X., Devos, A., Barbin, V. and Ollivier, P. 2019. Worldwide occurrence and origin of perchlorate ion in waters: A review. Sci. Total Environ. 661: 737–749. https://doi.org/10.1016/j.scitotenv.2019.01.107.

Chen, R., Luo, Y.H., Chen, J.X., Zhang, Y., Wen, L.L., Shi, L.D., Tang, Y., Rittmann, B.E., Zheng, P. and Zhao, H.P. 2016. Evolution of the microbial community of the biofilm in a methane-based membrane biofilm reactor reducing multiple electron acceptors. Environ. Sci. Pollut. Res. 23: 9540–9548. https://doi.org/10.1007/s11356-016-6146-y.

Choi, H. and Silverstein, J. 2008. Inhibition of perchlorate reduction by nitrate in a fixed biofilm reactor. J. Hazard. Mater. 159: 440–445. https://doi.org/10.1016/j.jhazmat.2008.02.038.

Chung, J., Nerenberg, R. and Rittmann, B.E. 2007. Evaluation for biological reduction of nitrate and perchlorate in brine water using the hydrogen-based membrane biofilm reactor. J. Environ. Eng. 133: 157–164. https://doi.org/10.1061/(ASCE)0733-9372(2007)133:2(157).

Coates, J.D., Fryman, J., Bruce, R.A., Michaelidou, U. and Achenbach, L.A. 2001. *Dechloromonas agitata* gen. nov., sp. nov. and *Dechlorosoma suillum* gen. nov., sp. nov., two novel environmentally dominant (per)chlorate-reducing bacteria and their phylogenetic position. Int. J. Syst. Evol. Microbiol. 51: 527–533. https://doi.org/10.1099/00207713-51-2-527.

Crespo, J.G., Reis, A.M., Fonseca, A.D. and Almeida, J.S. 1999. Ion exchange membrane bioreactor for denitrification of water (Reactor de membrana de permuta iónica para desnitrificação de água). Portuguese National Patent No. 102 385 N.

Crespo, J.G. and Reis, A.M. 2003. Treatment of aqueous media containing electrically charged compounds. Patent EP1246778.

Crespo, J.G., Velizarov, S. and Reis, M.A. 2004. Membrane bioreactors for the removal of anionic micropollutants from drinking water. Curr. Opin. Biotechnol. 15: 463–468. https://doi.org/10.1016/j.copbio.2004.07.001.

Dalsgaard, T. and Bak, F. 1994. Nitrate reduction in a sulfate-reducing bacterium, *Desulfovibrio desulfuricans*, isolated from rice paddy soil: sulfide inhibition, kinetics, and regulation. Appl. Environ. Microbiol. 60: 291–297.

Dudley, M., Salamone, A. and Nerenberg, R. 2008. Kinetics of a chlorate-accumulating, perchlorate-reducing bacterium. Water Res. 42: 2403–2410. https://doi.org/10.1016/j.watres.2008.01.009.

Ensano, B.M.B., Borea, L., Naddeo, V., Belgiorno, V., de Luna, M.D.G. and Ballesteros, F.C. 2016. Combination of electrochemical processes with membrane bioreactors for wastewater treatment and fouling control: a review. Front Environ. Sci. 4. https://doi.org/10.3389/fenvs.2016.00057.

Ensano, B.M.B., Borea, L., Naddeo, V., de Luna, M.D.G. and Belgiorno, V. 2019. Control of emerging contaminants by the combination of electrochemical processes and membrane bioreactors. Environ. Sci. Pollut. Res. 26: 1103–1112. https://doi.org/10.1007/s11356-017-9097-z.

Fonseca, A.D., Crespo, J.G., Almeida, J.S. and Reis, M.A. 2000. Drinking water denitrification using a novel ion-exchange membrane bioreactor. Environ. Sci. Technol. 34: 1557–1562. https://doi.org/10.1021/es9910762.

Fox, S., Oren, Y., Ronen, Z. and Gilron, J. 2014. Ion exchange membrane bioreactor for treating groundwater contaminated with high perchlorate concentrations. J. Hazard. Mater. 264: 552–559. https://doi.org/10.1016/j.jhazmat.2013.10.050.

Fox, S., Bruner, T., Oren, Y., Gilron, J. and Ronen, Z. 2016. Concurrent microbial reduction of high concentrations of nitrate and perchlorate in an ion exchange membrane bioreactor: Simultaneous nitrate and perchlorate reduction in an IEMB. Biotechnol. Bioeng. 113: 1881–1891. https://doi.org/10.1002/bit.25960.

Gal, H., Ronen, Z., Weisbrod, N., Dahan, O. and Nativ, R. 2008. Perchlorate biodegradation in contaminated soils and the deep unsaturated zone. Soil. Biol. Biochem. 40: 1751–1757. https://doi.org/10.1016/j.soilbio.2008.02.015.

Gao, M., Wang, S., Ren, Y., Jin, C., She, Z., Zhao, Y., Yang, S., Guo, L., Zhang, J. and Li, Z. 2016. Simultaneous removal of perchlorate and nitrate in a combined reactor of sulfur autotrophy and electrochemical hydrogen autotrophy. Chem. Eng. J. 284: 1008–1016. https://doi.org/10.1016/j.cej.2015.09.082.

Gardell, A.M., von Hippel, F.A., Adams, E.M., Dillon, D.M., Petersen, A.M., Postlethwait, J.H., Cresko, W.A. and Buck, C.L. 2017. Exogenous iodide ameliorates perchlorate-induced thyroid phenotypes in threespine stickleback. General and Comparative Endocrinology 243: 60–69. https://doi.org/10.1016/j.ygcen.2016.10.014.

Giblin, T. and Frankenberger, W.T. 2001. Perchlorate and nitrate reductase activity in the perchlorate-respiring bacterium perclace. Microbiological Res. 156: 311–315. https://doi.org/10.1078/0944-5013-00111.

Gu, B., Brown, G.M. and Chiang, C.C. 2007. Treatment of perchlorate-contaminated groundwater using highly selective, regenerable ion-exchange technologies. Environ. Sci. Technol. 41: 6277–6282. https://doi.org/10.1021/es0706910.

Guan, X., Xie, Y., Wang, J., Wang, J. and Liu, F. 2015. Electron donors and co-contaminants affect microbial community composition and activity in perchlorate degradation. Environ. Sci. Pollut. Res. 22: 6057–6067. https://doi.org/10.1007/s11356-014-3792-9.

Halling-Sørensen, B. and Jorgensen, S.E. 1993. The Removal of Nitrogen Compounds from Wastewater. Elsevier.

Han, K.R., Kang, T.H., Kang, H.C., Kim, K.H., Seo, D.H. and Ahn, Y.H. 2011. Autotrophic perchlorate-removal using elemental sulfur granules and activated sludge: batch test. J. Life Sci. 21: 1473–1480. https://doi.org/10.5352/JLS.2011.21.10.1473.

He, L., Zhong, Y., Yao, F., Chen, F., Xie, T., Wu, B., Hou, K., Wang, D., Li, X. and Yang, Q. 2019. Biological perchlorate reduction: which electron donor we can choose? Environ. Sci. Pollut. Res. 26: 16906–16922. https://doi.org/10.1007/s11356-019-05074-5.

King, G.M. and Schnell, S. 1994. Ammonium and nitrite inhibition of methane oxidation by *Methylobacter albus* BG8 and *Methylosinus trichosporium* OB3B at low methane concentrations. Appl. Environ. Microbiol. 60: 3508–3513.

Kluber, H.D. and Conrad, R. 1998. Inhibitory effects of nitrate, nitrite, NO, and N_2O on methanogenesis by *Methnoscarcina barkeri* and *Methanobacterium bryantii*. FEMS Microbiol. Ecol. 25: 331–339.

Lee, K.C. and Rittmann, B.E. 2002. Applying a novel autohydrogenotrophic hollow-fiber membrane biofilm reactor for denitrification of drinking water. Water Res. 36: 2040–2052.

Leung, A.M., Pearce, E.N. and Braverman, L.E. 2010. Perchlorate, iodine and the thyroid. Best Practice & Res. Clinical Endocrinology & Metabolism 24: 133–141. https://doi.org/10.1016/j.beem.2009.08.009.

Luo, Y.H., Chen, R., Wen, L.L., Lai, C.Y., Rittmann, B.E., Zhao, H.P. and Zheng, P. 2015. Complete perchlorate reduction using methane as the sole electron donor and carbon source. Environ. Sci. Technol. 49: 2341–2349.

Lv, P.L., Shi, L.D., Wang, Z., Rittmann, B. and Zhao, H.P. 2019. Methane oxidation coupled to perchlorate reduction in a membrane biofilm batch reactor. Sci. Total Environ. 667: 9–15. https://doi.org/10.1016/j.scitotenv.2019.02.330.

Lv, P.L., Shi, L.D., Dong, Q.Y., Rittmann, B. and Zhao, H.P. 2020. How nitrate affects perchlorate reduction in a methane-based biofilm batch reactor. Water Res. 171: 115397. https://doi.org/10.1016/j.watres.2019.115397.

Mahmudov, R. and Huang, C.P. 2010. Perchlorate removal by activated carbon adsorption. Sep. Purif. Technol. 70: 329–337. https://doi.org/10.1016/j.seppur.2009.10.016.

Matos, C.T., Velizarov, S., Crespo, J.G. and Reis, M.A.M. 2005. Removal of bromate, perchlorate and nitrate from drinking water in an ion exchange membrane bioreactor. Water Sci. Technol. Water Supply 5: 9–14. https://doi.org/10.2166/ws.2005.0033.

Matos, Cristina T., Fortunato, R., Velizarov, S., Reis, M.A.M. and Crespo, J.G. 2006a. Optimisation of the removal of toxic mono-valent anions from water supplies in the ion exchange membrane bioreactor. Desalination 199: 322–324. https://doi.org/10.1016/j.desal.2006.03.076.

Matos, C.T., Velizarov, S., Crespo, J.G. and Reis, M.A.M. 2006b. Simultaneous removal of perchlorate and nitrate from drinking water using the ion exchange membrane bioreactor concept. Water Res. 40: 231–240. https://doi.org/10.1016/j.watres.2005.10.022.

Matos, Cristina T., Fortunato, R., Velizarov, S., Reis, M.A.M. and Crespo, J.G. 2008. Removal of mono-valent oxyanions from water in an ion exchange membrane bioreactor: Influence of membrane permselectivity. Water Res. 42: 1785–1795. https://doi.org/10.1016/j.watres.2007.11.006.

Nagarale, R.K., Gohil, G.S. and Shahi, V.K. 2006. Recent developments on ion-exchange membranes and electro-membrane processes. Adv. Colloid. Interface Sci. 119: 97–130. https://doi.org/10.1016/j.cis.2005.09.005.

Nedwell, D.B. 1999. Effect of low temperature on microbial growth: lowered affinity for substrates limits growth at low temperature. FEMS Microbiol. Ecol. 30: 101–111. https://doi.org/10.1111/j.1574-6941.1999.tb00639.x.

Nerenberg, R., Rittmann, B.E. and Najm, I. 2002. Perchlorate reduction in a hydrogen-based membrane-biofilm reactor. Journal/American Water Works Association 94: 103–114.

Nerenberg, R., Kawagoshi, Y. and Rittmann, B.E. 2008. Microbial ecology of a perchlorate-reducing, hydrogen-based membrane biofilm reactor. Water Research 42: 1151–1159.

Nerenberg, R. 2016. The membrane-biofilm reactor (MBfR) as a counter-diffusional biofilm process. Curr. Opin. Biotechnol. 38: 131–136. https://doi.org/10.1016/j.copbio.2016.01.015.

Perez-Calleja, P., Aybar, M., Picioreanu, C., Esteban-Garcia, A.L., Martin, K.J. and Nerenberg, R. 2017. Periodic venting of MABR lumen allows high removal rates and high gas-transfer efficiencies. Water Res. 121: 349–360. https://doi.org/10.1016/j.watres.2017.05.042.

Ricardo, A.R., Carvalho, G., Velizarov, S., Crespo, J.G. and Reis, M.A.M. 2012. Kinetics of nitrate and perchlorate removal and biofilm stratification in an ion exchange membrane bioreactor. Water Research 46: 4556–4568. https://doi.org/10.1016/j.watres.2012.05.045.

Rittmann, B.E. 2007. The membrane biofilm reactor is a versatile platform for water and wastewater treatment. Environ. Eng. Res. 12: 157–175. https://doi.org/10.4491/eer.2007.12.4.157.

Sahu, A.K., Conneely, T., Nüsslein, K. and Ergas, S.J. 2009. Hydrogenotrophic denitrification and perchlorate reduction in ion exchange brines using membrane biofilm reactors. Biotechnol. Bioeng. 104: 483–491. https://doi.org/10.1002/bit.22414.

Saracco, G. and Zanetti, M.C. 1994. Ion transport through monovalent-anion-permselective membranes. Ind. Eng. Chem. Res. 33: 96–101.

Schaffer, R.B., Ludzack, F.J. and Ettinger, M.B. 1960. Sewage treatment by oxygenation through permeable plastic films. Journal (Water Pollution Control Federation) 32: 939–941.

Schmidt, F., Schnurr, S., Wolf, R. and Braunbeck, T. 2012. Effects of the anti-thyroidal compound potassium-perchlorate on the thyroid system of the zebrafish. Aquat. Toxicol. 109: 47–58. https://doi.org/10.1016/j.aquatox.2011.11.004.

Smith, Jr., C.V. 1968. The use of ultrafiltration membrane for activated sludge separation. Proc. 24th Annual Purdue Indus. Waste Conf. 1300–1310.

Song, W., Gao, B., Zhang, X., Li, F., Xu, X. and Yue, Q. 2019. Biological reduction of perchlorate in domesticated activated sludge considering interaction effects of temperature, pH, electron donors and acceptors. Process Safety and Environ. Protect. 123: 169–178. https://doi.org/10.1016/j.psep.2019.01.009.

Sun, J., Dai, X., Peng, L., Liu, Y., Wang, Q. and Ni, B.J. 2017. A biofilm model for assessing perchlorate reduction in a methane-based membrane biofilm reactor. Chem. Eng. J. 327: 555–563. https://doi.org/10.1016/j.cej.2017.06.136.

Urbansky, E.T. and Brown, S.K. 2003. Perchlorate retention and mobility in soils. J. Environ. Monitor. 5: 455–462. https://doi.org/10.1039/B301125A.

Van Ginkel, S.W., Ahn, C.H., Badruzzaman, M., Roberts, D.J., Lehman, S.G., Adham, S.S. and Rittmann, B.E. 2008. Kinetics of nitrate and perchlorate reduction in ion-exchange brine using the membrane biofilm reactor (MBfR). Water Res. 42: 4197–4205. https://doi.org/10.1016/j.watres.2008.07.012.

Velizarov, S., Matos, C., Reis, M. and Crespo, J. 2005. Removal of inorganic charged micropollutants in an ion-exchange membrane bioreactor. Desalination 178: 203–210. https://doi.org/10.1016/j.desal.2004.11.037.

Velizarov, S., Matos, C., Oehmen, A., Serra, S., Reis, M. and Crespo, J. 2008. Removal of inorganic charged micropollutants from drinking water supplies by hybrid ion exchange membrane processes. Desalination 223: 85–90. https://doi.org/10.1016/j.desal.2007.01.217.

Wan, D., Liu, Y., Wang, Y., Li, Q., Jin, J. and Xiao, S. 2019. Sulfur disproportionation tendencies in a sulfur packed bed reactor for perchlorate bio-autotrophic reduction at different temperatures and spatial distribution of microbial communities. Chemosphere 215: 40–49. https://doi.org/10.1016/j.chemosphere.2018.10.006.

Wu, M., Luo, J.H., Hu, S., Yuan, Z. and Guo, J. 2019. Perchlorate bio-reduction in a methane-based membrane biofilm reactor in the presence and absence of oxygen. Water Res. 157: 572–578. https://doi.org/10.1016/j.watres.2019.04.008.

Xie, T., Yang, Q., Winkler, M.K.H., Wang, D., Zhong, Y., An, H., Chen, F., Yao, F., Wang, X., Wu, J. and Li, X. 2018. Perchlorate bioreduction linked to methane oxidation in a membrane biofilm reactor: Performance and microbial community structure. J. Hazard. Mater. 357: 244–252. https://doi.org/10.1016/j.jhazmat.2018.06.011.

Yamamoto, K., Hiasa, M., Mahmood, T. and Matsuo, T. 1989. Direct solid-liquid separation using hollow fiber membrane in an activated sludge aeration tank. Water Science and Technology 21(4-5): 43–54. https://doi.org/10.2166/wst.1989.0209.

Yao, F., Zhong, Y., Yang, Q., Wang, D., Chen, F., Zhao, J., Xie, T., Jiang, C., An, H., Zeng, G. and Li, X. 2017. Effective adsorption/electrocatalytic degradation of perchlorate using Pd/Pt supported on N-doped activated carbon fiber cathode. J. Hazard Mater. 323: 602–610. https://doi.org/10.1016/j.jhazmat.2016.08.052.

Yoon, J., Yoon, Y., Amy, G. and Her, N. 2005a. Determination of perchlorate rejection and associated inorganic fouling (scaling) for reverse osmosis and nanofiltration membranes under various operating conditions. J. Environ. Eng. 131: 726–733. https://doi.org/10.1061/(ASCE)0733-9372(2005)131:5(726).

Yoon, Y., Amy, G., Cho, J. and Pellegrino, J. 2005b. Systematic bench-scale assessment of perchlorate (ClO_4^-) rejection mechanisms by nanofiltration and ultrafiltration membranes. Sep. Sci. Technol. 39: 2105–2135. https://doi.org/10.1081/SS-120039304.

Yoon, J., Amy, G., Chung, J., Sohn, J. and Yoon, Y. 2009. Removal of toxic ions (chromate, arsenate, and perchlorate) using reverse osmosis, nanofiltration, and ultrafiltration membranes. Chemosphere 77: 228–235. https://doi.org/10.1016/j.chemosphere.2009.07.028.

Youngblut, M.D., Tsai, C.-L., Clark, I.C., Carlson, H.K., Maglaqui, A.P., Gau-Pan, P.S., Redford, S.A., Wong, A., Tainer, J.A. and Coates, J.D. 2016. Perchlorate reductase is distinguished by active site aromatic gate residues. J. Biol. Chem. 291: 9190–9202.

Yu, L., Cañas, J.E., Cobb, G.P., Jackson, W.A. and Anderson, T.A. 2004. Uptake of perchlorate in terrestrial plants. Ecotoxicol. Environ. Safety 58: 44–49. https://doi.org/10.1016/S0147-6513(03)00108-8.

Zhang, Y., Chen, J.X., Wen, L.L., Tang, Y. and Zhao, H.P. 2016. Effects of salinity on simultaneous reduction of perchlorate and nitrate in a methane-based membrane biofilm reactor. Environ. Sci. Pollut. Res. 23: 24248–24255. https://doi.org/10.1007/s11356-016-7678-x.

Zhao, H.P., Van Ginkel, S., Tang, Y., Kang, D.W., Rittmann, B. and Krajmalnik-Brown, R. 2011. Interactions between perchlorate and nitrate reductions in the biofilm of a hydrogen-based membrane biofilm reactor. Environ. Sci. Technol. 45: 10155–10162. https://doi.org/10.1021/es202569b.

Zhao, H.P., Ontiveros-Valencia, A., Tang, Y., Kim, B.O., Ilhan, Z.E., Krajmalnik-Brown, R. and Rittmann, B. 2013. Using a two-stage hydrogen-based membrane biofilm reactor (MBfR) to achieve complete perchlorate reduction in the presence of nitrate and sulfate. Environ. Sci. Technol. 47: 1565–1572. https://doi.org/10.1021/es303823n.

Zhou, C., Ontiveros-Valencia, A., Nerenberg, R., Tang, Y., Friese, D., Krajmalnik-Brown, R. and Rittmann, B.E. 2019. Hydrogenotrophic microbial reduction of oxyanions with the membrane biofilm reactor. Front. Microbiol. 9: 3268. https://doi.org/10.3389/fmicb.2018.03268.

Ziv-El, M.C. and Rittmann, B.E. 2009. Systematic evaluation of nitrate and perchlorate bioreduction kinetics in groundwater using a hydrogen-based membrane biofilm reactor. Water Research 43: 173–181.

Zoeller, R.T. and Rovet, J. 2004. Timing of thyroid hormone action in the developing brain: clinical observations and experimental findings. J. Neuroendocrinol. 16: 809–818. https://doi.org/10.1111/j.1365-2826.2004.01243.x.

20
Nanobioremediation Technologies for Clean Environment

B. Chakrabarti,* P. Pramanik, S.P. Mazumdar and *R. Dubey*

1. Introduction

Pollution refers to an undesirable change in the physical, chemical and biological characteristics of air, water and soil that may harmfully affect life or create potential health hazard for any living organism. With the increase in population and growing industrialization and urbanization, pollution has become one of the biggest environmental challenges in modern times (Mehndiratta et al. 2013). Our natural resources like air, water and soil are subjected to pollution due to different anthropogenic activities. Sources of pollution include both point and nonpoint sources. Point sources include discharge of effluents and solid waste from industries, vehicular exhaust, metals released from mining and smelting, while agricultural activities like use of different insecticides/pesticides, and excessive use of chemical fertilizers comprise the nonpoint sources (Mohsenzadeh and Rad 2012). Various technologies are explored for remediation of pollution of the environment (Masciangoli and Zhang 2003). Different approaches can be followed while different materials can be used for remediation of environmental pollution (Guerra et al. 2018). In the 21st century, the most challenging task is to remove the pollutants from the environment using eco-friendly, economical and sustainable technologies (Yadav et al. 2017). "Remediate" means to remove the pollutant and "bioremediation" refers to the process in which different biological agents, like bacteria, fungi, protists, or their enzyme are used to degrade the pollutants (van Dillewijn et al. 2007). Among the various technologies practised nowadays, bioremediation technologies are efficient for controlling certain hazardous pollutants and are considered clean, green and environmentally friendly solution to pollution (Tripathi et al. 2018). Bioremediation refers to the use of naturally occurring microorganisms like bacteria, fungi as well as plants to breakdown or degrade toxic chemical compounds in the environment. Bioremediation can occur naturally with the help of indigenous microorganisms which is referred to as intrinsic bioremediation. Addition of nutrients through fertilizers to stimulate the indigenous microorganisms is known as biostimulation. Sometimes gaseous stimulants like oxygen (O_2) or methane (CH_4) are also used to stimulate the microbial activity. This process is known as bioventing. Bioremediation can be done both *in situ* and *ex situ*. *In situ* bioremediation is done in the location itself, while *ex situ* process requires excavation or removal of toxic pollutants prior to their treatment. The *in situ* process is less expensive but requires longer time and there is less uniformity in removal of the pollutant. On the other hand, *ex situ* process is rapid, more expensive and pollutant can be removed

---

CESCRA Indian Agricultural Research Institute, Pusa, New Delhi-110012, India.
* Corresponding author: bidisha2@yahoo.com

uniformly. Some of the common bioremediation technologies include bioleaching, bioventing, bioaugmentation, composting, biostimulation, bioreactor, phytoremediation, and rhizofiltration (Li and Li 2011). Although it is a good strategy for removing different types of contaminants, it may not hold good for high concentration of toxic chemicals which might even be harmful for the microorganisms (Rizwan et al. 2014). Best suitable method for remediation of contaminated sites depends on various factors like pollutant removal efficiency, cost effectiveness, complexity, hazards, availability of resources as well as the time required for remediation. It has been experienced that a single method of contaminant removal may not be appropriate. However, the problem related to single method can be overcome by the use of multiple technologies. Nanobioremediation is such an approach which uses physiochemical and biological methods together for removal of pollutants.

2. Nanobioremediation

Nano-bioremediation is the sum of three words, i.e., nano+bio+remediation. It refers to the use of nanoparticles and nanotechnology to aggravate the microbial activity or to enhance bioremediation process. Recently, studies have focused on development of technologies using nanomaterials for environmental remediation (Tratnyek and Johnson 2006). The concept of nanotechnology was given by Richard Feynman in 1959 (Feyman 1960) and presently it is one of the fastest growing areas of scientific research and technology development. Nanotechnology is an emerging branch of science and it is being used in various areas ranging from medicine to industrial applications. One of its application is for the remediation of environmental pollution (Agarwal and Joshi 2010). Nanomaterials will enhance the microbial activity and will be less toxic for the microorganisms, thereby reducing the overall cost as well as the time required for remediation.

3. Nanoparticles

Nanotechnology is characterized by the use particles between 1 and 100 nanometers (nm) of size. European Union (2011) has defined nano-materials as any material whether natural or artificial where more than 50% of the material is in the dimension of 1–100 nm. Nanoparticles are used in different fields like energy, medicine, agriculture, personal care products, electronics, textile, food processing and packaging, and environment (Figure 1).

In recent times, a lot of attention is being given to the application of nanomaterials for removing toxic substances from wastewater and air (Yang et al. 2006, Li et al. 2005). Nanoparticles (NP) can be classified in two groups, i.e., organic and inorganic. Organic nanoparticles comprise carbon nanoparticles, while inorganic are with magnetic, noble metal (palladium, gold and silver) and semiconductor nanoparticles (titanium dioxide and zinc oxide) (Tripathi et al. 2018). The materials used in nano-remediation are zeolites, metal oxides, carbon nanotubes, and many noble metals, out of which nano-valent iron is presently the most used material. Naturally occurring NPs are found in volcanoes and forest fires, and as by-products of combustion process, while engineered nanoparticles (ENPs) are manufactured for industrial use like black carbon, fumed silica, titanium

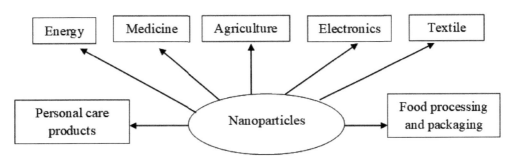

Figure 1. Use of nanoparticles in different fields.

Table 1. Characteristics and examples of different engineered nanoparticles.

Name	Characteristics	Examples
Carbon based nanomaterials	High specific surface area, specific adsorption capacity, high electrical and thermal conductivity, high strength, reusable	Carbon nanotube, fullerene, graphene
Metal based nanomaterials	Synthesised from metals, high surface area to volume ratio, crystalline and amorphous structures, high stability, low cost	Quantum dots (QD), nanogold, nanozinc, nano aluminum, and nanoscale metal oxides like TiO_2, ZnO and Al_2O_3
Organic nanoparticles	Biodegradable, non-toxic	Dendrimers, micelles, liposomes and ferritin
Nanocomposites	Nanoparticles with other nanoparticles or with larger bulk-type materials	Ceramic matrix nanocomposites, metal matrix nanocomposites, polymer matrix nanocomposites

Among these four types, uses of carbon and metal based nanoparticles are more frequent.

dioxide (TiO_2), iron oxide (FeOx), quantum dots (QDs), fullerenes, carbon nanotubes (CNTs), and dendrimers. Table 1 shows the characteristics and examples of different nanoparticles.

3.1 Nanobioremediation in pollution control

Nanoremediation methods are unique in the sense that it can treat persistent contaminants for which only few remediation techniques persist; it avoids formation of harmful intermediates and most importantly it remediates a contaminated site in lesser time which reduces the overall remediation cost of a site (Müller and Nowack 2010). Nanobioremediation techniques can be used in solid waste remediation, heavy metal pollution, groundwater and wastewater, uranium and hydrocarbon remediation.

At present time, atmospheric pollution caused due to the anthropogenic activities is one of the major concerns all over the world. In case of India, the problem of air pollution has increased due to the increasing population, industrialization and urbanization since the last few decades (Mina et al. 2013). Gases like carbon monoxide (CO), sulphur dioxide (SO_2), nitrogen oxides (NOx), as well as particulate matter (PM) are primary air pollutants. Primary air pollutants undergo chemical reactions in the air to form secondary air pollutants like ground level ozone (O_3), photochemical smog, acid rain, etc. Poor air quality affects vegetation, animals and also human beings. Scientists and engineers have suggested that nanotechnology promises less consumption of resources and energy and also produces less waste and hence causes less pollution (Fleischer and Grunwald 2008). Nanotechnology can be used in controlling air pollution by sensing and detection, remediation and treatment (Yunus et al. 2012). Nanomaterials have large specific surface areas and are highly reactive. There can be many methods through which remediation process can proceed of which some are absorption, absorption, chemical reactions, filtration and photo-catalysis.

3.1.1 Nano adsorbent

Nano-adsorbents have a high specific surface area with higher absorption rate and leave smaller footprint on the environment. It is mainly used for removal of organics, bacteria and heavy metals. Nanoadsorbents are broadly classified as oxide based nanoadsorbents, carbon nanotubes, and graphene based nanoadsorbents, depending on their role in adsorption process. Nanoscale adsorbents like carbon nano tubes (CNT) have property to increase adsorption of metal ions and can be used for treating polluted air (Sugunakala et al. 2017). CNTs and metal oxides are the most commonly used nanoparticles for the removal of heavy metals from aqueous solution (Ray and Shipley 2015). The adsorption capacity of CNTs is due to their pore structure, pore volume and existence of a broad spectrum of surface functional groups. These nanomaterials have good potential in removing different inorganic and organic pollutants, in both air and aqueous environment (Yunus et al. 2012). CNTs are powerful adsorbents for various organic compounds like dioxin, polynuclear aromatic hydrocarbons, chlorobenzenes and chlorophenols, dyes, etc. Zeolites are effective sorbents and ion-exchange media for metal ions. NaP1 zeolites can remove heavy metals from acid mine wastewaters (Moreno et al. 2001, Savage and Diallo 2005). Ordinary CNTs were modified into multiwall CNTs which helped in the removal of Pb (II) and Mn (II) (Tarigh and Shemirani 2013), and Cu (II) (Tang et al. 2012) from waste water. Mohamed et al. (2017) also reported the removal of Hg (II) by employing a functionalized-CNTs absorbent. Pharmaceutical and personal care products are source of some of the emerging pollutants which are difficult to detect, measure and remove from wastewater streams. Carbon nanomaterials are effective in their removal from water (Yu et al. 2014). Some compounds which can be effectively removed by carbon nanomaterials are ciprofloxacin (Li et al. 2014), ibruprofen (Cho et al. 2011), estradiols (Kumar and Mohan 2012), triclosan (Cho et al. 2011), etc.

Oxide based nanoadsorbents include titanium oxide/dendrimers' composites, zinc oxides, magnesium oxide, manganese oxides and ferric oxides. Dendrimers are mostly used for removal of organic compound and heavy metals from polluted water. However, dendrimers have found very little use commercially, since its synthesis is very complex. Some of the examples where dendrimers are used at batch scale to examine its effectiveness like silver dendrimers have been found effective to treat and remove bacteria such as *Pseudomonas aeruginosa, Escherichia coli* and *Staphylococcus aureus* (Balogh et al. 2001). Gupta et al. (2015) reported that the modified ferric oxide nano-adsorbents showed high affinity for the removal of different pollutants such as Cr^{3+}, Co^{2+}, Ni^{2+}, Cu^{2+}, Cd^{2+}, Pb^{2+} and As^{3+} simultaneously from wastewater. Modified ZnO nanoadsorbent, i.e., nano-assemblies were used for removal of different kinds of heavy metals (Singh et al. 2013). Kumar et al. (2013) reported that the use of mesoporous hierarchical ZnO nano-rods was very much effective in removing Pb (II) and Cd (II) from wastewater.

Graphene oxides and modification of graphene oxide or graphene with metal oxides or organics produce various nanocomposites which are useful in removal of pollutants from air and water (Wang et al. 2013). Various researchers reported that use of graphene and its other composite for the removal of heavy metals from wastewater was getting more attention due to its high surface area, mechanical strength, light weight, flexibility and chemical stability (Azamat et al. 2015, Dong et al. 2015, Vu et al. 2016, Zare-Dorabei et al. 2016, Ding et al. 2014). Besides these, different types of silicon nanomaterial like silicon nanotubes, silicon nanoparticles and silicon nanosheets are also used as nano-adsorbents. In addition to this, nanoclays, polymer-based nanomaterials, nanofibres, and aerogels are some of the nanomaterials used for adsorption of heavy metals from wastewater. Although use of CNTs for pollutant removal have several advantages, high cost of CNTs hinder their commercial use. Therefore, for production and use at commercial scale the technology needs to become economically viable.

3.1.2 Nanocatalyst

Nano-catalysts such as photocatalyst, electrocatalyst, fenton based catalyst, and chemical oxidant have immense potential for removing both organic and inorganic contaminants. Usually, these photocatalysts are comprised of semiconductor metals that can degrade variety of persistent organic pollutants in wastewater such as dyes, detergents, pesticides and volatile organic compound (Lin et al. 2014). Furthermore, semiconductor nano-catalysts are also highly effective for degradation of halogenated and non-halogenated organic compounds, and also heavy metals in specific situation (Adeleye et al. 2016). Some materials like titanium dioxide (TiO_2), zinc oxide (ZnO), iron oxide (Fe_2O_3) and tungsten oxide (WO_3) can be used as photocatalysts to oxidize the organic pollutants. The titanium dioxide nanoparticles (TiO_2) are capable of removing atmospheric pollutants like NOx, VOCs and other pollutants into less toxic species (Shen et al. 2015). Besides this, TiO_2 nanoparticles also possess antibacterial properties which are inversely proportional to the particle size (Mohamed 2017). Titanium dioxide (TiO_2) nanoparticles are also promising photocatalysts used for purification of water (Adesina 2004). TiO_2 nanoparticles can serve both as oxidative and reductive catalysts for organic as well as inorganic pollutants. Kabra et al. (2004) documented that TiO_2 nanoparticles degrade organic compounds like chlorinated alkanes and benzenes, dioxins, furans, PCBs and reduce toxic metal ions like Cr (VI), Ag (I) and Pt (II) in aqueous solutions. Nanotechnology can help in developing new nanocatalysts with enhanced surface area, thereby increasing the reaction efficiency for air pollution remediation (Özkar 2009). Nanogold based catalysts have excellent treating effect on different toxic air pollutants (Singh and Tandon 2014). ZnO photocatalyst is being developed and it is expected to have both detection and remediation functions (Yadav et al. 2017). When ZnO nanoparticles were used after modification with chitosan for removal of permethrin from contaminated water, results showed that more than 96% of permethrin was removed by using a small amount of nanoparticle (Dehaghi et al. 2014). Banerjee et al. (2012) used ZnO nanoparticles for reduction of Cr (VI) under the presence of sunlight. Results showed upto 35% reduction of Cr (VI) in 2 hours aqueous phase with ZnO-E nanoparticles as compared to only ZnO particles (19%).

3.1.3 Nanofiltration

Nano-filtration is the process wherein pressure of 3.5–16 bar is applied on the membrane and particles of size less than 0.5–1 nm are rejected by the membranes. These membranes are pressure-driven membranes and have pore sizes between 0.2 nm and 4 nm (Nowack 2008). These membranes are mostly used to remove colours, turbidity, microorganisms, organics, salts, hardness, and inorganic matter from the water. It is also helpful in removal of heavy metal from water (Kumar et al. 2014). Nanostructured membranes can also separate different air pollutants from the exhaust. These membranes remove multivalent ions, pesticides and heavy metals more effectively than conventional treatment methods. Nanofibre coated filter is used to remove dust from air at industrial plants (Muralikrishnan et al. 2014).

When inorganic fillers are added in polymeric nano membranes to improve their performance, they are known as nano-composite membranes. Nano composite membranes mainly find their application in water and wastewater treatment process. There are various nano fillers like carbon nanotubes, graphene, silver, titanium nanoparticles, etc. Advantages of these composites are that they enhance the selectivity process, improve thermal properties of material and moreover enhance the efficiency of filtration process (Ursino et al. 2018).

4. Nanosensors

Sensing and detection of pollutant in the environment is essential for the pollution control process. There are various conventional methods to detect pollutants in different media. Although these methods are accurate, some of them require expertise, and are more time consuming as well as expensive. Nanosensor is a minor device proficient of sensing and reacting to physical stimuli with at least one dimension in the order of nanometer scale. In recent times, nanotechnology helps in detecting several toxic pollutants at ppm and ppb levels in the environment (Zhou et al. 2015). Nanosensors have similar configuration as traditional sensors, but their fabrication is at nanoscale. These minute sensors are able to detect and respond to physicochemical as well as biological signal, conveying response into an output signal which is visualized by human beings. Nanosensors are evolving as capable tools for their use in agriculture and food production. They are superior to conventional chemical and biological method in terms of their selectivity, speed and sensitivity, detection time, economic feasibility and ease in handling. Presence of microbes, contaminants, pollutants and also the freshness of food items can be assessed through the use of these sensors (Joyner and Kumar 2015). At present, nanosensors are classified as:

Active nanosensors: These sensors are able to propel a signal that could be sensed distantly.

Passive nanosensors: These sensors rely on change in colour, opacity, or fluorescence (Liu 2003).

An ideal nanosensor should be of small size, highly specific to target types, super sensible to the changes in minute concentration in target species, should have rapid response time, lifetime should be long and production cost should be low. A nanosensor has a sensing unit for measurement of molecular composition of a matter, actuation unit for interaction of the sensor with the environment, power unit which converts mechanical, vibrational and hydraulic energies to electrical energy, processing unit like tiny transistors, storage unit, and communication unit.

Based on the material used for developing nanosensors, they can be classified as:

- Nanostructured materials like porous silicon based sensors
- Nanoparticles based sensors
- Nanoprobes
- Nanowire nanosensors
- Nanosystems: cantilevers, nano-electromechanical systems (NEMS)
- Electrochemical nanosensors
- Optical nanosensors

5. Why nanosensors?

Conventional sensors are usually quite large, from handheld devices up to mounted units like a small TV set, while nanosensors are much smaller. Due to extremely small size, surface to volume ratio rises significantly. As a result, nanoparticles show completely new properties than the same bulk material. The size of the sensing unit and transducer should be as small as possible for better configuration of the devices. Nanosensors can be used in hot wire anemometer for measuring fluid flow, thin film bolometers for detecting IR radiation, humidity sensors and in photodetectors.

Nanosensors are superior to traditional sensors because of greater sensitivity, lower detection range, smaller amount of sample required, direction detection without sample label and elimination of certain chemical use. The sensitivity or detection range of nanosensors can range down to just a few molecules of a specific inorganic, organic, or molecules of a specific inorganic, organic, or biological material.

Sensors combined with the knowledge of biology, chemistry and nanotechnology are called nanobiosensors. A nanobiosensor comprises of a biosensitive layer (biological element like antibody (Ab), enzyme, protein, DNA, etc.), covalently connected to the transducer. When the target interacts with bioreceptor, there is a change in physiochemical reaction and that change is converted into an electrical signal. Nano biosensors generate real-time data of the presence of antibodies, antigens, cell receptors, DNA and RNA, and nucleic acid. Where traditional sensors can detect in the range of 10^3 and 10^4 colony forming units (CFU)/ml, nanosensors can detect ~ 1 CFU/ml with less detection time.

Several types of nanobiosensors are available like mechanical, optical, nanowire biosensors, ion channel switch biosensor, electronic, Probes Encapsulated by Biologically Localized Embedding (PEBBLE) nanobiosensors and nanoshell biosensors. The advantages of nanobiosensors are that they are supersensitive and can even sense single virus or ultra-low amount of analyte that might cause possible damage, extremely capable with more surface to volume ratio. The main disadvantages associated with nano biosensors are that they are very delicate, complex and error prone and these are still in embryonic stage. The nano biosensors can be used in

- soil quality and disease assessment
- promotion of sustainable agriculture
- detection of contaminants and other molecules
- effective detection of biomolecules like DNA, RNA and protein
- analysis of food products

5.1 Challenges of nanobioremediation

Although nanotechnology provides opportunities to develop novel techniques for use in detection as well as remediation of environmental pollution, there is debate about the impact of the nanoparticles on the environment and biological organisms (Wiesner et al. 2006, Colvin 2003). The extensive use of manufactured nanoparticles for industrial, commercial, medical and agricultural purposes is leading to their release into the environment, eventually leading to the contamination of the environment (Nowack 2009, Paschoalino et al. 2010). Recently, there is a lot of discussion on nanotoxicology, which relates to the toxicity of nanomaterials (Nowack 2008). Nanoparticles can enter the environment during the production of raw material and products, during its use and also after disposal of the products (Bundschuh et al. 2018). After their release in the environment, the nanoparticles undergo changes like chemical transformation, aggregation, and disaggregation. Some scientists have suggested that nanoparticles form reactive oxygen species in the environment that could affect biological structures and cause harmful effects to biological organisms (Moore 2006, Hund-Rinke and Simon 2006). Certain toxicity mechanism of nanoparticles reported are oxidative stress (Mwaanga et al. 2014, Wu and Zhou 2012), reproductive failure due to the modifications in hormones or hatching enzymes (Muller et al. 2015, Nair et al. 2011) and alteration in photosynthetic pigments of algae and aquatic plants (Zou et al. 2016, Jiang et al. 2017). Besides this, there are reports that nanoparticles cause biochemical and physiological changes in terrestrial plants (Du et al. 2017). Certain metal or metal oxide nanoparticles were found to be highly toxic for soil microorganisms and also have impact on soil microbial species diversity (Fernández-Luqueño et al. 2017, He et al. 2016). Silver nanoparticles were found to inhibit the activity of dehydrogenase (Murata et al. 2005), phosphomonoesterase, arylsulfatase, b-D-glucosidase and leucineaminopeptidase (Peyrot et al. 2014) in soil, while iron nanoparticles could stimulate enzyme activities in soil (He et al. 2011). Kumar et al. (2012) reported that silver nanoparticles are toxic to the soil microbial community, especially to plant associated *Bradyrhizobium canariense*. Nanosized particles are reported to be taken up by different mammalian cells and are able to cross the cell membrane (Nel et al. 2006, Smart et al. 2006). Uptake of these particles is dependent on their size.

5.1.1 Health risk

Human beings can be exposed to nanoparticle toxicity either by direct exposure through air, soil or water or indirectly through intake of plants or animals which have accumulated nanoparticles. Because of small particle size and large surface area, nanoparticles are able to disperse easily and form bonds in the environment and human tissues. Besides this, NPs are biopersistent and can enter the food chain. Certain nanomaterials (NM) react with proteins and enzymes leading to oxidative stress and generation of reactive oxygen species (ROS), which can destroy mitochondria and thereby cause apoptosis. The six main entry routes of NM to the biological systems are inhalation, dermal, oral, subcutaneous, intraperitoneal and intra venous. Inhalation of NMs cause deposition of NPs in respiratory tract and lungs, which leads to lung related diseases like asthma, bronchitis, lung cancer, pneumonia, etc. NMs may penetrate into sweat glands and hair follicles. Skin exposure to cosmetics, sunscreens and dusts results in accumulation of nanoparticles, which are afterwards translocated to various parts of the body. These NMs remain in body in structurally unaltered, modified or metalized forms. They pass to the organs and remain in the cells for an indefinite period of time. They may move to other organs or get excreted. NMs can cause toxic effects like allergy, fibrosis, organ failure, cytotoxicity, swelling and inflation, tissue scar and damage, reactive oxygen species generation and DNA damage.

5.2 Eco-friendly nanomaterials

Synthesis of eco-friendly nanomaterials to reduce the environmental hazards is needed in future. Hence, the concept of green nanotechnology is coming up, which is a combination of nanotechnology and green chemistry with the goal of creating eco-friendly nanomaterials to reduce the environmental and human health hazards (Pandey 2018). Green synthesis of nanomaterials refers to that remediation process which will not pose any environmental risk. The method should be such that minimal waste is produced, products used can be recycled, materials used are renewable, nanomaterials produced should not get converted to some other more harmful secondary products after use and moreover it should be safe to use. In sum total, green synthesis of nanomaterial refers to avoiding production of undesirable, unsafe by-products and development of safe, sustainable and eco-friendly processes. In green synthesis, materials used are mostly microorganisms like algae, bacterial, fungi and plant materials. Researchers are using extracts from living organisms for synthesis of metallic nanoparticles. Most easy and readily available material for production of nanomaterial is plant extracts. Green nanoparticles could be produced by using reducing agents obtained from phytochemical extracts of different plants' parts (Boisselier and Astruc 2009, Shah et al. 2015). Different molecules like carbonyl groups, terpenoids, phenolic, flavones, amines, amides, proteins, pigments, and alkaloids, existing in both plant and microbial cells, may help in the synthesis of nanoparticles (Klaus et al. 1999). Microorganisms are also used for the synthesis of nanomaterials. Extracellular enzymes secreted by microbes are used for synthesis of green nanoparticles (Duran et al. 2011, Kupryashina et al. 2016). The metal binding ability of bacteria makes them useful for nanobioremediation. But fungi secrete more amount of protein, thereby producing larger amount of nanoparticles (Moghaddam et al. 2015). The methods of synthesizing nanoparticles involving microbes are slower than the methods involving plants. Some of the examples of biogenic nanomaterials are listed in Table 2.

Pollmann et al. (2006) reported that cells of bacteria *Bacillus sphaericus* JG-A12 (S-layer) were used for nano-remediation purposes as it was found that specifically these cells had high affinity for uranium and other heavy metals. A product called 'bioceramics' was prepared, where bacterial cells were encapsulated into silica gels. These bioceramics were effectively capable of eliminating copper and uranium from contaminated sites. Thus, green synthesis of nanoparticles has huge potential in the field of nanoscience with least environmental risk and will gain attention in future among researchers to focus on green part of nano science with more efficient result being derived from plant based products and their extracts.

Table 2. List of biogenic nanomaterials.

Nanoparticles	Plant, bacteria, yeast and fungi
Silver – Germanium nanoparticles	Freshwater diatom *Stauroneis* sp.
Gold and silver nanoparticles	*Saccharomyces cerevisiae, Citrus sinensis, Hibiscus rosasinensis*, Mushroom extract, *Verticillium* sp., *Fusarium oxysporum*
Silver nanoparticles	*Escherichia coli, Lactobacillus casei, Bacillus cereus, Aeromonas* sp., *Fusarium solani, Azadiracta indica, Brassica juncea*, Aloe vera extract
Gold nanoparticles	Banana peel, *Camellia sinensis, Chenopodium album, Rhodoseudomonas capsulate*
Copper, zinc, nickel nanoparticles	*Brassica juncea, Helianthus annuus, Medicago sativa*
Platinum nanoparticles	*Diopyros kaki, Ocimum sanctum*
Palladium nanoparticles	*Glycin max*
Lead nanoparticles	*Vitus vinifera, Jatropha carcus*

6. Conclusions

In recent times, nanotechnology has made enormous progress and it has useful applications in different fields. Nanobioremediation techniques can also help in remediating environmental pollution. However, there are certain risks associated with use of NMs in the environment.

Assessment of risks associated with use of nanoparticles requires an understanding of their mobility, bioavailability, toxicity and persistence. Development of eco-friendly NMs is required to reduce the environmental hazards. Hence, before recommendation of any technology for remediation of pollution, it is important to study the materials used for remediation and be sure that they are not a pollutant.

References

Adeleye, A.S., Conway, J.R., Garner, K., Huang, Y., Su, Y. and Keller, A.A. 2016. Engineered nanomaterials for water treatment and remediation: costs, benefits, and applicability. Chem. Eng. J. 286: 640–662.

Adesina, A.A. 2004. Industrial exploitation of photocatalysis progress, perspectives and prospects. Catal. Surv. Asia 8(4): 265–273.

Agarwal, A. and Joshi, H. 2010. Application of nanotechnology in the remediation of contaminated groundwater: a short review. Recent Res. in Sci. and Tech. 2(6): 51–57.

Azamat, J., Sattary, B.S., Khataee, A. and Joo, S.W. 2015. Removal of a hazardous heavy metal from aqueous solution using functionalized graphene and boron nitride nanosheets: insights from simulations. J. Molecul. Graphics Model. 61: 13–20.

Balogh, L., Swanson, D.R., Tomalia, D.A., Hagnauer, G.L. and McManus, A.T. 2001. Dendrimer–silver complexes and nanocomposites as antimicrobial agents. Nano Letters 1(1): 18–21.

Banerjee, P., Chakrabarti, S., Maitra, S. and Dutta, B.K. 2012. Zinc oxide nano-particles–sonochemical synthesis, characterization and application for photo-remediation of heavy metal. Ultrasonics Sonochemistry 19(1): 85–93.

Boisseliera, E. and Astruc, D. 2009. Gold nanoparticles in nanomedicine: preparations, imaging, diagnostics, therapies and toxicity. Chemical Society Reviews 38(6): 1759–1782.

Bundschuh, M., Filser, J., Lüderwald, S., McKee, M.S., Metreveli, G., Schaumann, G.E., Schulz, R. and Wagner, S. 2018. 2018.Nanoparticles in the environment: where do we come from, where do we go to? Environ. Sci. Eur. 30: 6. https://doi.org/10.1186/s12302-018-0132-6.

Colvin, V.L. 2003. The potential environmental impact of engineered nanoparticles. Nat. Biotechnol. 21: 1166–1170.

Cho, H.H., Huang, H. and Schwab, K. 2011. Effects of solution chemistry on the adsorption of ibuprofen and triclosan onto carbon nanotubes. Langmuir 27(21): 12960–12967.

Dehaghi, S.M., Rahmanifar, B., Moradi, A.M. and Azar, P.A. 2014. Removal of permethrin pesticide from water by chitosan–zinc oxide nanoparticles composite as an adsorbent. J. Saudi Chem. Soc. 18(4): 348–355.

Ding, Z., Hu, X., Morales, V.L. and Gao, B. 2014. Filtration and transport of heavy metals in graphene oxide enabled sand columns. Chem. Eng. J. 257: 248–252.

Dong, Z., Zhang, F., Wang, D., Liu, X. and Jin, J. 2015. Polydopamine mediated surface-functionalization of graphene oxide for heavy metal ions removal. J. Solid State Chem. 224: 88–93.

Du, W.C., Tan, W.J., Peralta-Videa, J.R., Gardea-Torresdey, J.L., Ji, R, Yin, Y. and Guo, H.Y. 2017. Interaction of metal oxide nanoparticles with higher terrestrial plants: physiological and biochemical aspects. Plant Physiol. Biochem. 110: 210–225.

Durán, N., Marcato, P.D., Durán, M. Yadav, A., Gade, A. and Rai, M. 2011. Mechanistic aspects in the biogenic synthesis of extracellular metal nanoparticles by peptides, bacteria, fungi and plants. Appl. Microbiol. Biotechnol. 90(5): 1609–1624. https://doi.org/10.1007/s00253-011-3249-8.

Fernández-Luqueño, F., Lopez-Valdez, F., Sarabia-Castillo, C.R., García-Mayagoitia, S. and Perez-Rios, S.R. 2017. Bioremediation of polycyclic aromatic hydrocarbons-polluted soils at laboratory and field scale: a review of the literature on plants and microorganisms. In: Anjum, N.a., Gill, S.s. and Tuteja, N. (eds.). Enhancing Cleanup of Environmental Pollutants Vol. 1: Biological Approaches. Springer. Switzerland, 43.

Feynman, R. 1960. There's plenty of Room at the Bottom. California Institute of Technology, December 29, 1959. Caltech. Eng. Sci. 23: 22–36.

Fleischer, T. and Grunwald, A. 2008. Making nanotechnology developments sustainable. A role for technology assessment? J. Clean. Prod. 16: 889–898.

Guerra, F.D., Attia, M.F., Whitehead, D.C. and Alexis, F. 2018. Nanotechnology for environmental remediation: materials and applications. Molecules 23: 1760. doi:10.3390/molecules 23071760.

Gupta, V.K., Tyagi, I., Sadegh, H., Shahryari-Ghoshekand, R., Makhlouf, A.S.H. and Aazinejad, B. 2015. Nanoparticles as adsorbent; a positive approach for removal of noxious metal ions: a review. Sci. Technol. Dev. 34: 195.

He, S., Feng, Y., Ni, J., Sun, Y., Xue, L., Feng, Y., Yu, Y., Lin, X. and Yang, L. 2016. Different responses of soil microbial metabolic activity to silver and iron oxide nanoparticles. Chemosphere 147: 195.

He, S.Y., Feng, Y.Z., Ren, H.X., Zhang, Y., Gu, N. and Lin, X.G. 2011. The impact of iron oxide magnetic nanoparticles on the soil bacterial community. J. Soils Sediments 11: 1408e1417.

Hund-Rinke, K. and Simon, M. 2006. Ecotoxic effect of photocatalytic active nanoparticles TiO_2 on algae and daphnids. Environ. Sci. Pollut. Res. 13(4): 225–232.

Jiang, H.S., Yin, L.Y., Ren, N.N., Zhao, S.T., Li, Z., Zhi, Y., Shao, H., Li, W. and Gontero, B. 2017. Silver nanoparticles induced reactive oxygen species via photosynthetic energy transport imbalance in an aquatic plant. Nanotoxicology 11(2): 157–167.

Joyner, J.R. and Kumar, D.V. 2015. Nanosensors and their applications in food analysis: a review. Vol. 1, Issue 4: 80–90.

Kabra, K., Chaudhary, R. and Sawhney, R.L. 2004. Treatment of hazardous organic and inorganic compounds through aqueous-phase photocatalysis: A review. Ind. Eng. Chem. Res. 43(24): 7683–7696.

Klaus, T., Joerger, R., Olsson, E. and Granqvist, C.G. 1999. Silver-based crystalline nanoparticles, microbially fabricated. Proc. Natl. Acad. Sci. USA 96: 13611–13614.

Kumar, A.K. and Mohan, S.V. 2012. Removal of natural and synthetic endocrine disrupting estrogens by multi-walled carbon nanotubes (MWCNT) as adsorbent: kinetic and mechanistic evaluation. Separation and Purification Technology 87: 22–30.

Kumar, A., Gayakwad, A. and Nagal, B.D. 2014. A review: nano membrane and application. International J. of Innovat. Res. in Science, Engineering and Technology 3(1): 8373–8381.

Kumar, N., Shah, V. and Walker, V.K. 2012. Influence of a nanoparticle mixture on an arctic soil community. Environ. Toxicol. Chem. 31: 131e135.

Kumar, K.Y., Muralidhara, H.B., Nayaka, Y.A., Balasubramanyam, J. and Hanumanthappa, H. 2013. Hierarchically assembled mesoporous ZnO nanorods for the removal of lead and cadmium by using differential pulse anodic stripping voltammetric method. Powder Technol. 239: 208–216.

Kupryashina, M.A., Vetchinkina, E.P. and Nikitina, V.E. 2016. Biosynthesis of silver nanoparticles with the participation of extracellular Mn-dependent peroxidase from Azospirillum. Appl. Biochem. Microbiol. 52: 384. https://doi.org/10.1134/S0003683816040104.

Li, Y., H., Di, Z.C., Ding, J., Wu, D.H., Luan, Z.K. and Zhu, Y.Q. 2005. Adsorption thermodynamic, kinetic and desorption studies of Pb^{2+} on carbon nanotubes. Water Res. 39: 605.

Li, Y. and Li, B. 2011. Study on fungi-bacteria consortium bioremediation of petroleum contaminated mangrove sediments amended with mixed biosurfactants. Advan. Materials Res. 183-185: 1163–1167.

Li, H., Zhang, D., Han, X. and Xing, B. 2014. Adsorption of antibiotic ciprofloxacin on carbon nanotubes: pH dependence and thermodynamics. Chemosphere 95: 150–155.

Lin, S.T., Thirumavalavan, M., Jiang, T.Y. and Lee, J.F. 2014. Synthesis of ZnO/Zn nano photocatalyst using modified polysaccharides for photodegradation of dyes. Carbohyd. Polym. 105: 1–9.

Liu, Y. 2003. Nanosensors, http://www.slideserve.com/kim-johnston/nanosensors, (3/10/2016).

Masciangoli, T. and Zhang,W. 2003. Environmental technologies. Environ. Sci. Technol. 37: 102–108.

Mehndiratta, P., Jain, A., Srivastava, S. and Gupta, N. 2013. Environmental pollution and nanotechnology. Environ. Poll. 2(2): 49–58.

Mina, U., Singh, R. and Chakrabarti, B. 2013. Agricultural production and air quality: an emerging challenge. Int. J. of Environ. Sci.: Develop. Monit. 4(2): 80–85.

Moghaddam, A.B., Namvar, F., Moniri, M., M.d Tahir, P., Azizi, S. and Mohamad, R. 2015. Nanoparticles biosynthesized by fungi and yeast: a review of their preparation, properties, and medical applications. Molecules 20: 16540–16565. doi:10.3390/molecules200916540.

Mohamed, E.F. 2017. Nanotechnology: Future of environmental air pollution control. Environ. Manage. Sustain. Develop. 6(2): 429–454.

Mohsenzadeh, F. and Rad, A.C. 2012. Bioremediation of heavy metal pollution by nano-particles of *Noaea mucronata*. Int. J. Biosci. Biochem. Bioinforma. 2(2): 85–89.

Moore, M.N. 2006. Do nanoparticles present ecotoxicological risks for the health of the aquatic environment? Environ. Int. 32: 967–976.

Moreno, N., Querol, X. and Ayora, C. 2001. Utilization of zeolites synthesized from coal fly ash for the purification of acid mines water. Environ. Sci. Technol. 35: 3526–3534.

Müller, N.C. and Nowack, B. 2010. Nano Zero Valent Iron the Solution for Water and Soil Remediation. Observatory NANO Focus Report.

Muller, E.B., Lin, S.J. and Nisbet, R.M. 2015. Quantitative adverse outcome pathway analysis of hatching in zebrafish with CuO nanoparticles. Environ. Sci. Technol. 49(19): 11817–11824.

Muralikrishnan, R., Swarnalakshmi, M. and Nakkeeran, E. 2014. Nanoparticle-membrane filtration of vehicular exhaust to reduce air pollution—A review. Int. Res. J. of Environ. Sci. 3(4): 82–86.

Mwaanga, P., Carraway, E.R. and van den Hurk, P. 2014. The induction of biochemical changes in *Daphnia magna* by CuO and ZnO nanoparticles. Aquat. Toxicol. 150: 201–209.

Nair, P.M.G., Park, S.Y., Lee, S.W. and Choi, J. 2011. Differential expression of ribosomal protein gene, gonadotrophin releasing hormone gene and Balbiani ring protein gene in silver nanoparticles exposed *Chironomus riparius*. Aquat. Toxicol. 101(1): 31–37.

Nel, A., Xia, T., Mädler, L. and Li, N. 2006. Toxic potential of materials at the nanolevel. Science 311: 622–627.

Nowack, B. 2008. Pollution prevention and treatment using nanotechnology. In: Harald Krug (ed.). Nanotechnology. Volume 2: Environmental Aspects. WILEY-VCH Verlag GmbH & Co. KGaA, Weinheim. ISBN: 978-3-527-31735-6.

Nowack, B. 2009. The behavior and effects of nanoparticles in the environment. Environ. Pollut. 157: 1063e1064.

Özkar, S. 2009. Enhancement of catalytic activity by increasing surface area in heterogeneous catalysis. Appl. Surface Sci. 256(5): 1272–1277. https://doi.org/10.1016/j.apsusc.2009.10.036.

Pandey, G. 2018. Prospects of nanobioremediation in environmental cleanup. Orient J. Chem. 34(6): 2828–2840.

Paschoalino, M.P., Marcone, G.P.S. and Jardim, W.F. 2010. Nanomaterials and the environment. Qumica Nova 33: 421e430.

Peyrot, C., Wilkinson, K.J., Desrosiers, M. and Sauve, S. 2014. Effects of silver nanoparticles on soil enzyme activities with and without added organic matter. Environ. Toxicol. Chem. 33: 115e125.

Pollmann, K., Raff, J., Merroun, M., Fahmy, K. and Selenska-Pobell, S. 2006. Metal binding by bacteria from uranium mining waste piles and its technological applications. Biotechnology Advances 24(1): 58–68.

Ray, P.Z. and Shipley, H.J. 2015. Inorganic nanoadsorbents for the removal of heavy metals and arsenic: a review. RSC Adv. 5: 29885–29907.

Rizwan, M., Singh, M., Mitra, C.K. and Morve, R.K. 2014. Ecofriendly application of nanomaterials: nanobioremediation. J. of Nanoparticles Article ID 431787, 7 pages. http://dx.doi.org/10.1155/2014/431787.

Savage, N. and Diallo, M.S. 2005. Nanomaterials and water purification: Opportunities and challenges. Journal of Nanoparticle Research 7: 331–342.

Shah, M., Fawcett, D., Sharma, S., Tripathy, S.K. and Poinern, G.E.J. 2015. Green synthesis of metallic nanoparticles via biological entities. Materials (Basel) 8(11): 7278–7308. Published 2015 Oct 29. doi:10.3390/ma8115377.

Shen, W., Zhang, C., Li, Q. Zhang, W., Cao, L. and Ye, J. 2015. Preparation of titanium dioxide nanoparticle modified photocatalytic self-cleaning concrete. J. of Cleaner Production 87: 762–765. https://doi.org/10.1016/j.jclepro.2014.10.014.

Singh, S.B. and Tandon, P.K. 2014. Catalysis: A brief review on nano-catalyst. Journal of Energy and Chemical Engineering 2(3): 106–115.

Singh, S., Barick, K.C. and Bahadur, D. 2013. Fe_3O_4 embedded ZnO nanocomposites for the removal of toxic metal ions, organic dyes and bacterial pathogens. J. Mater. Chem. A. 1: 3325–3333.

Smart, S.K., Cassady, A.I., Lu, G.Q. and Martin, D.J. 2006. The biocompatibility of carbon nanotubes. Carbon 44(6): 1034–1047. doi:10.1016/j.carbon.2005.10.011.

Sugunakala, S., Krishnaveni, K. and Neela, R. 2017. Applications of nanotechnology in water and air pollution treatment—review. Int. J. of I. Res. in Advan. Engineering 4(9): 76–79.

Tang, W.W., Zeng, G.M., Gong, J.L., Liu, Y., Wang, X.Y., Liu, Y.Y., Liu, Z.F., Chen, L., Zhang, X.R. and Tu, D.Z. 2012. Simultaneous adsorption of atrazine and Cu(II) from wastewater by magnetic multi-walled carbon nanotube. Chem. Eng. J. 211: 470–478.

Tarigh, G.D. and Shemirani, F. 2013. Magnetic multi-wall carbon nanotube nanocomposite as an adsorbent for preconcentration and determination of lead (II) and manganese (II) in various matrices. Talanta 115: 744–750.

Tratnyek, P.G. and Johnson, R.L. 2006. Nanotechnologies for environmental cleanup. Nano Today 1: 44–48.

Tripathi, S., Sanjeevi, R., Anuradha, J., Chauhan, D.S. and Rathoure, A.K. 2018. Nano-bioremediation: nanotechnology and bioremediation. pp. 202–219. In: Rathoure, A.K. (ed.). Biostimulation Remediation Technologies for Groundwater Contaminants. DOI: 10.4018/978-1-5225-4162-2.ch012.

Ursino, C., Castro-Muñoz, R., Drioli, E., Gzara, L., Albeirutty, M. and Figoli, A. 2018. Progress of nanocomposite membranes for water treatment. Membranes 8(2): 18.

United States Environmental Protection Agency (USEPA). 2005. Guidance on selecting age groups for monitoring and assessing childhood exposures to environmental contaminants. National Center for Environmental Assessment, Washington, DC; EPA/630/P-03/003F. Available from: National Technical Information Service, Springfield, VA, and online at http://epa.gov/ncea.

van, Dillewijn, P., Caballero, A., Paz, J.A., Gonz´alez-P´erez, M.M., Oliva, J.M. and Ramos, J.L. 2007. Bioremediation of 2,4,6-trinitrotoluene under field conditions. Environmental Sci. and Technol. 41(4): 1378–1383.

Vu, H.C., Dwivedi, A.D., Le, T.T., Seo, S.H., Kim, E.J. and Chang, Y.S. 2016. Magnetite graphene oxide encapsulated in alginate beads for enhanced adsorption of Cr (VI) and As (V) from aqueous solutions: role of crosslinking metal cations in pH control. Chem. Eng. J. http://dx.doi.org/10.1016/j.cej.2016.08.058.

Wang, S., Sun, H., Ang, H.M. and Tadé, M.O. 2013. Adsorptive remediation of environmental pollutants using novel graphene-based nanomaterials. Reviews in Chemical Engineering Journal 226: 336–347. https://doi.org/10.1016/j.cej.2013.04.070.

Wiesner, M.R., Lowry, G.V., Alvarez, P., Dionysiou, D. and Biswas, P. 2006. Assessing the risks of manufactured nanomaterials. Environ. Sci. Technol. 40: 4336.

Wu, Y. and Zhou, Q.F. 2012. Dose- and time-related changes in aerobic metabolism, chorionic disruption, and oxidative stress in embryonic medaka (*Oryzias latipes*): underlying mechanisms for silver nanoparticle developmental toxicity. Aquat. Toxicol. 124: 238–246.

Yadav, K.K., Singh, J.K., Gupta, N. and Kumar, V. 2017. Review of nanobioremediation technologies for environmental cleanup: a novel biological approach. J. of Materials and Environ. Sci. 8(2): 740–757.

Yang, K., Zhu, L.Z. and Xing, B.S. 2006. Adsorption of polycyclic aromatic hydro carbons on carbon nonmaterials. Environ. Sci. Technol. 40(6): 1855.

Yu Jin-Gang, Xiu-Hui Zhao, Hua Yang, Xiao-Hong Chen, Qiaoqin Yang, Lin-Yan Yu, Jian-Hui Jiang and Xiao-Qing Chen. 2014. Aqueous adsorption and removal of organic contaminants by carbon nanotubes. Sci. Total Environ. 482: 241–251.

Yunus, I.S., Harwin, Kurniawan, A., Adityawarman, D. and Indarto, A. 2012. Nanotechnologies in water and air pollution treatment. Environmental Technology Reviews 1(1): 136–148.

Zare-Dorabei, R., Ferdowsi, S.M., Barzin, A. and Tadjarodi, A. 2016. Highly efficient simultaneous ultrasonic-assisted adsorption of Pb (II), Cd (II), Ni (II) and Cu (II) ions from aqueous solutions by graphene oxide modified with 2, 20-dipyridylamine: central composite design optimization. Ultrason. Sonochem. 32: 265–276.

Zhou, R., Hu, G., Yu, R., Pan, C. and Wang, Z.L. 2015. Piezotronic effect enhanced detection of flammable/toxic gases by ZnOmicro/nanowiresensors. NanoEnergy 12: 588–59. https://doi.org/10.1016/j.nanoen.2015.01.036.

Zou, X.Y., Li, P.H., Huang, Q. and Zhang, H.W. 2016. The different response mechanisms of *Wolffia globosa*: light-induced silver nanoparticle toxicity. Aquat. Toxicol. 176: 97–105.

21

Biochar

An Imperative Amendment for Soil and Environment

Sumita Chandel,[1] *Ritika Joshi*[2,*] *and Ashish Khandelwal*[2]

1. Introduction

Continuous cultivation with a rice-wheat cropping system and large scale residue production leads to burning of surplus rice in the open field, which in turn leads to severe air pollution and greenhouse gases emission from field. The word "biochar" is a combination of 'bio' and 'char'. Bio means 'biomass' and char means 'charcoal'. Biochar is a carbon rich stable substrate obtained after pyrolysis (combusted under low or no oxygen conditions) of organic biomass or agricultural residues. During the production process of biochar, bio-oil and gases are also produced as a byproduct which can be used in other industries and for other purposes. Biochar is a form of carbon rich charcoal that is formed by the pyrolysis (thermal decomposition) of organic biomass or agricultural residues (raw materials) which is used as soil conditioner (Xiao et al. 2014). Biochar is also defined as charcoal produced from plant matter and is stored in the soil as a means of removing carbon dioxide from the atmosphere. Biochar is the carbon-rich product obtained when biomass, such as wood, manure, crop residues or leaves, is heated in a closed container with little or no available air. Biochar is charcoal used as a soil amendment. It is highly heterogeneous in nature and composed of stable and labile components. Major components of biochar are carbon (C), volatile matter, mineral matter and ash. These occur in different proportions according to biomass used for biochar preparations. Physically, it is dark black in color (Figure 1) and its highly porous structure makes it attractive. In biochar, most of the carbon is recalcitrant in nature; therefore, it helps in carbon sequestration by locking the carbon present in the plant biomass. From agriculture point of view, biochar has major role in soil management, carbon sequestration issues, and immobilization of pollutants (especially heavy metals). Recalcitrant carbon in biochar has multiple benefits ranging from soil improvement to waste management and mitigating climate change. Biochar is mainly used to improve soil nutrient content and to sequester carbon from the environment (Lehmann 2009). It has been estimated that through production of biochar, almost 12 per cent of the GHG emissions caused by human activities could be reduced (Woolf et al. 2010). Biochar acts as a carbon sink and it reduces decaying and

---

[1] Department of Soil Science, Punjab Agricultural University, Ludhiana.
[2] Centre for Environment Science and Climate Resilient Agriculture, ICAR-Indian Agricultural Research Institute, New Delhi.
* Corresponding author: ritikajoshi964@gmail.com, ritikajoshi90-coasoil@pau.edu

Figure 1. Biochar.

mineralization of the biological carbon cycle to establish a carbon sink. In recent years, the use of surplus organic matter or biomass (especially rice and wheat) for the production of biochar has yielded promising results in the reduction of CO_2 and mitigating climate change. Biochar is used for amendment of acidic soils (low pH soils), it increases basicity of acidic soils and also increase agricultural productivity. Recently, several researches were done in biochar, their properties and effects on crop yields. Currently, many projects are being carried out on biochar. In this chapter, we will discuss about biochar, its production, elemental composition, differences in biochar according to feedstocks used, effect of biochar on soil carbon sequestration, reducing heavy metal pollution, mitigation of climate change, etc.

Biochar production.

2. Production

Biochar production is thermo-chemical conversion of biomass under low or no oxygen environment. There are different steps (fast pyrolysis, slow pyrolysis, gasification, and carbonization) involved in the production of biochar, which are depicted in Figure 2.

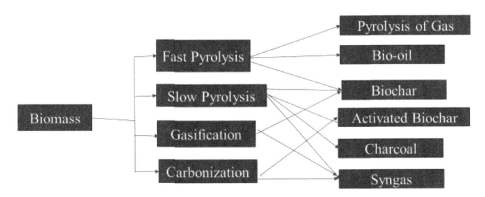

Figure 2. Step by step process for biochar production.

3. Effect of temperature and heating rate on biochar production

With increase in temperature, biochar production decreased, primarily due to depolymerisation of compounds like cellulose and hemicellulose as well as combustion of organic materials (Cooperman

2016, Cao and Harris 2010). The decrease in yield with increase in temperature is due to increase in volatilization rate of organic compounds (Muradov et al. 2012). Evolution of moisture occurs at 220°C. The high yield of biochar at low temperatures indicates that the material has been only partially pyrolysed (Katyal et al. 2003, Yang et al. 2004). The chemical properties of biochar are also greatly affected by temperature and heating rate. With the increase in the temperature, the pH values of the biochar also increased (Angin 2013). Similar results were also reported by Parmila and Saroha (2014) and Jindarom et al. (2007). Yuan et al. (2011) suggested that the increases in pH values might be attributed to the separating of alkali salts from organic materials due to increase in pyrolysis temperature. With the increase in the pyrolysis temperature, the carbon content in the biochar increases, while the oxygen and hydrogen content decreases with respect to carbon content (Chen et al. 2011, Chun et al. 2004). Cooperman (2016) suggested that cracking of weak bonds within the biochar structure might be attributed to loss in hydrogen and oxygen content. Lua et al. (2004) observed that with the increase in temperature, there is a decrease in the Brunauer-Emmett-Teller (BET) surface area. Lehmann (2007) suggested that the Cation Exchange Capacity (CEC) is directly proportional to temperature at which the biochar is produced, that is, CEC increases significantly with the increase in the temperature.

4. Biochar composition

Biochar composition is affected by raw materials which are used for biochar preparations. Feedstocks, which are rich in carbon and nitrogen, have more carbon and nitrogen concentration in biochar. Researches have been done for comparing different biochars (derived from different sources) regarding their elemental and functional properties. Researches were carried out on effects of different feedstocks (wheat straw and rice straw) on nutritional composition of biochar and it was found that rice straw biochar is rich in nutrient content with respect to wheat straw biochar.

5. Sources

Rural areas are the main sources of raw material for biochar production, which includes crop residue from the harvesting of crops, charcoal from chulhas (villages), rice husk by products, different forest wastes, weed biomass, etc.

6. Amendment of soil and environment through biochar

Burning of rice straw is a major problem all over the world, causing nuisance in the environment, soil and human health. In India, Punjab is the major rice-wheat growing state, where the area under paddy (rice) is approximately 3.0 million ha, which produces 20 million tons of paddy straw, out of which 75 per cent of the straw is burned in the fields only. Burning of rice straw generally gives rise to 1515 kg per tons of CO_2, 92 kg/tons of CO, 3.83 kg/tons of NO_2, 0.4 kg/tons of SO_2, 2.7 kg/tons of CH_4, 199 kg/tons of ash (PM) and 5.7 kg/tons of non methane volatized substances. These are the major greenhouse gases (GHG) contributing to global warming. So, its management becomes very imperative. To solve this problem, biochar production is one of the best option. Biochar is the key to many problems like waste management, energy production, soil fertility and productivity, carbon sequestration and mitigation of greenhouse gases (CO_2, CH_4 and N_2O). Biochar is being promoted as a way to initiate a "doubly green revolution" as it potentially address soil organic matter, GHG emission and food securities.

6.1 Waste management

The waste produced from various sectors like agriculture, industries, forest, animal and municipal solid waste can be managed by intelligent utilization, i.e., exploiting the pyrolysis technique for biochar production (Figure 2). The management of the waste in this way is not only sustainable to the environment, and cost effective, but also one of the best method to curb the environment pollution. In India, about 130–150 million tons of agricultural waste like acacia wood, coconut shell, etc. are dumped without any use. This clearly indicates that the biochar production potential is very high. Biochar produced from coconut shell increases the survival of microbial inoculants by up to six months, as it is free from toxic elements and is eco-friendly (Saranya et al. 2011). Biochar also has the potential to be used as an alternate carrier to lignite for the preparation of biofertilizers.

6.2 Soil management

Biochar (Figure 1) plays an important role in the management of soil either by increasing the fertility potential or by remediating the soil pollution (heavy metal pollution).

6.2a Improvement in soil fertility

Biochar can be extensively used as an amendment in the fields, which not only improves the overall quality of the soil, nutrient cycling, carbon sequestration but also helps in the curbing of soil pollution. It contains about 48 per cent carbon, 1 per cent nitrogen, 0.7 per cent phosphorous and 3.3 per cent potassium. There is no release of any harmful gases from the biochar. Soil biochar application offers the potential to stabilize carbon and recycling of nutrients present in the straw. Early trends suggest beneficial effects of biochar on crop yields (rice-wheat and potato-anion) and C content in soil. Biochar improves the soil water holding capacity and thus the water retention for a longer duration, which is attributed to the highly porous structure (Keech et al. 2005, Liang et al. 2006). So, the problem of water scarcity can be combated as it will reduce the amount and intensity of water requirement of any crop. The biochar addition will increase the soil pH and EC (Glaser et al. 2002). This is due to the presence of ash residue that is dominated by carbonates of alkali and alkaline earth metals and some amount of silica, heavy metal and organic and inorganic nitrogen. Thus, it will have a liming effect on the acidic soil. It will also suppress the enzymatic activities of microbes involved in the conversion of nitrite to nitrous oxide, thereby increasing nitrogen availability in the soil and suppressing nitrogen losses. Shenbagavalli and Mahimairaja (2012) produced the biochar from *Prosopis* sp. woods by pyrolysis material under high temperature. The characterization of this biochar revealed that it contains high carbon and was neutral in nature. When this biochar was added to soil with pH 8.42, it was found that the pH was decreased to 7.92 during the incubation with rise in CEC and organic carbon. However, the pH raised to 5.93 on application of biochar to acidic soil (pH 4.20) (Zwieten et al. 2010). Ibrahim et al. (2017) reported that soil pH, total nitrogen and total carbon, dissolved organic carbon and ammonium nitrogen were increased while available concentration of Cd, Pb and As decreased along with nitrate-nitrogen in biochar amended soils. Biochar also reduces the leaching losses of the nutrient to groundwater, so in addition to improving the economy it improves the ecology too. This is attributed to high CEC of the biochar because of more surface area and porous structure. Therefore, it holds the nutrients more strongly and improves the nutrient use efficiency of soils.

6.3 Remediation of heavy metal pollution

Biochar is an ideal candidate for remediation of organic and inorganic pollutants both in contaminated soil and water. This is because of its high specific surface area, micro-porosity and positively and negatively charged surface functional groups (Ahmad et al. 2014). It is used to sorb both organic compounds like pesticides and herbicides and heavy metal contaminants, but it reduces the ability of microbes to breakdown these substances. The sorption of both organic and heavy metals pollutants is influenced by pH, specific surface area, particle size, time of exposure of pollutants and soil moisture. The inorganic ions, or metals can be physically entrapped or chemically sorbed onto the biochar. The adsorptive competition occurs when multi-contaminants are adsorbed onto the biochar. Inyang et al. (2012) investigated the adsorption of Cu^{2+}, Cd^{2+}, Pb^{2+} and Ni^{2+} on the biochar. They showed that biochar demonstrated better capacity to remove Ni^{2+} and Cd^{2+}. Soil pH is an important parameter that affects both the surface charge density of the absorbent and the metal ion speciation (Chen et al. 2008). Biochar is alkaline in nature and therefore the increase in soil pH stabilizes metals, except arsenic (Ahmad et al. 2014). The alkalinity also causes some metals to precipitate from the solution on the surface of the biochar, therefore reducing the availability of theses metals to the plants. In addition to this, the heavy metals are less mobile in the soil with neutral pH or above, since the biochar increases soil pH, which in turn will decrease the mobility of the metals in the soil. Soil pH significantly affects the adsorption of Cu^{2+} and Zn^{2+} on the biochar.

Bioavailability and eco-toxicity of heavy metals in biochar were obtained from sludge of pulp and Parmila and Saroha (2014) studied paper mill effluent. They found that the matrix of biochar heavy metals gets enriched after pyrolysis, but their bioavailability and eco-toxicity were reduced because of relatively stable fraction. It reduces the mobility of the lead (Pb) due to the formation of insoluble Pb-phosphate as the biochar derived from manure is rich in phosphate (Chao and Harris 2010). The biochar, which is produced at 700°C, showed the residual form of heavy metals. So, there is no risk of utilization of biochar as the eco-toxicity of the heavy metals was reduced significantly due to pyrolysis.

6.4 Carbon sequestration

Carbon storing capacity of the soil is three times more than that of the carbon existing in the atmosphere. The storing capacity can be further increased if the rate of carbon inputs exceeds the rate of mineral decomposition. Carbon sequestration is the process of storing carbon in soil organic matter and thus removing it from the atmosphere. The biochar potential to sequester the carbon in the soil has received considerable attention in recent years and plays a big role in the climate smart agriculture. Approximately 70–80 per cent of the carbon present in the biomass gets trapped in the structure of the biochar that is stable in nature compared to the biomass, which on degradation releases the carbon back to the atmosphere. Therefore, it contributes more carbon to the soil compared to the plant residue. So, the production of the biochar and its subsequent addition in soil act as a carbon sink. Biochar generally slows down the decaying and mineralization of carbon cycle to establish a carbon sink. It is the home of beneficial microorganism and the large surface area provides the ability to absorb organic matter, gases and inorganic nutrients.

6.4.1 Biochar carbon persistence in soil

Biochar contains high level of carbon, so it always remains uncertain that for how much time it will remain persistent in the soil. The persistence level of biochar carbon in the soil depends upon the characteristics of the feedstock, which in turn depends upon the pyrolysis time and temperature, interaction of the biochar with the climate condition (precipitation and temperature) and soil environment. A shorter pyrolysis time and high temperature leads to recalcitrant biochar carbon (which remains persistence in soil for longer time). But this is a kind of trade-off between the quality and quantity of the biochar, as this leads to less production of biochar compared to per unit feedstock. Soil texture is another criteria controlling the persistence level of the biochar carbon in the soil. Clay particles have more surface area, so there is more interaction between clay particle and the biochar carbon. Therefore, it is more effective in stabilization of the biochar carbon (Cooperman 2016).

6.4.2 Priming effect

The increase in the rate of decomposition of organic matter with the addition of biochar or any other substance is known as the priming effect. This so-called priming effect complicates all the efforts to sequester carbon, as this leads to increase in microbial activity that results in faster rate of microbes decomposition compared to carbon input rates (Cooperman 2016).

6.5 Mitigation of climate change

Over the last few decades, the concentration of CO_2 and other GHGs has increased at a rapid rate, mainly because of industrialization and unsustainable development. This increase in CO_2 and GHGs generally leads to global warming, which needs to be controlled. So, it has become imperative to mitigate GHG emission by eco-friendly methods and biochar can be one of them. Biochar application to soils may increase carbon sequestration and it leads to increase in recalcitrant carbon pool. But the effect of biochar application on GHG flux is variable among many studies. So, there is uncertainty in carbon sequestration by using biochar in the fields. He et al. (2017) did the meta-analysis of 91 published research papers and did comparison to get the changes in GHG fluxes (i.e., CO_2, CH_4 and N_2O) in response to application of biochar. They showed that biochar application significantly increased CO_2 fluxes by 22.14%, decreased N_2O flux by 30.92% and did not affect the CH_4 fluxes. Therefore, according to them biochar application may significantly contribute to increase in global warming potential of total soil GHG fluxes due to high stimulation of CO_2 fluxes. However, the CO_2 flux from the soil was suppressed when biochar was added to fertilized soils, indicating that the CO_2 emission will be suppressed from the agricultural soils where fertilizers (particularly N) are added to the soils. The soil GHG flux varied with source of feedstock, soil texture and the pyrolysis temperature of biochar. To a limiting extent, soil and biochar pH, biochar application rate and latitude also influence the soil GHG flux. The biochar affects the soil microbial structure and function, which alters C and N cycling and subsequently changes the soil respiration and N_2O flux. Production of the biochar can reduce the GHG emission by 12 per cent (Woolf et al. 2010). Use of organic matter biochar in the soil can help in the reduction of CO_2 (Krishna et al. 2013). It is one of the best methods to sequester the CO_2 from the atmosphere. Biochar does not directly sequester the C from the atmosphere, but converts carbon into stable form, hence decreasing the emission of CO_2 due to decomposition (Kataki et al. 2012). Ibrahim et al. (2017) reported that the emission of greenhouse gases, i.e., carbon (CO_2) and methane (CH_4) was significantly reduced but nitrous oxide (N_2O) emission first increased and then decreased when amended with biochar.

7. Conclusion

Pyrolysis is the main process for biochar production and this process alters nutrient content and availability of nutrient in crop residues (used as raw material) with thermal degradation of raw materials, mainly due to volatile losses of nutrients. Biochar enhances several physical, chemical and biological properties of soil when it is mixed with soil. Cation exchange capacity of biochar is high, which helps in retaining nutrients on their surfaces and ultimately helps in increasing productivity of crops. Soil improvement in acidic soils (through increasing basicity, fertility), heavy metal remediation and carbon sequestration are alarming topics for encouraging research in these field with the use of biochar. Biochar can be used as an input in the field of conservation agriculture. However, the prime requirement is to optimize dose according to soil type and organic carbon status of the soil. In spite of its inert activity, it can act as a suitable amendment for microbial activates in the soil. The biochar can also be used as a suitable adsorbent for decreasing the pollution level in sewage water, big kilns and at contaminated sites. In a nutshell, potential use of biochar alone with suitable practices can help to enhance carbon sequestration and reduction of pollutants, which ultimately leads to the enhancement of soil productivity in the long term.

References

Ahmad, M., Rajapaksha, A.U., Lim, J.E., Zhang, M., Bolan, N., Mohan, D., Vithanage, M., Lee, S.S. and Ok, Y.S. 2014. Biochar as a sorbent for contaminant management in soil and water: A review. Chemosphere 99: 19–33.

Al-Wabel, M.I., Al-Omran, A., El-Naggar, A.H., Nadeem, M. and Usman, A.R.A. 2013. Pyrolysis temperature induced changes in characteristics and chemical composition of biochar produced from conocarpus wastes. Bioresour. Technol. 131: 374–379.

Angin, D. 2013. Effect of pyrolysis temperature and heating rate on biochar obtained from pyrolysis of safflower seed press cake. Bioresour. Technol. 128: 593–597.

Apaydın-Varol, E. and Pütün, A.E. 2012. Preparation and characterization of pyrolytic chars from different biomass samples. J. Anal. Appl. Pyrolysis. 98: 29–36.

Barrow, C.J. 2012. Biochar: Potential for countering land degradation and for improving agriculture. Appl. Geogr. 34: 21–28.

Cao, X. and Harris, W. 2010. Properties of dairy-manure-derived biochar pertinent to its potential use in remediation. Bioresour. Technol. 101: 5222–5228.

Chen, B.L., Zhou, D.D. and Zhu, L.Z. 2008. Transitional adsorption and partition of nonpolar and polar aromatic contaminants by biochars of pine needles with different pyrolytic temperatures. Envir. Sci. Technol. 42: 5137–5143.

Chen, X., Chen, G., Chen, L., Chen, Y., Lehmann, J., McBride, M.B. and Hay, A.G. 2011. Adsorption of copper and zinc by biochars produced from pyrolysis of hardwood and corn straw in aqueous solution. Bioresour. Technol. 102: 8877–8884.

Chun, Y., Sheng, G., Chiou, C.T. and Xing, B. 2004. Compositions and sorptive properties of crop residue-derived chars. Environ. Sci. Technol. 38: 4649–4655.

Chutia, R.S., Kataki, R. and Bhaskar, T. 2014. Characterization of liquid and solid product from pyrolysis of *Pongamia glabra* deoiled cake. Bioresour. Technol. 165: 336–342.

Cooperman, Y. 2016. Biochar and carbon sequestration. Solution center for nutrient management. Solution center blog published on 19 September 2019. Perspectives on nutrient management in California agriculture.

Demirbas, A. 2004. Effects of temperature and particle size on bio-char yield from pyrolysis of agricultural residues. J. Anal. Appl. Pyrolysis. 72: 243–248.

DeSisto, W.J., Hill, N., Beis, S.H., Mukkamala, S., Joseph, J., Baker, C., Ong, T.H., Stemmler, E.A., Wheeler, M.C., Frederick, B.G. and Heininen, A.V. 2010. Fast pyrolysis of pine sawdust in a fluidized-bed reactor. Energy & Fuels 24: 2642–2651.

Glaser, B., Lehmann, J. and Zech, W. 2002. Ameliorating physical and chemical properties of highly weathered soils in the tropics with charcoal—a review. Biol. Fert. Soils 35: 219–230.

He, Y., Zhou, X., Jiang, L., Li, M., Du, Z., Zhou, G., Shao, J., Wang, X., Xu, Z., Bai, S.H.5, Wallace Helen and Xu, C. 2017. Effects of biochar application on soil greenhouse gas fluxes: a meta-analysis. Bioenergy Res. 9: 743–755.

Inyang, M., Gao, B. and Yao, Y. 2012. Removal of heavy metals from aqueous solution by biochars derived from anaerobically digested biomass. Bioresour. Technol. 110: 50–56.

Ibrahim, M., Li, G., Khan, S., Chi, Q. and Xu, Y. 2017. Biochars mitigate greenhouse gas emissions and bioaccumulation of potentially toxic elements and arsenic speciation in *Phaseolus vulgaris* L. Environ. Sci. Pollut. Res. 24: 19524–19534.

Jones, D.L., Rousk, J., Edwards-Jones, G., DeLuca, T.H. and Murphy, D.V. 2012. Biochar-mediated changes in soil quality and plant growth in a three year field trial. Soil Biol. Biochem. 45: 113–124.

Jindarom, C., Meeyoo, V., Kitiyanan, B., Rirksomboon, T., Rangsunvigit, P. 2007. Surface characterization and dye adsorptive capacities of char obtained from pyrolysis/gasification of sewage sludge. Chem. Eng. 133: 239–246.

Kataki, R., Das, M., Chutia, R.S., Borah, M.s. Biochar for C-sequestration and soil amelioration. pp. 131–137. In: Shankar, G., Das, B. and Blange, R. (eds.). Renewable Energy Technology: Issues and Prospects. Excell India Publishers, New Delhi.

Katyal, S., Thambimuthu, K. and Valix, M. 2003. Carbonisation of bagasse in a fixed bed reactor: influence of process variables on char yield and characteristics. Renewable Energy 28: 713–725.

Keech, O., Carcaillet, C. and Nilsson, M.C. 2005. Adsorption of allelopathic compounds by wood-derived charcoal: The role of wood porosity. Plant and Soil. 272: 291–300.

Krishna kumar, S., Kumar, S.R., Mariappan, N. and Surendar, K.K. 2013. Biochar-boon to soil health and crop production. Agricultural Res. 8: 4726–4739.

Lee, J.W., Kidder, M., Evans, B.R., Paik, S., Buchanan, A.C., Garten, C.T. and Brown, R.C. 2010. Characterisation of biochar produced from cornstrovers for soil amendment. Environ. Sci. Technol. 44: 7970–7974.

Lee, Y., Jung, J., Park, J., Hyun, S., Ryu, C. and Gang, K.S. 2013. Comparison of biochar properties from biomass residues produced by slow pyrolysis at 500°C. Bioresour. Technol. 148: 196–201.

Lehmann, J. 2007. Bio-energy in the black. Front. Ecol. Soc. 5: 381–387.

Lehmann, J. 2009. Biological carbon sequestration must and can be a win approaches. Clim. Change 97: 459–463.

Liang, B., Lehmann, J., Solomon, D., Kinyangi, J., Grossman, J., O'Neill, B., Skjemstad, J.O., Thies, J., Luizao, F.J., Petersen, J. and Neves, E.G. 2006. Black carbon increases cation exchange capacity in soils. Soil Sci. Am. J. 70: 1719–1730.

Lua, A.C., Yang, T. and Guo, J. 2004. Effects of pyrolysis conditions on the properties of activated carbons prepared from pistachio-nut shells. J. Anal. Appl. Pyrolysis 72: 279–287.

Muradov, N., Fidalgo, B., Gujar, A.C., Garceau, N. and T-Raissi, A. 2012. Production and characterization of *Lemna minor* bio-char and its catalytic application for biogas reforming. Biomass Bio. 42: 123–131.

Novak, J.M., Lima, I., Xing, B., Gaskin, J.W., Steiner, C., Das, K.C., Ahmedna, M., Rehrah, D., Watts, D.W., Busscher, W.J. and Schomberg, H. 2009. Characterization of designer biochar produced at different temperatures and their effects on a loamy sand. Ann. Environ. Sci. 3: 195–206.

Parmila, D. and Saroha, A.K. 2014. Risk analysis of pyrolyzed biochar made from paper mill effluent treatment plant sludge for bioavailability and eco-toxicity of heavy metals. Bioresour. Technol. 162: 308–314.

Saranya, K., Krishnan, P.S., Kumutha, K. and French, J. 2011. Potential for biochar as an alternate carrier to lignite for the preparation of biofertilizers in India. Agric. Environ. Biotechnol. 4: 167–177.

Shenbagavalli, S. and Mahimairaja, S. 2012. Production and characterization of biochar from different biological wastes. Plant, Anim. Environ. Sci. 2: 197–201.

Volli, V. and Singh, R.K. 2012. Production of bio-oils from de-oiled cakes by thermal pyrolysis. Fuel 96: 579–585.

Woolf, D., James, E., Amonette, F., Street-Perrott, A., Lehmann, J. and Stephen, J. 2010. Sustainable biochar to mitigate global climate change. Nat. Commun. 1: 1–9.

Xiao, X., Chen, B. and Lizhong, Z. 2014. Transformation, morphology and dissolution of silicon and carbon in rice straw derived biochars under different pyrolytic temperatures. Environ. Sci. Technol. 48: 3411–3419.

Yang, H.P., Yan, R., Chin, T., Liang, D.T., Chen, H.P. and Zheng, C.G. 2004. Thermogravimetric analysis-Fourier transform infrared analysis of palm oil wastes pyrolysis. Energy Fuels 18: 1814–1821.

Yao, Y., Gao, B., Inyang, M., Zimmerman, A.R., Cao, X., Pullammanappallil, P. and Yang, L. 2011. Biochar derived from anaerobically digested sugar beet tailings: characterization and phosphate removal potential. Bioresour. Technol. 102: 6273–6278.

Yuan, J., Xu, R. and Zhang, H. 2011. The forms of alkalis in the biochar produced from crop residues at different temperatures. Bioresour. Technol. 102: 3488–3497.

22

Endophytic Microorganisms from Synanthropic Plants

A New Promising Tool for Bioremediation

*Olga Marchut-Mikolajczyk** and *Piotr Drozdzynski*

1. Introduction

Fast industrialization and urbanization are inseparable from the increasing pollution of the environment with organic compounds. Deterioration of the natural environment is a global problem that requires the development of new ecological and sustainable methods for removing pollution. Bioremediation technology, which uses the metabolic potential of microorganisms, is one of the most effective, economically viable and socially accepted method of cleaning up the environment. However, the progressive degradation of the natural environment indicates the deficient potential of microorganisms to effectively decompose hardly degradable pollutants and their mixtures. Regarding the search for new sources of microorganisms, showing high activity towards such contaminations as polychlorinated biphenyls, polyaromatic hydrocarbons, petroleum compounds, aromatic dyes, pesticides, etc., the interest of researchers in endophytic microorganisms has increased in recent years. It seems that synanthropic (ruderal) plants may be a particularly promising source of endophytes with high degradation activity. During the long course of the plant-endophyte relationship in a contaminated site, endophytes not only help in the mobilization of contaminants but also help in promoting plant growth, as well as in developing tolerance to various biotic and abiotic stress. Thus, these microorganisms can be a highly effective tool for modern bioremediation techniques (Gupta et al. 2020). In this chapter, the potential of endophytic microorganisms, mainly bacteria, in removing hydrocarbons from the environment is discussed. The potential of synanthropic plants as a source of microorganisms with high degradation activity and biosurfactant production will be reviewed.

2. Bioremediation—the right way for contamination removal

The state of pollution of the environment with organic compounds inclines to seek solutions that limit drastic and long-lasting changes in ecosystems, at the same time allowing for their effective

---

Institute of Molecular and Industrial Biotechnology, Faculty of Biotechnology and Food Science, Lodz University of Technology, Stefanowskiego 4/10, 90-924 Lodz, Poland.
Email: piotr.drozdzynski@dokt.p.lodz.pl
* Corresponding author: olga.marchut-mikolajczyk@p.lodz.pl

utilization. One of the main goals of environmental biotechnology is to develop highly effective biological processes using naturally occurring ones' catabolic potential of microorganisms to eliminate and detoxify pollutants. Bioremediation is one of the most effective strategies for cleaning up the environment.

For the first time, the concept of bioremediation appeared in the scientific literature in 1987. It describes a method of biological removal of various types of contamination chemical substances from the natural environment, mainly with the participation of microorganisms, to a level safe for the life of organisms. This technology is based on the use of microorganisms living in a polluted environment (natural bioremediation, biostimulation) and/or specially selected for microorganisms for the given contamination (bioaugmentation), which enzymes have the ability of transforming toxic organic compounds into simpler substances, harmless to the environment. Bioremediation techniques can be used for the removal of pollution with natural substances as well as those from the anthropogenic origin, including xenobiotics. Although bioremediation technology has many advantages, the main disadvantage is related to the long time of purification of the contaminated environment. To overcome this phenomenon, scientists are constantly seeking for new microorganisms, with extraordinary degradation activities. Over the last ten years, the attention of sourcing bioremediation agents has been directed to microorganisms living inside plant tissues, called endophytes.

3. Endophytic microorganisms

Until twentieth-century, scientists believed that plants are free of microorganisms (Waghunde et al. 2017). Galippe (1887) was the first researcher who reported for the first time that microorganisms can leave inside vegetable plants, e.g., carrot, onion, potato, cabbage, radish, etc. (Waghunde et al. 2017). Nowadays, it is known that in natural conditions, plants growth is highly affected by the presence of different microorganisms. These microbes are formed both by environmental factors and plant metabolic activity itself. The most favorable area for microbes to live is near the roots of plants which is rich in roots' secretion that can be utilized as a carbon, nitrogen and energy source. The area affected by root secretions is called the rhizosphere (Pisarska and Pietr 2014). Microorganisms inhabiting that zone are usually neutral for plants. However, some of them may exhibit non-antagonistic, favorable interactions with plants which may result in plant growth and promotion. Such interactions can be symbiotic, and even more. A special kind of symbiosis observed among microorganisms and plants is endosymbiosis. It occurs when cells of one organism (microorganism) are living inside other (plant). It is characterized by the fact that microorganisms do not show any harmful or disease symptoms for the plant host (Malfanova et al. 2013). Endophytic microorganisms are bacteria and fungi that are living in different plant tissues for a part, or all of their life. The most important feature of endophytes is causing no harm to the host. Usually, endophytes affect the host beneficially by inducing systemic resistance against phytopathogens and/or animals, promoting the plant's growth or increasing resistances for environmental stresses. In return, the plant host provides nutrients and safe residency for bacteria (Weyens et al. 2010, Afzal et al. 2014, Zhang et al. 2014).

There are a few types of endophytic microorganisms: obligate, competent, passenger and facultative. Microbial colonization depends on many different factors like the growth stage and plant genotype, the type of plant tissues and soil environmental conditions (Singh et al. 2009).

A large number of studies have shown that each plant can be the host to one or more kinds of endophytes. Therefore, single plant can be a host for several different species of endophytes, which can be isolated from different plant tissues, like roots, stems, leaves or even fruits (Bulgari et al. 2012, Stępniewska and Kuźniar 2013, Faegheh and Harighi 2018, Ekta et al. 2018, Maggini et al. 2019).

The relations between plants and bacteria are not the only since that kind of relations occurs between plants and fungi as well. They are described as a smooth change from mutualism through commensalism even to parasitism (Schulz and Boyle 2006, Aly et al. 2010).

This kind of symbiosis is documented even more than symbiosis with endophytic bacteria and occurs rather in difficult environments, where plants are exposed to lack of water, extreme temperature and pH (Rodriguez et al. 2008, Redman et al. 2011).

Presence of endophytic fungi inside tissues of plants usually has a positive effect, thus they have an ability to reduce water stress of the host plant, take a part into plant nutrition, and reduce some environmental stresses (Redman et al. 2002, Harris-Valle et al. 2009, Rodriguez et al. 2009, Vos et al. 2014, Johnson et al. 2012, Aschehoug et al. 2014). In return for so many positive effects, the plant supplies the symbiotic fungus with secretions that are a source of carbon and energy. Thanks to their presence, plants are better protected against some pathogens (Aschehoug et al. 2014). However, the reports on the use of endophytic fungi in bioremediation and biodegradation are rather scarce.

4. Synanthropic plants

Along with the development of the economy, the number of ecosystems characterized by the high expansion of anthropogenic disorders increases. One of the side effects of human activity is synantrophization of flora in natural ecosystems. This type of vegetation can be divided into two broad types: weed and ruderal (Lososova et al. 2006, Silc 2010). Weed plants can be found on arable land, while the ruderal vegetation is found in waste deposits, along transportation routes, settlements and semi-natural landscape (Silc 2010).

The term ruderal comes from Latin word *rudus* and defines disturbance-adapted species (Lachmud 2003). Thanks to the human activity, ruderal environments became nutrient-rich and thus are characterized by the presence of plants with high productivity and a variety of competitive relationships associated with continuous or occasional interference and rapid successive changes (Dietz et al. 1998). Ruderal plants encompass species from environments that are directly or indirectly influenced by human and may grow in artificial habitats, which are disturbed such as surroundings of houses, gardens, roadsides, railways sides, etc. (Johnson and Klemens 2005, Szafer 2013). The communities of synanthropic plants are composed of not only native species but also alien ones. Native plants may persist in their old habitat, which has been damaged by humans (*Pteridium aquilanum*, a cereal which grows on fields, which were previously covered by forests). Another mechanism involves settling after arriving from a near to far neighborhood, like *Urtica dioica* (originally a forest plant, which is now growing in ruderal environments). Due to the environment in which these plants grow, highly affected by human activity, very often they are exposed to different xenobiotics like herbicides, pesticides, hydrocarbons, and heavy metals. Nevertheless, they are capable of surviving in such an inconvenient environment, thanks to microorganisms that are inhabiting the interior of their tissues (Figure 1).

4.1 Endophytes from ruderal plants against environment pollution

Ruderal plants, growing in a difficult and/or extreme environment, exhibit strong adaptation skills, related to the expression of genes responsible for defense mechanisms (Andreolli et al. 2013, Anyasi and Atagana 2019). Therefore, these plants may serve not only as a source of unique biologically active compounds but also new microorganisms with great metabolic potential (Afzal et al. 2014). This approach can be applied in the remediation of a polluted environment. Recently, several studies have shown that endophytic microorganisms may demonstrate relatively high degradation activity towards organic contaminations (Table 1) (Stepniewska and Kuzniar 2013, Sun et al. 2014, Syranidou et al. 2018).

Research performed by Moore et al. (2006) indicates that endophytic bacteria like *Pseudomonas* sp., *Bacillus* sp., *Arthrobacter* sp. isolated from root, stem and leaves of poplar tree, which was growing on a site contaminated with BTEX compounds, have metabolic conditions to degrade this kind of pollution (Moore et al. 2006, Zhang et al. 2014).

Soleimani et al. (2010) reported that planting into a soil contaminated with hydrocarbons of two species of grass—*Festuca arundinacea Schreb.* and *Festuca pratensis Huds.* inoculated

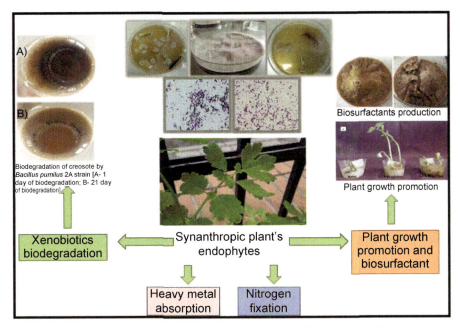

Figure 1. Environmental application of endophytes isolated from synantrophic plants.

with endophytic fungi *Neotyphodium coenophialum* and *Neotyphodium uncinatum*, respectively, caused increased biomass and achieved higher levels of dehydrogenases activity in the soil than uninoculated grass samples. They also reported that polyaromatic and total hydrocarbons reduction in soil was higher than in a control sample by about 30–40% (Soleimani et al. 2010).

An interesting study was performed by Germaine et al. (2006). Authors inoculated plants with endophytic bacteria, *Pseudomonas* sp., and used it for herbicides degradation. Noteworthy, they did not observe the herbicide accumulation in the plant tissue in an inoculated plant, unlike for the plant without endophytes (Germaine et al. 2006, Anyasi and Atagana 2019). Although there is some research on degradation ability of endophytic fungi towards different xenobiotics, the use of endophytic bacteria as degradation agents has not been demonstrated for *in situ* remediation (Anyasi and Atagana 2019).

4.2 Urtica dioica L.

Urtica dioica L. (stinging nettle) is a common plant growing in the countryside in northern Europe and the majority of Asia. The positive effect of *Urtica dioica* has been known for a long time worldwide. It was used for healing different diseases and disorders like eczema, rheumatism, hemorrhoids, hyperthyroidism, bronchitis, and cancer; thus, it is worth mentioning that it causes no adverse effect (Kavalali et al. 2003, Gaballu et al. 2015). Other authors claimed that a stinging nettle is a good tool for phytoremediation due to its hyperaccumulating ability (Hartley 2004, Viktorova et al. 2016).

Due to that fact, it was obvious that internal tissues of stinging nettle inhabit endophytes with special abilities.

Naoufal et al. (2018) isolated 54 endophytic bacterial strains from *Urtica dioica* L. from which authors chose gram-positive Bacilli characterized by different morphologies and abilities for further research, whose goal was to find the best phytopathogen inhibitors. Authors performed biochemical tests and on the basis of obtained results, they chose three isolates that exhibited the largest antagonistic activity against typical phytopathogens (*Fusarium oxysporum*, *Colletotrichum gloeosporioides*, *Rhizoctonia solani*, *Phytophthora parasitica*). The average size of phytopathogens' growth inhibition was 73.21% against the control sample which was uninoculated with endophytic

Table 1. Endophytic microorganisms with the ability for xenobiotics degradation (Feng et al. 2017).

Organic polluntants	Endophyte species	Host plant	References
PAHs (anthracene, naphtalene, benzene, toluene, xylene)	*Bacillus* sp. SBER3	*Populus deltoides*	Bisht et al. 2014
PAHs, petroleum hydrocarbons	*Pseudomonas* spp.	*Halimione portulacoides, Sarcocornia perennis*	Oliveira et al. 2014
	Neotyphodium coenophialum *Neotyphodium uncinatum*		
	Neotyphodium coenophialum	Tall fescue (*Festuca arundinaacea* Schreb.),	
	Neotyphodium uncinatum	Meadow fescue (*Festuca pratensis* Huds)	
Petroleum hydrocarbon	*Pseudomonas* spp., *Microbacterium* sp., *Rhodococcus* sp. *Bacillus pumilus* 2A *Bacillus* sp., *Pseudomonas* sp.	Ryegrass (*Lolium perenne* L.) *Chelidonium majus* L. *Azadirachta indica*	Kukla et al. 2014 Marchut-Mikolajczyk et al. 2018 Singh and Padmavathy 2015
Pyrene	*Staphylococcus* sp. BJ106	*Alopecurus aeqialis*	Sun et al. 2014
Phenanthrene	*Pseudomonas putida* PD1, *Pseudomonas* sp. Ph6-gfp, *Massilia* sp. Pn2, *Paenibacillus* sp. PHE-3	Poplar Ryegrass (*Lolium multiflorum* L.) *Alopecurus aequalis* Sobol, *Plantago asiatica* L., Ryegrass (*Lolium multiflorum* L.) Birdsfoot trefoil (*Lotus corniculatus*)	Khan et al. 2014, Sun et al. 2015a, Liu et al. 2014, Zhu et al. 2016
Mixture of aliphatic and aromatic hydrocarbons (diesel)	*Enterobacter ludwigii, Pseudomonas* sp., *Flavobacterium* sp., *Pantoea* sp.	Alfalfa (*Medicago sativa*)	Yousaf et al. 2011, Mitter et al. 2019
Crude oil	*Pseudomonas* sp. J4AJ, *Acinetobacter* sp. BRSI56	*Scripus triqueter* *Brachiaria mutica*	Zhang et al. 2014, Fatima et al. 2015
BTEX, trichloroethylene (TCE)	*Burkholderia cepacia* VM1468	*Populus trichocarpa*	Taghavi et al. 2005
Phenolic pollutants	*Achromobacter xylosoxidans* F3B	*Phragmites australis, Ipomoea aquatica*	Ho et al. 2013
TCE	*Pseudomonas putida* W-619-TCE	Poplar tree (*Populus deltoids*)	Weyens et al. 2015
2,4,6-Trinitrotoluene (TNT)	*Methylobacterium thiocyanatum* ES2, *Sphingomonas panni* ES4, *Pseduomonas* spp. ER9, *Stenotrophomonas* EL1, *Variovorax* ER18	Bent grass (*Agrositis capillaris*)	Thijs et al. 2014

bacteria. What is more, according to the authors, these endophytic Bacilli species may be valuable sources of different secondary metabolites and enzymes (Naoufal et al. 2018).

Victorova et al. (2016) assessed the impact of endophytic bacteria on stinging nettle's health and tolerance to toxic compounds during phytoremediation of soil contaminated with chlorinated biphenyls and heavy metals. Bacteria that were isolated by the authors mainly belong to the *Bacillus* and *Arthrobacter* genus. Authors noted that isolated endophytic *B. shackletonii* and *Streptomyces badius* represented all the beneficial abilities of endophytes-produced phytohormones, have ACC-deaminase and nitrogenase activity, perform solubilization of phosphate, and greatly increase the resistance of the host to heavy metals present in the environment. That is why authors claim that these microorganisms may constitute an effective tool for phytoremediation enhancement (Victorova et al. 2016).

4.3 *Lolium perenne* L.

Lolium perenne (perennial ryegrass) is a grass used in agriculture as a pasture all over the world, and because of an increasing demand of the public for meat and milk production, its role is significant. This grass usually grows in a moderate climate, so it is frequent to form an endosymbiosis with fungi like *Neotyphodium* and *Epichloe,* which provide an increased environmental stress tolerance and protection against pests (Saikkonen et al. 2006, Clay and Schardl 2002, Young et al. 2013). However, the reports about the usage of ryegrass and its endophytic fungi are scarce. Nevertheless, endophytic bacteria from that plant host are used more frequently. Jabeem et al. (2016) used *Mezorhizobium* sp. HN3 bacterial strain isolated from ryegrass for chlorpyrifos degradation (Jabeem et al. 2015).

4.4 *Chelidonium majus* L.

Chelidonium majus L. is a ruderal plant that naturally occurs in Europe, Asia, and South America. Although the plant is toxic, it produces many bioactive substances, e.g., alkaloids with high antimicrobial, antiviral and even anticancer properties (Goryluk et al. 2016, Marchut-Mikolajczyk et al. 2018). Therefore, it was essential to isolate endophytes from the plant and to examine the possibility of sharing the ability to produce the same biological active compounds. Goryluk et al. (2009) isolated 34 endophytic bacteria from internal stem tissues of *Chelidonium majus* L. and investigated their antifungal properties against six fungal species *Alternaria alternata*, *Paecilo-myces variotti*, *Aureobasidium pullulans*, *Byssochla-mysfulva*, *Chaetomium* sp. and *Exophiala mesophila*. The authors reported that eleven isolates exhibited the inhibition of growth of all fungi except *B. fulva*. Only one bacterial strain exhibited antifungal properties against all the tested fungi. The strain was classified on the basis of the API-20E, -50CHB tests and the analysis of 16S rDNA sequence as *B. amyloliquefaciens* (Goryluk et al. 2009). Studies on endophytic microorganisms from *Chelidonium majus* L. were extended by Marchut-Mikolajczyk et al. (2018). The authors isolated 11 endophytic bacteria from the tissues of *Chelidonium majus* L. plant growing in motorway neighborhood. They examined the ability of isolated bacteria for hydrocarbons degradation (diesel oil, waste engine oil) and biosurfactant production. All tested strains showed degradation activity. However, the strain marked as 2A had the highest degradation activity against both diesel and waste engine oil. The strain also exhibited the highest biosurfactant production. Based on the analysis of 16S rDNA sequence, the strain was classified as *Bacillus pumilus* 2A. Authors claim that produced biosurfactants not only enhance hydrophobic compounds degradation but also may act as plant-growth-promoting agents in a contaminated environment (Marchut-Mikolajczyk et al. 2018). Although the presented data is new and interesting, the application of the strain and obtained biosurfactant will be possible after acquired knowledge of the mechanism of an observed phenomenon, which will be important in evaluating the possibility of application of biosurfactants to promote plant growth, especially in contaminated areas.

5. Endophytes in phytoremediation

Due to the increasing contamination of the environment with xenobiotics (hydrocarbons, nitrogen compounds, pesticides, herbicides), researchers are more likely to use phytoremediation technology. Unfortunately, it has many disadvantages, e.g., very long cleaning time. To overcome these drawbacks, a promising option is to inoculate plants with suitable endophytes capable of degrading xenobiotics like *Burkholderia*, *Pseudomonas, Herbaspirillum,* and *Methanobacterium* (Mannisto et al. 2001, Barac et al. 2004, Van Aken et al. 2004). Many other endophytic microorganisms have ability to degrade pollutants that are present in the environment where these plants grow. The most known and described endophytes were isolated from poplar tissues, wheat and yellow lupine (Mannisto et al. 2001, Barac et al. 2004, Van Aken et al. 2004, Germaine et al. 2004).

Some of the endophytes can greatly increase resistance for the presence of heavy metals and thus may exhibit hyperaccumulating properties. These bacteria belong to the genus *Bacillus, Pseudomonas,* and *Achromobacter*. It was reported that *Bacillus thuringensis* GDB-1 strain significantly enhanced plant growth and improved metal absorption by *Pinus sylvestris* (Babu et al. 2013). Another researcher proved that *Microbacterium* and *Arthrobacter* inhabiting interior tissues of *Noccaea caerulenscens,* despite strong plant growth promoting abilities, also notably enhanced phytoextraction, translocation, and removal of heavy metals like Fe, Co, Ni and Cu from soil (Visioli et al. 2015).

Expanded enzymatic apparatus of endophytic microorganisms, related to the colonization of plant tissues, resulted in rising interest in using those enzymes as potential biocatalysts in chemical transformation of not only natural compounds but also drugs. The transformation of benzoxazinones, 2-benzoxazolinone (BOA) and 2-hydroxy-1,4-benzoxazin-3-one (HBOA) into different compounds was described for endophytic fungus isolated from *Aphelandra tetragona* roots and shoots (Gunjal et al. 2018). Also, endophytic fungi *Phomopsis* spp. from *Viguiera arenaria* were reported to catalyze transformation of a tetrahydrofuran lignan, (−)-grandisin to a new compound-3,4-dimethyl-2-(4-hydroxy-3,5-dimethoxyphenyl)-5-methoxy-tetrahydrofuran. What is interesting, this compound exhibits trypanocidal activity against *Trypanosoma cruzi,* which causes the Chagas disease. However, the studies on using endophytes for biotransformation processes are still scarce (Gunjal et al. 2019).

Despite remarkable abilities for promoting plant growth and increasing plants' resistance for environmental stresses related to the presence of pollution in the soil, endophytes inside plants' tissues may be not sufficient as a sole tool for cleaning the environment, especially in the presence of high concentrations of contaminants (Mitter et al. 2019).

6. Endophytes as a source of biosurfactants

Hydrophobic compounds present in the soil very often are bounded with soil matrix, thus their availability to microorganisms that are performing biodegradation process is limited. To increase the bioavailability of hydrophobic compounds, surface active agents may be applied. However, the chemical surfactants exhibit many hazardous features like toxicity and low biological compatibility and low biodegradability. However, it is a great alternative, namely, biological surface-active agents (biosurfactants) that demonstrate all advantages of synthetic surfactants and also do not have their disadvantages (Marchut-Mikolajczyk et al. 2018, Pinto et al. 2018, Sunkar et al. 2019).

Biosurfactants are surface active agents produced by microorganisms and they are very valuable during the biodegradation process; thus, they increase the solubility of hydrophobic compounds and heavy metals in soil matrix, which increase the bioavailability and mobility significantly (Marchut-Mikolajczyk et al. 2018, Pinto et al. 2018, Sunkar et al. 2019). Innovative sources of biosurfactants are endophytic bacteria.

Sunkar et al. (2019) isolated biosurfactant from endophytic *Bacillus cereus* strain, which is described as a glycolipid that contains rhamnose, fructose and glucose, and 9-hexadecanoic acid and 1-eicosanol. This compound has a good emulsifying activity, which predisposes its application for biodegradation enhancement. Also, this glycolipid has antibacterial activity against *Staphylococcus* sp. and *Serratia* sp., which is worth mentioning; nevertheless, it is not very effective against pathogenic bacteria (Sunkar et al. 2019).

7. Endophytes as plant growth-promoting agents

One of the most important abilities of endophytes is plant growth promotion. This can be achieved by such mechanisms as nitrogen fixation, production of antifungal compounds and phytohormones, regulating ethylene production in roots or Induced Systemic Resistance (Gunjal et al. 2018).

One of the most important abilities of endophytes is plant growth promotion. It can be achieved by the production of phytohormones. The most frequent compound that endophytic bacteria synthesizes is indoleacetic acid (IAA), which belongs to auxins, which induce main root elongation and lateral root production (Reinhold-Hurek and Hurek 1998, Strobel 2003, Klama 2004). The same effect was observed by Cacciari et al. (1989) during the growth of pearl millet (*Pennisetum americanum*) inoculated by *Arthrobacter giacomelloi* and *Azospirillum brasilense,* which have the ability to produce other phytohormones, namely, gibberellins and cytokinins (Cacciari et al. 1989). Also, positive modification of black nightshade roots caused by the production of IAA was observed by Long et al. (2008) (Long et al. 2008). Also, endophytic microorganisms can produce biocontrol agents and therefore, indirectly promote plant growth. The growth of such a plant pathogen as *Cryphonectria parasitica* (chestnuts blight) can be inhibited by endophytic Bacillus subtilis, isolated from the xylem sap of chestnut trees, thanks to its antifungal properties (Arthofer and Schafleitner 1997, Gunjal et al. 2018).

For plants, nitrogen deficiency is most adversely affected. It causes poor tillering, yellowing of leaves, weakness or stiffness of tissues (Chodań et al. 1980). However, this state can be easily resolved by the presence of endophytic microorganisms which have nitrogenase activity, namely, diazotrophs. Thanks to nitrogenase activity, microorganisms may fix the atmospheric nitrogen, due to its reduction to NH_3, thus significantly increasing nitrogen uptake by plants. Endophytes that have nitrogenase activity belong to genus *Azospirillum, Acetobacter, Herbaspirillum, Azoarcus* or *Azotobacter* (Pisarska and Pietr 2014). Rodrigues et al. (2008) in their research inoculated the *Oryza sativa* L. (seed rice) plants with diazotrophic endophytic bacteria *Azospirillum amazonense*. They noticed that the presence of diazotrophic bacteria increases the uptake of nitrogen by the plant from 3.5 up to 18.5%; also, it increased the biomass of tested plants (Rodrigues et al. 2008). Worth mentioning is the fact that sugar cane grows in some areas of Brazil without any nitrogen fertilization. A great amount, namely 50–80% of this element, is taken from the atmosphere, and this is possible due to endosymbiosis with the diazotrophic bacterium *Gluconacetobacter diazotrophicus* (Błaszczyk 2010, Chawla et al. 2014). The nitrogen-fixation by endophytic bacteria has been reported by Elbeltagy et al. (2001) for Enterobacter cloacae, Alcaligenes, *A. diazotrophicus, Azospirillum, Azoarcus, Herbaspirillum seropedicae* and Ideonella dechloratans (Elbeltagy et al. 2001, Gunjal et al. 2018).

Iron is one of the most important elements for plant health due to its participation in cellular respiration, intermediary metabolism, oxygen transport, DNA stability and repair, and photosynthesis. Iron is an element commonly found in soil; however, because of the form in which it occurs, it remains inaccessible to plants (López-Milan et al. 2013, Li and Lan 2017). It is dangerous for plants because deficiency of iron may cause adverse changes in root formation, changes in enzyme activity, and chlorosis. Nevertheless, the bioavailability of iron for plants can be increased by endophytic bacteria that can produce siderophores. Siderophores are low molecular mass compounds that have a high chelating affinity to Fe^{3+}. In this form, iron's bioavailability is increased significantly and plants can easily absorb it. Also, siderophores may exhibit static properties against different phytopathogens (Lacava et al. 2008).

8. Conclusions

There are numerous reports on the possibility of using endophytic microorganisms as degradation agents. However, there is still a great number of plants to explore, looking for microorganisms with special features. Discovery of new endophytic species isolated from a plant growing in ruderal environments may give an opportunity to source microorganisms with high metabolic activity towards many contaminants and biosurfactants production. Using endophytic microorganisms as a tool for bioremediation can help us to prevent our planet from ecological calamity.

References

Afzal, M. Khan, Q.M. and Sessitsch, A. 2014. *Endophyticbacteria*: Prospects and applications for the phytoremediation of organic pollutants. Chemosphere 117: 232–242.

Aly, A.H. Debbab, A. Kjer, J. and Proksch, P. 2010. Fungal endophytes from higher plants: a prolific source of phytochemicals and other bioactive natural products. Fungal Divers 41: 1–16.

Andreolli, M., Lampis, S., Poli, M., Gullner, G., Biro, B. and Vallini, G. 2013. Endophytic *Burkholderia fungorum* DBT1 can improve phytoremediation efficiency of polycyclic aromatic hydrocarbons. Chemosphere 92: 688–694.

Andria, V., Reichenauer, T.G. and Sessitsch, A. 2009. Expression of alkane monooxygenase (alkB) genes by plant-associated bacteria in the rhizosphere and endosphere of Italian ryegrass (*Lolium multiflorum* L.) grown in diesel contaminated soil. Environ. Pollut. 157: 3347–3350.

Anyasi, R.O. and Atagana, H.I. 2019. Endophyte: understanding the microbes and its applications department of environmental sciences. Institute for Science and Technology Education Pakistan Journal of Biological Sciences. ISSN 1028-8880.

Arthofer, W.E. and Schafleitner, R. 1997. *Bacillus subtilis* an endophyte of chestnut (*Castanea sativa*), as antagonist against chestnut blight (*Cryphonectria parasitica*). pp. 331–337. *In*: Cassells, A.C. (ed.). Pathogen and Microbial Contamination Management in Micropropagation. Dordrecht, Netherlands: Academic Publishers. doi:10.1007/978-94-015-8951-2_40.

Aschehoug, E.T., Callaway, R.M., Newcombe, G., Tharayil, N. and Chen, S. 2014. Oecologia, 175: 285–291. doi:10.1007/s00442-014-2891-0.

Babu, A.G., Kim, J.D. and Oh, B.T. 2013. Enhancement of heavy metal phytoremediation by *Alnus firma* with endophytic *Bacillus thuringiensis* GDB-1. J. Hazard Mater. 250: 477–483.

Barac, T., Taghavi, S., Borremans, B., Provoost, A., Oeyen, L., Colpaert, J.V., Add: Kumar, V. and Bisht, N.S. 2004. Engineered endophytic bacteria improve phytoremediation of water-soluble, volatile, organic pollutants. Nat. Biotechnol. 22(5): 583–588.

Bisht, S., Pandey, P., Kaur, G., Aggarwal, H., Sood, A., Sharma, S., Kumar, V. and Bisht, N.S. 2014. Utilization of endophytic strain *Bacillus* sp. SBER3 for biodegradation of polyaromatic hydrocarbons (PAH) in soil model system. Eur. J. Soil Biol. 60: 67–76.

Błaszczyk, M.K. 2010. Mikrobiologia środowisk, Wyd. PWN, War-szawa 81–85.

Bulgari, D., Bozkurt, A.I. and Casati, P. 2012. Endophytic bacterial community living in roots of healthy and 'Candidatus Phytoplasma mali'-infected apple (Malus domestica, Borkh.) trees. A Van Leeuw 102: 677.

Chawla, N., Phour, M., Suneja, S., Sangwaan, S. and Goyal, S. 2014. *Gluconacetobacter diazotrophicus*: An overview. Res. Environ. Life Sci. (1): 1–10.

Chodań, J., Grzesiuk, W. and Mirowski, Z. 1980. Zarys gleboznawstwa i chemii rolnej. PWN Warszawa, in polish.

Ciacciari, I., Lippi, D., Pietrosanti, T. and Pietrosanti, W. 1989. Phytohormone-like substances produced by single and mixed diazotrophic cultures of *Azospirillum* and *Arthrobacter*. Plant and Soil 115: 151–153.

Clay, K. and Schardl, C. 2002. Evolutionary origins and ecological consequences of endophyte symbiosis with grasses. Am. Nat. 160: 99–127.

Dietz, H.. Steinlein, T. and Ullmann, I. 1998. The role of growth form and correlated traits in competitive ranking of six perennial ruderal plant species grown in unbalanced mixtures. Acta Oecologica 19: 25–36.

Ekta, K., Jitendra, M. and Kumar, A.N. 2018. Multifaceted interactions between endophytes and plant: developments and prospects. Front. Microbiol. 9: 2732.

Elbeltagy, A., Nishioka, K., Sato, T., Suzuki, H., Ye, B., Hamada, T., Isawa, T., Mitsui, H. and Minamisawa, K. 2001. Endophytic colonization and in planta nitrogen fixation by a *Herbaspirillum* sp. isolated from wild rice species. Appl. Environ. Microbiol. 67: 5285–93.

Faegheh, E. and Harighi, B. 2018. Isolation and identification of endophytic bacteria with plant growth promoting activity and biocontrol potential from wild pistachio trees. Plant Pathol. 34: 208–217.

Fatima, K., Afzal, M., Imran, A. and Khan, Q.M. 2015. Bacterial rhizosphere and endosphere populations associated with grasses and trees to be used for phytoremediation of crude oil contaminated soil. Bull. Environ. Contam. Toxicol. 94: 314–320.

Feng, N.-X. Yua, J., Zhao, H.-M., Cheng, Y.-T., Moa, C.-H., Cai, Q.-Y., Li, Y.-W., Li, H. and Wong, M.-H. 2017. Efficient phytoremediation of organic contaminants in soils using plant–endophyte partnerships. Sci. Total Environ. 583: 352–368.

Gaballu, F. A., Abedi Gaballu, Y., Moazenzade Khyavy, O., Mardomi, A., Ghahremanzadeh, K., Shokouhi, B. and Mamandy, H. 2015. Effects of a triplex mixture of *Peganum harmala*, *Rhus coriaria*, and *Urtica dioica* aqueous extracts on metabolic and histological parameters in diabetic rats. Pharm. Biol. 53(8): 1104–1109.

Galippe, V. 1887. Note sur la présence de micro-organismesdans les tissus végétaux (2nd note). C R Hebd. Sci. MemSoc. Biol. 39: 557–560.

Germaine, K.J., Liu, X., Cabellos, G.G., Hogan, J.P., Ryan, D. and Dowling, D.N. 2006. Bacterial endophyte-enhanced phytoremediation of the organochlorine herbicide 2,4-dichlorophenoxyacetic acid. FEMS Microbiol. Ecol. 57: 302–310.

Goryluk, A., Piórek, M., Rekosz-Burlaga, H., Studnicki, M. and Blaszczyk, M. 2016. Identification and bioactive properties of bacterial endophytes isolated from selected European herbal plants. Polish J. Microbiol. 65: 369–75. https://doi.org/10.5604/17331331.1215617.

Gunjal, A., Waghmode, M., Patil, N. and Kapadnis, B.P. 2018, Endophytes and their role in phytoremediation and biotransformation process, 10.4018/978-1-5225-3126-5.ch015.

Gupta, A., Singh, S.K., Singh, V.K., Singh, M.K., Modi, A., Zhimo, V.Y., Singh, A.V. and Kumar, a. 2020. Endophytic microbe approaches in bioremediation of organic pollutants. pp. 157–174. In: Ajay Kumar and Vipin Kumar Singh (eds.). Woodhead Publishing Series in Food Science, Technology and Nutrition, Microbial Endophytes, Woodhead Publishing. ISBN 9780128187340.

Harris-Valle, C., Esqueda, M., Valenzuela-Soto, E.M. and Castellanos, A.E. 2009. Water stress tolerance in plant-arbuscular mycorrhizal fungi interaction: Energy metabolism and physiology. Article Revista Fitotecnia Mexicana 32(4): 265–271.

Hartley, L. 2004. Characterization of a Heavy Metal Contaminated Soil in Ohiofora Phytoremediation Project: University of Toledo.

Ho, Y.N., Hsieh, J.L. and Huang, C.C. 2013. Construction of a plant-microbe phytoremediation system: combination of vetiver grass with a functional endophytic bacterium, Achromobacter xylosoxidans F3B, for aromatic pollutants removal. Bioresour. Technol. 145: 43–47.

Jabeem, H., Iqbal, S., Ahmad, F., Afzal, M. and Firouds, S. 2016. Enhanced remediation of chlorpyrifos by ryegrass and a chlorpyrifos degrading bactieral endophyte Mezorizhobium sp. HN3. Int. J. Phytoremediation 18(2): 126–133.

Johnson, E.A. and Klemens, M.W. 2005. Nature in fragments: the legacy of spraw. Columbia University Press. p. 212.

Johnson, D., Martin, F., Cairney, J.W.G. and Anderson, I.C. 2012. The importance of individuals: intraspecific diversity of mycorrhizal plants and fungi in ecosystems. New Phytol. 194: 614–628.

Kavalali, G., Tuncel, H., Göksel, S. and Hatemi, H.H. 2003. Hypoglycemic activity of *Urtica pilulifera* in streptozotocin-diabetic rats. J. Ethnopharmacol. 84(2-3): 241–245.

Khan, Z., Roman, D., Kintz, T., delas Alas, M., Yap, R. and Doty, S. 2014. Degradation, phytoprotection and phytoremediation of phenanthrene by endophyte Pseudomonas putida, PD1. Environ. Sci. Technol. 48: 12221–12228.

Klama, J. 2004. Coexistence of bacterial endophytes and plants (review). Acta Sci. Pol. Agric. 3(1).

Kukla, M., Płociniczak, T. and Piotrowska-Seget, Z. 2014. Diversity of endophytic bacteria in Lolium perenne and their potential to degrade petroleum hydrocarbons and promote plant growth. Chemosphere 117: 40–46.

Lacava, P.T., Silva-Stenico, M., Araújo, Welington, L., Simionato, A.V.C., Carrilho, E., Tsai, S.M. and Azevedo, J.L. 2008. Detection of siderophores in endophytic bacteria Methylobacterium spp. associated with Xylella fastidiosa subsp. pauca. Pesquisa Agropecuária Brasileira 43(4): 521–528.

Lachmund, J. 2003. Exploring the city of rubble: botanical fieldwork in bombed cities in Germany after World War II. Osiris 18: 234–254.

Li, W. and Lan, P. 2017. The Understanding of the Plant Iron Deficiency Responses in Strategy I Plants and the Role of Ethylene in This Process by Omic Approaches 8(40): 1–15.

Long, H.H., Schmidt, D.D. and Baldwin, I.T. 2008. Native bacterial endo-phytes promote host growth in a specoes-specyfic manner, phytohormone manipulations do not result in common growth responses. PLoS ONE 3: e2702.

López-Milan, A.F., Grusak, M.A., Abadia, A. and Abadia, J. 2013. Iron deficiency in plants: an insight from proteomic approaches. Front. Plant Sci. 4(254): 1–7. doi: 10.3389/fpls.2013.00254.

Losovova, Z., Chytry, M., Kuhn, I., Hajek, M., Horakova, V., Pyrek, P. and Tichy, L. 2006: Patterns of plant traits in annual vegetation of man-made habitats in central Europe. Perspect. Plant Ecol. Evol. Syst. 8: 69–81.

Maggini, V., Mengoni, A., Gallo, E.R., Biffi, S., Fani, R., Firenzouli, F. and Bogani, P. 2019. Tissue specificity and differential effects on *in vitro* plant growth of single bacterial endophytes isolated from the roots, leaves and rhizospheric soil of Echinacea purpurea. BMC Plant Biol. 19: 284.

Malfanova, N., Lugtenberg, B.J.J. and Berg, G. 2013. Bacterial endophytes: who and where, and what are they doing there? Molecular Microbial Ecology of the Rhizosphere 391–403.

Mannisto, M.K., Tiirola, M.A. and Puhakka, J.A. 2001. Degradation of 2,3,4,6-tetrachlorophenol at low temperature and low dioxygen concentrations by phylogenetically different groundwater and bioreactor bacteria. Biodegradation 12(5): 291–301.

Marchut-Mikolajczyk, O., Drożdżyński, P., Pietrzyk, D. and Antczak, T. 2018. Biosurfactant production and hydrocarbon degradation activity of endophytic bacteria isolated from Chelidonium majus L. Microb. Cell Fact. 17: 171.

Mitter, E.K., Kataoka, R., de Freitas, J.R. and Germida, J.J. 2019. Potential use of endophytic root bacteria and host plants to degrade hydrocarbons. Int. J. Phytoremediation 21(9): 928–938.

Moore, F.P., Barac, T., Borremans, B., Oeyen, L., Vangronsveld, J., van der Lelie, D., Campbell, C.D. and Moore, E.R.B. 2006. Endophytic bacterial diversity in poplar trees growing on BTEX-contaminated site: The characterization of isolates with potential to enhance phytoremediation. Syst. Appl. Microbiol. 29: 539–556.

Naoufal, D., Amine, H., Ilham, B. and Khadija, O. 2018. Isolation and biochemical characterisation of endophytic *Bacillus* spp. from *Urtica dioica* and study of their antagonistic effect against phytopathogens. Annu. Res. Rev. Biol. 28(4): 1–7. https://doi.org/10.9734/ARRB/2018/43642.

Oliveira, V., Gomes, N.C.M., Almeida, A., Silva, A.M.S., Simões, M.M.Q., Smalla, K. and Cunha, Â. 2014. Hydrocarbon contamination and plant species determine the phylogenetic and functional diversity of endophytic degrading bacteria. Mol. Ecol. 23: 1392–1404.

Pinto, A.P., de Varennes, A., Dias, C.M.B. and Lopes, M.E. 2018. Microbial-assisted phytoremediation: a convenient use of plant and microbes to clean up soils. pp. 21–87. In: Ansari, A., Gill, S., Gill, R., Lanza, G. and Newman, L. (eds.). Phytoremediation. Springer, Cham.

Pisarska, K., Pietr, S. 2014. Bakterie Endofityczne – Ich pochodzenie i interakcje z roślinami, Post. Mikrobiol. 53(2): 141–151 in polish.

Redman, R.S., Sheehan, K.B., Stout, R.G., Rodriguez, R.J. and Henson, J.M. 2002. Thermotolerance conferred to plant host and fungal endophyte during mutualistic symbiosis. Science 298: 1581.

Redman, R.S., Kim, Y.O., Woodward, C.J., Greer, C., Espino, L., Doty, S.L. and Rodriguez, R.J. 2011. Increased fitness of rice plants to abiotic stress via habitat adapted symbiosis: a strategy for mitigating impacts of climate change. PLoS ONE 6(7): e14823.

Reinhold-Hurek, B. and Hurek, T. 1998. Life in grasses: diazotrophic endophytes. Trends Micorbiol. 6: 139–144.

Rodrigues, E.P., Rodrigues, L.S., Martinez de Oliveira, A.L., Bal-dani, V.L.D., Dos Santos Teixeira, K.R., Urquiaga, S. and Reis, V.M. 2009. Azospirillum amazonense inoculation: effects on growth, yield and N2 fixation of rice (Oryza sativa L.). Plant Soil 302: 249–261.

Rodriguez, R.J., Freeman, D.C., McArthur, E.D., Kim, Y.O. and Redman, R.S. 2009. Symbiotic regulation of plant growth, development and reproduction. Commun. Integr. Biol. 2: 1–3.

Saikkonen, K., Lehtonen, P., Helander, M., Koricheva, J. and Faeth, S.H. 2006. Model systems in ecology: Dissecting the endophyte–grass literature. Trends Plant Sci. 11: 428–433.

Schulz, B.J.E. and Boyle, C.J.C. 2006. What are endophytes? pp. 1–13. In: Schulz, B.J.E., Boyle, C.J.C. and Sieber, T.N. (eds.). Microbial Root Endophytes. Springer-Verlag, Berlin.

Silc, U. 2010. Synanthropic vegetation: pattern of various disturbances on life history traits. Acta Bot. Croat. 69(2): 215–227.

Singh, G., Singh, N. and Marwaha, T.S. 2009. Crop genotype and a novel symbiotic fungus influences the root endophytic colonization potential of plant growth promoting rhizobacteria. Physiol. Mol. Biol. Plants 15: 87–92.

Singh, M.J. and Padmavathy, S. 2015. Hydrocarbon biodegradation by endophytic bacteria from neem leaves. LS: Int. J. Life Sci. 4: 33–36.

Soleimani, M., Afyuni, M., Hajabbasi, M.A., Nourbakhsh, F., Sabzalian, M.R. and Christensen, J.H. 2010. Phytoremediation of an aged petroleum contaminated soil using endophyte infected and non-infected grasses. Chemosphere 81(9): 1084–1090. https://doi.org/10.1016/j.chemosphere.2010.09.034.

Stepniewska, Z. and Kuzniar, A. 2013. Endophytic microorganisms-promising applications in bioremediation of greenhouse gases. Appl. Microbiol. Biotechnol. 97: 9589–9596.

Strobel, G.A. 2003. Endophytes as sources of bioactive products. Microbes Infect. 5: 535–544.

Sun, K., Liu, J., Jin, L. and Gao, Y. 2014. Utilizing pyrene-degrading endophytic bacteria to reduce the risk of plant pyrene contamination. Plant Soil 374: 251–262.

Sun, K., Liu, J., Gao, Y., Sheng, Y., Kang, F. and Waigi, M.G. 2015. Inoculating plants with the endophytic bacterium Pseudomonas sp. Ph6-gfp to reduce phenanthrene contamination. Environ. Sci. Pollut. Res. 22: 19529–19537.

Sunkar, S., Nachiyar, C.V., Sethia, S., Ghosh, B., Prakash, P. and Devi, R.K. 2019. Biosurfactant from endophytic Bacillus cereus: Optimization, characterization and cytotoxicity study. Malaysian J. Microbiol. 15(2): 120–131.

Syranidou, E., Thijs, S., Avramidou, M., Weyens, N., Venieri, D., Pintelon, I., Vangronsveld, J. and Kalogerakis, N. Responses of the endophytic bacterial communities of Juncus acutus to pollution with metals, emerging organic pollutants and to bioaugmentation with indigenous strains. Frontiers in Plant Science 18 October 2018.

Szafer, W. 2013. The Vegetation of Poland: International Series of Monographs in Pure and Applied Biology: Botany, Elsevier. 116–137.

Taghavi, S., Barac, T., Greenberg, B., Borremans, B., Vangronsveld, J. and van der Lelie, D. 2005. Horizontal gene transfer to endogenous endophytic bacteria from poplar improves phytoremediation of toluene. Appl. Environ. Microbiol. 71: 8500–8505.

Thijs, S., Van Dillewijn, P., Sillen, W., Truyens, S., Holtappels, M., D'Haen, J., Carleer, R., Weyens, N., Ameloot, M., Ramos, J.L. and Vangronsveld, J. 2014. Exploring the rhizospheric and endophytic bacterial communities of Acer pseudoplatanus growing on a TNT-contaminated soil: towards the development of a rhizocompetent TNT-detoxifying plant growth promoting consortium. Plant Soil 385: 15–36.

Van, A., Yoon, J.M. and Schnoor, J.L. 2004. Biodegradation of nitro-substituted explosives 2, 4, 6-trinitrotoluene, hexahydro-1, 3, 5-trinitro-1, 3, 5-triazine, and octahydro-1, 3, 5, 7-tetranitro-1, 3, 5-tetrazocine by a phytosymbiotic Methylobacterium spp. associated with poplar tissues (Populus deltoids x nigra DN34). Appl. Environ. Microbiol. 70(1): 508–517.

Viktorova, J., Jandova, Z., Madlenakova, M., Prouzova, P., Bartunek, V., Vrchotova, B., Lovecka, P., Musilova, L. and Macek, T. 2016. Native phytoremediation potential of Urtica dioica for removal of PCBs and heavy metals can be improved by genetic manipulations using constitutive CaMV35S promoter. PLoS ONE 11(12): e0167927. doi:10.1371/journal.pone.0167927.

Visioli, G., Vamerali, T., Mattarozzi, M., Dramis, L. and Sanangelantoni, A.M. 2015. Combined endophytic inoculants enhance nickel phytoextraction from serpentine soil in the hyperaccumulator Noccaea caerulescens. Front Plant Sci. 6: 638.

Vos, C.M., Yang, Y., De Coninck, B. and Cammue, B.P.A. 2014. Fungal (-like) biocontrol organisms in tomato disease control. Biol. Control 74: 65–81.

Waghunde, Rajesh, Shelake, Rahul, Shinde, Manisha and Hayashi, Hidenori. 2017. Endophyte microbes: a weapon for plant health management. 10.1007/978-981-10-6241-4_16.

Weyens, N., Croes, S., Dupae, J., Newman, L., van der Lelie, D., Carleer, R. and Vangronsveld, J. 2010. Endophytic bacteria improve phytoremediation of Ni and TCE co-contamination. Environ. Pollut. 158: 2422–2427.

Weyens, N., van der Lelie, D., Taghavi, S. and Vangronsveld, J. 2009. Phytoremediation: plant-endophyte partnerships take the challenge. Curr. Opin. Biotechnol. 20: 248–254.

Weyens, N., Beckers, B., Schellingen, K., Ceulemans, R., Van der Lelie, D., Newman, L., Taghavi, S., Carleer, R. and Vangronsveld, J. 2015. The potential of the Ni-resistant TCE-degrading Pseudomonas putida W619-TCE to reduce phytotoxicity and improve phytoremediation efficiency of poplar cuttings on a Ni-TCE co-contamination. Int. J. Phytorem. 17: 40–48.

Young, C.A., Hume, D.E. and McCulley, R.L. 2013. Forages and pastures symposium: Fungal endophytes of tall fescue and perennial ryegrass: Pasture friend or foe? J. Anim. Sci. May 2013, 91(5): 2379–2394.

Yousaf, S., Afzal, M., Reichenauer, T.G., Brady, C.L. and Sessitsch, A. 2011. Hydrocarbon degradation, plant colonization and gene expression of alkane degradation genes by endophytic Enterobacter ludwigii strains. Environ. Pollut. 159: 2675–2683.

Zhang, X., Liu, X.,Wang, Q., Chen, X., Li, H., Wei, J. and Xu, G. 2014. Diesel degradation potential of endophytic bacteria isolated from Scirpus triqueter. Int. Biodeterior. Biodegrad. 87: 99–105.

Zhu, X., Jin, L., Sun, K., Li, S., Ling,W. and Li, X. 2016. Potential of endophytic bacterium Paenibacillus sp. PHE-3 isolated from Plantago asiatica L. for reduction of PAH contamination in plant tissues. Int. J. Environ. Res. Public Health 13: 633.

23

Bioremediation of Chlorinated Organic Pollutants in Anaerobic Sediments

Archana V.,[1] *Salom Gnana Thanga Vincent*[1,2] *and Thava Palanisami*[2,]*

1. Introduction

Chlorinated organic compounds are designated as priority pollutants, and as persistent organic pollutants (POPs), because they are highly toxic, can bioconcentrate and persist in the environment. Various sources of chlorinated compounds include improper disposal of chemical wastes, agricultural run-off, use of herbicides, insecticides, fungicides, solvents, hydraulic and heat-transfer fluids, plasticizers, cleaning agents, fumigants, aerosol propellants, gasoline additives, degreasers, and intermediates for chemical syntheses (Rawford et al. 2005). In addition to anthropogenic sources, there are various natural sources of chlorinated organic compounds in the environment. Chlorinated compounds also occur naturally in the environment in low concentrations, such as volcanic eruption. Chlorobenzene, carbon tetrachloride, phenolic compounds (phenol, chlorophenols, nitro phenol, polychlorinated biphenyls (PCBs)) and dyes are some major chlorinated organic compounds prevalent in sediments. For example, in contaminated places, concentration of PCBs in sediments was reported as high as 104 mg/Kg, which is several orders of magnitude higher than the permissible level (Vasilyeva and Strijakova 2007).

Complete removal of POPs from the environment is difficult. Some POPs are still persistent, can accumulate in fatty tissues, and are present in higher concentrations at higher levels in the food chain, with long-range mobility (Bhatt et al. 2007). Chlorinated pollutants are recalcitrant under aerobic conditions because oxidative stress associated with biodegradation of chlorinated pollutants is a significant barrier for the evolution of aerobic pathways for chlorinated compounds, thereby allowing for the emergence of anaerobic counterparts (Nikel et al. 2013). Although anaerobic dechlorination does not result in complete degradation of xenobiotics, it contributes to the detoxification of the environment by forming less toxic products (Wiegel and Wu 2000). Aquatic sediments are typical examples of anaerobic conditions since oxygen diffusion is limited and microbial respiration depletes the oxygen when organic substrates are available. Aquatic sediments also act as sinks for recalcitrant and hazardous organic pollutants entering from various sources, most of which are anthropogenic. The problem is worldwide and this becomes alarming when these compounds, through various biogeochemical processes, can be available to the benthic organism as well as to the organisms in the water column through the sediment-water interface (Perelo 2010).

---

[1] Department of Environmental Sciences, University of Kerala.
[2] Global Innovative Centre for Advanced Nanomaterials (GICAN), University of Newcastle, Australia.
* Corresponding author: thava.palanisami@newcastle.edu.au

This chapter explores the anaerobic degradation of chlorinated organic pollutants in anaerobic sediments by natural attenuation.

2. Microbial degradation of chlorinated compounds

Although many of the chlorinated compounds escape from the degradation and persist in the environment, certain microorganisms exposed to these synthetic chemicals have evolved the ability to utilize some of them. Bacteria can degrade chlorinated aromatic compounds in four ways (Häggblom 1992, Commandeur and Parsons 1994): (1) dehalogenation after ring cleavage, (2) oxidative dehalogenation, (3) hydrolytic dehalogenation and (4) reductive dehalogenation. Biodegradation of chlorinated compounds is of two types: aerobic degradation and anaerobic degradation. Degradation of a compound depends on the structure of the compound (either oxidized or reduced), the number of chlorine substituents in the structure, and the position of chlorine in the molecules. In aerobic degradation, molecular oxygen serves as the electron acceptor (e.g., chloroaliphatic compounds). Bhatt et al. (2006) reported that 4-chlorophenol (4-CP) can be partially or completely degraded aerobically by *Pseudomonas, Alcaligenes, Rhodococcus, Azotobacter*, etc. In anaerobic degradation, the electron acceptors are NO_3^-, SO_4^{2-}, Fe^{3+}, H^+, S, fumarate, trimethylamine oxide, an organic compound, or CO_2. In anaerobic conditions, chlorinated aromatic compounds become reductively dehalogenated in a process known as halo respiration, resulting in the accumulation of lower chlorinated congeners.

Degradation of most of the chlorinated compounds are not based on microbial energy metabolism. When microorganisms can transform one substrate to different substrates that is not associated with that organism's energy production such as carbon assimilation, or any other growth process, that particular mode of activity is called "cometabolism". During cometabolism, in the presence of one organic material (primary energy source), degradation of another compound takes place. Several chlorinated hydrocarbons are transformed cometabolically by bacteria, and degrade the chlorine unsubstituted hydrocarbons. Detoxification and complete mineralization of chlorinated compounds can be easily attained by using a sequential treatment process, that is, anaerobic followed by aerobic treatments (Bhatt et al. 2007). However, the difficulty in assessing the degree of biodegradation of compounds like PCBs in the natural environment like sediments is due to the great diversity of congeners as well as the difficulty of their chemical analysis. A slow decrease in higher-chlorinated content and accumulation of lower-chlorinated PCB congeners like ortho-substituted PCBs was observed in river and marine sediments by Vasilyeva and Strijakova (2007).

3. Terminal electron acceptors and electron donors during degradation

In sediments, a series of redox zones is prevalent in which microbial respiration proceeds based on the availability of terminal electron acceptors (TEA). Hence, the type of microorganisms and their metabolisms depend on the availability of electron acceptors and their corresponding energy yield. Once oxygen in the sediment is depleted by aerobic respiration, nitrate is used as a TEA by denitrifying bacteria, followed by manganese and iron reduction. The degradation of halogenated aromatic compounds and carbon tetrachloride has been reported under these conditions (Kazumi et al. 1995, Boopathy 2002). After iron, sulphate anions are used as TEA. Several chlorinated compounds have been reported in the sulphate reducing conditions (Drzyzga et al. 2001, Savage et al. 2010, van Eekert and Schraa 2001). In the absence of other inorganic electron acceptors, methanogenesis has an important role for the biodegradation of organic matter by using CO_2 as TEA. Degradation of several chlorinated organic compounds has been reported in methanogenic conditions.

There exists a wide source of electron donors or substrates for dechlorination by microorganisms, and several lines of evidences show that once the electron donor is depleted, dechlorination does not proceed further (Villarante et al. 2001). Nies and Vogel (1990) observed that methanol was the most effective electron donor, followed by glucose, acetone and acetate to reductively dechlorinate polychlorinated biphenyls in an anaerobic sediment enrichment. In addition, other substrates like lactate for dechlorination of trichloroethylene to ethane, butyrate, propionate, and ethanol for dechlorination of trichlorophenoxyacetic acid were also observed (Gibson and Sulfita 1990). Ammonia was an effective electron donor for *Nitrosomonas* sp. capable of degrading several chlorinated aliphatic compounds.

4. Mechanism of anaerobic degradation: dechlorination or dehalorespiration

Dechlorination is a more promising anaerobic degradation mechanism for halogenated compounds. Since chlorinated compounds are electron deficient, they act as electron acceptors, thus generating energy in the anaerobic respiration process. The chlorinated organic compounds are used as TEA and the specific microorganisms gain energy during the reductive chlorination, wherein, a chlorine atom is replaced by a hydrogen atom. The energy available from reductive dechlorination reactions is similar, irrespective of the position of the chlorine atoms or its oxidation state and the free energy obtained varies from –130 to –171 kJ/Cl atom removed (Dolfing and Harrison 1992).

Several chlorinated aromatic compounds and pesticides have been reported to be degraded under anaerobic conditions. Under anaerobic conditions, chlorinated volatile organic compounds (VOCs) can be degraded by the reductive dehalogenation pathways to less chlorinated products. For example, perchloroethylene (PCE) is degraded sequentially by reductive chlorination to vinyl chloride (VC) and ethylene/ethane (Lorah and Olsen 1999) (Figure 1). The chlorinated organic carbon compounds are used as electron acceptors by the microorganisms, and the available organic matter in the sediments would serve as a continuous supply of electron donors, which is required to drive the reaction. The strictly anaerobic conditions promote reductive dechlorination by cleavage of ether bonds, which are more prevalent (Ghattas et al. 2017).

It is difficult to generalize the dechlorination pathways because the dechlorinating microbial species depend on the position of the chlorine atom. Several mechanisms were proposed for the dechlorination of chlorobenzene and pentachlorophenol. For chlorobenzenes, chlorine atoms with two or one adjacent chlorine atoms were preferentially cleaved under sulphate reducing conditions in a contaminated river sediment (Susarla et al. 1998). However, for chlorophenols, ortho-chlorine removal was preferred by both sulphate reducers and methanogens (Takeuchi et al. 2000). Contrastingly, preferential para- and meta-chlorine removal was observed by Häggblom et al. (1993) under both sulphate reducing and methanogenic conditions, respectively. Halorespiring bacteria are a group of microorganisms which grow independently of the above inorganic electron acceptors. They are highly relevant in the bioremediation of chlorinated organic pollutants as they utilize these compounds itself as TEA. They utilize an array of chlorinated organic compounds including chlorinated biphenyls and dioxins (Cutter et al. 2001, Wu et al. 2002, Bunge et al. 2003). In natural sediments, microbes have different types of dechlorinating enzymes based on the type of

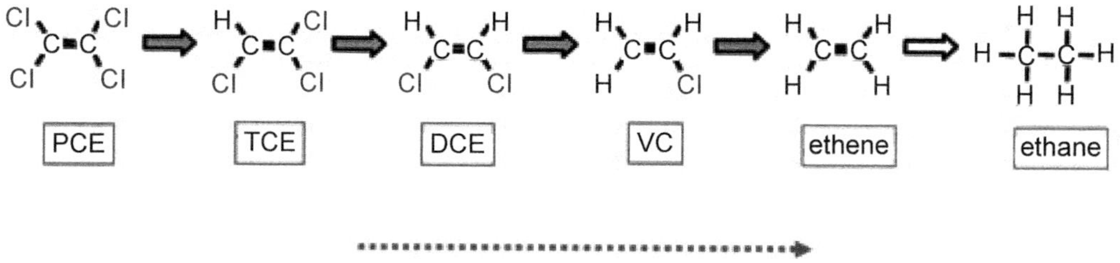

PCE : Perchloroethylene/Tetrachloroethene
TCE : Trichloroethylene/Trichloroethene
DCE : *1,2-dichloroethylene/1,2-dichloroethene/cis-1,2-dichloroethene/trans-1,2-dichloroethene*
VC : Vinyl Chloride

Figure 1. Anaerobic dechlorination pathways of perchloroethylene (PCE).

the chlorinated organic compounds as well as their congeners. For example, the chlorine atoms in PCBs are replaced one by one by hydrogen atoms and are removed as chloride ions by dehalogenase activity. Moreover, the rate of dechlorination in the sediments decreases from higher chlorinated congeners to less chlorinated congeners (Vasilyeva and Strijakova 2007). Around 70% of arachlor introduced in pure sediments of Lake Michigan was found to be dechlorinated after six months. It was also observed that low-chlorinated congeners were completely dechlorinated (Natarajan et al. 1998).

4.1 Degradation in sulphate reducing conditions

During anaerobic degradation of halogenated compounds or dehalogenation in sulphate reducing conditions, the most important electron acceptor is sulphate. Sulphate was found to be an important electron acceptor for transformation of TCE and chlorinated benzenes than oxygen or nitrate (Cobb and Bower 1991). The degradation capacity of chlorinated aromatic compounds coupled to sulphate reduction is considerably more favourable than degradation coupled with methane production (Häggblom and Young 1995). Desulfitobacterium chlororespirans gains energy from the reductive ortho-dechlorination of 3-chloro-4-hydroxy benzoate and 2,3-di and polychloro-substituted phenols (Loeffler et al. 1996). Reductive dechlorination was found to related to respiratory growth in *Desulfitobacterium multivorans* (Neumann et al. 1994).

It was observed that 4-chlorophenol was used as a sole source of carbon in a sulfidogenic consortium enriched with an estuarine sediment (Häggblom and Young 1995). After reductive dechlorination, 4-chlorophenol was dechlorinated to phenol under sulphate-reducing conditions, and mineralization of the phenol ring to CO_2 occurred only with the coupling of sulphate reduction. The most common mechanisms of dehalogenation are hydrogenolysis and hydrolysis. *Dehalococcoides* sp. was found to degrade chlorinated ethenes, PCB and dioxins (Bedard et al. 2007, Bunge et al. 2003). Certain species of sulphate reducers are strictly halorespiring like *Dehalococcoides* and *Dehalobacter. Desulfomonas michiganensis*, an acetate-oxidizing anaerobic bacteria, can utilize perchloro ethylene as electron acceptor (Sung et al. 2003).

4.2 Degradation in methanogenic conditions

Under methanogenic conditions also, the common biotransformation pathway is reductive dechlorination or reductive dehydrogenolysis for chloroaliphatics containing one or two carbon atoms. Numerous chlorinated aromatic compounds have been revealed to be degraded by reductive dehalogenation under methanogenic conditions (e.g., 2,4,5-trichlorophenoxyacetate, 3-chlorobenzoate, 2,4-dichlorophenol, 4-chlorophenol, 2,3,6-trichlorobenzoate, and dichlorobenzoates) (Häggblom 1992, Mohn and Tiedje 1992, Commandeur and Parsons 1994). Bower and McCarty (1983) observed that under methanogenic conditions, numerous 1- and 2-carbon halogenated aliphatic organic compounds present at low concentrations (< 100 ug/liter) were degraded in batch bacterial cultures and also in a continuous-flow methanogenic fixed-film laboratory-scale column. Acetate was used as a primary substrate. Within two-day detention time, more than 90% degradation was observed. Although trichloroethylene (TCE) is degraded under different types of facultative anaerobic and anaerobic conditions such as nitrate-reducing, iron-reducing, and sulphate-reducing, the degradation is fast and complete under methanogenic conditions, where the degradation of TCE forms non-toxic end products such as ethane and ethylene (McCarty and Semprini 1994).

Anaerobic degradation of tricolosan was reported under both sulphate reducing and methanogenic conditions, where triclosan was converted to catechol and phenol by dechlorination. Phenol and catechol were also further anaerobically degraded to reduce the aromatic ring (Veetil et al. 2012). As methanogens cannot directly degrade complex organic compounds, they depend on other organisms such as hydroloytic, fermentative and syntrophic acetogenic microbes for the supply of electron donors and substrates like H_2, CO_2, formate and acetate. Hence, methanogens co-exist as a microbial consortium (Stams et al. 2005).

5. Conclusions

Anaerobic degradation of organic compounds occurs frequently in environments such as in sediments which are contaminated. These compounds are used as electron acceptors under varying electron accepting conditions such as sulphate reducing and methanogenic conditions, which are strictly anaerobic. The availability of organic matter serves as the electron donor, which makes the biodegradation of contaminants possible. The type of microorganisms and metabolisms available for biodegradation in the sediments depend on the environmental conditions as well as the type of electron donors and acceptors. The degradation of chlorinated organic compounds, which are persistent in aerobic conditions, proceeds through the anaerobic dechlorination pathways in the sediments. Many of the studies are focused at the laboratory level on this topic and several studies underscore the variations in bioremediation performance in field compared to laboratory. In many instances, laboratory results do not reflect or predict field results, as a multitude of environmental factors influence the biodegradation process in the environment. Hence, in the natural environment, although the remediating organisms may be present, bioremediation may be slow. In addition, most of the studies are related to temperate ecosystems and the studies pertaining to tropical conditions are lacking. It is well known that the environmental conditions, particularly temperature, favour the growth and proliferation of more microbes in tropical than temperate conditions. Further research is warranted focusing on the enhancement of bioremediation in the real field conditions by making use of the natural bioremediation potential of the native microbes.

References

Arora, P.K., Sasikala, C. and Ramana, C.V. 2012. Degradation of chlorinated nitroaromatic compounds. Appl. Microbiol. Biotechnol. 93(6): 2265–2277.

Bedard, D.L., Ritalahti, K.M. and Loeffler, F.E. 2007. *Dehalococcoides* population in sediment-free mixed cultures metabolically dechlorinates the commercial polychlorinated biphenyl mixture Aroclor 1260. Appl. Environ. Microbiol. 73: 2513–2521.

Boopathy, R. 2002. Anaerobic biotransformation of carbon tetrachloride under various electron acceptor conditions. Bioresour. Technol. 84(1): 69e73.

Bouwer, E.J. and McCarty, P.L. 1983. Transformations of 1- and 2-carbon halogenated aliphatic organic compounds under methanogenic conditions. Appl. Environ. Microbiol. 45(4): 1286–1294.

Bunge, M., Lorenz, A., Kraus, A., Opel, M., Lorenz, W.G. and Andreesen, J.R. 2003. Reductive dehalogenation of chlorinated dioxins by an anaerobic bacterium. Nature 421: 357–360.

Chaudhry, G.R. and Chapalamadugu, S. 1991. Biodegradation of halogenated organic compounds. Microbiol. Mol. Biol. Rev. 55(1): 59–79.

Cobb, G.D. and Bouwer, E.J. 1991. Effects of electron acceptors on halogenated organic compound biotransformations in a biofilm column. Environ. Sci. Technol. 25: 1068–1074.

Colberg, P.J. 1990. Role of sulfate in microbial transformations of environmental contaminants: chlorinated aromatic compounds. Geomicrobiol. J. 8(3-4): 147–165.

Crawford, R.L. and Crawford, D.L. (eds.). 2005. Bioremediation: Principles and Applications (Vol. 6). Cambridge University Press.

Cutter, L.A., Watts, J.E.M., Sowers, K.R. and May, H.D. 2001. Identification of a micro-organism that links its growth to the reductive dechlorination of 2,3,5,6-chlorobiphenyl. Environ. Microbiol. 3(11): 699e709.

Das, N. and Chandran, P. 2011. Microbial degradation of petroleum hydrocarbon contaminants: an overview. Biotechnol. Res. Int.

Dolfing, J. and Harrison, B.K. 1992. Gibbs free energy of formation of halogenated aromatic compounds and their potential role as electron acceptors in anaerobic in environments. Environ. Sci. Technol. 26: 2213–2218.

Drzyzga, O., Gerritse, J., Dijk, J.A., Elissen, H. and Gottschal, J.C. 2001. Coexistence of a sulphate-reducing Desulfovibrio species and the dehalorespiring Desulfito-bacterium frappieri TCE1 in defined chemostat cultures grown with various combinations of sulphate and tetrachloroethene. Environ. Microbiol. 3(2): 92e99.

Ennik-Maarsen, K. 1999. Degradation of chlorobenzoates and chlorophenols by methanogenic consortia. Saff Publications Wageningen University and Research.

Farhadian, M., Vachelard, C., Duchez, D. and Larroche, C. 2008. *In situ* bioremediation of monoaromatic pollutants in groundwater: a review. Bioresource Technol. 99(13): 5296–5308.

Ghattas, A.K., Fischer, F., Wick, A. and Ternes, T.A. 2017. Anaerobic biodegradation of (emerging) organic contaminants in the aquatic environment. Water Res. 116: 268–295.

Gibson, S.A. and Sulfita, J.M. 1990. Anaerobic biodegradation of 2,4-5-trichloro-phenoxyacetic acid in samples from a methanogenic aquifer: Stimulated by short chain organic acids and alcohols. Appl. Environ. Microbiol. 56: 1825–1832.

Guerin, T.F. 2001. A biological loss of endosulfan and related chlorinated organic compounds from aqueous systems in the presence and absence of oxygen. Environ. Pollut. 115(2): 219–230.

Häggblom, M.M., Rivera, M.D. and Young, L.Y. 1993. Influence of alternative electron acceptors on the anaerobic biodegradability of chlorinated phenols and benzoic acids. Appl. Environ. Microbiol. 59(4): 3255–3260.

Häggblom, M.M. 1998. Reductive dechlorination of halogenated phenols by a sulfate-reducing consortium. FEMS Microbiol. Ecology 26(1): 35–41.

Jianlong, W. and Yi, Q. 1999. Microbial degradation of 4-chlorophenol by microorganisms entrapped in carrageenan-chitosan gels. Chemosphere 38(13): 3109–3117.

Kazumi, J., H€aggblom, M. and Young, L. 1995. Diversity of anaerobic microbial processes in chlorobenzoate degradation: nitrate, iron, sulfate and carbonate as electron acceptors. Appl. Microbiol. Biotechnol. 43(5): 929–936.

Kim, C.J. and Maier, W.J. 1986. Acclimation and biodegradation of chlorinated organic compounds in the presence of alternate substrates. J. Water Pollut. Control. Fed. 157–164.

Kouznetsova, I., Mao, X., Robinson, C., Barry, D.A., Gerhard, J.I. and McCarty, P.L. 2010. Biological reduction of chlorinated solvents: Batch-scale geochemical modeling. Adv. Water Resour. 33(9): 969–986.

Loeffler, F.E., Sanford, R.A. and Tiedje, J.M. 1996. Initial characterization of a reductive dehalogenase from *Desulfitobacterium chlrorespirans* Co 23. Appl. Environ. Microbiol. 62: 3809–3813.

Lorah, M.M. and Olsen, L.D. 1999. Natural attenuation of chlorinated volatile organic compounds in a freshwater tidal wetland: Field evidence of anaerobic biodegradation. Water Res. 35: 3811–3827.

Mao, X., Polasko, A. and Alvarez-Cohen, L. 2017. Effects of sulfate reduction on trichloroethene dechlorination by Dehalococcoides-containing microbial communities. Appl. Environ. Microbiol. 83(8): e03384–16.

McCarty, P.L. and Semprini, L. 1994. Ground-water treatment for chlorinated solvents. pp. 87–116. *In*: Norris, R.D. et al. (eds.). Handbook of Bioremediation. A. F. Lewis, New York.

Natarajan, M.R., Wu, W.M., Wang, H., Bhatnagar, L. and Jain, M.K. 1998. Dechlorination of Spiket PCBs in lake sediment by anaerobic microbial granules. Water. Res. 32: 3013–3020.

National Research Council. 1993. *In situ* bioremediation: When does it work?. National Academies Press.

Neumann, A., Scholtz, M.-H. and Diekert, G. 1994. Tetrachloroethene metabolism of *Dehalospirillum multivorans*. Arch. Microbiol. 162: 295–301.

Nies, L. and Vogel, T.M. 1990. Effects of organic substrates on dechlorination of aroclor 1242 in anaerobic sediments. Appl. Environ. Microbiol. 56: 2612–2617.

Nikel, P.I., Pe´rez-Pantoja, D. and de Lorenzo, V. 2013. Why are chlorinated pollutants so difficult to degrade aerobically? Redox stress limits 1,3-dichloprop-1-ene metabolism by Pseudomonas pavonaceae. Philosophical Transactions of The Royal Society B 368: 20120377. http://dx.doi.org/10.1098/rstb.2012.0377.

Perelo, L.W. 2017. *In situ* and bioremediation of organic pollutants in aquatic sediments. J. Hazard. Mater. 177: 81–89.

Savage, K.N., Krumholz, L.R., Gieg, L.M., Parisi, V.A., Suflita, J.M., Allen, J., Philp, R.P. and Elshahed, M.S. 2010. Biodegradation of low-molecular-weight alkanes under mesophilic, sulfate-reducing conditions: metabolic intermediates and community patterns. FEMS Microbiol. Ecol. 72(3): 485–495.

Stackelberg, P.E. 1997. Presence and distribution of chlorinated organic compounds in streambed sediments, New jersey 1. Jawra J. of the American Water Res. Association 33(2): 271–284.

Stams, A.J.M., Plugge, C.M., De Bok, F.A.M., Van Houten, B.H.G.W., Lens, P., Dijkman, H. and Weijma, J. 2005. Metabolic interactions in methanogenic and sulfate-reducing bioreactors. Water Sci. and Technol. 52(1-2): 13–20.

Sung, Y., Ritalahti, K.M., Sanford, R.A., Urbance, J.W., Flynn, S.T. Tiedje, J.M. and Loeffler, F.E. 2003. Characterization of two tetrachloroethene reducing acetate-oxidixing anaerobic bacteria and their description as *Desulfuromonas michiganensis* sp. nov. Appl. Environ. Microbiol. 69: 2964–2974.

Susarla, S., Yonezawa, Y. and Masunaga, S. 1998. Reductive transformations of halogenated aromatics in anaerobic estuarine sediment: kinetics, products and pathways. Water Res. 32: 639–648.

Takeuchi, R., Suwa, Y., Yamagishi, T. and Yonezawa, Y. 2000. Anaerobic transformation of chlorophenols in methanogenic sludge unexposed to chlorophenols. Chemosphere 41: 1457–1462.

Thapa, B., Kc, A.K. and Ghimire, A. 2012. A review on bioremediation of petroleum hydrocarbon contaminants in soil. Kathmandu University J. of Science, Engineering and Technol. 8(1): 164–170.

van Eekert, M.H.A. and Schraa, G. 2001. The potential of anaerobic bacteria to degrade chlorinated compounds. Water Sci. Technol. 44(8): 49–56.

Vasilyeva, G.K. and Strijakova, E.R. 2007. Bioremediation of soils and sediments contaminated by polychlorinated biphenyls. Microbiology 76: 639–653.

Veetil, P.G.P., Nadaraja, A.V., Bhasi, A., Khan, S. and Bhaskaran, K. 2012. Degradation of triclosan under aerobic, anoxic, and anaerobic conditions. Appl. Biochemistry and Biotechnology 167(6): 1603–1612.

Vidali, M. 2001. Bioremediation an overview. Pure and Appl. Chemistry 73(7): 1163–1172.

Villarante, N.R., Armenante, P.M., Quibuyen, T.A.O., Favak, F. and Kafkewitz, D. 2001. Dehalogenation of DCE in a contaminated soil: Fatty acids & alcohols as e donors and an apparent requirement for tetrachloroethene. Appl. Microbiol. Biotech. 55: 239–247.

Wackett, L.P. 1996. Biodegradation of chlorinated aliphatic compounds. Biotechnol. Res. Ser. 6: 300–311.

Wiegel, J. and Wu, Q.Z. 2000. Microbial reductive dehalogenation of polychlorinated biphenyls. FEMS Microbiol. Letts. 32: 1–15.

Wu, Q., Watts, J.E.M., Sowers, K.R. and May, H.D. 2002. Identification of a bacterium that specifically catalyzes the reductive dechlorination of polychlorinated bi-phenyls with doubly flanked chlorines. Appl. Environ. Microbiol. 68(2): 807–812.

24

Bioremediation of Wastewater by Sulphate Reducing Bacteria

Panchami Shaji,[1] *Salom G.T. Vincent*[1,2] *and Thava Palanisami*[2,]*

1. Introduction

Sulphate reducing bacteria (SRB) are a diverse group of microorganisms that have a vital role in the biogeochemical cycle of sulphur in anaerobic environments. Sulphate reducing bacteria are prokaryotic microorganisms that belong to bacteria and archaea domains. During the energy metabolism in SRB, sulphate acts as the terminal electron acceptor resulting in dissimilatory sulphate reduction. Due to their importance in the environmental remediation and ecosystem functioning, research interests in SRB have been ascending for the past few years. SRB can be considered as a problem due to the formation of hydrogen sulphide, as well as due to biocorrosion. However, there are various advantages of SRB in wastewater treatment, particularly the less sludge production and also the potential for removal of hazardous materials like heavy metals. Moreover, treatment of wastewater using SRB can be considered as a pretreatment for anaerobic digestion. Given the suitable environmental conditions such as presence of sulphate, temperature, and anoxic conditions, SRB has immense applications in wastewater treatment. Another highlight of SRB is their versatility. Most SRB are mesophilic, although there are some thermophilic and psychrophilic species available in the nature. Regarding pH, most SRB prefer acidic environment; however, there are SRB which thrive and prefer alkaline and hyper saline environments like marine sediments. Such vast adaptation makes SRB preferred candidates to treat various types of wastewater containing sulphate. This chapter explores the various applications of SRB in the treatment of different types of wastewater containing sulphate, heavy metals and chlorinated organic compounds.

2. Ecological significance of SRB

Sulphate reduction is one of the most commonly occurring and extensive microbiological processes on the earth. SRB are physiologically unique among living organisms in being able to reduce sulphate to sulphide. SRB was first discovered by Bejerneck (1895), and around 220 species of 60 genera have been identified until now. There are five divisions within the bacteria, and two divisions among the archea. These include: the spore-forming *Desufotomaculum*, *Desulfosporomusa* and *Desulfosporosinus* species within the Firmicutes division; *Deltaproteobacteria* and

---

[1] Department of Environmental Sciences, University of Kerala, India.
[2] Global Innovative Centre for Advanced Nanomaterials, University of Newcastle, Australia.
* Corresponding author: thava.palanisami@newcastle.edu.au

Thermodesulfovibrio species within the Nitrospira division; and two phyla represented by *Thermodesulfobium narugense* and *Thermodesulfobacterium/ Thermodesulfatator* species, and two divisions within the archaea (euryarchaeotal genus *Archaeoglobus*, and the two crenarchaeotal genera *Thermocladium* and *Caldivirga*, affiliated with the Thermoproteales) (Castro et al. 2000, Itoh et al. 1998, 1999, Mori et al. 2003, Muyzer and Stams 2008, Ollivier et al. 2007, Rabus et al. 2006).

Representatives of sulphur reducing bacteria are found in environments ranging from brackish, super cooled Antarctic waters to hot artesian springs and deep Pacific sediments (Fauque 1995, Loubinoux et al. 2002, Muyzer and Stams 2008, Ollivier et al. 2007, Rabus et al. 2006). Generally, SRB prefer sulphate rich habitat (Cypionka 2000, Fareleria et al. 2003, Sass et al. 1992) like marine sediments, where sulphate reduction is a predominant terminal electron accepting process during the mineralisation of organic matter (Vincent et al. 2017). However, they are also prevalent in freshwater habitats, anaerobic digesters and gastrointestinal tract of humans and animals (Postage 1984). In these environments, the SRB influences overall biogeochemistry by the oxidation of organic matter and concomitant sulphide production and/or metal reduction (Hao et al. 1996).

$$SO_4^{2-} + \text{organic matter} \xrightleftharpoons{\text{SRB}} HS^- + H_2O + HCO_3^-$$

SRB are versatile in nature, and have the ability to use electron acceptors other than sulphates in facultative anaerobic conditions, including elemental sulphur (Bottcher et al. 2005, Finster et al. 1998), fumarate (Tomei et al. 1995), dimethyl sulfoxide (Jonkers et al. 1996), Fe (III) (Lowely et al. 1993, 2004), and nitrate (Krekeler and Cypionka 1995). Earlier, SRB was believed to utilize only a limited range of substrates as energy source such as lactate, molecular hydrogen, pyruvate, and ethanol; however, later studies revealed a wide range of electron donors used by SRB (Rabus et al. 2006) such as fructose, glucose, amino acids, monocarboxylic acids, oxalic acids, alcohols and aromatic compounds (Fauque et al. 1991, Rabus et al. 2006).

3. SRB in domestic wastewater treatment

The application of SRB in domestic wastewater treatment is limited compared to industrial wastewater. This is because the concentration of sulphate in domestic wastewater is comparatively less (500 mg L^{-1}) than industrial wastewater. The advantage of SRB to treat municipal wastewater is the low sludge yield that significantly reduces the amount of excess sludge produced. However, the challenge in application of SRB is the limitation of sulphate in wastewater. At higher sulphate levels, a maximum of 75% chemical oxygen demand (COD) is removed by SRB (van der Brand et al. 2018).

4. SRB in industrial wastewater treatment

The sulphate concentration is high in industrial wastewater, which originates from pharmaceuticals and chemical units, edible oils, and paper, pulp and molasses based units (Visser et al. 1993, Fang et al. 1997, Percheronetal. 1997, Knobel et al. 2002, Silva et al. 2002, Venkata Mohan and Sharma 2002). The sulphate containing wastewater is normally treated by means of various physicochemical and biological methods, which although are effective, have disadvantages such as separation and appropriate disposal of solid phases, high cost compared to the biological methods, and high energy consumption (Silva et al. 2002). Among the various biological methods, anaerobic process is generally used for the treatment of sulphate containing wastewater (Visser et al. 1993, Fang et al. 1997, Knobel et al. 2002, Omil et al. 1996, Dries et al. 1998, Venkata Mohan et al. 2005). In anaerobic process, the various organisms act in a sequential process during which organic compounds are converted to simple molecules and finally to CO_2 and CH_4. SRB grow by the splitting of thiosulfate, sulphite and sulphur, which results in the formation of sulphate and sulphide (Muyzer and Stams 2008). Two key enzymes, adenylylsulfate reductase and bisulfite reductase, are involved in the process of reduction of sulphate to sulphite, and to elemental sulphur.

The presence of high concentration of sulphate in the wastewater will affect the process due to problems such as corrosion (Vincke et al. 2001), reduced treatment efficiency and biogas production due to the production of H_2S. Anaerobic treatment can be more effectively done by integration of sulphate reduction and methanogensis (Lens et al. 2002). Treatment of cotton delinting wastewater using halophilic SRB *Desulfovibrio halophilus* achieved a removal of 78.4% sulphate within a 24 h retention time (Torbaghan and Torghabeh 2019). SRB has also been applied for the remediation of acid mine wastewater. The successful application at the industrial scale has been patented (BioSulphide and Thiopaq technologies) for removal of sulphate and heavy metals from acid mine drainage (Ayangbenro et al. 2018). Ito et al. (2002) observed *Desulfobulbus, Desulfovibrio* and *Desulfomicrobium* to be the numerically important members of a wastewater biofilm growing in microaerophilic conditions. Some of the reactor configurations previously described for biological reduction of sulphate are sequential batch reactors, anaerobic filters, fluidized bed reactors, membrane bioreactors, hybrid anaerobic reactors and upflow anaerobic sludge reactor.

5. Heavy metal remediation by SRB

Heavy metal pollution is one of the serious environmental issues faced presently. Heavy metals can cause neurological problems due to the toxic effect, and causes damage to the nerves, liver, bones and enzymes in humans. They can also enter into the food chain, and their accumulation leads to serious health and ecological problems (Malik 2004, Sanyal et al. 2005). These toxic metal ions generated from mining operations, metal-plating facilities, power generation facilities, electronic device manufacturing units, and tanneries commonly exist in process waste streams (Liu et al. 2009), which have increased dramatically during the last few decades in ambient environment. Thus, the removal of such toxic metal ions from effluent is a crucial issue. Chemical precipitation, chemical oxidation, chemical reduction, electrocoagulation, electrodeposition, solvent exchange and membrane separation are the various physical and chemical methods widely used for the heavy metal ion removal from industrial wastewater. However, these processes might be ineffective or expensive when the heavy metal ions are present in solution in the order 1–100 mg dissolved heavy metal ions/L (Meunier et al. 2003, Volesky and Holan 1995). In this context, biological methods become a viable alternative for the removal of heavy metal ions.

During the past few decades, SRB have been identified as one of the primary bacteria for the remediation of heavy metals in wastewater. In anaerobic condition, SRB oxidizes organic compounds by using sulphate and a terminal electron acceptor. Thus, the sulphate is reduced to sulphide (Postgate 1984, White et al. 1997). As a result of generation of sulphide, heavy metal ions from the aqueous phase can be removed (Utgikar et al. 2003, Jalali et al. 2000). The process consists of two stages: (1) the production of H_2S by SRB, and (2) the precipitation of metals by the biologically produced H_2S, a reaction in which insoluble metal sulphides are produced, and can be easily separated (Jiang et al. 2008).

Organic matter + SO_4^{2-} → $2CH_3COO^-$ + HS^- + HCO_3^- + Me^{2+} + HS^- → MeS + H^+

SRB can also reduce the heavy metals in an enzymatic way (Goulhen et al. 2006, Pagnanelli et al. 2010, Sahinkaya et al. 2011, Sheng et al. 2011). But there are a few drawbacks of using SRB for the remediation, such as when the metal concentration increases, the time of reaction also increases (Jiang et al. 2008). Also, the processing time is longer than the chemical or physical precipitation (Zhou et al. 2013). The functions of SRB can be greatly affected by the physicochemical parameters such as pH, ionic strength and temperature (Zhou et al. 2013). Guha and Bhargava (2005) reported that the removal efficiency of Cr in wastewater by SRB was limited by the formation of an oxidized layer and deposition of sulphide on Fe^0 surface. Sulphate reduction was found to be an effective process in removing toxic lead PB (II) and mercury Hg (II) ions from wet flue gas desulfurization wastewater through anaerobic sulphite reduction (Zhang et al. 2016).

6. Bioremediation of chlorinated organic compounds

Contamination of the environment by solvents is an important aspect resulting from improper handling and disposing of chloroethenes such as tetrachloroethene (PCE) or trichloroethene (TCE) (Barton et al. 2007). Specific strains of SRB can use chloroethene as electron acceptor in sulphate limited environment (Table 1).

The H_2 produced by the *Desulfovibrio* sp. from fermentation of lactate under the sulphate limiting condition was used by dehalogenating bacteria (Drzyzga et al. 2001). The reductive release of chloride from TCE is shown in Figure 1.

Mixed culture of SRB showed greater dehalogenation than pure culture of *D. frappieri* (Drzyzga et al. 2001). Pure culture of *D. frappieri* metabolizes PCE, which results in the accumulation of cis-dichloroethane. However, dehalogenation of PCE by mixed culture results in the production of 55% ethene and 45% ethane (Drzyzga et al. 2002).

Table 1. SRB that metabolize chlorinated organic compounds.

Substrate	Organism	Reference
Tetrachloroethene and/or polychloroethenes	*Desulfitobacterium frappieri* TCE1 *Desulfitobacterium* sp. PCE-1	Gerritse et al. 1999 Gerritse et al. 1996
	Desulfitobacterium frappieri PCP1	Dennie et al. 1998
	Desulfitobacterium frappieri TCE1 *Desulfitobacterium* sp. Viet1	Drzyzga et al. 2001 Loeffler et al. 1999
	Desulfitobacterium sp. Y51 *Desulfitobacterium hafniense* Y51	Suyama et al. 2001 Nonaka et al. 2006
	Desulfitobacterium sp. KBC1	Tsukagoshi et al. 2006

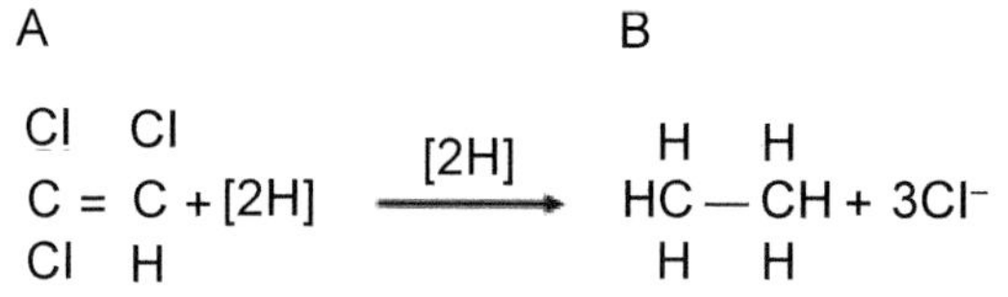

$$\text{A} \qquad \qquad \text{B}$$
$$\begin{array}{c} \text{Cl} \quad \text{Cl} \\ \text{C} = \text{C} + [2H] \\ \text{Cl} \quad \text{H} \end{array} \xrightarrow{[2H]} \begin{array}{c} \text{H} \quad \text{H} \\ \text{HC} - \text{CH} + 3\text{Cl}^- \\ \text{H} \quad \text{H} \end{array}$$

Figure 1. Microbial reductive dehalogenation (A) trichloroethane (B) ethane, [2H] indicate proton from the cell.

7. Future applications: energy recovery as a dual advantage

Yun et al. (2019) conducted a study to recover energy during waste water treatment using SRB by designing a SRB-based wastewater treatment system integrated with sulphide fuel cell (SFC). The ratio of COD and sulphate was a critical parameter in this process. They could achieve sulphate reduction coupled with electricity production, which is highlighted as a green and effective method of wastewater treatment.

8. Conclusions

Concern regarding the fate of sulphate and heavy metals in various types of industrial wastewater and sewage water has been growing. Biological method is the most common method used for the treatment of xenobiotics in the effluents released from industrial sites. Biological treatments are more cost effective than physical and chemical treatments, and help to maintain the ecological balance and reestablishment of polluted environments. SRB are one of the most common groups of microorganisms used for the biological anaerobic treatment of sulphate and heavy metal containing wastewater. Clearly, the synergism among SRB and other anaerobes is important in the bioremediation process, and future investigations are expected to enhance these ecological frameworks.

References

Ayangbenro, A.S., Olanrewaju, O.S. and Babalola, O.O. 2018. Sulfate-reducing bacteria as an effective tool for sustainable acid mine bioremediation. Front. Microbiol. 9: 1986.

Barton, L.L. and Hamilton, W.A. (eds.). 2007. Sulphate-Reducing Bacteria: Environmental and Engineered Systems. Cambridge University Press.

Böttcher, M.E., Thamdrup, B., Gehre, M. and Theune, A. 2005. 34S/32S and 18O/16O fractionation during sulfur disproportionation by Desulfobulbus propionicus. Geomicrobiol. J. 22(5): 219–226.

Castro, H.F., Williams, N.H. and Ogram, A. 2000. Phylogeny of sulfate-reducing bacteria. FEMS Microbiol. Ecol. 31(1): 1–9.

Cypionka, H. 2000. Oxygen respiration by Desulfovibrio species. Annu. Rev. Microbiol. 54(1): 827–848.

Dries, J., De Smul, A., Goethals, L., Grootaerd, H. and Verstraete, W. 1998. High rate biological treatment of sulfate-rich wastewater in an acetate-fed EGSB reactor. Biodegradation 9(2): 103–111.

Drzyzga, O., Gerritse, J., Dijk, J.A., Elissen, H. and Gottschal, J.C. 2001. Coexistence of a sulphate-reducing Desulfovibrio species and the dehalorespiring Desulfitobacterium frappieri TCE1 in defined chemostat cultures grown with various combinations of sulphate and tetrachloroethene. Environ. Microbiol. 3(2): 92–99.

Fang, H.H., Liu, Y. and Chen, T. 1997. Effect of sulfate on anaerobic degradation of benzoate in UASB reactors. J. Environ. Eng. 123(4): 320–328.

Fareleira, P., Santos, B.S., António, C., Moradas-Ferreira, P., LeGall, J., Xavier, A.V. and Santos, H. 2003. Response of a strict anaerobe to oxygen: survival strategies in Desulfovibrio gigas. Microbiology 149(6): 1513–1522.

Fauque, G., LeGall, J. and Barton, L.L. 1991. Sulfate-reducing and sulfur-reducing bacteria. Variations in autotrophic life, 271–337.

Fauque, G. D. 1995. Ecology of sulfate-reducing bacteria. Sulfate-Reducing Bacteria. Springer, Boston, MA. 217–241.

Finster, K., Liesack, W. and Thamdrup, B.O. 1998. Elemental sulfur and thiosulfate disproportionation by Desulfocapsa sulfoexigens sp. nov., a new anaerobic bacterium isolated from marine surface sediment. Appl. Environ. Microbiol. 64(1): 119–125.

Guha, S. and Bhargava, P. 2005. Removal of chromium from synthetic plating waste by zero-valent iron and sulfate-reducing bacteria. Water Environ. Res. 77(4): 411–416.

Goulhen, F., Gloter, A., Guyot, F. and Bruschi, M. 2006. Cr (VI) detoxification by Desulfovibrio vulgaris strain Hildenborough: microbe–metal interactions studies. Appl. Microbiol. Biotechnol. 71(6): 892–897.

Hao, O.J., Chen, J.M., Huang, L. and Buglass, R.L. 1996. Sulfate-reducing bacteria. Crit. Rev. Environ. Sci. Technol. 26(2): 155–187.

Hsu, H.F., Jhuo, Y.S., Kumar, M., Ma, Y.S. and Lin, J.G. 2010. Simultaneous sulfate reduction and copper removal by a PVA-immobilized sulfate reducing bacterial culture. Bioresour. Technol. 101(12): 4354–4361.

Ito, T., Okabe, S., Satoh, H. and Watanabe, Y. 2002. Successional development of sulfate-reducing bacterial populations and their activities in a wastewater biofilm growing under microaerophilic conditions. Appl. Environ. Microbiol. 68(3): 1392–1402.

Itoh, T., Suzuki, K.I. and Nakase, T. 1998. Thermocladium modestius gen. nov., sp. nov., a new genus of rod-shaped, extremely thermophilic crenarchaeote. Int. J. Syst. Evol. Microbiol. 48(3): 879–887.

Itoh, T., Suzuki, K.I., Sanchez, P.C. and Nakase, T. 1999. Caldivirga maquilingensis gen. nov., sp. nov., a new genus of rod-shaped crenarchaeote isolated from a hot spring in the Philippines. Int. J. Syst. Evol. Microbiol. 49(3): 1157–1163.

Jalali, K. and Baldwin, S.A. 2000. The role of sulphate reducing bacteria in copper removal from aqueous sulphate solutions. Water Res. 34(3): 797–806.

Jiang, W. and Fan, W. 2008. Bioremediation of heavy metal–contaminated soils by sulfate-reducing bacteria. Ann. N. Y. Acad. Sci. 1140(1): 446–454.

Jonkers, H.M., van Der Maarel, M.J., van Gemerden, H. and Hansen, T.A. 1996. Dimethylsulfoxide reduction by marine sulfate-reducing bacteria. FEMS Microbiol. Lett. 136(3): 283–287.

Knobel, A.N. and Lewis, A.E. 2002. A mathematical model of a high sulphate wastewater anaerobic treatment system. Water Res. 36(1): 257–265.

Krekeler, D. and Cypionka, H. 1995. The preferred electron acceptor of Desulfovibrio desulfuricans CSN. FEMS Microbial. Ecol. 17(4): 271–277.

Lens, P., Vallerol, M., Esposito, G. and Zandvoort, M. 2002. Perspectives of sulfate reducing bioreactors in environmental biotechnology. Rev. Environ. Sci. Biotechnol. 1(4): 311–325.

Li, H., Liu, T., Li, Z. and Deng, L. 2008. Low-cost supports used to immobilize fungi and reliable technique for removal hexavalent chromium in wastewater. Bioresour. Technol. 99(7): 2234–2241.

Liu, X., Hu, Q., Fang, Z., Zhang, X. and Zhang, B. 2009. Magnetic chitosan nanocomposites: a useful recyclable tool for heavy metal ion removal. Langmuir 25(1): 3–8.

Loubinoux, J., Bronowicki, J.P., Pereira, I.A., Mougenel, J.L. and Le Faou, A.E. 2002. Sulfate-reducing bacteria in human feces and their association with inflammatory bowel diseases. FEMS Microbiol. Ecol. 40(2): 107–112.

Lovley, D.R., Giovannoni, S.J., White, D.C., Champine, J.E., Phillips, E.J.P., Gorby, Y.A. and Goodwin, S. 1993. Geobacter metallireducens gen. nov. sp. nov., a microorganism capable of coupling the complete oxidation of organic compounds to the reduction of iron and other metals. Arch. Microbiol. 159(4): 336–344.

Lovley, D.R., Holmes, D.E. and Nevin, K.P. 2004. Dissimilatory Fe (III) and Mn (IV) reduction. Advan. in Microbial Physiol. 49(2): 219–286.

Malik, A. 2004. Metal bioremediation through growing cells. Environ. Int. 30(2): 261–278.

Meunier, N., Laroulandie, J., Blais, J.F. and Tyagi, R.D. 2003. Cocoa shells for heavy metal removal from acidic solutions. Bioresour. Technol. 90(3): 255–263.

Mori, K., Kim, H., Kakegawa, T. and Hanada, S. 2003. A novel lineage of sulfate-reducing microorganisms: Thermodesulfobiaceae fam. nov., Thermodesulfobium narugense, gen. nov., sp. nov., a new thermophilic isolate from a hot spring. Extremophiles 7(4): 283–290.

Muyzer, G. and Stams, A.J. 2008. The ecology and biotechnology of sulphate-reducing bacteria. Nat. Rev. Microbiol. 6(6): 441–454.

Ollivier, B., Cayol, J.L. and Fauque, G. 2007. Sulphate-reducing bacteria from oil field environments and deep-sea hydrothermal vents. Sulphate-Reducing Bacteria: Environ. and Engineered Syst. 305–328.

Omil, F., Lens, P., Pol, L.H. and Lettinga, G. 1996. Effect of upward velocity and sulphide concentration on volatile fatty acid degradation in a sulphidogenic granular sludge reactor. Process Biochem. 31(7): 699–710.

Pagnanelli, F., Viggi, C.C. and Toro, L. 2010. Isolation and quantification of cadmium removal mechanisms in batch reactors inoculated by sulphate reducing bacteria: biosorption versus bioprecipitation. Bioresource Technol. 101(9): 2981–2987.

Percheron, G., Bernet, N. and Moletta, R. 1997. Start-up of anaerobic digestion of sulfate wastewater. Bioresour. Technol. 61(1): 21–27.

Postgate, J.R. 1984. The Sulphate-reducing Bacteria, second ed. Cambridge University Press, New York.

Rabus, R.A.L.F., Hansen, T.A. and Widdel, F.R.I.E.D.R.I.C.H. 2006. Dissimilatory sulfate-and sulfur-reducing prokaryotes. The Prokaryotes 2: 659–768.

Sanyal, A., Rautaray, D., Bansal, V., Ahmad, A. and Sastry, M. 2005. Heavy-metal remediation by a fungus as a means of production of lead and cadmium carbonate crystals. Langmuir 21(16): 7220–7224.

Sahinkaya, E., Gunes, F.M., Ucar, D. and Kaksonen, A.H. 2011. Sulfidogenic fluidized bed treatment of real acid mine drainage water. Bioresour. Technol. 102(2): 683–689.

Sass, H., Steuber, J., Kroder, M., Kroneck, P.M.H. and Cypionka, H. 1992. Formation of thionates by freshwater and marine strains of sulfate-reducing bacteria. Arch. of Microbial. 158(6): 418–421.

Sheng, Y., Cao, H., Li, Y. and Zhang, Y. 2011. Effects of sulfide on sulfate reducing bacteria in response to Cu (II), Hg (II) and Cr (VI) toxicity. Chinese Sci. Bull. 56(9): 862.

Silva, A.J.D., Varesche, M.B., Foresti, E. and Zaiat, M. 2002. Sulphate removal from industrial wastewater using a packed-bed anaerobic reactor. Process Biochem. 37(9): 927–935.

Torbaghan, M.E. and Torghabeh, G.H.K. 2019. Biological removal of iron and sulfate from synthetic wastewater of cotton delinting factory by using halophilic sulfate-reducing bacteria. Heliyon 5(12): e02948.

Tomei, F.A., Barton, L.L., Lemanski, C.L., Zocco, T.G., Fink, N.H. and Sillerud, L.O. 1995. Transformation of selenate and selenite to elemental selenium by *Desulfovibrio desulfuricans*. J. of Ind. Microbiol. 14(3-4): 329–336.

Utgikar, V.P., Tabak, H.H., Haines, J.R. and Govind, R. 2003. Quantification of toxic and inhibitory impact of copper and zinc on mixed cultures of sulfate-reducing bacteria. Biotechnol. Bioeng. 82(3): 306–312.

van den Brand, T., Snip, L., Palmen, L., Weij, P., Sipma, J. and van Loosdrecht, M. 2018. Sulfate reducing bacteria applied to domestic wastewater. Water Pract. Technol. 13(3): 542–554.

Venkata Mohan, S. and Sarma P.N. 2002. Anaerobic treatment of pharmaceutical wastewater. Biomed. Pharmacol. J. 2: 101–108.

Venkata Mohan, S., Chandrasekhara Rao, N., Krishna Prasad, K., Murali Krishna, P., Sreenivas Rao, R. and Sarma, P.N. 2005. Anaerobic treatment of complex chemical wastewater in a sequencing batch biofilm reactor: Process optimization and evaluation of factor interactions using the Taguchi dynamic DOE methodology. Biotechnol. Bioeng. 90(6): 732–745.

Vincent, S.G.T., Reshmi, R.R., Hassan, S.J., Nair, K.D. and Varma, A. 2017. Predominant terminal electron accepting processes during organic matter degradation: Spatio-temporal changes in Ashtamudi estuary, Kerala, India. Estuar. Coast. Shelf Sci. 198: 508–517.

Vincke, E., Boon, N. and Verstraete, W. 2001. Analysis of the microbial communities on corroded concrete sewer pipes–a case study. Appl. Microbiol. Biotechnol. 57(5-6): 776–785.

Visser, A., Beeksma, I., Van der Zee, F., Stams, A.J.M. and Lettinga, G. 1993. Anaerobic degradation of volatile fatty acids at different sulphate concentrations. Appl. Microbiol. Biotechnol. 40(4): 549–556.

Volesky, B. and Holan, Z.R. 1995. Biosorption of heavy metals. Biotechnol. progress, 11(3): 235–250.

White, C., Sayer, J.A. and Gadd, G.M. 1997. Microbial solubilization and immobilization of toxic metals: key biogeochemical processes for treatment of contamination. FEMS Microbiol. Rev. 20(3-4): 503–516.

Ye, J., Yin, H., Mai, B., Peng, H., Qin, H., He, B. and Zhang, N. 2010. Biosorption of chromium from aqueous solution and electroplating wastewater using mixture of Candida lipolytica and dewatered sewage sludge. Bioresour. Technol 101(11): 3893–3902.

Yun, Y.M., Lee, E., Kim, K. and Han, J.I. 2019. Sulfate reducing bacteria-based wastewater treatment system integrated with sulfide fuel cell for simultaneous wastewater treatment and electricity generation. Chemosphere 233: 570–578.

Zhang, L., Lin, X., Wang, J., Jiang, F., Wei, L., Chen, G. and Hao, X. 2016. Effects of lead and mercury on sulfate-reducing bacterial activity in a biological process for flue gas desulfurization wastewater treatment. Sci. Rep. 6: 30455.

Zhou, Q., Chen, Y., Yang, M., Li, W. and Deng, L. 2013. Enhanced bioremediation of heavy metal from effluent by sulfate-reducing bacteria with copper–iron bimetallic particles support. Bioresour. Technol. 136: 413–417.

Index

A

Accumulators 82, 83
Adsorbents 246, 249
Aerobic bioremediation 173
Agrochemicals 228, 235
Aquatic sediments 330
Aquatic systems 241–243
Aromatic plants 138, 140, 147–149
Arsenic toxicity 240, 241, 246, 248

B

Bio concentration factor 83
Bioaccumulation 93, 95
Bioadsorption 95, 96
Bioaugmentation 61, 66, 69
Biochar 310–315
Bioconcentration 182, 183
Biodegradation 154, 157, 159–165, 215–217, 220, 223
Biofilm 280–282, 284–286, 289, 291–293
Bioremediation 15–30, 37–52, 57, 59–62, 64, 65, 67–72, 92, 93, 95–99, 227–229, 231, 232, 234, 235, 252, 253, 256, 257, 259, 261–263, 330, 332, 334
Biostimulation 186, 194, 196
Biosurfactants 62, 69–71, 217–219, 222, 223, 318, 321, 323–325
Biotechnology 61, 64

C

Chlorinated hydrocarbons 171
Chlorinated organic compounds 337, 340
Chlorinated pollutants 330
Composting 202, 206, 208, 209
Contamination 267–270, 273
Conventional membrane bioreactor 279

D

Denitrification 253, 259–263

E

Endophytes 318–325
Environment 37–39, 42–44, 46–52, 157, 159, 160, 165, 240–243, 246–249
Excluders 82, 87

F

Fluoride 252–259, 261, 262
Fossil-based polymers 165

G

Genetically modified microorganism 192
Greenhouse gases 310, 312, 315

H

Heavy metal 15–17, 19–22, 24–27, 79–85, 87, 88, 92–98, 103–105, 108–115, 337, 339, 340
Hydrocarbons 318, 320–323
Hyperaccumulator 82, 83, 85

I

In situ and *Ex situ* bioremediation techniques 4
In situ and *ex situ* treatment 278
Indicators 82
Industrial waste 228
Inorganic pollutant 57–60
In-situ technologies 219

M

Marine sediments 337, 338
Microbes 155–157, 159–165
Microbioremediation 4
Microorganism 38–44, 49, 173–175, 204–206, 208–210, 257, 259–262
Micropollutant 279–281
Mineralization 124, 126, 130, 131

N

Nanoparticles 299–306
Nutrients 173–176

O

Olive mill wastes 267, 268, 270, 271, 273
Organic carbon 122, 123, 125, 133
Organic pollutant 58, 63, 67, 68
Organic solids 206

P

Particulate organic matter 122
Pesticide 180–197
Photocatalysts 301
Phycoremediation 3, 4
Phytodecontamination 82, 83
Phytoextraction 81–86, 232–234
Phytoremediation 3–5, 7, 16, 18, 19, 22, 24, 27, 28, 62, 64, 67, 79, 81, 82, 84, 86–88, 97, 98, 138, 140, 143–147, 149, 150
Phytostabilization 82, 86, 87
Phytovolatilization 82, 85, 86
Plant growth-promoting bacteria 233
Plastic waste 154, 157
Pollution 318–320, 324
Polymerization 205, 209
Polymers 154, 157–165

R

Reductive dehalogenation 331–333
Rhizofiltration 82, 85

S

Salt tolerant crops 138
Sequestration 310–315
Soil 92, 93, 96–99
Soil desertification 270
Soil enzymes 125, 127–129, 132, 133
Soil heavy metal pollutants 310, 313, 315
Soil indicators 122, 123, 130
Soil management 310, 313
Soil microbes 17, 20, 26
Soil pollution 15, 17
Soil properties 267, 271, 273
Soil salinity 138, 142, 145, 146, 148
Soil's microbes 229
Sorbents 258
Storage 111–113
Synanthropic 318, 320, 321

T

Translocation factor 83, 84, 87

U

Uptake 104, 108–112, 114, 115

W

Wastewater 281, 283, 286–289, 291, 337–340
Wastewater irrigation 235

About the Editors

Amitava Rakshit, an IIT Kharagpur alumnus, is presently the faculty member in the Department of Soil Science and Agricultural Chemistry at Institute of Agricultural Sciences, Banaras Hindu University (IAS, BHU). Dr. Rakshit worked in the Department of Agriculture, Government of West Bengal in administration and extension roles. He has visited Scandinavia, Europe and Africa pertaining to his research work and presentation. He was awarded with TWAS Nxt Fellow (Italy), Biovision Nxt Fellow (France), Darwin Now Bursary (British Council) Young achiever award and Best Teacher Award at UG and PG level. He is serving as review college member of British Ecological Society, London since 2011, member of Global Forum on Food Security and Nutrition of FAO, Rome and Commission on Ecosystem Management of IUCN. He has published 70 research papers, 35 book chapters, 28 popular articles and one manual and co-authored twenty books.

Dr. Manoj Parihar is currently working at ICAR-VPKAS, Almora as a Soil Scientist in Crop Production Division. He did his graduation from SKRAU, Bikaner and was selected as ICAR-JRF fellow for post-graduation in BHU, Varanasi. He has been awarded doctorate from the same university in the year 2018. He has received various recognitions such as ICAR-SRF, UGC-BSR, UGC-RGNF, etc.

Binoy Sarkar is a Lecturer at the Lancaster Environment Centre of Lancaster University. Previously he was a Research Associate at University of Sheffield, and a Research Fellow at University of South Australia from where he also received his Ph.D. His research works extend to remediation of conventional and emerging contaminants in soil and water environments, and atmospheric carbon dioxide capture and carbon sequestration in soils. He has edited two books, and published more than 20 book chapters and more than 115 peer-reviewed journal papers including in Nature. He teaches Soil Science, Biogeochemical Cycles, and Earth's Interior subjects, and has supervised seven Ph.D. students and four post-doctoral fellows. He is an alumnus of Indian Agricultural Research Institute, awardee of the Australian Endeavour Research Fellowship pursued at Indiana University, and recipient of the Geof Proudfoot Award, and Desai-Biswas Medal.

Harikesh Bahadur Singh is presently a retired Professor of Mycology and Plant Pathology from IAS, BHU. Over the past 35 years, Professor Singh has served at Central Universities and CSIR Institutes with his teaching and research. Based on his scientific contribution and leadership in the field of plant pathology, Professor Singh has been honoured with prestigious awards, such as the CSIR technology award, M.S. Swaminathan award, Mundkur Memorial award, and BRSI Industrial Medal. His research has resulted in more than 300 publications and 17 books.

**Leonardo Fernandes Fraceto,** Ph.D., has a bachelor's degree in Chemistry from the State University of Campinas (1997), a master's in Functional and Molecular Biology from the State University of Campinas (2000) and a PhD in Functional and Molecular Biology from the State University of Campinas (2003). He is currently an Associate Professor at the Institute of Science and Technology of São Paulo State University (UNESP), Campus Sorocaba at the Environmental Engineering undergraduate course and at the postgraduate in Environmental Sciences. He coordinated the Postgraduate Program in Environmental Sciences of Unesp/Sorocaba (from 2012 to 2016). He is also Fellow from the Royal Society of Chemistry. He has experience in environmental nanotechnology, with an emphasis on health and environmental nanotechnology and applications of nanotechnology in agriculture. He has published about 200 papers in international peer reviewed journals and has supervised undergraduate, graduate (master and Ph.D.) and supervised Pos-docs. Dr. Fraceto has been active in the development of nanocarrier systems for the encapsulation of compounds with bioactive properties. In addition, from these studies, Dr. Fraceto has obtained grants from fellow research agencies, as well as scientific/technological development partnerships with companies from different areas. It also highlights the effective performance in the training of human resources, through the supervision of students of undergraduation, master's, Ph.D. and Post-docs. Many of his former students now hold positions at public and private universities and research centers. As such, Dr. Fraceto has developed high-level research and contributed to scientific and technological development at the national and international levels, with numerous internationally renowned scientific collaborators.

9780367343965